ELECTRONIC IMAGING IN ASTRONOMY
Detectors and Instrumentation

WILEY–PRAXIS SERIES IN ASTRONOMY AND ASTROPHYSICS
Series Editor: John Mason, B.Sc., Ph.D.

Few subjects have been at the centre of such important developments or seen such a wealth of new and exciting, if sometimes controversial, data as modern astronomy, astrophysics and cosmology. This series reflects the very rapid and significant progress being made in current research, as a consequence of new instrumentation and observing techniques, applied right across the electromagnetic spectrum, computer modelling and modern theoretical methods.

The crucial links between observation and theory are emphasised, putting into perspective the latest results from the new generations of astronomical detectors, telescopes and space-borne instruments. Complex topics are logically developed and fully explained and, where mathematics is used, the physical concepts behind the equations are clearly summarised.

These books are written principally for professional astronomers, astrophysicists, cosmologists, physicists and space scientists, together with post-graduate and undergraduate students in these fields. Certain books in the series will appeal to amateur astronomers, high-flying 'A'-level students, and non-scientists with a keen interest in astronomy and astrophysics.

ROBOTIC OBSERVATORIES
Michael F. Bode, Professor of Astrophysics and Assistant Provost for Research, Liverpool John Moores University, UK

THE AURORA: Sun–Earth Interactions
Second Edition
Neil Bone, British Astronomical Association and University of Sussex, Brighton, UK

PLANETARY VOLCANISM: A Study of Volcanic Activity in the Solar System, Second edition
Peter Cattermole, formerly Lecturer in Geology, Department of Geology, Sheffield University, UK, now Principal Investigator with NASA's Planetary Geology and Geophysics Programme

DIVIDING THE CIRCLE: The Development of Critical Angular Measurement in Astronomy 1500–1850
Second edition
Allan Chapman, Wadham College, University of Oxford, UK

TOWARDS THE EDGE OF THE UNIVERSE: A Review of Modern Cosmology
Stuart Clark, Lecturer in Astronomy, University of Hertfordshire, UK

ASTRONOMY FROM SPACE: The Design and Operation of Orbiting Observatories
John K. Davies, Joint Astronomy Centre, Hawaii, USA

THE DUSTY UNIVERSE
Aneurin Evans, Department of Physics, University of Keele, UK

MARS AND THE DEVELOPMENT OF LIFE, Second edition
Anders Hansson, Ph.D., Senior Science Consultant, International Nanobiological Testbed

ASTEROIDS: Their Nature and Utilization, Second edition
Charles T. Kowal, Allied Signal Corp., Applied Physics Laboratory, Laurel, Maryland, USA

ELECTRONIC IMAGING IN ASTRONOMY: Detectors and Instrumentation
Ian. S. McLean, Department of Physics & Astronomy, UCLA, and Director, Infrared Imaging Detector Lab, USA

THE PLANET NEPTUNE: An Historical Survey Before Voyager, Second edition
Patrick Moore, CBE, D.Sc.(Hon.)

ACTIVE GALACTIC NUCLEI
Ian Robson, Director, James Clerk Maxwell Telescope, Director Joint Astronomy Centre, Hawaii, USA

EXPLORATION OF TERRESTRIAL PLANETS FROM SPACECRAFT, Second edition
Yuri Surkov, Chief of the Planetary Exploration Laboratory, Russian Academy of Sciences, V. I. Vernadsky Institute, Moscow, Russia

THE HIDDEN UNIVERSE
Roger J. Tayler, Astronomy Centre, University of Sussex, Brighton, UK

Forthcoming titles in the series are listed at the back of the book.

ELECTRONIC IMAGING IN ASTRONOMY

Detectors and Instrumentation

Ian McLean
Department of Physics & Astronomy, UCLA
and Director, Infrared Imaging Detector Lab, USA

JOHN WILEY & SONS
Chichester • New York • Weinheim • Brisbane • Singapore • Toronto

Published in association with
PRAXIS PUBLISHING
Chichester

Copyright © 1997 Praxis Publishing Ltd
The White House,
Eastergate, Chichester,
West Sussex, PO20 6UR, England

Published in 1997 by
John Wiley & Sons Ltd
in association with Praxis Publishing Ltd

Based on *Electronic and Computer-Aided Astronomy*
first published in 1989.

Wiley Editorial Offices

John Wiley & Sons Ltd, Baffins Lane,
Chichester, West Sussex PO19 1UD, England

John Wiley & Sons, Inc., 605 Third Avenue,
New York, NY 10158-0012, USA

VCH Verlagsgesellschaft mvH, Pappelallee 3
D-69469 Weinheim, Germany

Jacaranda Wiley Ltd, G.P.O. Box 859, Brisbane,
Queensland 4001, Australia

John Wiley & Sons (Asia) Pte Ltd, 2 Clementi Loop #02-01,
Jin Xing Distripark, Singapore 129801

John Wiley & Sons (Canada) Ltd, 22 Worcester Road,
Rexdale, Ontario M9W 1L1, Canada

Library of Congress Cataloging-in-Publication Data
McLean, Ian S., 1949–
    Electronic imaging in astronomy : detectors and instrumentation / Ian McLean.
    p. cm. — (Wiley–Praxis series in astronomy and astrophysics)
    Includes bibliographical references and index.
    ISBN 0-471-96971-0 (cloth : alk. paper). – ISBN 0-471-96972-9 (pbk. : alk. paper)
    1. Imaging systems in astronomy—Instruments. 2. Astronomical instruments
    I. Title.
QB51.3.I45M36 1997
522'.2—dc20
                                                            97-32366
                                                            CIP

A catalogue record for this book is available from the British Library

ISBN 0-471-96971-0 Cloth   ISBN 0-471-96972-9 Paperback

Printed and bound in Great Britain by Hartnolls Ltd, Bodmin

*to my father, for
showing me the stars*

Table of contents

The colour plate section appears between pages 306 and 307

List of illustrations, plates and tables

Chapter 3

Chapter 4

Chapter 5

Chapter 6

Chapter 7

Chapter 8

Chapter 9

Chapter 13

Chapter 14

Appendix 5

Colour plates

(Positioned between pages 306 and 307.)

Tables

Preface

Astronomical discoveries often make news headlines. Little is said, however, about the methods and apparatus used to reach these conclusions, other than perhaps a reference to the observatory or satellite which was used. Moreover, visitors to a modern research observatory are surprised by the panoply of complex electronic equipment and computers attached to, or surrounding the telescope. What is the purpose of the multitude of electrical cables, and the plumes of white vapour emerging silently from instrumentation being cooled with liquid nitrogen? Usually there isn't even an eyepiece to enable one to look through the telescope! Of course, the well-read amateur astronomer knows that professional astronomy is done with "electronic imaging" devices, like CCDs, and that these devices are now available to amateur astronomers as well. In fact, all of modern astronomy relies heavily on recent advances in technology, and in some cases, astronomers have been the driving force behind those developments. Modern astronomy is therefore as exciting and challenging for the professional engineer and applied physicist as it is for the astronomer. Today, would-be astronomers must reckon on acquiring a wide range of skills, or on working as a member of a multidisciplinary team.

The idea for the substance of this book, and its precursor, *Electronic and Computer-Aided Astronomy: from Eyes to Electronic Sensors*, grew out of a desire to explain to others just how much applied physics and engineering goes into the seemingly "pure" science of astronomy. The first book, published in 1989, was also stimulated by the remarkable impact which one small "silicon microchip"—the CCD—had on astronomical imaging methods. Since then, the rapid pace of technology, has wrought many other fundamental changes in observational astronomy which have dated that account. For example, the launch of the Hubble Space Telescope, and the new developments in infrared astronomy, adaptive optics, laser guide stars and very large telescopes. In this text I have tried to concentrate more on general principles and techniques, as well as take an historical perspective. Of course, some "current" performance levels of astronomical detectors and instruments are inevitably included, but these will soon be surpassed. Readers can come up-to-date by consulting articles in manufacturers catalogues, professional journals, conference proceedings and popular magazines, but at least you will have broken the "jargon" barrier.

This book has been written on several levels in the hope that a wide range of people will be able to find something in it for them. Chapters 1 and 2 provide a general introduction to electronic imaging and the methods and instruments of modern ground-based astronomy. The impact of the charge-coupled device (or CCD) becomes apparent. Chapters 3 and 4 contain basic descriptions of fundamental instrumentation—photometers and spectrographs—for "optical" astronomy, and further develops the underlying theme of the book, which is the synergy between advances in technology and our best images of the Universe. Chapter 5 is a lengthy tutorial on designing astronomical instruments. Chapters 6 and 7 explain in detail the basic principle of the amazing charge-coupled device and cover all the practicalities of its use. Chapters 8 and 9 treat the revolutionary "infrared array" detector in a similar manner and illustrate how the advent of this new imaging device has even changed the way telescopes are designed. Computers and image processing is discussed in Chapter 10 and Chapter 11 deals with a wide range of calibration issues common to most electronic imaging devices. Electronic imaging at other wavelengths is treated in Chapter 12. The ubiquitous CCD appears again as a potential detector for X-ray astronomy. In Chapter 13 we see the impact of a new generation of telescopes on electronic imaging in astronomy, including the Hubble Space Telescope and several huge ground-based telescopes with 8–10-metre apertures. Finally, Chapter 14 covers the technology of "adaptive optics" with its goal to undo the blurring effects of atmospheric turbulence and achieve unprecedented resolution.

Unlike a typical college text, this book includes historical perspectives and describes the role that modern astronomers and technologists have played in the development of electronic imaging in astronomy. Mathematical expositions are controlled to encourage a wider audience, especially the rapidly growing community of amateur astronomers who now own CCD cameras. The book can be used at the college level for a one-semester introductory course on modern astronomical detectors and instruments, and as a supplement for a practical or laboratory class. Many college observatories have CCD cameras and some also have infrared cameras. By selecting more challenging exercises, de-emphasizing the historical accounts and perhaps supplementing the book with classic material on optics and detector physics, this book provides the core of a one-semester course on astronomical instrumentation for new graduate (PhD) students who may very soon be faced with using, or even building, electronic imaging systems. Too often these sophisticated and elegant instruments are viewed as a "black box". Finally, it is hoped that the text will form a useful reference book for professionals in the scientific imaging field. I hope this book will encourage more college courses on detectors and instrumentation for astronomy, but equally important, I hope it will also encourage an even greater appreciation of the remarkable link between astronomy and technology.

I would like to take this opportunity to thank all who helped and encouraged this work. I am particularly indebted to the Series Editor John Mason for his advice and to Clive Horwood of Praxis for his patience and support. Many people kindly supplied me with new information, reference material and photographs. In addition to all who helped with the earlier book, I especially wish to thank Jim Janesick, formerly at JPL and now with PixelVision, Paul Jorden of the Royal Greenwich Observatory, Tony Tyson of AT&T Bell Labs (now Lucent Technology) and Al Fowler of the National Optical Astronomy Observatories. It is also a pleasure to acknowledge Richard Aikens, Roger Angel, Jacque Beckers, Eric Becklin, Morley Blouke, Todd Boroson, George Carruthers, David Clarke, Martin Cullum, Sandra Faber, John Fordham, Bob Fugate, John Geary, Tom Geballe, James

Graham, Jim Gunn, Andrea Ghez, Arne Henden, Jeff Hester, Alan Hoffman, Chuck Joseph, Pierre-Olivier Lagage, Gerry Luppino, Craig Mackay, Claus Madsen, David Malin, Claire Max, Craig McCreight, Jerry Nelson, Ian Parry, Rick Puetter, Harold Reitsema, Lloyd Robinson, Francois Roddier, Brad Smith, Kadri Vural, Fred Vrba, Fred Watson, Jim Westphal and Ned Wright. It is a particular pleasure to acknowledge the many fine engineers and technologists that I have been privileged to work with over the years including my current team at the UCLA Infrared Imaging Detector Lab (George Brims, John Canfield, Woon Wong, Frank Henriquez, Nick Magnone and Fred Lacayanga) as well as my former colleagues at the Royal Observatory Edinburgh, most especially Donald Pettie. I am also very grateful to Melinda Laraneta for preparation of several diagrams. Perhaps most of all I thank all my students and post-docs over the years from whom I have learned so much, including Mark McCaughrean, John Rayner, Colin Aspin, Tim Liu, Bruce Macintosh, Don Figer, Suzanne Casement, Sam Larson and Harry Teplitz. Finally, I am most appreciative of the unswerving support of my wife and family. I look forward to hearing from readers and teachers.

Thousand Oaks, California
August 1996 Ian McLean

Introduction

All astronomers speak about "going observing", but what does this mean? If you are an amateur enthusiast then it may mean going no further than your backyard or your local astronomy club to use a telescope to view your favourite objects. For professional astronomers, however, the phrase means much more. Implicit in the phrase is the fact that to understand the Universe we must observe it, and to do so we will need more than our human eyes, more than a telescope in the backyard. We will need all that modern electronic technology can offer. The largest ground-based telescopes in the world are located at relatively remote, pristine sites, high above sea level where the air is thin and the skies are astonishingly clear—so "going observing" can also mean going far away from home.

Access by professional astronomers to national ground-based optical/infrared observatories, as well as at most university or privately owned facilities, is on a highly competitive basis. To obtain an allocation of "observing time" an astronomer must submit in writing a well-argued scientific case for permission to carry out his or her observational experiment. Deadline dates are set typically twice or three times per year. Selection is done by **peer review**, that is, by a committee formed from the body of scientists who actually use the facility. Unfortunately, all of the major telescopes are heavily oversubscribed, so disappointment is a fact of life. To maximize the progress of scientific experiments at each facility, and to make the optimum use of weather conditions, the astronomical community world-wide is expending considerable effort on technology. As we shall see, this means highly automated observatories with much reliance on well-engineered instrumentation and computers, and it also implies new cost-effective solutions for the design and management of telescopes and measuring equipment.

Observing time on large telescopes is therefore difficult to obtain and is very valuable; it is important that no time be wasted. Also, since the telescope and instrumentation is quite complex, guest astronomers who may visit the observatory only twice per year cannot be expected to learn the myriad of operational details. To solve this problem all large observatories provide one or more highly trained personnel to support the visitor. Usually a Night Assistant/ Telescope Operator is provided; he or she will be responsible for control of the telescope and dome, ensuring efficient operation, keeping an observatory logbook and the preparation of observatory equipment. The night assistant has the final word regarding safety matters such as closing the telescope dome if the wind speed becomes too high. Sometimes a Support Scientist, who is a professional research astronomer on the observatory staff familiar with the application of the instrumentation in question, is also available to assist first-time or irregular users of the observatory.

A guest observer (or G.O.) planning to use a modern, computer-controlled electronic imaging camera or spectrograph at one of these major facilities might encounter the following pattern of work. The visiting astronomers will probably arrive by air a few days before their allocated time to ensure that they are not travel-weary and to discuss their plans with observatory staff. They may have travelled from North America or Europe to Mauna Kea, Hawaii, or from Europe and North America to Chile or Australia, or any of several other destinations. By mid-afternoon on the day of the first night on the telescope, the observatory staff, which may include the support scientist and an engineer experienced with the system to be used that night, will be in the telescope dome making sure that the telescope control system and the instrument are both functioning correctly. The visiting astronomer(s), often including graduate students receiving training in observational methods or seeking data for a thesis topic, may well elect to be present for these checks, and may wish to practise using the instrument. This may mean becoming familiar with a control panel, or with the operation of a computer console on which the experimental modes can be displayed and changed by typing at a keyboard. To feel confident that they understand the operation of the instrument, the visiting astronomers will carry out some tests of their own such as a "noise check" on the detector, or a calibration image or spectrum; all of these terms and procedures will be described later. With everything in readiness for the evening, they return to the observatory residential lodge where a meal is prepared for them. This is usually a great chance to meet people from all over the world, and the dinner conversation is often buzzing with astronomical jargon! Between one and two hours prior to sunset the "observers" go back to the telescope dome, usually in the company of the Night Assistant, to complete their preparations. This entails discussing the "observing plan" with the Night Assistant, providing him/her with a list of object names and positions or "coordinates" in the sky which get typed into the telescope "control" computer, and taking some important calibration data such as "dark" frames with the detector covered and "flat fields" or "arc lamps" which are exposures using special artificial sources (lamps).

As the twilight fades and the sky becomes dark enough to commence work, the Night Assistant will "call-up" the first object on the target list and a computer will instruct electrically driven motors on the telescope's rotation axes to move or slew to that position. Using a special TV-type camera at the focus of the telescope, the guest astronomer examines the field of view to confirm that the telescope is pointing at the object of interest—usually by reference to an existing star chart. Sometimes of course, nothing can be seen because the object(s) are too faint and require a long exposure; in that case the field must be confirmed by checking the pattern of brighter non-target objects in the vicinity. When the object is correctly centred, the observation begins. Having put the camera or spectrograph to the required settings by typing command words or letters into the "instrument" computer (rather than the telescope control computer), all that is required next is to issue a "start" command. The total time for which the measurement lasts is called the "integration time" and this may be anything from a fraction of a second to hours depending on the brightness of the object, the efficiency of the instrument plus detector, the wavelength, and the nature of the experiment. If the integration time is fairly long then it is essential to ensure that the telescope continues to track the object very accurately. In principle, this can be done manually by viewing the object or a nearby star with the TV camera, and pushing buttons on a "hand-set" electronically connected to the telescope drive motors in such a way as to counteract any drift of the image. More likely, this action which is known as guiding, will be performed automatically by the telescope control computer which analyses the image of the guide star on the TV screen, computes any motion and issues a correction to the drive motors of the telescope. When the exposure is complete the image or spectrum will be displayed on a computer screen, an adjustment to the setting of the instrument might be made and another exposure started. Meanwhile, some rapid analysis of the first result is undertaken using the observatory computing facilities. Analysis and display of the measurements the moment they are obtained is crucial to the optimum use of telescope time. The same pattern of work is used repeatedly throughout the night.

The night can be long, from before dusk until well after dawn, typically twelve hours non-stop. Considerable concentration and often a degree of patience is required—sometimes the latter is in relation to the other observers rather than with the experiment—and so some sustenance or "night lunch" might be taken "on the job". Some look forward to opening up the little brown bag collected at dinner, others would just as soon not watch! Depending on how smoothly the experiment has progressed or on what has been found, tactical decisions may be required to optimize the use of the night. Certainly, as dawn approaches, an extra effort is made to get the most out of the remaining time. A golden rule of observing, which many newcomers forget to follow, is to assume that "every night is your last" and never leave a crucial measurement or calibration until tomorrow. Finally, with the last exposure complete, the Night Assistant is given the go-ahead to close the dome, a few more calibration frames are made and then the mirror covers are closed and the telescope is returned to its parked position. At last the equipment is shut down or placed in standby mode, logbooks or fault reports are filled out, and the weary group rally round for the walk or drive back to the lodge. The observers sleep until early afternoon and then rise to prepare and review for the next night. Several days of this activity constitutes the "observing run". The visitors will then spend a few days at the offices of the observatory perhaps to obtain copies of their data on magnetic tapes, discuss their observations with observatory staff and then fly back to their home institute to analyse the astrophysical content of their data in detail and write a scientific paper.

If you plan to use single-dish radio telescope facilities then you may well follow the same pattern as optical and infrared astronomers, particularly for submillimetre observations. Interferometer and aperture synthesis telescopes such as the Very Large Array (VLA) near Socorro, New Mexico, are too automated to have visiting astronomers present for interactive sessions. Usually one submits an "observe" file and the observations are taken as part of a larger preprogrammed sequence. It may, however, be necessary for the astronomer to visit the facility during the data analysis phase to make use of computers and software not available at their home institute. Of course, the same process of competitive application for telescope time based on scientific merit is used.

A similar situation is also encountered for telescopes located in space. While it may be necessary to visit the "home" institute of the satellite, such as the Space Telescope Science

Institute in Baltimore or the ST-European Coordinating Facility near Munich, in the case of the Hubble Space Telescope, it is rarely necessary to visit the ground-station and operations centre. Time on the Hubble Space Telescope and many other satellites is very competitive and is awarded in cycles on the basis of scientific proposals reviewed by a peer committee. Once again when time is awarded it is necessary to submit a detailed observing request that is programmed into a larger sequence by observatory staff. You will definitely not get to "play" with the telescope! Often, your data will simply arrive in the mail on magnetic tape and usually in a reduced form ready for you to begin scientific analysis.

Of course, there are numerous privately owned observatories throughout the world, mostly associated with universities and research consortia. Some of these facilities are quite large and others are fairly small. At many of these places there is a much greater degree of "do it yourself". Nevertheless, the pattern of preparation and work is essentially the same.

Whatever the circumstances, professional astronomers usually work in the comfort of a warm control room, rather than in the darkened telescope dome or the open air, while electronic imaging devices and computer systems gather data. Occasionally, forays are made into the chilly mountain air outside to check on the weather, or as in my own case, simply to look up at the star-studded canopy of the night sky, marvel at its awe-inspiring beauty and remind oneself that *this* is what it is all about!

1

The development of electronic imaging in astronomy

1.1 OBSERVATIONAL ASTRONOMY

Even the earliest civilizations realized that careful astronomical observations were important to survival because such observations enabled them to predict seasonal events, such as when to plant and when to harvest. Observational astronomy was also one of the earliest scientific activities. The Greek astronomer Hipparchus (c. 127 BC) used astronomical observations to determine the lengths of the four seasons and the duration of the year to within 6.5 minutes. He also derived the distance to the Moon and the Sun, but his most amazing feat was to notice a small westward drift of the constellations which we now call the **precession of the equinoxes**. This effect causes the current Pole Star (Polaris) to move away from the North point and circle back after 26 000 years! Oriental astronomers recorded the appearance and fading of an exceptionally bright star in 1054 AD in the constellation we now call Taurus, but it was not until the twentieth century that Edwin Hubble associated this event with the supernova explosion which gave rise to the Crab Nebula. With the invention of the telescope in the early 1600s however, Galileo Galilei and others were finally able to enhance the sensitivity of the only light detector available to them—the human eye—and to resolve details such as craters and mountains on the Moon, the rings of Saturn, moons orbiting Jupiter and the individual stars in the Milky Way. By making careful drawings (Fig. 1.1) of what their eyes could detect during moments of minimum atmospheric turbulence—what astronomers today call moments of "good seeing"—the early seventeenth century scientists were able to convey pictorially those observations to others.

Better telescopes led to more astronomical discoveries, which in turn stimulated the development of even bigger and better telescopes. Opticians developed colour-corrected lenses for telescopes and then, following Isaac Newton, telescopes using reflections from curved mirrors instead of lenses were gradually introduced. William Herschel (1738–1822), a prolific observer and discoverer of the planet Uranus, pioneered the construction of many reflecting telescopes with long "focal lengths" and large magnifications; in later years the emphasis would move to larger diameter mirrors rather than longer focal lengths. With the invention of the prism **spectroscope** by Joseph Fraunhofer

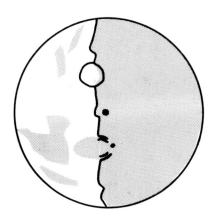 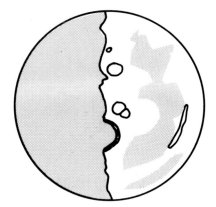

Fig. 1.1. A typical hand-drawn sketch of features on the surface of the Moon made by the first astronomers to use telescopes to enhance the power of the eye. Prior to photography, a (subjective) sketch was the only way to preserve a permanent record of the images seen.

(1787–1826), the chemical constitution of the Sun and stars became amenable to physical study. In Fraunhofer's early experiments a beam of sunlight was passed through a narrow slit and then through a glass prism to produce a coloured spectrum in the manner similar to Newton and others (Fig. 1.2). The critical addition made by Fraunhofer was a small telescope mounted on a moveable arm which could be set to precise angles to view the spectrum. Initially, the detector was still the human eye. Fraunhofer found that the normal band of colours from violet to red was crossed by numerous dark vertical lines. Eventually the pattern of these Fraunhofer or "absorption lines" (actually images of the entrance slit partially devoid of light) were shown to be characteristic of individual chemical elements. The elements hydrogen, calcium, sodium and iron were recognized in the spectra of the Sun, and later, the stars. Further spectroscopic observations of the Sun soon led to the discovery by Janssen and Lockyer (in 1868) of an unknown element which we now know to be a major constituent of the universe. This new element, helium, was named after the Greek word for the Sun—helios; helium was not discovered on Earth until 1895.

When dry, gelatine-based photographic emulsions became routinely available in the late nineteenth century, astronomers such as Henry Draper (1880) lost no time in putting them to use to catalogue the appearance and properties of a wide range of objects in the night sky. The photographic process was unarguably more accurate and more sensitive than the keenest human eye and the most artistic hand. From planets to stars to galaxies, the new observational tools were applied. Still larger telescopes were constructed, each a technical feat for its era, reaching a mirror diameter of 100 inches or 2.5 metres in 1917 with the completion of the Hooker Telescope on Mount Wilson by George Ellery Hale. Just one of the great discoveries which followed was the expansion of the universe by Edwin Hubble and Milton Humason in 1929.

The history of astronomy is marked by such sporadic progress. Each improvement in scientific apparatus, each new development in technology, helps to provide answers to old questions. Inevitably, the new observational methods uncover a host of new questions, which in turn, drive the quest for even better measuring equipment! Progress in studying

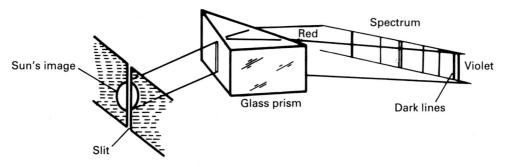

Fig. 1.2. The principle of Joseph Fraunhofer's spectroscope and the dark lines in the spectrum of the sun which now bear his name. This instrument combined with the photographic plate rather than the human eye opened the way to a physical understanding of the universe.

the universe has always been related to "deeper" surveys of the cosmos reaching to ever fainter objects, or higher resolution yielding more and more fine detail, or larger statistical samples from which generalizations can be made, or broader spectral response to sample *all* the energy forms passively collected by the Earth. That trend has continued since the Renaissance of the sixteenth century to the present day in a kind of ever-increasing spiral, with new tools or technologies leading to new discoveries which in turn drive the development of better tools.

A key feature of observational astronomy has been record-keeping; maintaining archives of observations, usually in some pictorial form, for future investigators to compare and consider. In terms of its ability to convert light into a measurable quantity, the photographic plate is actually less sensitive than the human eye. The great advantage of the photographic plate however, is that it can "build-up" a picture of a faint object by accumulating light on its emulsion for a long period of time. It is therefore called an "integrating" detector. The eye cannot do this. Moreover, the plate provides a permanent record which may be saved for future comparison and study by others.

By using a photographic plate as the recording device in a spectrometer astronomers could extend their investigations effectively and efficiently into the domain of quantitative astrophysics. Initially, of course, the flood of photographic material was analysed by human eye, and those eyes were mostly those of a dedicated group of female assistants hired by the director of the Harvard Observatory College, Edward Charles Pickering, toward the end of the last decade of the nineteenth century. Over forty women were employed by the observatory during the period of Pickering's tenure as director, and their efforts in handling the torrent of new astronomical data laid the foundations of modern astrophysics. Stellar spectral classifications led to the understanding that the colours of stars was largely a temperature sequence and that stars shine by the energy released in thermonuclear fusion reactions brought about spontaneously by the enormous temperatures and pressures at their centres. Among the most well-known of the Harvard ladies is Henrietta Leavitt whose work on the class of stars called Cepheid variables, which pulsate in brightness with a period which is proportional to their true or absolute average brightness, led to a distance-estimator and an appreciation of the true size and shape of our galaxy. During the first half of the twentieth century, these tools inevitably led to more discoveries (per year) and also to a massive increase in the "data rate", that is, the amount of information being collected, scrutinized and archived for posterity. But these advances were only the beginning.

Even as the 100-inch Hooker telescope was discovering the expansion of the universe, plans were being laid to build the great 200-inch (5-m) reflecting telescope on Mount Palomar in southern California. That telescope, named after George Ellery Hale, went into operation in 1946 and remains a powerful telescope to this day. Observational astronomy received another boost in the 1960s partly by the construction of several new major optical observatories with 4-metre class telescopes, that is, with mirror diameters of approximately 4 metres. These new facilities were located on excellent but somewhat more remote mountain sites in different parts of the world including the Arizona desert, the mountains of northern Chile, Australia and Hawaii. Another part of the expansion was stimulated by the exciting "new look" at the universe which accompanied the rise of radio astronomy and the discovery of completely new phenomena such as the incredibly luminous and distant quasars—thought to be supermassive black holes at the centre of large galaxies—and the remarkable pulsars—spinning neutron stars embedded in the remnants of a supernova explosion. All of this occurred during the successful development of the Soviet and American space programs which led to satellite astronomy and the opening up of the X-ray and ultraviolet regions. Other, more subtle, transformations began to occur around this time through the introduction of electronic computing machines and electronic devices which could be used as detectors of light. Photocells, photomultiplier tubes and sensitive "night-vision" TV cameras came first, but the steep rise of consumer microelectronic products through the seventies was to accelerate the changes rippling through astronomy. Even the telescopes themselves could be improved by the use of electronically encoded computer-controlled drive systems, thereby enabling much faster set-up times and more reliable tracking across the sky. The new radio and optical telescopes were remotely controlled, and the concept of converting measurements into an electronic form readily suitable for a computer to accept became standard practice. Computer power expanded exponentially, and astronomers eagerly used those capabilities to the full. And so the caricature of the "gentleman astronomer" peering through the eyepiece of some giant telescope was no longer even approximately true. Astronomy became more technological!

The explosion in astronomical data-gathering in recent years has been enormous. This flood of quantitative information is due to strides in the sensitivity of the detection devices used by astronomers. It is the impact of **semiconductor** (also called **solid-state**) electronic light-sensors attached to the new generation of telescopes which has had an effect as dramatic as was the introduction of the photographic plate itself over one hundred years ago.

There can be little doubt that we are living in a time of rapid technology development. This is the Semiconductor Age—the age of the "microchip". Semiconductor technology, of which the "silicon chip" found in computers is by far the most widely known example, has touched almost every aspect of our daily lives. The mass production of silicon chips has brought personal computers (PCs) of incredible power, at relatively low cost, to almost every environment—homes, schools, offices and industry. The Semiconductor Age is also the age of global electronic communication. There can be few people who haven't at least heard of the Internet and the World Wide Web! School kids can "download" images from the Hubble Space Telescope or send "email" messages to friends half-way around the world by typing at a computer keyboard.

What is a semiconductor? A semiconductor is a crystalline material with some of the properties of a good conductor of electricity (like copper metal), and some of the properties of an electrical insulator (like glass, for example). Because of its crystalline (solid-state) structure, a slab of such material behaves the same at all points. Semiconductor crystals can be "grown" in a controlled way from a melt, and moreover, the electrical properties can be tailored by introducing so-called impurity atoms into the crystal structure at the atomic level, so that by microscopic sculpting of the semiconductor material, all sorts of tiny electrical components and circuits can be constructed. The final piece—often not much larger than a thumbnail—is referred to as an **integrated circuit** or more commonly, as a "chip". Besides silicon, there is germanium, gallium arsenide, indium antimonide, and several other materials with these properties. Semiconductors can be used to manufacture a host of low-power microelectronic components including amplifiers, electronic counters, all sorts of logic units, computer memory, very complex chips called **microprocessors** capable of many computer functions, and tiny imaging devices of remarkable sensitivity. Silicon is the most well-developed semiconductor so far, but even for silicon the potential for yet smaller and smaller microchips still exists.

Astronomy has benefited in this semiconductor revolution because the apparatus needed for scientific experiments and for complex calculations, which were completely impossible before, have now become viable with the aid of new electronic imaging devices and powerful, extremely compact high-speed electronic computers.

Almost all modern astronomical research is carried out with photoelectronic equipment, by which we mean instrumentation which converts radiant energy (such as light) into electrical signals which can be **digitized**, that is, converted into numerical form for immediate storage and manipulation in a computer. Usually highly automated, and often remotely controlled, these instruments, and the telescopes to which they are attached, necessarily rely heavily on electronics and computers. Computers play an equally crucial role in helping astronomers assimilate, analyse, model and archive the prodigious quantity of data from the new instruments. The miniaturization of computers and the ever-increasing availability of large amounts of relatively cheap computer "memory", means that astronomers can employ fairly complex electronic and computer systems at the telescope which speed up and automate data-gathering and reduction. As a result, those astronomical facilities—which may be costly initially—and the data they produce can be available to a much wider range of scientists than would otherwise be possible. Surprisingly to some, a modern observatory requires an enormous breadth of engineering, scientific and managerial skills to operate efficiently and produce the very best results.

Most readers will be familiar with sources of current and topical astronomical results, whether these are professional journals (e.g. *Nature*, the *Astrophysical Journal*) or popular magazines (e.g. *Sky & Telescope*) or any of the numerous astronomical sites accessible on the World Wide Web. How are such remarkable observations obtained? Most press releases and/or scientific papers do not describe in detail the apparatus or the technology used in making the discovery. Of course, it would not be easy to do so because of the "jargon barrier" and the complexity of the technology itself. This is unfortunate, because it under-emphasizes an important link between modern technology and the quest for pure knowledge embodied in astronomy, a search for answers to the most fundamental questions about the universe we live in.

1.2 THE OBSERVABLES

Astronomy is an observational science. Unlike in a laboratory experiment, the conditions cannot be changed. That is, we on Earth are passive observers (so far) in almost all astronomical experiments, and we can do nothing other than intercept the various forms of energy which reach the Earth from the depths of space. Of course, there have been a few exceptions for solar system studies involving manned and unmanned spacecraft, and from time to time we can retrieve rocks from space which have survived passage through the Earth's atmosphere in the form of meteorites. The energy forms that we can intercept can be summarized as:

> **electromagnetic radiation** (gamma-rays through radio waves)
> **cosmic rays** (extremely energetic charged particles)
> **neutrinos** (tiny neutral particles with almost immeasurably small mass)
> **gravitational waves** (disturbances in a gravitational field)

Of these, the study of electromagnetic radiation which, as shown by the great Scottish mathematical physicist James Clerk Maxwell in 1865, incorporates visible light, is still the most dominant. Gravitational waves have not yet been detected directly, but neutrino detectors and cosmic ray experiments have been developed successfully. In this book we are concerned only with electromagnetic radiation and astronomical imaging techniques.

The nature of electromagnetic radiation is essentially characterized by oscillations of the electric and magnetic energy which gives the radiation the property of a wave motion. Different regions of the **spectrum** correspond to different **wavelengths** (denoted by the Greek letter lambda, λ; see Appendix 1 for the Greek alphabet) and the energy in the wave moves through empty space at a speed of approximately 3×10^8 metres per second (m/s), which is of course the speed of light (c); actually Maxwell derived this number from two electrical constants. The frequency of the oscillations (denoted by the Greek letter nu, ν) is related to the wavelength by the very simple equation

$$\nu \, \lambda = c \tag{1.1}$$

and the average intensity of the light is proportional to the square of the amplitude (or swing) of either the electric or magnetic part of the wave; electromagnetic waves are described in any good college physics text. The rate at which the energy flows is called the "power", and the power which falls on one square metre is the **irradiance** (W/m^2). Since the oscillations are transverse to the direction of propagation of the energy, these waves can be "polarized", that is, they have an associated "plane of vibration". As shown in Fig. 1.3, all the well-known forms of radiant energy are part of this spectrum. Radio waves are characterized by wavelengths of metres to kilometres, whereas X-rays have wavelengths around 1 nanometre (nm) or one billionth (10^{-9}) of a metre; other units such as the micrometre (micron) (μm, 10^{-6} m) and the angstrom (Å, 10^{-10} m) are commonly used. Visible light, with wavelengths from 400 to 700 nm, occupies only a very small portion of this enormous spectrum.

The measurements which can be made on electromagnetic radiation are also limited. Basically, we can determine:

- the *direction* of the radiation
- the *intensity* or energy spectrum
- the *polarization* or degree of alignment of the electric and magnetic fields in the radiation
- the *phase* or relation between waves

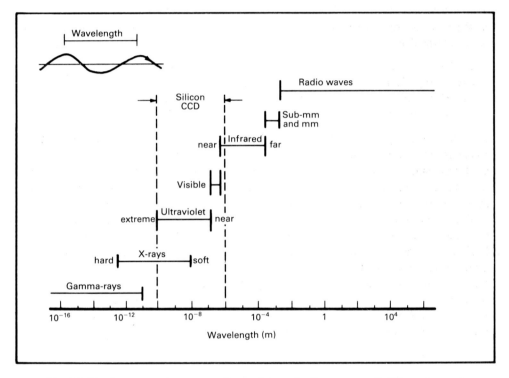

Fig. 1.3. The electromagnetic spectrum. X-rays, light and radio waves are all different forms of electromagnetic radiation. In the vacuum of empty space each of these forms of radiation travel in straight lines with the same speed— the speed of light.

All that we know about the universe must be extracted from measurements of these energy forms. Naturally, astronomy began as an optical science (rather than a radio or X-ray science) because human beings have optical sensors (our eyes) which detect the band of electromagnetic radiation which we call visible light. Since much of our technology is concerned with enhancing human vision, then it is hardly surprising that optical astronomy expanded rapidly with the advent of photography. Photoelectric devices and television cameras accelerated the pace, and now tiny solid-state imaging systems have pushed optical astronomy even further.

1.3 FROM EYES TO ELECTRONIC SENSORS

The blackening of silver halides when exposed to light gave birth to astronomical photography in the later half of the nineteenth century. Even today, photography is still a vital component in the astronomers' arsenal of observational tools. Modern photographic emulsions are sensitive to a wider range of wavelengths than the human eye; from the ultraviolet to the near infrared. Because it is difficult to relate the sensitivity to illumination of one part of a plate to another, brightness cannot usually be measured to high accuracies, compared to the corresponding exercise for electronic imaging devices. In addition, the ever-present, albeit faint light from the night-sky eventually saturates the emulsion rendering it insensitive. This behaviour is illustrated by the "characteristic curve" of the emulsion which is a plot of the "density" of the developed plate versus the logarithm of the "exposure". In this context, the density of a plate is a measure of how exposed (blackened) it is. Density (D) is just the logarithm (see Appendix) of the "opacity" (O) of the plate, which in turn is the reciprocal of the transmittance (T), i.e. the ratio of the amount of light transmitted (I_{out}) by the plate to the amount that is incident (I_{in}).

$$D = \log(O) = -\log(T) = -\log\left(\frac{I_{out}}{I_{in}}\right) \tag{1.2}$$

The term "exposure" measures the *total* amount of energy falling on a unit area of the plate and is found by multiplying the irradiance (E) of the light (in W/m^2) by the exposure time (t). A typical characteristic curve is shown in Fig. 1.4. Note that a minimum density occurs even for no exposure; this is the "fog" level of the plate. Then there is a short curved region in which the response suffers **reciprocity failure** which means that the response of the plate is not linearly proportional to the total energy. As we will see, this non-linear behaviour is in stark contrast to that of modern electronic imaging devices. Beyond a certain threshold in exposure the characteristic curve does become almost a straight line and the response of the plate to light is almost linear; the slope of this line defines the "gamma" for the plate ($\gamma = \tan \theta$). As gamma increases the contrast improves for a properly developed plate. Once all the grains are blackened there can be no further increase in density and so the curve "turns over" at large exposures. Unfortunately, the characteristic curve can differ even between two emulsions of the same type. Clever techniques, called hypersensitization, have been developed for treating photographic plates. For example, Kodak IIIaJ plates (Eastman Kodak, 1987) are "soaked" in nitrogen and hydrogen gas, in order to detect much fainter objects. In addition, highly automated computer-processing of large photographic plates has enabled astronomers to carry out significant statistical studies of many classes of objects. Nevertheless, it is the very low efficiency with which light is converted to blackened grains that limits the sensitivity, if not the utility, of photographic plates in astronomy.

From the seventeenth century, when Christiaan Huygens and Isaac Newton carried out fundamental experiments on the nature of light, evidence existed that light was a form of energy which tended to interact with matter in a way which suggested that the energy was transported as a "wave". For example, two beams of light could be made to "interfere" and produce a pattern of light and dark regions, very similar to the high crests and flat calms produced when two identical water waves meet; the effect is easily observed by dropping

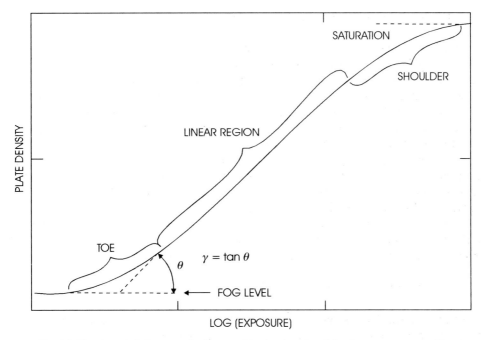

Fig. 1.4. The characteristic curve of a photographic plate is a plot of density versus exposure. The plate is linear over a limited range of exposure.

two pebbles simultaneously into a calm pond. Eventually, however, situations in which light behaved more like a stream of particles were encountered. It was not until the end of the nineteenth century with the work of Max Planck that it was understood that light can carry energy only in specific amounts, as though it came in individual packets; Planck called these packets **quanta**. For light (or any electromagnetic wave) with a frequency of vibration ν, the energy of one quantum is simply given by

$$E = h\nu = \frac{hc}{\lambda} \tag{1.3}$$

where h is a constant of nature called Planck's constant (= 6.626×10^{-34} J s). Other physical constants are listed in Appendix 3.

In the early 1900s Einstein's work on a phenomenon called the **photoelectric effect** showed that a quantum of radiant energy (now called a **photon** of light) could eject a negatively charged electron from the atoms in certain materials, and that the photon of light behaves like a particle—with energy and momentum—and yet has a "wavelength" associated with it. It was also discovered that a beam of electrons impacting on the surface of certain materials could cause the ejection of photons of light (or phosphorescence) and these developments led to the invention of **cathode ray tubes**, valve amplifiers, radio and ultimately to television. In the late 40s and early 50s, a photoelectric device called a **photomultiplier tube** became widely available for accurate brightness measurements. In such a device, a photon or quantum of radiation strikes the surface of a certain type of "photoemissive" material which responds by emitting an electron, provided that the

photon energy ($h\nu$) exceeds a minimum energy (ϕ) called the **work function** of the material. If the photon energy is less than this value, then no emission occurs irrespective of
how intense the light source is. Within the glass-encapsulated photomultiplier tube, the
negatively charged electron is accelerated by an electric field and made to impact on another photoemissive surface called a **dynode** (Fig. 1.5). The electric field is established by
a voltage potential of about 150 V between succeeding dynode stages; note that this implies about 1500 V (or 1.5 kV) across a tube of 10 stages. On impacting the first dynode
surface the photoelectron generates two or three new electrons which are all accelerated
and directed at yet another dynode surface where each electron generates two or three
more. This cascading process is continued several more times until a final surface collects
what is now a huge pulse of electric charge generated by just one photon of light. All that

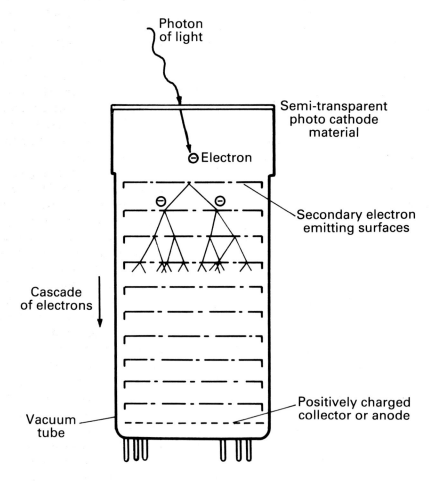

Fig. 1.5. The photomultiplier tube produces a large cascade of electrical current when illuminated
with faint light. It employs the photoelectric effect in which very thin slabs of certain materials
emit a negatively charged electron when bombarded by a photon of sufficient energy. The
electrons are drawn toward the positively charge anode.

needs to be done is to "count" the pulses for a given (arbitrary) time interval and determine the count rate; hence the process is called pulse counting or **photon counting**.

The beauty of this detector lies in the fact that the precision of the measurement depends only on the total number of photons counted. In fact, if N is the number of photons counted, then the random error or "noise" in N is given by $\sqrt{N}$; this is called Poisson or photon counting statistics and represents a fundamental limit (see Appendix for more information on statistics). Generally, it will be necessary to employ an additional amplifier and a circuit called a "discriminator" which suppresses pulses caused by unwanted effects. In some cases, the efficiency of photomultiplier tubes can be ten times that of a photographic emulsion and indefinitely long measurements (equivalent to exposures) can be used to detect extremely faint objects. The photomultiplier tube (or PMT) was a major innovation because it cheaply and easily allowed astronomers to establish reproducible and accurate brightness measurements of stars. Initially, the most common photocathodes were the S-11 (Cs_3Sb on MnO) and the $(Cs)Na_2KSb$ tri-alkali S-20 which responded from the UV to 600 and 700 nm respectively. Although S-1 material (Ag-O-Cs) went beyond 1000 nm into the near infrared, the response was very low (<1%). Most photomultiplier tubes in use in astronomy today have gallium arsenide (GaAs) photocathodes because these provide very good response from the UV to about 900 nm; these PMTs are also known as "negative electron affinity" devices. For best results, these detectors must be cooled to temperatures below about $-25°C$. The real drawback, of course, is that only one tiny patch of the sky can be observed at a time; a photomultiplier tube has no "panoramic" or camera-like advantage. It is like a single cell on the retina of the eye.

Alongside this development came the TV tube and work on electronic **image intensifiers** for very low light-level applications. The aim was to combine the attributes of accuracy and unlimited exposure time of the photomultiplier tube with the extended field of view of the TV camera. Military applications for ultra-low light-level camera systems helped to stimulate this technology and an astonishing variety of complex television-type image-intensified schemes were proposed and tested for astronomy in the 1960s and 1970s. Most of the devices begin with the light striking a photoemissive surface to release an electron, but a variety of ingenious methods are then employed to amplify the flow of electrons. Some image intensifier stages use electric fields (electrostatic amplification), others use magnetic fields and some use an intermediate stage to generate brighter light by means of a phosphorescent screen. Detection of the final amplified stream of negatively charged electrons, or the amplified light emission, is by some form of TV tube. The characteristic feature of all the various types of TV imaging devices is the use of a beam of electrons to "scan" across a region called the "target" on which there is an electrical charge pattern representing the optical image; the generic term for these electron-scanned imaging devices is a **vidicon** (see Fig. 1.6). In essence, the electron beam completes the circuit and provides a source of current to reset the target after exposure to light; the amount of current required for this action is the "video signal". TV tubes themselves were greatly improved during this time with the introduction of a range of new variants called the SEC, SIT and silicon vidicons and the Plumbicon or lead-oxide vidicon, which was the primary "studio" camera for many years. All of the TV-based image-intensified camera systems were complex, physically large, high-voltage (tens of kilovolts) systems. Some found instant success as faint-object TV track-and-guide cameras on large telescopes, but

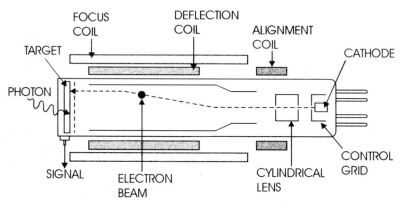

Fig. 1.6. A schematic representation of a vidicon showing its electron scanning beam.

most turned out to be difficult to use for accurate brightness measurements in astronomical cameras and spectrometers. Problems with the stability and reproducibility of the detector, critical for scientific research, often seemed to preclude effective techniques for the removal of systematic errors in long exposures. The most successful of the image-tube devices were those which appealed to photon counting rather than attempting to produce an amplified copy or "analogue" image of the scene. Two of the best-known of these systems were the Wampler Scanner and the Boksenberg Image Photon Counting System (IPCS). In the Wampler Scanner, developed at the University of California's Lick Observatory, the amplified light emission from the intensifier stage was fed to a unit called an "image dissector" which has a photocathode and an electromagnetic focusing mechanism which directs photoelectrons to a small rectangular aperture behind which is an electron multiplier stage just as in a normal photomultiplier tube. The area of the photocathode seen by the electron multiplier section can be "scanned" by controlling the electromagnetic focusing coils. IPCS systems were used for many years at the Anglo-Australian Telescope and the Isaac Newton Group on La Palma. Also, the Faint Object Camera (FOC) on the Hubble Space Telescope is an IPCS system. Photon counting arrays are especially well-suited to faint-object spectroscopy. Every fraction-of-a-second a frame of the TV output of the IPCS can be digitized and sent to a computer which can immediately analyse the pattern in "real-time". The main limitation of the IPCS is that it saturates around one event per pixel per second, which is a very low light-level.

Ironically, the perfection in TV tube construction was reached just about the time that the major breakthrough came in semiconductor devices. Things first began to take a different trend when in 1958 the first multi-transistor integrated circuit (IC) was demonstrated only 10 years after the invention of the transistor itself at Bell Labs[†], New Jersey, USA, in 1948. In the first transistors a small crystal of germanium was used; by adding other materials into the crystal structure it was possible to create a small sandwich which amplified electrical currents just like a (much larger) glass-encapsulated tube or valve.

Soon, silicon surpassed germanium as the best material to build transistors, the devices themselves got smaller as fabrication methods improved and then in 1958 the integrated

---

[†]In 1996 Bell Labs and other AT&T research centres were incorporated into a new organization called Lucent Technologies.

circuit was demonstrated. During the 1960s better and better manufacturing processes led to smaller and smaller integrated circuit units still capable of electrical functions. Today, silicon chips the size of a thumbnail can be constructed which contain millions of transistors (amplifiers). Dozens of these silicon chips could fit inside the glass vacuum casing of a single valve amplifier!

One simple device easily manufactured from silicon technology is the **photodiode**, in which the strength of the electric current flowing through the device is proportional to the amount of light falling on it. Small, cheap and low-powered, silicon photodiodes also had the immense benefit of very high **quantum efficiency**, that is, 80–90% of the incoming photons of light were converted to electrical charge. Compare this to 1–2% for photographic emulsions (perhaps 4% with hypersensitization) and 10–20% efficiency for the tube-type systems. Moreover, by the nature of the properties of silicon, this kind of sensor could be used far into the red, well beyond the limit of sensitivity of the human eye and much better than any existing detector. As imaging devices, silicon photodiodes were limited primarily by structural awkwardness in making a large densely packed two-dimensional (i.e. having width and length) assemblage or array of photodiodes with no gaps between them. Nevertheless, "linear arrays" of diodes (just one or two lines wide), used directly (Reticons) and behind image intensifiers (Digicons), found applications in astronomical spectrometers and several systems like this came into operation in the 1970s; Reticon and Digicon are both trade names.

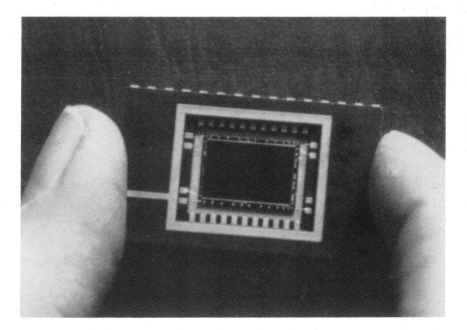

Fig. 1.7. A typical charge-coupled device or CCD. These tiny imaging devices are made from a small slab of the crystalline form of silicon containing upwards of 250 000 individual imaging sites called pixels. The one shown here is made by EEV. Photo courtesy Royal Greenwich Observatory.

In the late 1970s however, a new device challenged and virtually vanquished all contenders. It was the **charge-coupled device** or CCD (Fig. 1.7). The CCD is an array of microscopic square-shaped light-sensitive regions arranged in a checkerboard pattern. Tiny light-sensitive squares, usually called **pixels**—for picture elements—are formed directly in a slab of silicon. These pixels are so small that they cannot be seen simply by examining the surface of the CCD by eye. The introduction of these devices into astronomy has been revolutionary, and has sent repercussions through not only optical astronomy, but X-ray to infrared astronomy as well!

1.4 THE IMPACT OF SOLID-STATE IMAGING

Toward the end of 1969, two researchers, Willard S. Boyle and George E. Smith (Fig. 1.8) at the same Bell Labs in Murray Hill, New Jersey, where the transistor was invented, were investigating new ways of imaging with solid-state, silicon methods in an effort to develop a Picturephone! This concept involved having a tiny, inexpensive solid-state camera built into the telephone receiver to enable callers to see each other. Market research later failed to convince Bell Telephone of the worth of the Picturephone concept at that time; Picture-

Fig. 1.8. Willard S. Boyle (left) and George E. Smith inventors of the charge-coupled device at the Bells Labs AT&T research centre at Murray Hill, New Jersey, in December 1974 when they received their patent for the CCD. Photo courtesy Bell Labs AT&T.

phone is a registered trademark of Bell Labs AT&T. As luck would have it, the method devised by the two scientists turned out to be an incredible innovation, destined to change the whole philosophy of imaging away from vidicon-type TV tubes and even from cellu-loid movie film. Research on magnetic bubble memory, three-phase plasma display panels and silicon vidicons was in full swing at Bell Labs when executive director of the semicon-ductor division, Bill Boyle, and his close friend George Smith—who was department head in charge of developing a silicon diode array camera tube for Picturephone applica-tions—got together in front of a blackboard one afternoon and began musing about the idea, as George put it, of "friendly competition" from silicon technology. Bill asked George what about an "electric bubble". The obvious analogy to passing a magnetic do-main from one site to another was to pass charge from one site to the next. It was already known that charge could be stored by insulating a small metal plate placed on the surface of a silicon crystal, but it was the concept of stringing these storage sites together and using voltage differences between them to pass the charge along that constituted the new idea. Devising the basic concept took only a couple of hours and within a few weeks George had a six-element device under test! They began their paper, published in the *Bell System Technical Journal*, Vol. 49, No. 4, in April 1970, with the following words, "A new semiconductor device concept has been devised which shows promise of having wide application". Their invention was of course the charge-coupled device or CCD, a name chosen by Bill Boyle who recalls the tremendous reception given to the idea following its announcement. As Mike Tompsett, who came to Bell Labs from English Electric Valve (EEV) in the UK where he had worked on the development of the pyroelectric vidicon and himself an author of a classical text on CCDs recalls, "CCDs were an idea of the time. Within a few months all sorts of applications were listed—and many were actually of relevance to the phone company". Bill and George recall reactions to their idea as ranging from "I should have thought of that" to "it will never work"! But it did work. Experimental verification of the charge-coupled device concept was published in the same issue of the *Bell System Technical Journal* by Gil Amelio, Mike Tompsett and George Smith.

CCDs have a great many advantages over other electronic and photographic imaging devices. They are small, linear, stable, low-power devices with excellent sensitivity over a very wide range in wavelengths (see Fig. 1.9) and a huge range in light levels. Moreover, there is the possibility of relatively low cost due to silicon manufacturing processes and mass production. Commercial applications of CCDs now range from home video cameras to professional broadcast cameras, to remote-sensing and surveillance, to robotics, to medicine and to science. For astronomical applications, CCDs are nearly perfect and rep-resent such a dramatic improvement over all other imaging techniques, except perhaps at the very faintest levels in spectroscopy, that they were promptly embraced by astronomers. Indeed, it is fair to say that astronomers and space scientists were in the forefront of the pioneering days of CCDs. One of those very early developments was an effort by Texas Instruments to provide large CCDs under contract to NASA Jet Propulsion Lab for inclu-sion in instruments being proposed for the Jupiter Orbiter Probe (later re-named the Galileo mission) and for the Space Telescope (later renamed the Hubble Space Tele-scope). Further details of this remarkable development are given in Chapter 6.

CCDs have found such astonishingly widespread use in science in general, and astron-omy in particular, that it is really essential for all newcomers to understand their operation

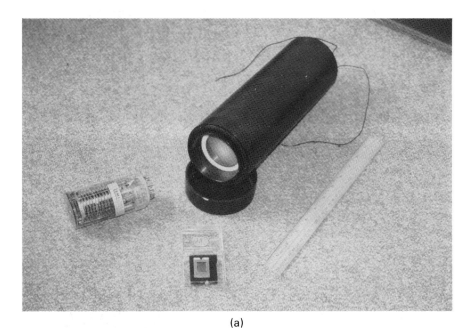

(a)

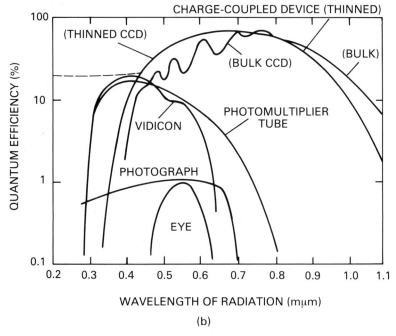

(b)

Fig. 1.9. (a) Comparing the sizes of a CCD chip, a photomultiplier tube and a large (black tube) image intensifier. The ruler is 12 inches long. (b) The sensitivity or quantum efficiency of certain CCDs to light of different wavelengths compared to other forms of light detectors. Note that the scale on the left increases by factors of ten! One of the CCD's main attributes is its extensive range in wavelength coverage.

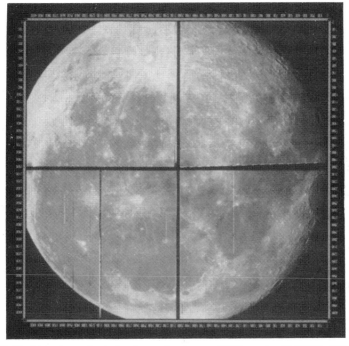

Fig. 1:10. The Moon imaged onto four large, closely spaced CCDs each with 2048 by 2048 pixels. Additional streaks in the image are due to bad regions of the detector; these effects can be removed by appropriate procedures. Courtesy NOAO/Todd Boroson.

and applications. In astronomy, CCDs are routinely used for all kinds of imaging. Today, no major observatory in the world lacks a suite of CCD cameras! In fact, CCD cameras are now widespread among amateur astronomers too, so much so that popular magazines dedicated to CCDs, such as Sky Publishing Corporation's *CCD Astronomy*, can be supported.

In an introductory paper to the conference on "Solid State Imagers for Astronomy" organized by John Geary and David Latham and held June 10–11 1981 at the Harvard-Smithsonian Center for Astrophysics in Cambridge, Massachusetts, Herbert Gursky of that institute, opened the meeting with the following remark, "There is by now enough information, in the form of laboratory data on a variety of devices and published scientific results, to confirm the original assessment that solid state arrays will become a permanent entry in the repertory of astronomical instruments". He was right!

The impact of CCDs has been phenomenal. Charge-coupled devices have been installed in cameras of every description, in spectrometers and in numerous ingenious, though more specialized, instruments, such as speckle cameras, coronagraphs, polarimeters and velocity meters. Astronomers now "design their own" CCDs and collaborate to organize runs at "silicon foundries" to make devices better suited for astronomy applications. Although the first CCDs used in astronomy had 10 000 pixels in an array of 100×100, devices are now available with typically 4–16 million pixels, and astronomers try to package more than one CCD into an instrument, i.e. they form a "mosaic" of detectors (Fig. 1.10). Forthcoming chapters will explain how CCDs work and what astronomers

Fig. 1.11. A deep, four-hour exposure of a cluster of galaxies called Abell 2390 obtained by Gary Bernstein and Tony Tyson using a CCD camera on the 4-m telescope at Kitt Peak. The short streaks on bright stars is called blooming and is explained in later chapters. Courtesy Tony Tyson.

have to do in practice to achieve remarkable images, such as Fig. 1.11 and others through-out the book. For more advanced treatments on a variety of topics the reader will be guided to publications in the technical literature.

CCDs have been used for observing everything from the Sun to the most distant objects in the universe. Fig. 1.12 shows one rendition of the impact of CCDs on faint-object as-tronomy. The curve shows how the limiting brightness in faint galaxy surveys improved with the introduction of the photographic plate and then remained fairly steady until the invention of the CCD.

1.5 OTHER TECHNOLOGIES

The CCD is not alone. Astronomers have incorporated many other technologies into their instruments in an effort to make better measurements. In addition to powerful electronic computers of many kinds, we will also encounter sophisticated electronics, optical fibres, new methods of fabricating and coating optics, applications of lasers, holography and su-perconductivity.

An in-depth study of CCDs will make it easier to understand how similar techniques applied to infrared astronomy have dramatically impacted that field. Silicon CCDs turn out to be excellent detectors at X-ray wavelengths too. The new light detectors on the world's largest existing telescopes have now attained limits of sensitivity close to the theo-

MODERN LIMITS FOR GALAXY PHOTOMETRY

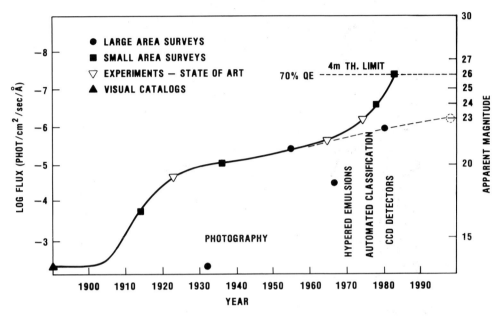

Fig. 1.12. The history of the faint limit of galaxy photometry has been dominated by advances in detector technology. The diagram shows the faintness levels detected in optical surveys as a function of time. The lower dashed line charts the large-area photographic surveys by Schmidt telescopes, but it is the CCD which has pushed the efficiency per detector area of modern telescopes to their theoretical maximum. Courtesy Tony Tyson, Bell Labs.

retical maximum at some wavelengths! Any further improvements in sensitivity will require telescopes with much larger collecting mirrors and/or observatories in space.

Throughout the world there is currently an intensive effort under way to develop a new generation of very large ground-based optical/infrared telescopes. At the time of writing, fourteen telescopes with diameters of 6.5 metres or larger are planned, under construction or already completed. Several of these facilities include plans to link individual telescopes as interferometers, in much the same way as radio astronomers have been doing for years. To build these huge telescopes, new mirror-making technologies have had to be developed. Computer-controlled, multiple segmented mirrors are one of the innovations, another is the use of "spin-casting" glass to remove most of the material before polishing.

In addition, the combination of high-speed image data processing and recent CCD technology has made it possible to implement strategies which "optically correct in real time" the distortions in an image caused by passage of the light through the Earth's atmosphere. The promise of this new field, called "adaptive optics", is that the very large telescopes now under construction around the world will perform as if they were located in outer space, instead of at the base of our turbulent atmosphere.

Bigger and better detectors, remote operation, overcoming atmospheric turbulence with mirrors that change their shape at high speeds, employing laser beams to generate artificial guide stars in the sky, and achieving "super resolution" are among the many technical innovations ahead. The magnificent Hubble Space Telescope and the gigantic, twin 10-metre Keck telescopes already in operation, represent just the beginning of a new wave of technological innovations which will propel astronomical imaging well into the twenty-first century.

SUMMARY

We have shown that progress in our understanding of the universe is intimately linked with developments in technology. Larger telescopes and more sensitive detectors of light, together with advances in electronics and computers, have enabled astronomers to observe and measure natural phenomena in a quantitative way. New technologies lead to new discoveries and the quest for more knowledge drives the development and search for better techniques. Electronic imaging devices which convert photons to electrical charge are the primary tools of modern astronomy.

EXERCISES

1.   Compare and contrast the following three classes of astronomical detectors: photographic emulsions, photomultiplier tubes and CCDs. What are the advantages and disadvantages of each detector?

2.   Cite some examples of how the development of technology has impacted astronomy in (a) the last 100 years (b) the last 25 years.

3.   Summarize the innovations to practical astronomy made by Galileo Galilei, Joseph Fraunhofer, Isaac Newton, William Herschel and Henry Draper.

4.   Why is astronomy called an observational science? List all the possible "observables" that can be used to study the cosmos.

5.   (a) What is meant by the "characteristic curve" of a photographic plate? (b) Draw and label all parts of a characteristic curve with $\gamma = 2$. (c) What does this curve imply for precise quantitative measurements? (d) How can photographic emulsions be made more sensitive?

6.   What is the wavelength interval in nanometres occupied by visible light?

7.   (a) Calculate the frequency of a radio wave which has a wavelength of 1 mm. (b) Calculate the wavelength of a radio wave with a frequency of 1427 MHz.

8.   List all the properties of electromagnetic radiation which can be measured?

9.   What is meant by a "quantum" of energy? Calculate the quantum energy associated with a photon of light of wavelength 500 nm.

10.   What important invention occurred in 1970 that was to change the way modern astronomy is done at visible-light wavelengths? List some advantages of the new device.

11.   What is meant by an "image photon counting system"? Why are they advantageous for very low light levels? What is their primary disadvantage?

12.   In a given photomultiplier tube, each dynode surface generates three electrons for the impact of one electron. Show that the "gain" of a photomultiplier tube with 10 dynodes is 3^{10}. Given that the charge on the electron is 1.6×10^{-19} coulomb, how much charge is this?

13. A photon-counting experiment collects 1 million counts in a given time. (a) What is the error in this measurement assuming Poisson statistics? (b) What is the fractional error expressed as a percentage? (c) If the measurement time is increased by a factor of 4, by how much does the percentage error improve (assuming a constant flux)?

14. What other technologies do astronomers use?

REFERENCES

Amelio, G.F., Tompsett, M.F., and Smith, G.E., (1970) Experimental verification of the charge coupled device concept, *The Bell System Technical Journal*, Vol. 49, No. 4, 593–600.

Boyle, W.S. and Smith, G.E. (1970) Charge coupled semiconductor devices, *The Bell System Technical Journal*, Vol. 49, No. 4, 587–593.

Csorba, I.P. (1985) *Image Tubes*, Howard Sams, Indianapolis, Indiana.

Eastman Kodak Co. (1987) *Scientific Imaging with Kodak Films and Plates*, Rochester, New York.

Geary, J.C. and Latham, D.W. (Eds) (1981) *Solid State Imagers for Astronomy*, SPIE, Vol. 290, Bellingham, WA.

Tyson, J. A. (1986) Low-light level charge-coupled device imaging in astronomy, *J. Opt. Soc. Am.*, Vol. 3, No. 12, 2131–2138.

SUGGESTIONS FOR ADDITIONAL READING

The following books provide additional material relevant to this chapter, either in terms of historical developments, general astronomy or more quantitative treatments.

Bode, M.F. (Ed.) (1995) *Robotic Observatories*, Wiley–Praxis, Chichester, England.

Birney, D. Scott (1991) *Observational Astronomy*, Cambridge University Press, Cambridge, England.

CCD Astronomy, Sky Publishing Corp., Cambridge, MA, USA; quarterly magazine.

Computers and the Cosmos (1988) by the editors of Time-Life Books, USA.

Gingerich, O. (Ed.) (1975) *The Nature of Scientific Discovery*. Proceedings of a symposium in honour of Copernicus, Smithsonian Institution Press, Washington.

Hearnshaw, J. B. (1986) *The Analysis of Starlight: One Hundred and Fifty Years of Astronomical Spectroscopy*, Cambridge University Press, Cambridge, England.

Kaufmann, W.J. and Comins N.F. (1996) *Discovering the Universe,* 4th edn, W.H. Freeman, New York.

Krisciunas, K. (1988) *Astronomical Centers of the World*, Cambridge University Press, Cambridge, England.

Learner, R. (1981) *Astronomy through the Telescope*, Van Nostrand Reinhold, New York.

Moore, P. (Ed.) (1987) *The International Encyclopedia of Astronomy*, Orion Books, New York.

Pasachoff, J.M. (1992) *Journey through the Universe*, Saunders College Publishing.

Rieke, G.H. (1994) *Detection of Light from the Ultraviolet to the Submillimeter*, Cambridge University Press, Cambridge, England.

Roy, A.E. and Clarke, D. (1988) *Astronomy: Principles and Practice*, 3rd edn, Adam Hilger, Bristol, UK.

Verschuur, G.L. (1987) *The Invisible Universe Revealed—the Story of Radio Astronomy*, Springer Verlag, New York.

Wall, J.V. and Boksenberg, A. (Eds) (1990) *Modern Technology and its Influence on Astronomy*, Cambridge University Press, Cambridge, England.

Zeilik, M. (1994) *Astronomy: the Evolving Universe*, 7th edn, John Wiley, New York.

2

Automation and computers

This chapter continues our introduction to modern astronomical instrumentation with a brief discussion of the impact that technology has had on the way that astronomical research is organized, particularly in terms of the development and support of ever more complex instrumentation. Large telescopes and multi-purpose detection equipment, together with all their attendant electronic control and data-gathering systems, are quite expensive and time-consuming to construct. Consequently, these facilities must be widely shared to be cost-effective. On the other hand, there is perhaps an increased risk of inefficient use of the instruments and telescopes because many of the visitors will lack detailed technical knowledge about the equipment. To counteract this it is essential to design and develop the instrumentation and the observatory to a much higher standard of reliability and ease-of-use. Achieving this "user-friendliness" requires project management, teamwork between astronomers and engineers, closer ties with industry and significant levels of funding. Training in astronomy should include exposure to this kind of interdisciplinary experience.

2.1 COMMON-USER INSTRUMENTS AND AUTOMATION

Many individual astronomers in university departments and research institutes throughout the world design and develop innovative instrumentation which they transport to various telescopes, but there has also been a strong trend toward shared or **common-user** facilities. Fig. 2.1 shows a typical Cassegrain cluster of instruments attached to a modest-sized telescope at a national facility. Apart from making economic sense by preventing needless duplication, common-user instruments have the advantage that their properties become well-known and more widely known to many people and so questions of systematic errors or optimum methods of calibration become less contentious. Expertise becomes distributed and better scientific results ensue. Of course, it is essential that individual creativity is not stifled in this process, and there must always be opportunities for specialized, "one-off" experiments in which the apparatus is not intended to have a long life, or to be used by anyone other than the builders.

Most often, a consortium of university groups and a national laboratory or observatory will get together to define or specify the new instrument. Usually, some effort is made to

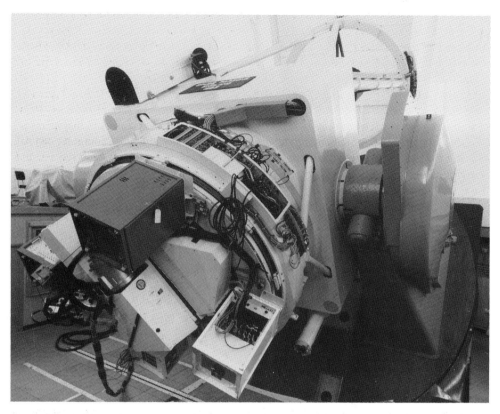

Fig. 2.1. A compact cluster of electronic, computer-controlled instruments attached to the Cassegrain focus of the 2.5-m Isaac Newton Telescope. Courtesy Royal Greenwich Observatory.

ensure that the new instrument is "modular" in design and reasonably flexible to allow innovative additions at a later time. It is particularly important to ensure that new detectors can be readily substituted. To be competitive, each common-user instrument must be very sensitive, must be easy to use correctly at the telescope and must provide computer-compatible data in a form which can be readily handled by the guest astronomer at his home institute. The instrument must have a high degree of reliability, not only because the original builders/designers may not be around to participate in every experiment done with it, but also because such instruments tend to become workhorses, and get used night after night on the cold mountain-tops where the world's major observatories reside. Inevitably therefore, common-user instruments must be well-engineered.

Modern astronomical instruments are highly automated and computer-controlled. Observational sequences can be pre-programmed and linked to the motion of the telescope, and in some cases the entire system can be operated remotely from the other side of the world.

Perhaps one of the most contentious questions that frequently arises is, how much automation? For instance, having invested in a modern electronic imaging system with a CCD camera, does it make any sense to have hand-operated mechanisms for changing filters? The answer really depends on the way the instrument is to be used. One concern at major observatories with large telescopes is safety, since the instrumentation clustered at the focus soon becomes quite inaccessible once the telescope moves away from the vertical position. It is also true that there is much interest in "remote observing" so as to permit more participants in a project yet avoid incurring large increases in travel costs and logistics problems from having too many collaborators all at the remote site simultaneously. Finally, by placing the instrument under computer control, observing sequences can be pre-programmed which not only saves time, but also reduces the chances of human error.

Given that the electronic imaging device being used is already under computer control, it is not difficult to ensure that all other mechanisms which are routinely required can also be computer-controlled. Trying to reach instruments attached to large modern telescopes by standing in the dark on tall step-ladders when the telescope is pointing at a source well away from the zenith is simply impracticable, and dangerous!

Astronomers have found that automation pays dividends in scientific flexibility also. By providing a modular package of control programs and other software options, modern astronomical instruments can be "reconfigured" to optimally match the needs of the most ingenious researcher without recourse to manual adjustments. It is this feature of re-programmability which ensures that the extra cost and extra effort which must be expended to design and construct a common-user instrument is worthwhile. Wherever possible, then, the trend has been toward having all the key features of a modern, common-user astronomical instrument under remote computer control.

2.2 DATA MANAGEMENT

A major problem facing all of astronomy as a result of these technological innovations is data "management". Ground-based optical and near-optical observational astronomy alone, to say nothing of space and radio astronomy, now produces data in the form of images or spectra, in such voluminous quantities as to make interactive analysis by human beings difficult. How much data? Typically, the information in an image is digitized to 16 bits or $2^{16} = 65\,536$ levels. Recall that eight bits (0s or 1s) make a byte, and 2 bytes (16 bits) make a word; 32 bits is a double word (corresponding to $2^{32} = 4\,294\,967\,296$ pieces of information). A CCD with 1024×1024 pixels and 16-bit digitization requires 2 megabytes (Mbyte) to store a single frame, and a typical 2048×2048 camera needs 8 Mbyte/frame. At 50 images per night (a lower limit at most observatories) the storage requirement becomes 400 Mbyte/night. At any one major facility, where several telescopes and cameras operate each night (see Fig. 2.2), a data collection rate of about 1000 Mbyte (1 gigabyte) per night is easily obtained. In only three years (3×365 nights), that single observatory would have amassed over 1 terabyte (1000 Gbyte) of digital data.

Again, a great reliance must be placed on computers. Unfortunately, the burden of cost on astronomy is considerable, and due to the rapid pace of developments in the computer industry, it is hard to keep up. Most observatories are still under-equipped with suitable computer hardware, assorted peripheral equipment and sufficiently powerful computer programs (the "software"). As a result, the "turn-around" time in getting observations (data recorded at the telescope) into the form of published scientific results in a professional journal can be displeasingly long. Getting the numbers needed from the experimental apparatus right there at the telescope requires what is known as "on-line" data reduction, and this is becoming more and more feasible. Even more serious is the task of **archiving** all the digital data being produced. A photographic plate has a big advantage in terms of long-term data storage over magnetic tape. Fortunately, bulky reels of 9-track magnetic tapes have given way to more compact and robust storage media, and there is a worldwide effort to establish "astronomical data centres" where the digital data from a wide variety of astronomical sources—both ground-based and space—are not only safe-guarded, but can be accessed electronically using facilities available on the Internet and the World Wide Web.

Fig. 2.2. The busy mountain top of Kitt Peak, Arizona, home of the US National Optical Astronomy Observatories (NOAO) where terabytes of digital data are being produced every few years by electronic imaging systems. Courtesy NOAO.

2.2.1 Data archives

Several electronic astronomical data archives now exist around the world. Among these are the Canadian Astronomy Data Center (CADC), the Strasbourg Astronomical Data Archive and the Astrophysics Data System (ADS).

ADS is a distributed data system chartered to provide access to information gathered from NASA space missions. The roots of the development of the ADS began in the late eighties with the realization that the "data volume" from projected NASA missions was going to be enormous and that some action to gather and archive the data was going to be essential. Actually, the data volume was underestimated! The first operational release came out in June 1991 and was updated to a Graphical User Interface in 1992/1993 and in 1994 the service became available on the World Wide Web. The ADS provides access to several hundred catalogues, several data archives and hundreds of thousands of abstracts. Major development centres are, the Center for Astrophysics & Space Astronomy (CASA), University of Colorado, Boulder, CO, the Infrared Processing and Analysis Center (IPAC), Caltech, Pasadena, CA, and the Smithsonian Astrophysical Observatory (SAO), Center for Astrophysics, Cambridge, MA. Many other data nodes also participate, including the Automated Plate Scanning Project (APS), University of Minnesota, Center for EUV Astrophysics (CEA), University of California, Berkeley, and the Space Telescope Science Institute in Baltimore. The ADS also allows access to several other database systems such as SIMBAD and the NASA Extragalactic Database (NED). Two image display tools are provided: SAOimage and IPAC's Skyview. An excellent review is given by Guenther Eichhorn (1994).

2.2.2 The Internet and its facilities

If you visit any bookstore or computer shop you will find many excellent books which explain and introduce the Internet and the World Wide Web for any level required. Most readers of this text will have encountered the Internet and the Web in some way. Here is a brief discussion for completeness.

Most introductions to the Internet begin by stressing that it is not a singular entity. Indeed there is no "it". The term Internet refers to a vast set of **inter**connected **net**works of computers. There is no centre or hub, it is not a "star-system" with one almighty powerful computer somewhere, it is much more like a spider-web with multiple links to many computers and computer networks, called domains. Each domain is linked by special high-speed telephone lines operating at rates in excess of 56 kbps. The genesis of the Internet was a distributed, web-like network of computer systems established in 1969 by the United States Defense Advanced Research Project Agency known as the ARPANET with the idea that whatever happened to one site, it would still be possible to route information to everywhere else, because of the multiple connections approach. In the 1980s the academic community built on this idea and developed its own network called NSFNET with funding from the US National Science Foundation. Similar developments occurred in other nations too. Eventually, everything was merged into the current Internet. There is no central control, no central ownership and no central data storage. There is, however, an Internet Society (ISOC) which is a non-profit organization formed in 1992, based in Reston, Virginia, USA, and composed of representatives of all the users and providers of the Internet with the goal of maintaining and broadening Internet availability. A subgroup of ISOC called IANA (Internet Assigned Numbers Authority) supervises allocation of Internet addresses.

Electronic mail

Electronic mail is simply a message typed at a keyboard which is sent electronically over the Internet to the recipient instead of via paper mail. Usually you will prepare the message within an "email environment" or program and provide an address and a subject. To send electronic mail over the Internet you need to know the "username" of the person, the name of their computer—the "host"—and the network or "domain" to which the machine is connected. Often, electronic mail has to make its way through several computers called "gateways" to get where it is going. A standard of communications called **TCP/IP** (Transmission Control Protocol/Internet Protocol) is generally observed and the domains are grouped into the following categories:

com—commercial and business
edu—educational/research institutions
gov—civilian government institutions
mil—military installations
net—network resources
org—other organizations (typically non-profit)

Individual computers are assigned a numerical address of four numbers separated by periods within a hierarchical system, e.g. 128.xx.yy.zz. The first number (128 in this case) identifies the individual network and the other numbers specify the subnet, site and individual computer. These numeric codes correspond to alphanumeric descriptions (usually saved in host tables) which are of the general form

username@machine.subnet.site.domain

For addresses outside the USA there is usually a top-level domain with an abbreviation for the country (e.g. au for Australia, ca for Canada, fr for France and uk for the United Kingdom).

Wide area information servers (WAIS)
This tool (usually pronounced "wayz") is intended for searching rather than browsing. WAIS servers contain electronic databases on specific topics, such as literature, molecular biology or astronomy, and present the user with a query form into which you enter criteria for your search. Essentially, WAIS is for searching the Internet when you know what you are looking for and where to find it, at least in a general sense.

File transfer protocol (FTP)
If you want to "log on" to another computer and copy over or "download" some files to your computer then you need FTP (pronounced by saying the letters). The term FTP is used in two ways, very similar to the way we use the word "telephone". You can say "the data are available by FTP" or "ftp-ing the file to you will be easy". FTP is both a program which enables file transfers and a protocol or set of rules which ensures that the exchange of information is independent of the original machine and/or the original file format. There are many windows-based programs and Internet browsers which embody FTP facilities and make it easy to perform downloads. Usually, you will need to have an account, username and password on the machine that you are trying to access. Typically, you would type "FTP machine name or code" (without the quotes) and that computer would reply with a request for your name and password. Simple commands like "put filename" and "get filename" are then used to transfer specific files, by substituting their names for "filename", from that machine to your machine and vice versa. NASA and other similar organizations often wish to make files (e.g. Hubble Space Telescope images) freely available for anyone to download without having an account or password. In this case you can use "anonymous FTP", that is, you log in to the remote computer using the account/ user-

name "anonymous" and there will either be no password or you must give your own Internet address as a password.

Telnet

Telnet is a service which allows one computer to contact another by knowing its Internet address. For example, if you have the Telnet software, then you simply type (without the quotes) "telnet hostname" (or substitute the numeric code mentioned above) and you will be connected to that machine. Of course, you will need an account/ username and a password to get logged into the machine you have reached, but once connected, you can then access all of that computer's features just as if it was your own.

Gopher

Originally developed in 1991 by the Computer & Information Services at the University of Minnesota—whose mascot is the gopher—to provide an answer service on computers to all campus users. Gopher uses a "client" program on the user's machine to obtain information from a "server" program running on a remote computer. Gopher is more than FTP and more than Telnet. You select items from a menu or enter text. Text files can be displayed immediately and bookmarks can be inserted for easy return to that item. A Gopher server can also run a local program and then send results to the user.

The World Wide Web

In 1989, scientists at the CERN—the European Particle Physics Laboratory in Geneva, Switzerland—set out to bring more organization and hospitality to the Internet by gathering all the tools and features in one place and providing better ways of linking from one topic to another. Their new concept, the World Wide Web (WWW) was introduced in 1990 and resulted in an instant "explosion" in the use of the Internet. Within the WWW you can find ftp, Gopher, WAIS gateways, email and virtually everything else! Although the Web is part of the Internet, you need a Web "browser" to gain access. There are many such programs (e.g. Netscape Navigator, NCSA Mosaic) and many providers (e.g. America-on-Line, Compuserve, Prodigy), and frequent reviews occur in popular computing magazines. A Web browser usually starts with its "home page" and gives you many options on where to go from there. Each web "page" consists of text and graphics (sound and video can also be included) with several areas of the page marked with so-called "hyperlinks". If you click on a hyperlink you are transferred to another document or page, or perhaps to another computer half way around the world. From there, depending on which hyperlink you click, you can be whisked away to yet another computer. If you were to do this for a few hours and plot lines on a map of the world showing your path, it might indeed look like a "web"!

A few other important terms are the "URL", or Uniform Resource Locator, which is the specific address of each page, article, text or image on the Web. The URL begins with **http** which stands for HyperText Transfer Protocol, followed by a colon and two slashes "://" and sometimes, but not always, "www." and then an Internet address. Subdocuments are separated by slashes, e.g. *http://www.astro.ucla.edu/irlab/irhome* would connect you to the "home page" for the Infrared Lab at UCLA. Each Web page is created using a programming language called HTML which means HyperText Markup Language.

Astronomers make good use of the Internet and the World Wide Web, and astronomical sites are very popular among browsers. Astronomical images are among the most frequently downloaded images from the Web.

2.3 GETTING FEEDBACK TO THE ASTRONOMER

A major concern, with all forms of electronic imaging is how to handle the electronic signals from the detector in such a way that the astronomer can appreciate what is going on at the telescope. Simply recording images on magnetic tape for later analysis is no longer adequate. Modern computers, called **workstations**, with large colour monitors can display images and spectra just seconds after the observation has ended, and programs can perform numerical manipulations of the digital images known as "reductions". For example, the object of interest may be faint relative to other "background" sources of light such as moonlight, and so this background must be subtracted. Likewise, no matter how good the manufacturing process is, each pixel in a CCD will differ slightly from its neighbours in its response to light and this variation must be corrected by comparing every frame with a calibration image. Performing thousands of repeated subtractions and divisions is just what a computer is good at doing. Sometimes this is not enough, however. In astronomical measurements of faint sources the contrast between the interesting part of the image and the rest of it is often poor. Computers can be used to perform other types of numerical

Fig. 2.3. The control room of a modern telescope (the WIYN telescope, located on the far left in Fig. 2.2). This is an altitude-azimuth mounted telescope with an actively controlled primary mirror. Courtesy NOAO.

manipulations which help to pick out the significant information, and present the result in a form more useful to the astronomer such as a contour map, or a picture in "false-colour" in which shades of grey are replaced by unusual colour hues to make low-contrast features more prominent (see the colour plate section). All of these techniques are described in later chapters.

The ability to assess the quality of the data being obtained at the telescope in so-called "real-time" is in no sense a luxury. It requires, of course, considerable provision for computers and associated equipment called "peripherals" at the telescope, and it may require additional staff to support those facilities, but in the end it is more cost-effective and more efficient in the use of telescope time because it gives the astronomer the (only) opportunity to properly control the experiment.

Therefore, at most observatories, the visitor will encounter a telescope "control room" such as the one shown in Fig. 2.3, containing several TV-type displays and several computer screens and keyboards. The status of the telescope and its instrumentation will be on display, together with normal images of the sky seen by "acquisition and guide" TV cameras—which also may be CCDs—and computer-enhanced images or spectra of the particular region under study.

2.3.1 Performance, operating systems and programming languages

Benchmarks for computer performance are difficult to standardize. In the mid-eighties, the Digital Equipment Corporation (DEC) VAX11/780 computer was common throughout astronomy and had a MIPS (millions of instructions per second) rating of 1. Typical "desk-top workstations" (e.g. Sparc 10/51) now achieve about 155 MIPS and a 133 MHz Pentium gives 188 MIPS. Computers designed for commercial applications do not normally rely on good floating-point performance but, since this quantity is independent of the computer architecture, it is a useful measure for the scientific user. Many computer and trade magazines carry articles from time to time which explain the standard performance "benchmarks" and which compare leading computers and new products. SPEC95 is a standard which includes two test suites, one for integer and the other for floating-point performance, and delivers a performance index relative to a baseline system of a 40 MHz Sun SparcStation 10 with a score of 1.0 (Yager, 1996).

Vector computers can perform floating-point operations on vectors (quantities with a magnitude and a direction) as well as scalars (ordinary numbers), and parallel computers can perform simultaneous operations on several different scalars and vectors. This kind of computer is often applied by radio astronomers to perform the intensive data reduction associated with aperture synthesis, and in supercomputer architectures. The Cray Y-MP8I/8128 supercomputer can achieve a peak performance of 2.7 gigaflops (2.7×10^9 floating-point operations per second) and typically has 128 million 64-bit words of memory.

Of equal importance in many applications will be the efficiency with which a computer can perform input and output actions—referred to as I/O—such as "writing" to or "reading" from magnetic disk.

Four broad classes of computer systems found in astronomy can be distinguished as follows. Personal computers (PCs and Macs), often networked and certainly connected to the Internet. Powerful desk-top image processing workstations, perhaps with limited magnetic and/or optical disk storage, but linked to larger disks elsewhere; small departmental systems with multiple workstations supported over Ethernet; large departmental systems involving "mini-supercomputers" with ample vector processing used, for example, for complex numerical inversions. For computing work beyond this capability it is necessary to form large collaborative efforts and establish "supercomputer" centres.

Vendors most popular with astronomers for image processing workstations include Sun Microsystems and Digital Equipment Corporation, with Silicon Graphix, Hewlett-Packard, IBM and several others also being used. Decstations and Sparcstations are likely to be found in the office, lab and even "at the telescope". The usual approach is to "distribute" the computing power by linking several machines over Ethernet to a "server". All of these workstation computers have large, built-in, high-quality colour monitors and each runs some kind of "windows" software for image display.

All computers employ special software supplied under licence by the manufacturer to allow the machine to operate and execute programs—this is called the "operating system". For PCs this might be DOS, Windows 3.1, Windows 95, Windows NT or OS/2 Warp. For image-processing workstations, one would like to use an operating system which is independent of the actual computer hardware. In practice this never seems to be perfectly achievable. The operating system which comes closest to this ideal in portability is called UNIX—a development and registered trademark of Bell Labs AT&T.

Each computer program is written in a "high-level language" and then "compiled" into an "executable" form for the machine. For many years the mainstay of scientific programming has been FORTRAN, originally developed by IBM in 1957. The COmmon Business Oriented Language (COBOL) is not used in astronomy. Other languages have found specific applications and variable popularity over the years including Pascal, Turbo Pascal, C and Forth. Of these, C is extremely popular and most astronomical instrumentation today is controlled by programs written in C. Most of the applications required for data analysis are also written in C, and of course almost all commercially available software is in C. There are several "extensions" to the C language, but ANSI C is recommended to ensure compatibility.

2.4 DEALING WITH THE GROWING COMPLEXITY

Although the technological innovations demanded by modern astronomy are as diverse as they are difficult, it is interesting to note that over the years, industrial manufacturers of many sorts have happily participated in these developments, for reasons of prestige and subsequent spin-off rather than for the relatively small market-value of astronomy itself. The charge-coupled device is a fine example of a technological innovation driven to near ultimate perfection by the demands of ultra-low light-level astronomical studies of distant sources in the cosmos, as much as by the mass market for home video cameras. Yet, high-performance CCDs (HCCDs) as Richard Aikens of Photometrics termed them, now find application in fluorescence microscopy, industrial radiology, underwater imaging, high-energy physics and general medical imaging.

Instrumentation developed for a national facility, such as those in mainland North America, the Canary Islands, Hawaii, Chile and Australia, is expected to be used by many different scientists and to be supported by astronomers and engineers other than those who designed and built it. To make this truly feasible it is absolutely essential to carefully design and build the instrument to a very high standard of engineering and to a specification agreed with the astronomers. Drawings, documentation, and test procedures need to be excellent. In a phrase, the instrument must be "user-friendly". It can be argued that it is too costly for astronomy to attend to such "details" as accurate drawings and parts schedules, documentation, test procedures, software design and modelling tools such as CAD (computer-aided-design). Clearly, if the correct balance is struck, the cost savings in terms of the number of breakdowns and the inefficiency of repair and maintenance once the instrument is finally located at the observatory can be immense—as any observatory director will testify. Astronomers cherish "telescope-time" jealously and are not at all tolerant of inoperative equipment during their block of assigned time!

If an instrument, CCD or otherwise, is to be rapidly and successfully deployed it makes sense to deliver it to the telescope with a full suite of computer programs (simply referred to as software) which enables the instrument to take data and the astronomer to analyse and archive those data as conveniently as possible. It is easy to over-emphasize the hardware component at the expense of software development. Too often this occurs because of faulty assumptions such as,

(1) the software requirements are too hard to define until the hardware is built
(2) the software is easier to design and implement than hardware
(3) it is up to the astronomer anyway!

It has gradually become accepted that it is the job of the instrument-building team to provide a complete system, including software. The required software can be specified and designed alongside the hardware, and it can be coded as the hardware is built. Moreover, the software package can and should provide for data simulations which can test the software, and "exercise" routines for assisting engineers to check the hardware. When such an approach is taken the result can be very powerful. One example from my own experience in which this strategy was followed was the development from 1984 to 1986 of the first state-of-the-art, common-user infrared electronic imaging system (called IRCAM) by my colleagues and me at the Royal Observatory, Edinburgh. This new and fairly complex system went straight into successful operation on the United Kingdom Infrared Telescope on the 14 000 ft summit of Mauna Kea in Hawaii with only a few months of laboratory trials because of a planning technique which emphasized the following:

(1) design reviews—including software
(2) sub-assembly testing
(3) system simulation and modelling
(4) simultaneous development of on-line data reduction and analysis software

Therefore, the way to deal with the growing complexity of electronic computer-controlled instrumentation in astronomy is to adopt high standards in developing apparatus for general use. Project planning and management is needed, and close interactions between astronomers and engineering specialists is required. The appropriate standards are already familiar to professional engineers and industrial technologists.

2.5 APPLICATIONS SOFTWARE—A UNIFIED APPROACH

Clearly, computers are an integral part of all modern astronomical instrumentation. The computer provides the immense degree of flexibility and reliability demanded by astronomical researchers, and it is an essential tool in handling the digital information from a wide range of electronic imaging detectors, including CCDs. Computers are also needed for later "analysis and modelling" of data once the astronomer has left the mountain observatory. Fortunately, the widespread use of the CCD in astronomy and its growing availability to the amateur market, has coincided with extremely rapid advances in computers and a downward cascade in the cost of these machines. Astronomers need to be aware of progress in all aspects of computer technology.

Of paramount concern to many groups of astronomers in university departments and observatories throughout the world is the question of "portability" of good software for astronomical image processing and data analysis from one institute and one computer system to another. It may seem strange now, but when CCDs first entered astronomy in 1976, observational astronomers had no option but to develop their own image-processing software which dealt with the typical "data reduction" problems found in astronomical instruments, e.g. correcting for non-uniform detector response, enhancing images, photometry, spectrum extraction and wavelength calibration. Unlike in the commercial world, this kind of software is usually freely available within a nation or an international partnership—regardless of the cost of programming effort—because it is often developed under the auspices of the national science funding agency for the benefit of all. Major software packages are also exchanged between nations on a reciprocal basis.

As in other branches of science and space research, several countries find it much more cost-effective to collaborate to construct and operate major astronomical facilities—often in relatively remote areas of the world—and so, more than ever, compatibility is important. Astronomers and national institutions around the world have become aware that computing facilities and good software is of the utmost importance to the successful use of modern detectors such as CCDs. The lesson has been learned slowly, and there is still much more that could be done. It can be a serious problem for scientific development if observations are made but yet their publication, in final form, is delayed because subsequent, so-called "off-line", computing facilities are inadequate. Some of this reluctance can certainly be traced to a genuine concern about capital expenditure on items which have a very short lifetime before they are superseded!

Some excellent attempts to reach common standards in astronomical computing have been made. Credit must go to Don Wells who from 1974 to 1978 at the Kitt Peak National Observatory (NOAO) was responsible for the development of a system known as the Interactive Picture Processing System or IPPS which was to begin the move toward workstations and to common-user software.

In the late seventies an ambitious project called STARLINK was initiated in the UK when in mid-1978 a committee, chaired by Professor Mike Disney (who invented the name STARLINK), began to look into urgent complaints by the optical/UV astronomers of a dreadful lack of image-processing machinery, and then struck it lucky when, within 6 months, an underspend in the Science Research Council budget allowed the project to get off the ground and go into operation in early 1980. From the outset STARLINK was to be a nationwide linked network of VAX computers from Digital Equipment Corporation covering all of the major astronomy centres in the UK as well as at overseas observing sites in Australia, Hawaii and La Palma. Similar peripherals were established at each site and identical software was provided to each site. Moreover, applications programmers at all sites contribute to the common store of software. Headquarters for the project was established at the Rutherford-Appleton Labs (RAL) near Oxford, from where each new software package—all written to the same standard specification—is scrutinized before general release. Everyone contributes and everyone benefits. STARLINK supports a large community of users in the UK, covers every wavelength regime in astronomy, and has been widely exported. STARLINK later underwent a metamorphosis with major hardware upgrades and the widespread adoption of a software "environment" called ADAM, and then another change when the code was ported to Unix from VMS.

STARLINK applications software is accessed through an "environment" which overlays the commercial operating system. This environment is known as ADAM—for Astronomical Data Acquisition Monitor—a system originally developed for data-taking and the control of instruments at the telescope, but now extended in a grand unification to data reduction; in other words, the observer at the telescope will "see" the same computer environment as he does at his home institution. The ADAM environment offers menu-style interaction as well as the more familiar single-line command input, and a programmable command language. Critically important in the design of this kind of software is that it satisfies the following goals:

- must run on the users computer and display system without any difficulties
- must be reasonably easy to use and learn
- must be well-equipped with a wide range of computational procedures which are properly documented and supported

Conceptually, the degree of generality offered by the STARLINK approach has many advantages if the community using these facilities is large. For example, the reduction process looks the same to all, the system used at the telescope can also be used "back home", more complex algorithms can be acquired at no extra cost and standards of disk and tape storage formats become established such as the FITS or Flexible Image Transport System. One disadvantage of the STARLINK approach is that generalized software packages tend to "run" slower than more specific programs, which means that more powerful (costly) machines are required. But in this era of decreasing cost for increasing computing power, this does not seem to be a major problem.

Of more concern perhaps, is the time needed to develop and document an astronomical data-analysis system which has universal approval. But the alternative of do-it-yourself schemes leads to an immense amount of duplication and no lasting product, which may be just as costly!

A similar general-purpose scheme for astronomical data reduction has been set up in Europe under the auspices of the European Southern Observatory (ESO) which operates several telescopes on La Silla in Chile on behalf of a large consortium of European nations; the ESO headquarters are in Garching near Munich in Germany. The ESO scheme is called MIDAS and as the name implies—Munich Image Data Analysis System—it is principally intended for imaging and spectroscopic data. As with ADAM, within the

MIDAS environment there is a command language (MCL) and a large set of general software packages for image processing, plotting, spectral analysis and input/output functions.

In the USA a major initiative, supported by the National Optical Astronomy Observatories (NOAO), called the IRAF (for Image Reduction and Analysis Facility), has become the accepted standard system. Headed by its originator Doug Tody, the IRAF project began formally in late 1981 at Kitt Peak National Observatory. IRAF was implemented initially on a VAX 11/750 computer running Berkeley UNIX—a development of the AT&T UNIX undertaken by a group at Berkeley (some of whom formed SUN Microsystems)—but from its inception was intended to be transferable to all major scientific computers. In December 1983 the Space Telescope Science Institute in Baltimore selected the IRAF Command Language to host their Science Data Analysis System (SDAS), and an initial transfer or "port" of the IRAF to VAX/VMS was made in 84/85. The first public release of IRAF was in 1986. IRAF is now extremely widely used and NOAO currently supports the IRAF on an incredible variety of different computers. It is also available through STARLINK and MIDAS.

There was a time when most software systems developed in academic research environments were of such poor engineering that they barely survived upgrades of the host operating system and usually died with the departure of the original author or graduate student! This may be a little severe, but resources were being wasted. Efforts like IRAF and STARLINK, have ensured that this will no longer be true. Despite this progress, however, experience indicates that "software engineering" is still not taken seriously enough in astronomy. It is often under-manned and under-funded. Software development is a much more sophisticated and complex engineering task than any other. What is being overlooked is the fact that it is very hard to deal with the hundreds of small details in a complex system unless time is allotted for a thorough design and evaluation analysis—just as must be done for hardware. Therefore, the best strategy is:

- plan and review the software development process
- obtain a complete specification of requirements
- model the system in several ways
- provide good programming tools and a support environment

To do these tasks in a rigorous way requires planning and management. Therefore, it is worth including some recommendations on general project planning.

2.6 PROJECT PLANNING AND MANAGEMENT

Building complex, highly automated common-user astronomical instruments requires much more than a good idea, a knowledge of basic physics and a lot of enthusiasm. The key is "project planning". While it is true that a small group of good experimental scientists developing apparatus for their own use can, and do, often avoid the formal project planning approach, it is dangerously cavalier for a national or international facility to take this approach. Actually, in reality there is much to be gained by the small experimental group if they too endeavour to follow some simple rules of project management. For example, they can end up with a more reliable and better documented instrument, win more telescope time and provide better postgraduate training for their students.

When setting up a project to develop a complex common-user instrument for astronomy, the most commonly adopted approach is to begin by breaking up the project into clearly identifiable areas requiring specialist expertise. These might be, for example,

(1)   optics
(2)   mechanics
(3)   electronics
(4)   cryogenics
(5)   software
(6)   detector physics
(7)   science applications

The next step is to further divide each of these areas if required, for example—(a) structural mechanics, compact mechanisms, (b) digital electronics, analogue electronics, microcomputers, (c) control software, data acquisition software, user interface software, data reduction and data analysis software.

Having prepared a detailed scientific Case-for-Support and a draft Specification for the proposed new instrument, it is wise for the scientist or science team to conduct a "preliminary" design study—sometimes referred to in space-mission jargon as a "pre-Phase A" study—by consulting professional engineers and instrument physicists with expertise in each appropriate field. In this way the Specification can be strengthened, detailed and finalized. These steps are largely self-evident and standard practice in industry. Remember, however, that the specification is not absolutely sacrosanct, compromises may be possible and even necessary for purely practical reasons. It is the role of the Project Scientist to "point the direction" and to mediate in any design reviews in which there is a request for a change. Once there is a specification, the next step is to obtain a budget and define a Project Team.

At this stage it becomes essential to have a Project Engineer, and for very large projects one must also have a Project Manager, who can deal with day-to-day management, scheduling and financial activities. The Project Team should be formed so as to cover the required technical areas of expertise (e.g. as outlined above). It is important to ensure that a lot of work can proceed in parallel and that a given person is not overloaded. Management and distribution of personnel, their responsibilities and the chain of reporting, can be shown in a simple ORG (organizational) Chart and the timetable for the project should be given in a Planning Chart which indicates the expected start date and durations of major parts of the project, as well as the expected dates of reviews.

Once the project team has been formed then a full design study (Phase A) is carried out in all areas, including software. Before any metal is cut, before any wires are soldered and before any computer code is written it is important to verify that a clear and comprehensive design has emerged. Detailed "design reviews" must be held to critically discuss proposals. Whenever possible, an accurate replica or "space model" should be constructed and small demonstration parts can be made to verify a concept. Extensive computer simulations should be used, for example for optical ray tracing, mechanical structural stress/deformation analysis, thermal modelling, detector modelling, and so on. Usually, a two-step review process is adequate. The first review is called the Preliminary Design Review (PDR) and should be sufficiently detailed to enable any potentially serious problem areas (sometimes called "show-stoppers") to be identified. Towards the end of the planned design phase the Critical Design Review (CDR) is held to assess whether or not all remaining design problems have been solved and the project can proceed to the Fabrication Phase. It is extremely important at this stage to understand the budget and timetable for the project. Time to respond to issues raised at CDR should be allowed in the overall plan. Changes to the scope of the project (called "de-scoping") may be necessary, and such changes have to be worked out carefully between astronomers and technical experts.

When the design is complete the next phase (Phase B) that of Fabrication & Procurement can begin. Several commercial software packages, such as SuperProject Plus (from Computer Associates) or Microsoft Project (from Microsoft), are available for PCs to help plan project activities in detail, and they can also yield graphical displays (e.g. PERT and GANTT charts) for progress evaluation and analysis of the network of overlapping jobs which must be accomplished in time, and in the correct order, to complete the project in the most efficient manner. In every project there will be at least one item, one subset of the project, which if not completed on-schedule will seriously delay and disrupt the delivery of the instrument. These items are said to lie on the **critical path**, and it is usually possible to accurately identify items on the critical path.

Of paramount importance is Phase C (Assembly & Test; sometimes called System Integration). If care has gone into the design of test procedures and test rigs—as well as the instrument itself—then the design and performance of sub-systems can be verified more easily and any unforeseen problems can be caught at the sub-assembly stage. It is a well-known maxim of the professional engineer that "if you can't test it you can't build it"! A thorough Phase C results in a much easier final phase, which in astronomy is usually called the Commissioning Phase (Phase D).

Of course, there are always problems. What seems like an elegant experimental set-up on paper rarely transposes into a straightforward exercise in practice. Everything is harder than expected, everything costs more and everything takes longer. And, even worse, one actually has to deal with real people to make any progress! Some of the difficulties encountered in any project are:

- limitations on funding
- limitations on manpower and technical resources
- personnel factors such as
  —personality clashes
  —incompetence
  —occasional personal matters
  —demarcation rules—who does what?
  —conflicts of interest
  —poor communication between personnel

The last item is certainly an area of responsibility for the Project Scientist (and/or the Project Manager) and, at times, can be very difficult. If the team is allowed to splinter into non-interacting factions, or if there is no means to ensure that everyone's viewpoint is heard and understood, or if the Project Scientist fails to encourage discussion among the team, then there is a real danger of a design flaw being overlooked. Regular monthly "progress and review" meetings, properly recorded with written Minutes and Action Lists, are essential, as are occasional meetings of special-interest sub-groups. Well-known rules of good management, such as "catching people in the act of doing well and praising them accordingly", and "motivating people by your own enthusiasm for their involvement", are just as valid for astronomy projects as for any other activity.

SUMMARY

We have learned that there have been changes in the way astronomers carry out their observational research and changes in the way that observatories responsible for supporting a large community of "users" must react and plan. Automation, remote observing, common-user instrumentation and international collaborations imply that astronomers need a wider training in technology and in management skills.

EXERCISES

1. What is meant by a "common-user" or "facility" instrument and why might you expect to find several such instruments at national facilities?
2. List some advantages of fully automating an astronomical instrument. Can you think of any disadvantages?
3. What is IRAF and STARLINK? Why do you think these projects were initiated and pursued?
4. In what sense is a photographic plate both a detector and an archival medium for data storage? Compare this with a digital camera system such as a CCD. Describe what steps are being taken to construct digital archives of astronomical data.
5. Investigate several Web sites and see how easy it is to access astronomical data.
6. Explain what is meant by the term "critical path" in a project.

REFERENCES AND SUGGESTED FURTHER READING

Blanchard, K. and Johnson, S. (1982) *The One-Minute Manager*, William Morrow, USA.
Blanchard, K. and Lorber, R. (1984) *Putting the One-Minute Manager to Work*, William Morrow, USA.
Brooks, F.P., Jr (1982) *The Mythical Man Month*, Addison-Wesley, Reading, MA, USA.
Bode, M.F. (Ed.) (1995) *Robotic Observatories*, Wiley–Praxis, Chichester, England.
Computers and the Cosmos (1988) by the editors of Time-Life Books, USA.
Eichorn, G. (1994) An overview of the Astrophysics Data System, *Experimental Astronomy*, **5**, 205.
Yager, T. (1996) Bringing benchmarks up to SPEC, *BYTE*, March, 145–146.

SOME FACILITIES ON THE WORLD WIDE WEB

ADS—Astrophysics Data System—archival retrieval system. Track down publications and abstracts. Search by authors, titles, keywords.
http://adswww.colorado.edu/adswww/adshomepg.html

NED–NASA/IPAC Extragalactic Database–a database for extragalactic objects only. IPAC is the NASA Infrared Processing and Analysis Center.
http://www.ipac.caltech.edu/ned/ned.html   for access to NED
http://www.ipac.caltech.edu/   for general access to IPAC
http://www.ipac.caltech.edu/2mass/images/2mass.gif   for info on the 2MASS project

CDS—Centre Donnes Strasbourg—the Strasbourg Data Centre at the Strasbourg Observatory. Operates SIMBAD—Set of Identifications, Measurements and Bibliography for Astronomical Data. This database covers all classes of astronomical objects. You need to get an account.
http://cdsweb.u-strasbg.fr/CDS.html   in Europe
http://hea-www.harvard.edu/SIMBAD/simbad.home.html   in USA

SkyView is a "virtual observatory" on the Internet; an electronic database from gamma rays to radio. Developed by the High Energy Astrophysics group at the Goddard Space Flight Center (GSFC). Type in the name of an object or its coordinates and get a GIF or FITS image from the multi-wavelength database.
http://skyview.gsfc/nasa.gov/skyview.html   for access to SkyView

DSS—Digital Sky Survey—provides GIF or FITS images of any field in the sky using digitizations of the first and second epoch Schmidt surveys. Operated by the Space Telescope Science Institute. Go to the Web address below, enter object name or coordinates (get these from SIMBAD or NED if necessary), select first or second survey, select size of image, e.g. 10×10 arcmin and file format—usually GIF for an on-screen display.
http://stdatu.stsci.edu/dss/

CADC—Canadian Astronomical Data Centre—provides access to all of the above and to the Canada–France–Hawaii Telescope (CFHT). Located at the Dominion Astrophysical Observatory (DAO).
http://cadcwww.dao.nrc.ca/CADC-homepage.html

Other useful astronomy resources:
http://www.stsci.edu/   main access to Hubble Space Telescope resources
http://marvel.stsci.edu/net-resources.html   an extensive listing of resources
http://www.gsfc.nasa.gov/nasa_homepage.html   provide links to all NASA centres
http://www.tuc.noao.edu/noao.html   access to NOAO and many links to other sites
http://iraf.noao.edu/   information about IRAF
http://www.ast.cam.ac.uk/   access to Cambridge University (UK) and Royal Greenwich Observatory (RGO)

3

The power of modern astronomical instrumentation

We began (in the Introduction) by following a visiting astronomer through a night at a large national observatory and we have now reviewed the development of astronomical imaging from eyes to electronic sensors, including the changes in how astronomical research is carried out. Any introductory text on astronomy will describe the basic "tools" of the astronomer and will certainly mention telescopes, spectrographs and photometers. Before going into more detail on the design of modern electronic imaging instruments, this chapter provides an overview and some illustrations of use. Subsequent chapters explain the underlying design principles and concentrate on detector properties in detail.

3.1 TELESCOPES—MORE THAN LIGHT-GATHERING POWER

3.1.1 Telescope designs

Telescopes are of course the means by which light from distant objects is collected and focused, but telescopes must do more than gather light. Providing excellent telescope optics and proper alignment is the first critical step in obtaining good images. Modern astronomical telescopes are identified by the diameter (D) of the primary collecting surface. Most readers will be familiar with the fact that all large astronomical telescopes are "reflectors" and therefore use curved mirrors rather than lenses to achieve light collection and focus (Fig. 3.1). The large, "primary" mirror is usually a paraboloid rather than a simple spherical surface (because a spherical surface cannot bring a parallel beam of light to a single, sharp focus); if you spin a bowl of water on a turntable, the shape taken up by the surface of the water is a parabola. In normal two-mirror telescopes, the "secondary" mirror can have a variety of surface shapes depending on the telescope design. One exception to this standard approach is the Schmidt design in which the primary mirror has a spherical surface, but a thin refracting plate with a complex shape is placed at the entrance tube of the telescope to correct for the poor focusing ability of the sphere. In this case, the focal surface is significantly curved and lies between the **corrector plate** and the spherical primary mirror. Photographic plates or CCD detectors can be placed at this focus.

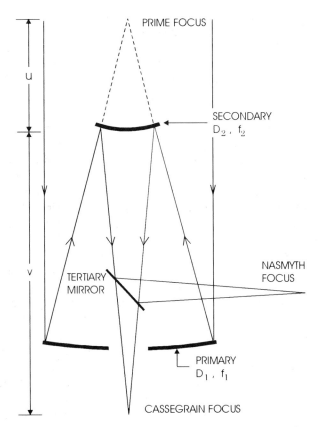

Fig. 3.1. A schematic representation of a typical reflecting telescope showing the relative location of the primary and secondary mirrors and the optional tertiary mirror. The symbols D_1, f_1 and D_2, f_2 refer to the diameters and focal lengths of the primary and secondary mirrors respectively. The distance $u + v$ is greater than f_1 by an amount between $0.5D_1$ and $1.0D_1$ (the back focal distance) to enable the focus to be accessible at Cassegrain or Nasmyth. Also, $1/u + 1/v = 1/f_2$.

Light reflected from the primary mirror will come to a focus—called the **prime focus**—near the top of the telescope tube. Provided the obstruction is small, it is possible to place a detector at that focus. If the telescope is large enough, like the 5-m Hale telescope on Mt. Palomar, it is perfectly possible to install a cage large enough to carry a human being at the prime focus. A flat or plane mirror placed in the centre of the tube to intercept the beam before the prime focus yields the well-known Newtonian design. Usually the plane mirror is at 45° to deflect the beam through a right-angle and out the side of the tube, but it can also be at right-angles to the beam to send it back towards the primary; the 61-inch US Naval Observatory astrometric reflector at Flagstaff, Arizona, is designed this way. Many large reflecting telescopes are of the Cassegrain design and contain a central hole in the primary mirror. The light reflected from the secondary passes through the Cassegrain hole, exits the telescope tube at the back of the primary and comes to a focus—the **Cassegrain focus**. This arrangement produces a much longer focal length and a more restricted field of view than the prime focus. In the classical Cassegrain telescope the primary mirror is a parabola and the secondary mirror is a hyperbola. The hyperbolic secondary is placed to intercept the beam before it reaches the prime focus. Image quality is limited by the off-axis effect called "coma". An alternative design is the Gregorian in which the converging rays from the primary are allowed to go through the prime focus and reflect from an elliptical secondary mirror back through the "Cassegrain" hole in the primary; the Gregorian has more "field curvature" than the classical Cassegrain. In either case, instruments can be attached at the Cassegrain focus. A variation of this approach is called the Ritchey–Chrétien (or RC) design in which both the primary and secondary are hyberbolic surfaces. This design gives good performance over a larger field of view. Hybrid designs such as the Maksutov and Schmidt–Cassegrain which involve refractive corrector elements are very popular among manufacturers of small telescopes in the 4–14 inch range, but watch out for poor performance at the edge of a large field.

It is also possible to locate a third or "tertiary" mirror to direct the light along an axis of rotation of the telescope. Depending on the type of telescope mounting used (see below), additional mirrors may be required but the final focus is a "stationary point", independent of telescope pointing. For classical telescopes constructed on an **equatorial mount**, this focus is called the **coudé** (koo-day), and for telescopes using **altitude-azimuth** mounts, this is the **Nasmyth** (nay-smith) focus. A change of secondary mirror is required with the coudé arrangement and the result is a very long focal length. In general, the Nasmyth focus does not require a change of secondary mirror and thus the image scale is the same as at the Cassegrain focus.

Each focus is characterized by a **plate scale** which measures the number of arcseconds on the sky corresponding to 1 mm at the focus of the telescope; the longer the focal length (F), the smaller the number of arcseconds per millimetre and the greater the magnification. The shortest focal length corresponds to the prime focus of the telescope and there the scale is largest in arcseconds per millimetre. The **focal ratio** or "f-number" is given by f/D and the smaller this number, the "faster" the beam converges and the shorter the telescope for a given diameter. Hence the smaller the dome or enclosure. Modern telescopes tend to employ very fast primary mirrors. Note that the rate of energy collection or "speed" of an optical system is proportional to $1/(f\text{-number})^2$.

For direct imaging, CCD cameras will be placed at either the prime focus or the Cassegrain focus. Prime focus cameras have the largest field of view and are ideal for surveys, whereas Cassegrain cameras are usually designed to better sample the blurred image of a star (caused by the Earth's atmosphere) called a "seeing disk".

3.1.2 Telescope mounts

Since the rotation of the Earth on its axis causes the daily rising and setting of the stars, all telescopes require a means to continually "update" the direction in which they are pointing. Otherwise, a star would simply drift through the field of view and a lengthy exposure with a CCD camera would be hopelessly "trailed". No telescope points perfectly and no telescope can maintain its pointing accuracy from one part of the sky to another without numerous corrections. The sidereal rate is 360° in approximately 23 hours and 56 minutes or about 15.04 arcseconds per second of time. The main problems are:

- alignment of the rotation axes
- mechanical flexure of the telescope structure

- cyclic errors in the gear trains
- atmospheric refraction (which depends on wavelength, pressure and temperature)

Refraction makes the star appear nearer the zenith. For zenith distance $\zeta < 50°$ the displacement is given by $R \approx 58.2''$ tan ζ. These effects are removed or minimized by a computer model of the telescope derived from hundreds of pointings spread over the sky. Atmospheric dispersion, that is, the correction for the wavelength-dependence in the refractive index of air, requires an optical device, which some telescope now have. At sites with exceptional seeing, or in adaptive optics applications (Chapter 14), the differences in refraction between a "blue" object and a "red" guide star may result in light loss at the slit of a spectrograph unless dispersion is taken into account.

Telescope mounts fall into three basic categories depending on the motion of the telescope axes: (i) **equatorial** mounts, (ii) **altitude-azimuth** (alt-az) mounts and (iii) **transit** mounts.

There are several different implementations of the equatorial mount (German mount, English yoke, horseshoe), but all have the same fundamental purpose. In the equatorial mount, one axis of the telescope is parallel to the Earth's axis, i.e. it is pointed at the North Celestial Pole. This is called the **polar axis**, and it is rotated in the counter-direction from the Earth's motion at precisely the same rate. The other axis, the **declination** (dec) axis, moves at right angles to the first. Both axes must be moved to locate the star, but only the polar axis needs to be turned continuously to "track" the star. Since the telescope is sweeping along the same arc in the sky as traversed by the star, the image at the telescope focus maintains the same orientation at all times. This is not so for the alt-az style of mounting (Fig. 3.2) which is identical to that used for radars. The telescope rotates about a vertical axis to point to any compass or **azimuth** direction and it also rotates about a horizontal or **elevation** axis to point from the horizon to the point overhead (the **zenith**). To track the curved path of a rising and setting star both axes must be rotated, but the rate of rotation varies with the position of the object in the sky and becomes impossibly large at the zenith which leads to a "blind spot" overhead. In addition, the image at the focus rotates with time; this is called **field rotation**. The basic equation for field rotation is

$$\omega = \Omega \cos A \frac{\cos \phi}{\sin z}$$

where ω is the field rotation rate in radians per second, Ω is the sidereal rate (7.2925×10^{-5} rad/s, A and z are the azimuth and zenith distance respectively in degrees and ϕ is the latitude of the telescope. Spherical trigonometry formulas are available to convert to (A,z) from right ascension (or hour angle) and declination; see Appendix 5. Field rotation can be compensated by counter-rotating the entire instrument at a variable rate, or rotating an optical compensator such as a K-mirror. Without compensation there will be image smear. The amount of image movement in arcseconds is given by ω (rad/s) × plate scale ($''$/mm) × radial distance from rotation axis (mm). Despite these complications, which are easily handled by computers, all large modern telescopes are built this way because it aids the mechanical design and reduces the total weight and load on the bearings. The final mounting scheme is the **transit** telescope. In this case, the telescope is fixed to point only in the plane of the north–south meridian line, but it can select any point from zenith to horizon, and the stars are allowed to drift across the field of view.

Fig. 3.2. The WIYN (Wisconsin–Indiana–Yale–NOAO) telescope, a relatively small modern telescope with an altitude-azimuth (alt-az) mounting. Courtesy NOAO.

3.2 IMAGING THE SKY

Mapping the distribution of celestial sources on the sky at the wavelength of interest, serves not only to locate the position of the source precisely—a practice called **astrometry**—but also to provide information on its form or structure, and that of its environment.

Much can be learned from the appearance or "morphology" of certain objects. Images of the sky at different wavelengths are essential for classification of objects. Accurate brightness measurements, or **photometry**, of each source in the image at different wavelengths yields basic physical information. Usually, some form of electronic camera is used.

3.2.1 Photography and computers

Although the classical photographic plate is not an electronic detector and cannot match the CCD for sensitivity, it would be quite wrong to ignore its utility in modern astronomy. Specially designed telescopes, called Schmidt telescopes—such as the UK Schmidt Telescope (now operated by the Anglo Australian Observatory) at Siding Spring in Australia or the older Palomar Schmidt Telescope which produced the original Palomar Observatory Sky Survey (POSS)—have been constructed to accept very large glass-based plates about 14 inches on a side. Schmidt telescopes are named for the German optician, Bernhard Schmidt, who pioneered the technique in the 1930s of grinding a thin lens into a complicated shape to compensate for the shortcomings of spherical shaped mirrors. These telescopes are designed to provide images of very large areas of sky ($6.5° \times 6.5°$). Recall that the full Moon is only $0.5°$ across. A typical photographic plate might easily contain 2.5 billion (potential) picture elements—the grains of emulsion—but only 1 out of 50 incident photons of light will be detected (i.e. the quantum efficiency is 0.02 or 2%). The product of the number of pixels and the quantum efficiency of a typical (hypersensitized) Schmidt plate such as Kodak IIIa-J emulsion is about 300 times better than that of a 512×512 silicon CCD with 70% quantum efficiency, simply because the CCD chip is so small. Another way to look at this is to compare the product of the quantum efficiency and area of sky covered. Even for the large mosaic of CCDs shown in Fig. 1.10 in Chapter 1, only 0.25 deg^2 of sky is viewed compared to the 42 deg^2 on a Schmidt plate. When the quantum efficiencies are included however, the plate is only about five times more efficient than the CCD mosaic, and the CCD provides very large gains in the ultraviolet and far red parts of the spectrum. Higher quantum efficiency implies that fainter limits are reached in the same time. Alternatively, the same detection limit can be reached in a much shorter time, hence allowing several other patches of sky to be measured. In summary, when area-coverage rather than sensitivity is the issue, photographic plates are extremely useful but, as we have implied, the gap is closing as larger CCDs and arrangements of multiple CCDs become available. The all-sky survey produced by the Palomar, UK and ESO Schmidt telescopes is in use at every observatory, and a new or "second epoch" Palomar survey is under way. A special Schmidt survey to find guide stars for the Hubble Space Telescope is widely available on CD-ROM.

Having obtained one of these 14-inch plates containing tens of thousands of images, the next step is to find some way to extract useful statistical information such as the coordinates or positions of objects, their shapes, sizes and orientations, and of course, their magnitudes or brightnesses. It is at this point that even photographic cameras must rely on computers. At a number of establishments throughout the world special "plate-measuring" machines have been constructed. In such a machine the original photographic plate is "scanned" under computer control by a special beam of light and the transmitted signal—which depends on how densely the emulsion has been blackened—is detected by a photoelectric sensor to produce an electrical signal as shown in Fig. 3.3. This signal is converted to a numerical value by some electronics and stored in the memory of a computer—the plate is said to have been "digitized". Since these plates are so large, an immense amount of digital data is produced and stored or "archived" on magnetic tape, disk or CD-ROM. The digital data can now be analysed by a high-speed computer to search automatically, for instance, for galaxies rather than stars.

Inevitably, interesting features are found on Schmidt plates which demand "follow-up" images of much smaller patches of the sky, but with 100 times more sensitivity. Astronomers call this "going deep". To achieve fainter limits we must turn to much larger telescopes and to the more sensitive photoelectric imaging devices such as the CCD.

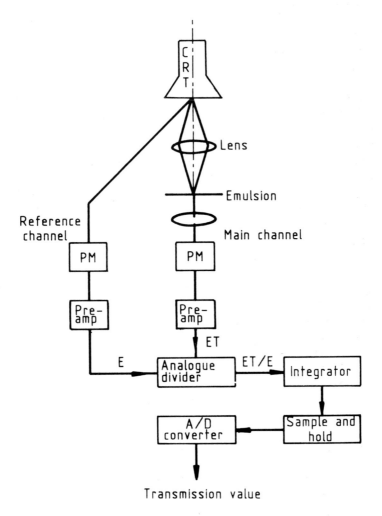

Fig. 3.3. A schematic drawing of the computer-controlled plate scanning and detection system used by the COSMOS measuring machine to digitize images recorded on the large photographic emulsion plates used with Schmidt telescopes. Photo courtesy of Royal Observatory, Edinburgh.

3.2.2 Electronic imaging

Electronic imaging covers all wavebands, but we begin with what can be done from the ground using visible light detectors such as CCDs. The most obvious application of the CCD is in a straightforward "camera" for direct imaging at the focus of a telescope. Among the advantages of CCDs over photographic plates are:

(1)   very high sensitivity—a much larger fraction of the light falling on the CCD is converted to a measurable quantity, in this case, an electronic voltage
(2)   greater coverage of the spectrum—the silicon material of the CCD is sensitive to light of many different colours
(3)   immediate compatibility with computers—no developing or processing of film is required, the "output" from a CCD is a stream of electronic pulses immediately suitable for transfer to a computer
(4)   instant display of the image

Compared to TV-type cameras, CCDs offer the additional very important advantages of precision and stability which are essential for quantitative brightness measurements. CCDs are therefore excellent photometers too.

Every major observatory in the world now possesses a CCD camera system of some sort. At most of the larger observatories several different kinds of CCDs are offered and CCDs feature in almost all forms of modern instrumentation. Many smaller astronomy groups have transportable or "travelling" CCD systems. CCD cameras have now reached the amateur astronomy community (see, for example, the magazine *CCD Astronomy* by

Fig. 3.4. A set of liquid-nitrogen-cooled CCD cameras used at the University of Hawaii telescopes on Mauna Kea. Chips of various sizes can be seen along with associated electronic boxes.

Sky Publishing Corp.). Smaller telescopes are astonishingly re-vitalized by the addition of CCD cameras and other CCD-based instruments. For the most part, CCD cameras have also replaced TV tubes for acquiring and guiding on the object of interest. In most CCD cameras the chip is mounted in a small vacuum chamber because it must be cooled to a low temperature. Every CCD camera requires a "box of electronics" with circuits to operate the CCD. One of these circuits converts the voltage signal from the CCD into a digital number which can be stored by a computer. Several CCD cameras are shown in Fig. 3.4; see the colour plate section for images.

In all cameras there will be some small amount of light which gets into the detector after being scattered from the slender struts or vanes which support the secondary mirror of the telescope, or from the edges of the primary mirror. Scattered light is particularly

Fig. 3.5. The discovery image by Terrile and Smith of the disk around the star Beta Pictoris using a CCD camera and an occulting spot (a stellar coronagraph). Courtesy Brad Smith.

serious when you are looking for very faint sources close to a very bright source. One way to overcome this is to place a small "occulting finger" or disk in the focal plane to blot out the bright object in much the same way as the Moon covers up the Sun during a total eclipse to reveal the outer layer or corona of the Sun's atmosphere. To cut down scattering from the secondary mirror support vanes a blackened mask is constructed with a shape which imitates the size and orientation of the vanes and this mask is placed very accurately inside the camera at a position where there is an optical image of the secondary mirror. [If there is no such image then one can be created by means of a "field lens", i.e. a weak lens placed exactly at the focus of the telescope.] This arrangement is called a **coronagraph**. Fig. 3.5 shows a remarkable result from an instrument using this technique. The object is the disk of proto-planetary material around the star Beta Pictoris.

As we will see in later chapters, a remarkable property of the CCD is that charge can be transferred (along a column of pixels) one row at a time at any desired rate. In particular, the CCD can be operated at the "sidereal rate", that is, the rate at which the Earth turns on its axis relative to the most distant stars. Suppose a sensitive CCD camera is placed at the focus of a moderate-sized telescope which is pointing at the zenith (or anywhere in the meridian), but the telescope drive motors are switched off. Stars will drift across the CCD pixels, at the sidereal rate of about 15 arcseconds per second of time, producing trails. Now, initiate the electronic process of reading out the CCD charge pattern, at the sidereal rate and in the same direction as the star moves, and then open the shutter. There will no

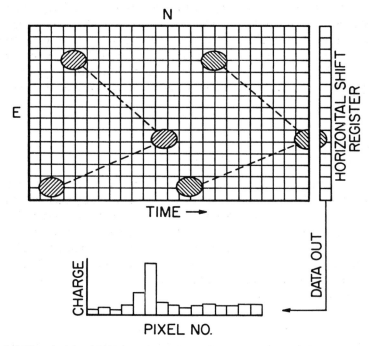

Fig. 3.6. The principle of "drift-scanning" using the unique charge-coupling property of a CCD which allows the image charge to be moved along columns from pixel to pixel at any rate. In this way, all pixels in a given column contribute to the signal.

longer be star-trails. Instead, the charge image from previous pixels is added to the next one and the current position of the charge pattern will move along the column so as to keep up with the current optical image position (see Fig. 3.6), more and more photons will be collected and, ultimately, the entire column of pixels will be read out and will have contributed to the detection process. Carrying on in this way a huge strip of sky can be surveyed systematically to a deep level without actually moving anything except electrons! This technique is called **drift scanning**.

3.3 SPECTROSCOPY—ATOMIC FINGERPRINTS

Astronomers rely on spectrometers to provide them with fundamental physical information on the chemical composition, temperatures, densities and velocities of remote objects in the universe. To record an image of the spectrum, almost all astronomical spectrometers now use CCD detectors.

The term, "spectrometer", is a generic term for any device which "measures" the spectrum. Astronomers generally always use **spectrographs**, that is, spectrometers which contain an imaging device to make a record of the spectrum. Note that this is not the same as an image of the scene at the telescope focus. Instruments which can record both the normal two-dimensional image of the scene *and* the one-dimensional wavelength spectrum are called **imaging spectrometers** or sometimes "3-d spectrometers". When a non-imaging detector such as a PMT is used, the spectrum can only be recorded by either scanning the detector along the spectrum in small steps or scanning the image of the spectrum over the detector. Such instruments are usually called **scanning monochromators**. Spectrographs fall into several classes depending on the amount of fine detail or spectral "resolution" achieved in the spectrum. The resolving power (R) is defined by the ratio of the wavelength (λ) divided by the smallest discernible change in wavelength ($\Delta\lambda$); $R = \lambda/\Delta\lambda$. Values of R range from a few hundred to a few hundred thousand. These classes are:

- faint object spectrographs—low resolution ($R \sim 1000$)
- intermediate dispersion spectrographs ($R \sim 10\,000$)
- high resolution spectrographs ($R > 50\,000$)
- imaging spectrometers (depends on technique)

To "disperse" or spread the incident light beam into a spectrum, most astronomical spectrographs use a **diffraction grating** rather than a glass prism. The principle of the diffraction grating, pioneered by Joseph Fraunhofer, relies on the wave nature of light. Thousands of very fine parallel lines or grooves are cut in an optical substrate and reflected (or transmitted) light waves "interfere" to produce a spectrum. The more the light is spread out the less bright any part of the spectrum becomes, and the more difficult it is for a detector to make an accurate measurement of that brightness. Hence the more sensitive the detector needs to be.

3.3.1 Classical spectrographs

By "classical" we really mean single-slit spectrographs. Of course, if the slit is long enough and the field dense enough then there may be more than one object contained in the slit length. This was the classic configuration for many years until the schemes de-

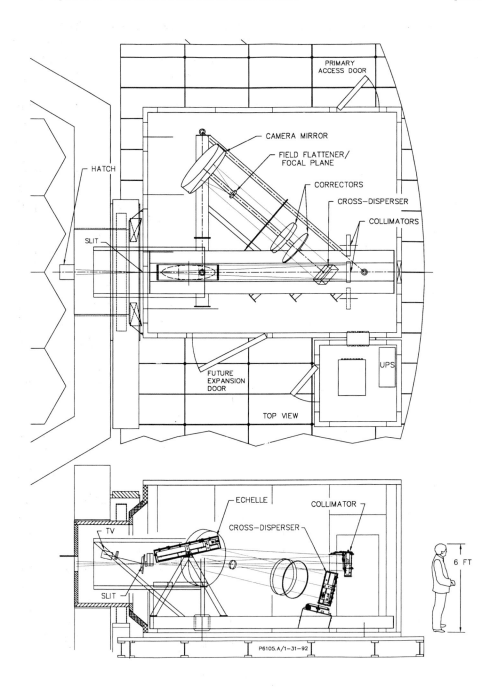

Fig. 3.7 The layout of the High Resolution Echelle Spectrograph (HIRES) on the W.M. Keck 10-m telescope. This instrument, which is contained in a large room, is so big that you can walk inside it. The detector is a 2048 × 2048 pixel CCD. Courtesy Steve Vogt.

scribed below were devised to observe many more objects simultaneously. The high-resolution class of spectrographs are still mainly of this type. Such spectrographs tend to be
very large instruments. They are usually located at a stationary focus of the telescope (e.g.
coudé or Nasmyth). Some of these instruments are so large that they effectively occupy a

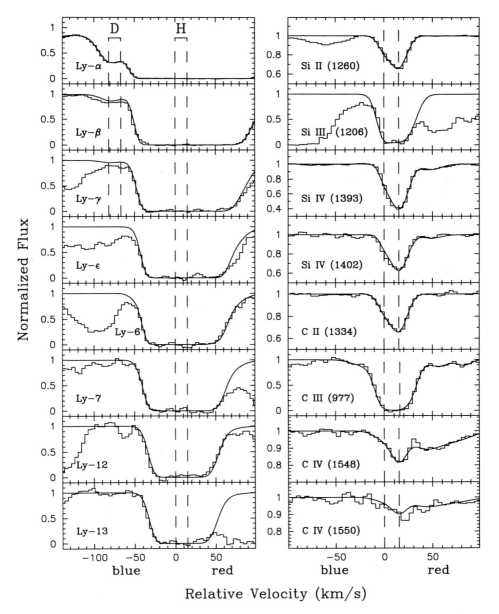

Fig. 3.8. A very high resolution spectrum obtained by HIRES on the Keck I telescope which
reveals the presence of primordial deuterium in the light from a distant quasar. Courtesy David
Tytler.

small room, and you can literally walk "inside" the instrument—although you would not normally be allowed to do so in order to keep the optics safe and clean. One example of this class of instrument is the HIRES (High Resolution Echelle Spectrograph) on the W.M. Keck Telescope, Mauna Kea, Hawaii, which was developed by Steve Vogt and colleagues of the University of California, Santa Cruz (Fig. 3.7). Among the many applications of such an instrument is the study of the subtle spectral differences between chemical isotopes, in particular hydrogen and deuterium, in distant galaxies in order to determine their primordial abundance (Fig. 3.8). Measurements like this are fundamentally important to our theories of the origin and evolution of the universe from the Big Bang. An excellent review of high-resolution spectrographs designed or built for very large telescopes, including the Keck, Gemini, Hobby-Eberly, MMT and the ESO-VLT, is given by Pilachowski *et al.* (1995).

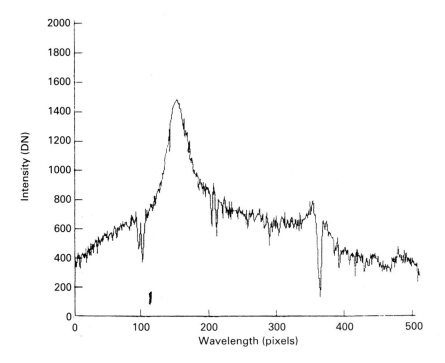

Fig. 3.9. A typical moderate-resolution blue spectrum of a faint object, in this case the quasar 1331 + 170. This drawing is performed by the computer which examines the CCD image of the spectrum and plots the brightness against the pixel number or wavelength along the spectrum.

Intermediate resolution spectrographs are the workhorses of astronomy and are in use all over the world. Most telescopes larger than about 2-m will have an intermediate resolution spectrograph. A major fraction of the "dark time", that is the period of the month around new Moon when the sky is darkest, is devoted to obtaining spectra of faint objects with these instruments. Many objects are of such low brightness that the light of the darkest night sky still dominates the raw picture. Computer processing is required to measure and remove the signature of the background sky from the spectrum of the faint object.

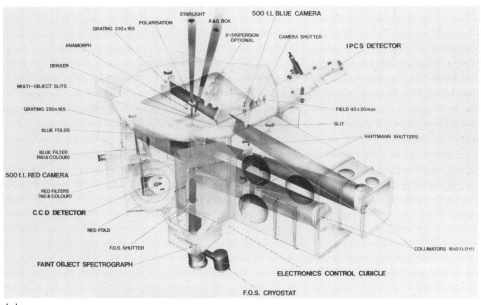

(a)

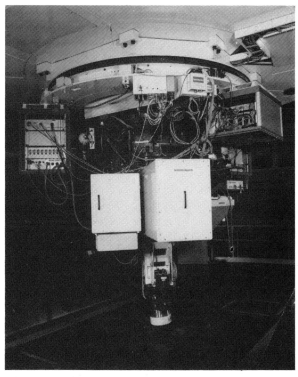

(b)

Fig. 3.10. ISIS is a highly automated triple spectrograph for the 4.2-m William Herschel Telescope. CCD cameras are attached at three places. The lower camera constitutes a faint object spectrograph. Courtesy Royal Greenwich Observatory.

In 1982 Bev Oke pioneered the concept of a "double" spectrograph for the 5-m Hale Telescope at Palomar, driven in large part by the huge spectral range of the CCD. Many Cassegrain spectrographs are now "double" systems. For example, the Kast spectrograph for the 3-m Shane telescope at Lick Observatory, the MOS/SIS double spectrograph on the CFHT and the LRIS (Low Resolution Imaging Spectrograph) on the W.M. Keck telescope. Fig. 3.9 shows the blue spectrum of the quasar $1331 + 170$. A trace like this one is produced easily from the digital CCD image of the spectrum by a computer program

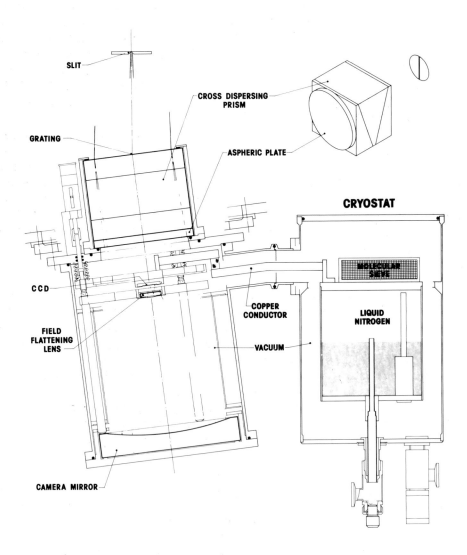

Fig. 3.11. A drawing of the basic design of a faint-object spectrograph (FOS) which employs a Schmidt camera and corrector plate, together with a CCD detector. The one shown here was designed by Charles Wynne. Courtesy Royal Greenwich Observatory.

which plots the signal along a row of pixels. The strong emission feature protruding upwards is actually due to triply-ionized carbon atoms and occurs at a wavelength of 1550 Å in the far ultraviolet; it is "redshifted" into the normal visible part of the spectrum by the expansion of the universe during the time it has taken this light to reach us.

Even more complex is the ISIS "triple" spectrograph (Fig. 3.10) on the 4.2-m William Herschel Telescope, La Palma. The ISIS is essentially a double intermediate spectrograph (for red and blue wavelengths) with a quick-change mode to send the light to a "built-in" faint-object low-resolution CCD spectrograph.

The "faint object" class of spectrographs is basically a CCD camera capable of imaging a spectrum of a faint object with relatively low resolution of the spectral detail so as not to "spread out" the available light too much. The less the spectrum is spread out, the more light there will be on any given pixel of the CCD and the fainter the source that can be detected. Obviously a compromise is needed which ensures that enough spectral resolution is retained otherwise features or lines in the spectrum which could yield the velocity of the object, for example, will be hopelessly smeared out. Examples include FORS at the AAT, and LDSS on the WHT.

One approach is to use an arrangement like that in Fig. 3.11. Light rays diverging from the telescope focus pass through a transmission diffraction grating and a prism; the prism is oriented to spread or disperse light at right angles to the grating. A tiny Schmidt telescope forms an image of the spectrum onto the CCD, and the entire system is in a vacuum. This combination is three times more efficient than the equivalent intermediate dispersion spectrographs. An instrument of this sort is particularly appropriate for the study of objects like quasars which have prominent broad emission lines in their spectra which stand out above the underlying and much fainter general or continuum spectrum.

3.3.2 Multi-object spectroscopy

Spectroscopic measurements can be painfully slow, even with very sensitive CCD detectors. This is because the accuracy of the measurement depends on the amount of light (the number of photons) falling on the CCD pixel and, as already stated, the greater the spectral resolution the fewer the number of photons on any pixel. Typically, the signal at any point in the spectrum of an astronomical source will be several hundred times fainter than the signal in a direct image of the source obtained with a CCD camera. Thus the possibility of recording spectra from several objects at once is very attractive.

One approach to this problem is the multi-object spectrograph which employs an entrance slit composed of multiple sub-sections which can be positioned by computer to pick up many different objects in the field of view. The Low Dispersion Survey Spectrograph (LDSS-2) developed by Durham University and the Royal Greenwich Observatory for the WHT in 1992 is a very efficient multislit faint object spectrograph which covers a large 11.5 arcminute field of view. A commercial engraving machine is used to produce the multislit aperture masks at the telescope during the course of an observing run. Exceptionally good optics designed by Charles Wynne and Sue Worswick yield a transmission of about 30% in the V band for a typical CCD detector.

Many observatories now have spectrographs with multislit designs, one of the most powerful is the LRIS spectrograph on the Keck 10-m telescope. Fig. 3.12 shows a typical result with about 30 slits after extraction of the spectrum.

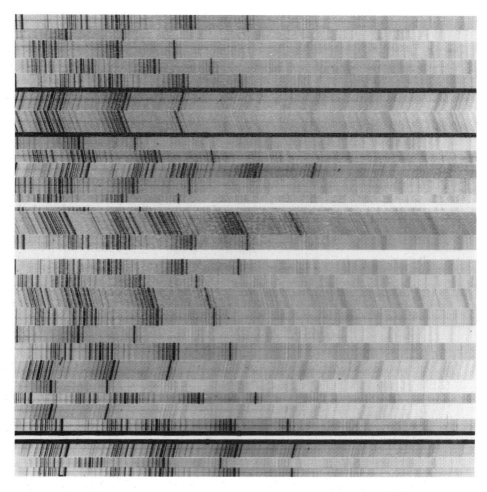

Fig. 3.12. Multislit spectroscopy is illustrated here by results from the LRIS spectrograph on the Keck telescopes. The spectral range on the 2048 × 2048 CCD is about 400–660 nm and the spectrum shows numerous "night-sky" lines (long). Different science targets can be observed simultaneously. For example, line-of-sight MgII absorption in an active galaxy or the distribution of 3727 Å emission in an inclined-disk galaxy at $z = 0.4$ or the Lyman alpha emission at $z = 3.2$ in a very distant source. Courtesy Sandra Faber and Drew Phillips.

Of even greater impact on "multi-object spectroscopy" has been the new technology of transparent **optical fibres**. Developed mainly for the telecommunications industry, these slim, flexible glass conduits—which resemble normal electrical cables—can be used as "light-pipes" to transmit light over very long distances with only slight losses or attenuation. The basic idea is to position one end of each of several optical fibres in the focal plane of the telescope at points corresponding to interesting objects, such as a cluster of distant galaxies, and to stack up the other ends along the entrance slit of the spectrograph; the spectrum of each source is therefore recorded simultaneously. Multi-fibre spectroscopy can allow over 100 simultaneous spectra to be obtained.

The first of the fibre optic coupled systems—appropriately named MEDUSA—was developed at the Steward Observatory in Arizona by John Hill, Roger Angel and co-workers in 1979. Such systems are now fairly common on telescopes all over the world, and have drastically transformed many branches of astronomy where statistical properties on many sources are required. Examples include WYFFOS/AUTOFIB2 on WHT, 2dF on the AAT, HYDRA on the KPNO 4-m, ARGUS on CFHT. More than 50% of all spectroscopic time on the Anglo-Australian Telescope is now multi-object in nature according to Richard Ellis and Ian Parry of Cambridge University (and formerly Durham University) England, who have been pioneers of the use of optical fibres. With techniques like this, a complete redshift (distance) limited survey of quasars and faint galaxies becomes viable thus affording us the opportunity to visualize accurately the universe in three dimensions.

All of the earlier fibre-positioning schemes were manual "plug-plate" systems in which an aperture plate is prepared in advance from the astronomer's target list using a milling machine, and then each fibre is inserted by hand—paying careful attention to which fibre goes to which hole! More automated systems such as the Durham/AAT AUTOFIB systems soon followed (Fig. 3.13). Obviously, these new "robot-like" systems depend on a control computer to do all the accurate positioning and keep all the necessary logbooks. Perhaps the most ambitious system to date is called 2dF on the Anglo-Australian Tele-

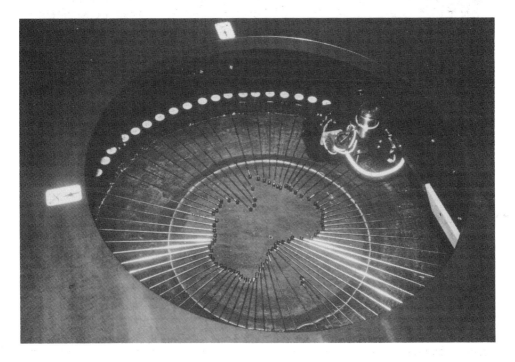

Fig. 3.13. A robotic positioner for optical fibres (AUTOFIB 1) on the Anglo-Australian Telescope. Photo Ian Parry.

scope which operates at the prime focus and enables one set of fibres to be positioned robotically while observations proceed with another set (Fig. 3.14). The primary astronomical concerns with optical fibre systems are accurate positioning and consistent sky subtraction.

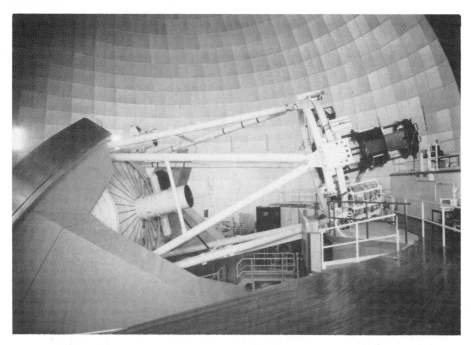

Fig. 3.14. 2d-F, perhaps the most ambitious multi-fibre system to date. A robotic mechanism can position hundreds of fibres across the prime focus field of the AAT and begin recording their spectra, while an identical system gets "configured" under computer control. When ready, the two massive units exchange positions by mechanically tumbling over and the newly prepared fibres start taking data. Courtesy, Anglo-Australian Telescope.

Light transmits down an optical fibre by total internal reflection. Step index fibres are used. These have an inner core and an outer cladding. Let n_f and n_c be the refractive index of the fibre core and the cladding respectively, then $\sin\theta_{max} = \sqrt{\left(n_f^2 - n_c^2\right)}/n_0$ where n_0 is the index of the external medium. Typically, $\theta_{max} \lesssim 40$ degrees which corresponds to $f/2.2$. For practical focal ratios slower than $f/2.2$ (e.g. $f/8$) the output focal ratio is always faster than the input (e.g. $f/8$ would emerge as $f/5$); this effect is called focal ratio degradation. Fibre optic coupling is therefore most efficient at prime focus. Silica fibres come in different types depending on the OH ion contribution. High OH⁻ implies better UV transmission and ultra low OH⁻ gives transmission from 500 nm to 2000 nm. In the AUTOFIB and 2dF systems, the fibre tips are small units only 6 mm long with tiny microprisms to deflect the light into the fibre and strong NdFeB magnet "buttons" in their base to keep them completely stationary when deposited by the robotic positioning system on the flat magnetic field plate. Positioning accuracy is about 0.15 arcseconds. Fibre core diameters vary, but 140 μm is common; the goal is to match to an appropriate image size such as 2 arcseconds. The 2dF system provides about 400 fibres and a "tumbler" which allows a second set of fibres to be positioned by the robotic gantry while the other is in actual use.

Optical fibres have also re-vitalized the photographic Schmidt telescopes in a novel way. At the Anglo-Australian (formerly UK) Schmidt Telescope in Australia, Fred Watson (then of the Royal Observatory, Edinburgh) pioneered an instrument called FLAIR which is a fibre optic system which allows simultaneous spectroscopic observations of objects scattered over the huge $6.5° \times 6.5°$ field of view; Fred was also responsible for WYFFOS on the WHT. The location of the fibre ends with the target objects is achieved with a specially made warped-to-shape 1 mm thick positive glass plate, rather than the usual negative plate, of the field (Fig. 3.15). The fibres are glued to this plate with an ultraviolet-curing optical cement and the assembled unit is loaded into the position normally occupied by the photographic plate. The bundle of fibre cables is led away from the telescope, stacked one above the other in a vertical line and fed into the slit of a standard grating spectrometer which is set up on an optical bench on the dome floor; the spectrometer is therefore completely stationary at all times, which greatly enhances the stability of

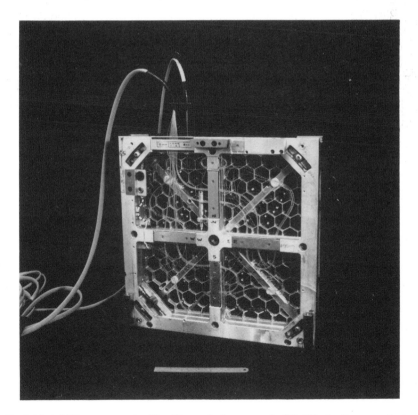

Fig. 3.15. A CCD spectrograph with a fibre-optical system called FLAIR; the fibres are attached to where the 14-inch photographic plate would be in the wide-field UK Schmidt Telescope. Courtesy Fred Watson, RGO.

the apparatus. A cooled CCD camera is used to take images of these spectra. The current version (FLAIR II) has over 90 fibres and two plate holders.

Similar techniques have been extended into the infrared part of the spectrum by Ian Parry and his team at Cambridge, including the development of a fibre-fed high-resolution spectrograph from which they remove or suppress the unwanted emission lines from the night sky and then recombine the "clean" spectrum to produce a lower resolution version free of contamination.

3.3.3 Precision velocities

Several astronomers are engaged in the extremely difficult task of searching for evidence of massive planets orbiting around relatively nearby stars and with orbital periods in the range of 3 to 10 years. Of course, the planets themselves are too faint and too close to their parent stars to be seen directly. The basis of the measurement is the reflex motion on the star being orbited; the star itself must orbit about the common centre of mass—called the "barycentre"—of the star–planet system. For example, the sun orbits its centre of mass in common with Jupiter with a speed of only 12.7 m/s (metres per second). This speed is many times smaller than the errors of conventional velocity measurements of the Doppler shift ($\Delta v = \Delta\lambda \, c/\lambda$, where Δv is the change in velocity of the source corresponding to the wavelength shift $\Delta\lambda$) from stellar spectra. How can such a small effect be observed? In fact, the best results give 3 m/s (Butler *et al.*, 1996).

Precise calibration of the wavelength scale is obtained by allowing the incoming starlight to pass through a chamber containing an almost translucent gas (usually iodine) which absorbs a small amount of light at a few very specific places in the spectrum.

Other astronomers use this technique to search for and study small pulsation patterns in stars. The nature of these vibrations, which also occur in our Sun, yields clues to the *internal* structure of stars and so has been called, "stellar seismology".

3.3.4 Imaging spectrometers

Even the multi-object spectrograph suffers from the fact that in order to record the dispersed spectrum of any object in the field of view of the telescope the remainder of the field, which is most of it, must be hidden from the spectrometer by the slit mask. An image of the scene cannot be recorded with high spectral resolution unless the measurements are repeated many times with slight displacements of the telescope. This strategy was pioneered at the Anglo-Australian Telescope with software written by Pat Wallace (STARLINK). Scanning or stepping the telescope in a sufficiently smooth uniform way to produce a two-dimensional image of the scene composed of numerous strips, with every pixel in every strip providing a spectrum of that point in the image, places stringent requirements on the telescope motion control. For spectroscopic studies of extended objects such as galaxies and nebulae, this is a significant disadvantage.

There are two basic approaches in trying to achieve both spatial and spectral information simultaneously. One is to use interferometric techniques such as the Fabry–Perot interferometer or an imaging Michelson interferometer. The other method involves innovative optical designs called image slicers or more generally, **integral field** units. The Fabry–Perot (described in the next section) allows only a very small range of wavelength space to be covered, but has the advantage of producing an excellent image of the field.

Slit stepping or scanning can provide a large wavelength coverage but it is hard to obtain good quality images along the scanning direction for most telescopes. Integral field spectroscopy combines some of the advantages of each of these; the field of view is less than the Fabry–Perot method but spectral coverage is greater.

One way to implement an integral field mode is to subdivide the focal plane image into numerous very small segments using an array of tiny lenses. This is the basic idea behind the TIGER instrument at the CFHT (Bacon *et al.*, 1995). First, the normal image is greatly magnified and fed to a "microlens" array, which effectively slices up the image before it goes into the spectrograph. The pattern on the CCD detector is complicated, representing both spatial and spectral information for every small segment in the field. Variations of the microlens approach have been developed for other instruments too. An alternative strategy pioneered in the ARGUS instrument at CFHT is to subdivide the focal plane with numerous, closely packed optical fibres in a two-dimensional pattern, but collect all the fibres to "reformat" this image into a one-dimensional stack which can be fed directly to the long slit of a conventional spectrograph.

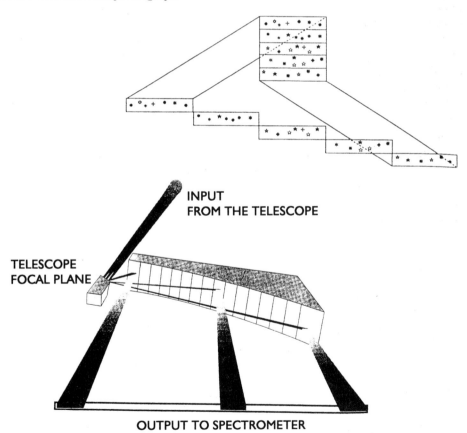

Fig. 3.16. The image slicing concept used in the "3D" near infrared imaging spectrograph developed at the Max Planck Institute to obtain images and spectra of a field simultaneously. Courtesy Alfred Krabbe.

A completely different approach is employed in the "3D" integral field spectrometer developed at the Max Planck Institute (Krabbe *et al.*, 1995). This near infrared instrument uses a complex mirror with many tilted facets to subdivide the image in the focal plane of the telescope into narrow strips and then another similar mirror to "stack" these parts one beside the other along the length of a spectrograph slit (Fig. 3.16). Thus, every region of the image produces a spectrum. The field of view that can be covered is relatively small ($8'' \times 8''$) and once again the image on the array detector is a peculiar-looking combination of spectral and spatial information.

3.4 POLARIZATION —TRANSVERSE WAVES

Polarization is the term used to describe the phenomenon in which a fraction (or all) of the electromagnetic waves in a beam of light are vibrating in the same plane, such as the vertical or horizontal planes (or any orientation in between) along the line of sight to the source. If there is no preferred orientation and the beam contains electromagnetic waves having a random jumble of all planes of vibration then the beam is said to be "unpolarized". When there is a preferred plane of vibration which does not change no matter at which point in the wave cycle we sample, the beam is said to be "linearly" polarized. On the other hand, if the plane of vibration rotates by 360° through a wave cycle then the beam is said to be "circularly" polarized. The light emitted by an ordinary tungsten lamp is essentially unpolarized, whereas the scattered sunlight which constitutes "blue sky" is very strongly linearly polarized.

Many important astrophysical phenomena produce polarization either through the interaction of unpolarized light with matter, or by the generation of polarized light by the atoms themselves. For example, reflection from solid surfaces, scattering of photons by electrons, molecules and small grains, and absorption by certain materials in the interstellar medium all cause polarization; atoms emitting light in the presence of a magnetic field (as in sunspots) suffer the Zeeman effect and the emitted radiation is polarized; very high energy (relativistic) electrons spiralling in a magnetic field around a neutron star will emit polarized light called synchrotron radiation. Polarization spectra and polarization images therefore contain additional information about physical processes and source geometry which cannot be discerned from brightness measurements alone.

Polarimetry depends on relative brightness measurements, on the same source, at different "settings" of a polarization sensitive element—often called a polarization modulator. Because of the systematic, but unpredictable, intensity variations in the light due to poor seeing or poor tracking, it is essential that the rate at which one moves from one setting of the polarimeter modulator to another be fairly rapid. Alternatively, some means must be found to measure two polarization positions simultaneously, so that both are affected in the same way by any systematic errors. One solution is to use the charge-coupling attributes of the CCD to build up the alternating (polarization) signal *on the chip* until enough signal has accumulated to swamp any electronic readout noise. This idea was first suggested to me by Pete Stockman in an imaging context using a device called a Pockels cell when we were both at Steward Observatory in Arizona in 1978/79. This led me to the idea of developing an instrument called the Imaging Spectro-Polarimeter (or ISP) based on the unique properties of a 3-phase CCD detector. Although basically a CCD camera, the ISP could be converted to a spectrometer by placing a "grism"—small right-angled prism with a transmission grating deposited on the hypotenuse face—in one of the filter positions, and converted to a polarimeter (imaging or spectro-) by inserting a polarization modulator in front of the entire optical system. In the spectropolarimetry mode, a special polarizer made of calcite was placed under the slit of the spectrograph which produced two oppositely polarized spectra on the CCD; actually, two tiny slits were used, one for the object and one for the sky, and so four spectra appeared on the CCD. The 3-phase structure of the CCD—described and explained in Chapter 6—allows so-called "bi-directional" charge-transfer, therefore alternating images (or spectra) corresponding to orthogonal polarization states of the modulator could be stored in the top and bottom thirds of the CCD array while only the central part was used for actual light collection (see Fig. 3.17). The collected charge image in the centre is transferred by charge coupling to the appropriate (upper or lower) storage area during the dead-time required to move to the other polarimeter setting. No other detector can perform this trick!

One result from this instrument was the first very high resolution images of the synchrotron polarization from the Crab Nebula and the consequent disentangling of the nebu-

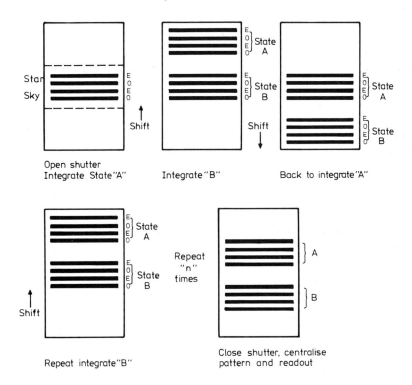

Sequence of CCD frames for shift image polarimetry

Fig. 3.17. The unique bi-directional charge-shifting principle employed in the CCD Imaging Spectro-polarimeter (or ISP) developed at the Royal Observatory Edinburgh by the author.

lar and pulsar polarizations (Fig. 3.18). Other CCD-based spectropolarimeter systems include one developed by Jaap Tinbergen for the ISIS spectrograph on the WHT, one by Joe Miller of University of California, Santa Cruz, for the 3-m Lick Observatory telescope and a similar system is in use with the LRIS instrument on the Keck Telescope. Several national facilities offer imaging- and spectropolarimetry.

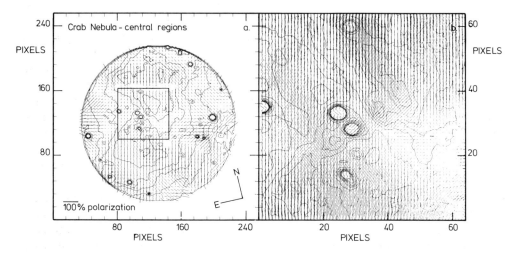

Fig. 3.18. A contour map of the bright emission associated with the Crab Nebula supernova remnant overlaid with tiny segments which represent the amount of polarization and the orientation of the magnetic field in the nebula. Observations obtained with the ROE ISP.

3.5 INTERFEROMETERS—STAYING IN PHASE

A different approach to combining imaging and spectroscopy is the Fabry–Perot interferometer. Developed initially by Charles Fabry and Albert Perot as long ago as 1896, it remained dormant until the 1950s because it was considered too difficult to manufacture and use. The Fabry–Perot "etalon" is essentially a cavity formed by two face-to-face circular plates of high reflectivity and low absorption which are held parallel and flat to a tiny fraction (typically 1/200) of the wavelength of light, i.e. a mere 2.5 nanometres (nm) in the mid-visible. This device has a much greater efficiency or throughput for a given spectral resolution than the normally used diffraction grating. [The throughput of a spectrometer is the "solid angle" it can accept multiplied by its "aperture".] As the spacing or gap between the pair of reflective surfaces is changed, so the wavelength transmitted by the etalon changes. The beauty of the Fabry–Perot etalon lies in the fact that it behaves essentially like an ordinary wavelength selection filter and can be placed directly into the beam of an imaging system to yield a picture of the entire field of view, with the spectral purity and detail that could otherwise only be obtained by a fairly powerful spectrometer. Fabry–Perot etalons are most useful for studies of specific spectral lines and have very limited wavelength span compared to a classical spectrograph.

The Fabry–Perot (or FP for short) can be "tuned" to different wavelengths by careful adjustment of the gap between the reflective plates. This can be accomplished under computer control and a whole series of images can be stacked up in computer memory corresponding to slightly different wavelengths or, equivalently, slightly different velocities. The resulting collection of images is known in the jargon as a "data cube" because it has three "dimensions" namely, the ordinary two spatial dimensions of a picture, and a third dimension which is the wavelength at which the picture was recorded. A plot of the signal values along the wavelength dimension for any spatial position yields the spectrum at that point in the source. Another way to visualize this data cube is to create and project a hologram. One example of a Fabry–Perot instrument is called TAURUS, developed originally for the AAT by Keith Taylor and Paul Atherton, in which the etalons were made by Queensgate Instruments, England (Atherton and Hicks) and use very small position sensors whose electrical "capacitance" is employed to monitor parallelism and spacing of the reflective plates. An electronic control system uses these signals to drive tiny piezoelectric actuators—devices which produce a small movement in response to an electric voltage—to maintain perfect alignment. FPs are also used at infrared wavelengths.

In practice, large telescopes cannot achieve their theoretical angular resolution because of the motions and turbulence of the Earth's atmosphere—even at the very best sites. There is one way around this problem however, at least for sufficiently bright objects, and that is to take numerous very short snapshots, of such short duration that the image motion which leads to blurring is essentially "frozen". Using a camera with a high magnification in order to see the tiny diffraction-limited images, a movie of these snapshots shows a small compact image which changes its position randomly with time; this is called a **speckle** pattern. If a single long integration had been used then the resulting image would be the normal blurred seeing disk. A simple strategy sometimes used is to record each of the snapshots and then use computer processing to "shift-and-add" each image so that the brightest speckles coincide with each other. From a more sophisticated treatment involving intensive mathematical manipulation of the entire speckle pattern and the phase relationships involved, it is possible to reconstruct a single image with diffraction-limited resolution.

One other instrument worth mentioning is the scanning Michelson Interferometer. This is a powerful although relatively rare form of instrument in which no dispersive element, such as a diffraction grating is used, and yet it is possible to obtain a spectrum of the light. The technique relies on decoding a complex signal called an **interferogram** using the mathematical technique of Fourier transforms; the other name for this instrument is a Fourier transform spectrometer or FTS. This class of spectrometers is capable of very high spectral resolution and has been used to obtain the spectrum of the Sun over a huge wavelength range well into the infrared.

SUMMARY

Astronomical telescopes and instruments have been developed which can measure all the properties of electromagnetic radiation. Cameras, spectrometers, polarimeters and interferometers are all applied to study astronomical phenomena. In almost all cases, charge-coupled devices dominate as the detector of choice and these instruments are relatively large, complex and computer-controlled.

EXERCISES

1.   Explain the difference between an equatorial mount and an altitude-azimuth mount. Give an advantage and disadvantage of each.
2.   The 3-m reflector at Lick Observatory has an f/5 prime focus and an f/17 Cassegrain (Cass) focus. (a) Which focus gives the fastest exposure on an extended source and by what factor? (b) Which mode gives the fastest exposure on an unresolved point source?
3.   If you have a CCD camera with a large field of view attached to a typical small telescope and you see images with small "tails" toward the edges of the field, what could be wrong?
4.   How is the data on large photographic plates handled and made available for analysis?
5.   Explain the difference between a Faint Object Spectrograph and a High Resolution Spectrograph. Give an example of a project that would require one or other of these instruments.
6.   What techniques are used to produce spectra of many objects simultaneously? Discuss the advantages of multi-object spectroscopy over conventional spectroscopy.
7.   Cite some examples of modern instruments and telescopes and the discoveries made with them.
8.   Suppose you wish to search the entire sky for relatively faint, point-like objects called quasars whose spectra show strong emission lines at high redshifts. What technique would you use? Why is a CCD camera not so useful in this context?
9.   Suppose you wish to study the gravitational lensing of a distant quasar caused by a massive foreground galaxy. You need deep images and low resolution spectroscopy. What would be the best choice of detector in this application?
10.   What instrument could you use to prove that the light from the Crab Nebula supernova remnant was due to non-thermal emission from the process called synchrotron radiation?
11.   What is meant by "speckle interferometry" and why is it a very useful technique for ground-based imaging?
12.   Describe a Fabry–Perot interferometer.
13.   What is meant by "integral field spectroscopy"?
14.   Explain the technique of drift-scanning.

REFERENCES AND SUGGESTIONS FOR ADDITIONAL READING

Bacon, R. *et al.* (1995) *Astronomy and Astrophysics*, **113**, 347.
Baranne, A. (1972) in ESO/CERN Conference on Auxiliary Instrumentation for Large Telescopes, S. Lausten and A. Reiz (Eds) (Geneva), p.227.
Butler, R.P., Marcy, G.W., Williams, E., McCarthy, C., Dosanjh, P., and Vogt, S.S. (1996) Attaining Doppler precision of 3 m/s, *Pub. Astron. Soc. Pacific*, **108**, 500–509.

Clarke, D. and Grainger, J.F. (1971) *Polarized Light and Optical Measurement*, Pergamon Press, Oxford.

Crawford, D.L. (Ed.) (1986) Instrumentation in Astronomy VI, Proc. SPIE, Vol. 627.

Coyne, G.V., Magalhaes, A.M., Moffat, A.F.J., Schulte-Ladbeck, R.E., Tapia, S. and Wickramasinghe, D.T. (Eds) (1988) *Polarized Radiation of Circumstellar Origin*, University of Arizona Press, Tucson, AZ, USA.

Krabbe, A. *et al.* (1995) *SPIE*, **2475**, 172.

Kovalevsky, J., (1995) *Modern Astrometry*, Springer-Verlag, Berlin.

Oke, J.B., Cohen, J.G., Carr, M., Cromer, J., Dingizian, A., Harris, F.H., Labreque, S., Lucinio, R., Schaal, W., Epps, H. and Miller, J. (1995) The Keck Low-Resolution Imaging Spectrometer, *Pub. Astron. Soc. Pacific*, **107**, 375–385.

Pilachowski, C., Dekker, H., Hinkle, K., Tull, R., Vogt, S., Walker, D.D., Diego, F. and Angel, R. (1995) High-resolution spectrographs for large telescopes, *Pub. Astron. Soc. Pacific*, **107**, 983–989.

Rieke, G. H. (1994) *Detection of Light from the Ultraviolet to the Submillimeter*, Cambridge University Press, Cambridge, UK.

Robinson, L.B. (Ed.) (1988) *Instrumentation for Ground-based Optical Astronomy; Present and Future*, Springer-Verlag, New York.

Tinbergen, J. (1996) *Astronomical Polarimetry*, Cambridge University Press, Cambridge, UK.

Walker, G. (1988) *Astronomical Observations: an Optical Perspective*, Cambridge University Press, Cambridge, UK.

4

Instrumentation—basic principles

In previous chapters we have introduced many different astronomical instruments and techniques without providing much explanation of the underlying physical principles. In this chapter each class of instrument is examined in more detail. Some typical layouts are shown and the basic relationships involving spatial and spectral resolution are given. Chapter 5 expands on this theme by providing additional information useful to the novice designer, and important in understanding the detailed descriptions of solid-state detectors that follow.

4.1 CLASSIFICATION SCHEMES

Broadly speaking, there are four classes of instruments used in optical astronomy namely, (1) **photometers/cameras**, which measure the brightness and direction of radiation, (2) **spectrometers**, which measure the distribution of brightness (or energy) as a function of wavelength, (3) **polarimeters**, which determine the degree of alignment of wave vibrations in a beam, and (4) **interferometers**, which rely on coherent phase relationships to achieve interference effects. Not all cameras are good photometers, and good photometers based on single-cell detectors such as the PMT are not cameras. Analogues of these instruments exist from X-ray wavelengths to radio wavelengths, although the methods of implementation differ considerably. The descriptions which follow refer to visible and near-visible wavelengths. Other wavebands are discussed later.

4.2 PHOTOELECTRIC PHOTOMETERS

A photometer is a device for measuring the brightness of a source. Usually one does not measure the total energy integrated over all wavelengths, but instead the brightness in a limited band of wavelengths which has been selected by means of an "optical filter" such as a red or blue glass plate. Coloured glass is not always sufficient to define the band of wavelengths properly for scientific applications and so special methods are required to make these filters. For photometry of individual stars, a detector with a single cell can be used, e.g. a photomultiplier tube (PMT) or a silicon photodiode. Several systems of brightness measurements have been in use since the introduction of PMTs. The most familiar of these is the UBV system (U = ultraviolet, B = blue and V = visual or yellow). While the underlying principle of the photometer is simple, it is quite difficult to measure the brightness of astronomical sources reliably to even 1 part in 100 (a level of 1% accuracy) unless extreme care is taken. Several dedicated observers have established very stable systems and set up sequences of "standard" stars which other astronomers can use to calibrate their own photometry. Initially, coloured glass filters and the detector's own wavelength-dependent response to light determined the wavebands observed, but it is now possible to design and make an optical filter to pass any specific band of wavelengths desired. These filters are known as **interference filters** because they utilize destructive interference in multiple, very thin, dielectric (non-conducting) layers deposited on the glass substrate (see Chapter 5).

The essential features of a simple photoelectric photometer are shown in Fig. 4.1. The instrument is a light-tight box attached to the telescope by means of a flange. At the telescope focus, which falls inside the box, a circular aperture (or diaphragm) isolates a given star. Usually, these apertures will be interchangeable by constructing a series of them in a wheel or slide. The size of the diaphragm needs to be larger than the image of the star (the "seeing" disk), but not excessively so, otherwise too much "sky" background is included. Another wheel or slide carries a selection of filters such as UBVRI. The detector is usually a PMT (e.g. a thermoelectrically cooled GaAs tube) or sometimes a silicon photodiode. A **field lens**, located near to the focal plane, produces an image of the telescope primary mirror—the "collecting aperture"—onto the detector and NOT an image of the star. All light rays from the star *must* pass through this image no matter where the star is located in the aperture. This design prevents movement of the illuminated image on the detector which might occur due to drifting of the star image across the diaphragm from poor tracking. Consequently, the signal is stable and insensitive to variations in detector response over the photocathode surface. As shown in the figure, solid lines trace rays when the star is centred in the focal plane aperture and the dashed lines indicate the light path when the star is at the edge of the aperture.

Many small telescopes all over the world are equipped with simple photoelectric photometers using pulse-counting PMTs. These systems are routinely used to monitor variable stars of all types.

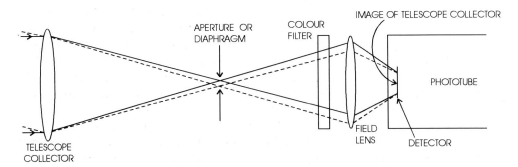

Fig. 4.1. Shows the basic layout of a photometer. Note that the image of the star is not focused onto the detector. A field lens produces an image of the telescope collecting aperture on the detector. In this way the signal strength is independent of the position of the star within the aperture.

One of the great advantages of the photomultiplier tube is its speed of response to a change in brightness, typically one thousandth of a second. There are many useful and important applications of this so-called "high-speed photometry". For example, objects such as cataclysmic variables, and pulsars suffer rapid changes in brightness on short time scales. Also, when stars are occulted by the Moon (or a planet) passing in front of them, or satellites of planets are occulted by the planet itself, there is a very rapid dimming which yields the physical dimensions of the sources.

4.3 CAMERAS—MATCHING THE PLATE SCALE

Fig. 4.2 shows the basic layout of a camera. In the simplest design, the detector (CCD or other array detector) is placed directly in the focal plane of the telescope behind a light-tight shutter. Band pass filters are therefore located in a wheel or slide in the converging beam from the telescope. Care is required to ensure that all filters have the same "optical path", i.e. the product of refractive index and thickness, in order to avoid refocusing the telescope after each filter change. An alternative approach, shown in the figure, is to **colli-mate** the beam by placing a lens after the focal plane at a distance equal to its focal length

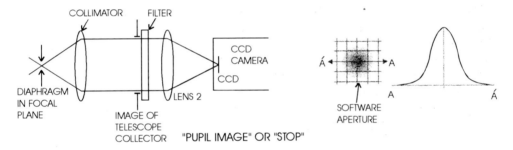

Fig. 4.2. The basic layout of a camera system in which optics are used to collimate the diverging beam from the telescope focus and re-image the field at a different magnification. Filters can be placed in the collimated beam.

and to re-image the field onto the detector with a camera lens (or mirror). This design has many advantages. First, by selecting the focal lengths of the collimator and camera sections one can either magnify or reduce the plate scale. Filters of arbitrary thickness can be located in the parallel (collimated) beam and the filters can be placed near the so-called "pupil" image of the primary mirror. In addition, a circular aperture or "stop" can be placed at the pupil image to reject stray light from outside the beam. This is particularly important in infrared cameras where this mask is at cryogenic temperatures and so becomes a "cold" stop (see Chapter 8). Of course, in this design, the image of the star may drift by a small amount due to tracking or pointing errors, but photometry is performed "after the fact" on the digital image by selecting an appropriately sized "software aperture" and summing up all the signal. An annulus around the summed region is used to construct an estimate of the sky flux contained in the summed aperture. Thus no "separate" measurement of the sky is required. Since the star image is spread over many pixels, and since different pixels are used for the sky image, it is essential to have a good procedure to normalize all the pixels to the same sensitivity or gain. This is a general requirement with array detectors which is covered in considerable detail later.

In most astronomical instruments (UV, visible, IR), the best optical component is usually the telescope itself. For imaging applications it is therefore very attractive to place the detector directly at the focal plane of the telescope, but this is not always appropriate. Of primary concern is "matching" the image quality to the pixel size on the detector, and there may be additional issues such as the need for a pupil image to baffle scattered moonlight or a cold stop to reduce thermal backgrounds in the case of infrared instruments.

There are two issues to be considered when matching the spatial or spectral resolution element to the physical size of the detector pixels: (1) maximizing observing efficiency, meaning more light onto a pixel and therefore keeping the required integration time to a minimum, and (2) accomplishing this task without compromising the ability of the camera system to obtain very accurate brightness measurements (photometry). The spatial resolution element may be determined by seeing conditions or by optical constraints. In general, the image is either **critically sampled**, meaning that there will be about two pixels (also known as the Nyquist limit) across the resolution element, or it will be over-sampled which implies that there may be about five pixels across the resolution element. It is very rare to design a system which is under-sampled deliberately.

In a spectrometer, the width of the entrance slit is usually the determining factor. A narrow slit implies higher spectral resolution, but the highest efficiency is achieved when the slit is wide enough to accept the full image diameter.

Consider first the plate scale of the telescope which is given in arcseconds per millimetre ("/mm) by:

$$(ps)_{tel} = \frac{206\,265}{F_{tel}} \tag{4.1}$$

where F_{tel} is the focal length of the telescope in millimetres ($F_{tel} = D_{tel} \times f$/number) and the numerical factor is the number of arcseconds in 1 radian; 2π radians $= 360°$. Plate scales vary considerably. For instance, at the prime focus of the 3.6 m Canada–France–Hawaii (CFHT) telescope the scale is 13.70 "/mm, whereas at the Cassegrain focus the scale is 7.33 "/mm. With an infrared telescope however, the focal ratio is usually larger (slower) so that at the Cassegrain focus of the 3.8 m UK Infrared Telescope (UKIRT) the scale is only 1.52 "/mm. For our f/16 24-inch reflector at UCLA we get 21.1"/mm. For direct imaging, the angle on the sky subtended by the detector pixel is,

$$\theta = (ps)_{tel}\, d_{pix} \tag{4.2}$$

where d_{pix} is the physical pixel size in milimetres. For CCDs and near-infrared array detectors, values range from about 0.009 mm (9 μm) up to about 0.030 mm (30 μm); detector pixels on mid-infrared arrays may be significantly larger. For 20 μm detector pixels we would get 0.27 and 0.15 "/pixel at the prime and Cass foci of the CFHT respectively, 0.42 "/pixel on the 24-inch at UCLA and only 0.03 "/pixel on UKIRT. We need to compare these values with the image quality to determine whether or not some optical magnification is required. For example, for our "roof-top" conditions on the UCLA campus we use 3" for the average seeing disk, whereas for the instruments on the CFHT and other telescopes on Mauna Kea, Hawaii, one might adopt 0.3–0.5″ for the seeing! Calculating the magnification factor can proceed as follows:

- choose a value for the diameter of the seeing in arcseconds
- decide on the sampling (2–5 pixels)
- divide the seeing diameter by the sampling factor to give the required angular resolution θ_{pix} in arcseconds
- given the size of the detector pixels, derive the plate scale at the detector from $(ps)_{det} = \theta_{pix}/d_{pix}$
- the required magnification (m) is then

$$m = \frac{(ps)_{tel}}{(ps)_{det}} \tag{4.3}$$

Note that m also defines an effective focal length (EFL = mF_{tel}) for the entire optical system. If $m > 1$ then the optics are a magnifier; if $m < 1$ (the usual case) then the optics are called a **focal reducer**. We can also relate the pixel size in arcseconds to the f-number of the focal reducer optics (or simply, "the camera") by

$$\theta_{pix} = 206\,265\,\frac{d_{pix}}{D_{tel}(f / number)_{cam}} \tag{4.4}$$

where $(f/number)_{cam} = F_{cam}/D_{cam}$.

Example: If $d_{pix} = 27$ µm and $D_{tel} = 10$ m, then $\theta_{pix} = 0.56″/(f/number)_{cam}$. Assuming seeing of 0.5″ (on Mauna Kea) and two-pixel sampling, this implies $\theta_{pix} = 0.25″$ which leads to $(f/number)_{cam} = 2.2$. Remembering that the optics must be well-corrected, this is quite a fast camera (more on this issue in Chapter 5). For 18.5 µm pixels, however, we would need an $f/1.5$ camera! As CCD pixels get smaller and telescope mirrors get larger, it becomes more challenging to invent an optical re-imaging (or matching) system. Over-sampling to three or four pixels, or having a smaller image size in the first place (due to a system which eliminates atmospheric turbulence effects; see Chapter 11), makes things much easier.

The image of a distant point-source object produced by a perfect telescope with a circular entrance aperture (D_{tel}) should have a bright core surrounded by fainter rings. This pattern is called the **Airy diffraction disk** (after Sir George Airy) and the first dark minimum between the bright core and the first faint ring corresponds to an angular radius (in radians) of

$$\theta = 1.22\,\frac{\lambda}{D_{tel}}\ \text{radians} \tag{4.5}$$

where λ is the wavelength. Again, to convert this angle to seconds of arc (″) multiply by 206 265, the number of arcseconds in one radian.

Example: $\lambda = 0.5$ µm, $D_{tel} = 0.5$ m then $\theta = 0.25″$. The same limit is reached at $\lambda = 1$ µm for a 1-m telescope and again at 10 µm for a 10-m telescope. Alternatively, at 0.5 µm on a 10-m telescope the diffraction limit is only 0.0125″, which is much smaller than the typical seeing disk (even on Mauna Kea). The size of the seeing disk can be predicted as λ/r_0, where the parameter r_0 is the length over which the incoming wavefront is not significantly disturbed by motions in the Earth's atmosphere (see Chapter 11). For $r_0 = 20$ cm at a wavelength of 0.5 μm the seeing would be at least 0.5″.

The angular size of the diffraction-limited image can be related to a physical size by multiplying by the appropriate focal length (F_{tel} in this case) and using the definition of f/number,

$$r_{diff} = 1.22\lambda\left(f / number\right)_{tel} \tag{4.6}$$

which shows that the diameter of the diffraction-limited image from an f/15 telescope is 36.5 μm at $\lambda = 1$ μm. The same formula can also be used to find the physical size of the image spot in a camera system. For example, an f/3 camera has a diffraction-spot of only 7.3 μm at a wavelength of 1 μm, half this size in the mid-visible and double this in the near-infrared bands.

Optical matching to a CCD in spectroscopic mode is similar to the normal camera mode except that it will usually be relevant to choose the pixel size to correspond to the spectral resolution of the spectrometer, and this is partially determined by how narrow the entrance slit is made. Sometimes this will be smaller than the seeing disk, although designers try to ensure that all of the light is accepted without loss of spectral resolution; the consequence of this is that the spectrograph has to be physically very large (see section 4.4). As in the case of spatial resolution, it is best to oversample the spectral resolution with two, three or more pixels.

4.4 SPECTROMETERS—DISPERSION AND RESOLVING POWER

All spectrometers, whether high, medium or low resolution, have essentially the same basic design, but many different implementations are possible depending on the constraints and choice of spectral disperser. Fig. 4.3 shows the essential features of the layout.

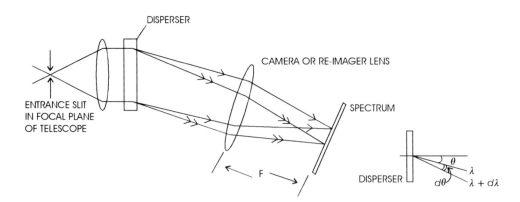

Fig. 4.3. The essential features in the optical layout of a spectrometer. The beam is collimated before intersecting the dispersive element and then the spectrum is re-imaged with camera optics onto the detector.

Instead of a wide field aperture at the focal plane of the telescope, a mask with a narrow slit is used. The slit width must be matched to either the seeing conditions or the diffraction disk depending on the design and application. As the beam diverges from the slit it is first collimated and then directed to the dispersing system (one or more prisms or diffraction gratings) after which the spectrally dispersed beam is finally collected by the camera optics and re-imaged onto the CCD or other array detector. The important quantities to determine when beginning the design are (1) the resolving power (R), (2) the slit width, (3) the diameter of the collimated beam, (4) the sampling or matching of the slit width to the detector pixels, and (5) the resulting f/number of the camera system. Several key terms and properties are as follows:

Angular dispersion (AD): the rate of change of the dispersed angle (θ) of the beam with respect to wavelength (λ).

$$AD = \frac{d\theta}{d\lambda} \tag{4.7}$$

[Although expressed here as a derivative, $d\theta$ and $d\lambda$ can also be taken as small intervals.]

Linear dispersion (LD): relates an interval of length (dx in millimetres) along the spectrum to a wavelength interval ($d\lambda$ in Å)

$$LD = \frac{dx}{d\lambda} = \frac{dx}{d\theta}\frac{d\theta}{d\lambda} = F\frac{d\theta}{d\lambda} \tag{4.8}$$

where F is the focal length of the spectrograph camera and the units are usually expressed as mm/Å. A more useful form is the **reciprocal linear dispersion** which is simply the inverse of the above expression in Å/mm.

Resolving power: is the ability to distinguish two wavelengths separated by a small amount $\Delta\lambda$.

$$R = \frac{\lambda}{\Delta\lambda} \tag{4.9}$$

Note that the "resolution" is often stated as

$$\frac{1}{R} = \frac{\Delta\lambda}{\lambda} = \frac{v}{c} \tag{4.10}$$

where the non-relativistic Doppler formula is used to relate the smallest detectable wavelength shift to the velocity v of the source which could cause that shift. For example, for $R = 10\,000$ or 0.01% resolution ($\Delta\lambda = 0.0001\lambda$) then $v = 0.0001c = 30$ km/s.

The usual dispersing element is a diffraction grating and the general grating equation is

$$m\lambda = d(\sin i + \sin\theta)\cos\gamma \tag{4.11}$$

where d is the spacing of adjacent grooves, i is the angle of incidence of the collimated beam, θ is the angle of the emergent diffracted beam, γ is the angle out of the normal plane of incidence (usually 0°, hence $\cos\gamma = 1$) and m is an integer called the "order" of interference. The angular dispersion of a grating is therefore given by

$$\frac{d\theta}{d\lambda} = \frac{m}{d\cos\theta\cos\gamma} \tag{4.12}$$

Substituting for m/d gives

$$\frac{d\theta}{d\lambda} = \frac{\sin i + \sin\theta}{\lambda\cos\theta\cos\gamma} \tag{4.13}$$

Usually $\cos\gamma \sim 1$, and therefore the angular dispersion is determined entirely by i and θ for a given λ. Many combinations of m and d yield the same AD provided the grating angles remain unchanged. Typical "first order gratings" ($m \sim 1$) have 300–2400 grooves or lines/mm. Coarsely ruled reflection gratings (large d) can achieve high angular dispersion by making i and θ very large, typically 60°. Such gratings, called **echelles**, have groove densities from 20 to 200 lines/mm with values of m in the range 10–100.

A grating produces a different magnification in the dispersion direction than at right angles to the dispersion. The **anamorphic magnification** factor describes this effect and is found by determining the change in θ for a change in i.

$$\frac{d\theta}{di} = \left| \frac{\cos i}{\cos\theta} \right| \tag{4.14}$$

and the size of the slit image (Δx) at the detector becomes

$$\Delta x = w\frac{\cos i}{\cos\theta}\frac{F_{cam}}{F_{coll}} \tag{4.15}$$

where w is the true slit width and F_{cam} and F_{coll} are the focal lengths of the camera and collimator optics; this ratio is the normal magnification factor.

If the grating is to accept all the light from the collimator then it follows that the ruled width of the grating (W) must be $W = D_{coll}/\cos i$. In the diffraction-limited case the resolving power of a grating spectrometer is then

$$R = mN = \frac{mW}{d} = \frac{W(\sin i + \sin\theta)}{\lambda} \tag{4.16}$$

where N is the total number of grooves illuminated. In practice however, spectrometers are usually slit-width or seeing-limited. If the slit is matched to the angular size (ϕ) of the seeing disk then, $\phi = \lambda/D_{tel}$ and

$$R = \frac{W(\sin i + \sin\theta)}{\phi D_{tel}} \tag{4.17}$$

As the telescope diameter increases, R decreases, unless W gets larger too. Hence the effort to produce larger reflection gratings and echelles as the size of telescopes has increased.

Taking ϕ to be $p \times \theta_{pix}$, where p is the number of pixels across the slit image, and converting to arcseconds, gives the form which shows explicitly the tradeoffs of size versus resolution:

$$R = \left(\frac{\sin i + \sin \theta}{\cos i}\right) \frac{D_{coll}}{D_{tel}} \frac{206\,265}{p\theta_{pix}} \tag{4.18}$$

Blaze angle: By tilting the facets of a reflection grating through an angle θ_B with respect to the plane of the grating surface it is possible to maximize the grating efficiency in the direction in which light would have been reflected in the absence of diffraction. Grating efficiency is a maximum when the angle of incidence is $\theta_B + \theta_0$, and the angle of diffraction is $\theta_B - \theta_0$, where θ_0 is measured with respect to the normal to the facet (not the grating surface). There is a special case when $\theta_0 = 0$, then the incident ray enters along the normal to the facet and the diffracted ray leaves along the same direction. This is the **Littrow** condition and the incident and diffracted angles measured relative to the grating normal are now equal to each other and to the blaze angle. The grating equation simplifies to $m\lambda = 2d \sin \theta_B$ and the resolving power is given by

$$R = \frac{2 D_{coll} \tan \theta_B}{\phi D_{tel}} \tag{4.19}$$

The only way to work in the Littrow condition is with a central obscuration in the optics. Alternatively one can use the "near" Littrow condition by moving off by a few degrees or the "quasi" Littrow condition by going out of the plane ($\gamma > 0°$). Grating efficiency drops rapidly as the angle away from Littrow grows, whereas the drop is very slow for the quasi-Littrow mode but the adverse effect is that the slit images are tilted. The tilt angle χ is given by

$$\tan \chi = \tan \gamma \, \frac{(\sin i + \sin \theta)}{\cos \theta}$$

which for a tan $\theta_B = 2$ echelle gives tan $\chi = 4 \tan \gamma$. For example, for $\gamma = 5°$, $\chi = 19.3°$ and there is also a change $\Delta\chi$ in this angle across an order; the higher the order the smaller the change.

Free spectral range: For a given pair of incident and diffraction angles the grating equation is satisfied for all λ for which m is an integer. There are two wavelengths in successive orders, λ and λ', for which $m\lambda' = (m+1)\lambda$. The wavelength difference $\lambda' - \lambda$ is called the **free spectral range** (FSP) therefore

$$\Delta\lambda_{FSP} = \frac{\lambda}{m} \tag{4.20}$$

The two wavelengths are diffracted in the same direction and require either an "order sorter" filter or a cross-disperser element (when m is large) which is another grating or a prism at right angles to the first one.

A very popular way to convert a camera into a spectrograph is to deposit a transmission grating on the hypotenuse of a right-angled prism and use the deviation of the prism to bring the first order of diffraction on axis. Such a device is called a **grism** and Fig. 4.4 shows the basic geometry. The advantage of a grism is that it can be placed in a filter wheel and treated like another filter. The basic relationships required to design a grism are

$$m\lambda_c T = (n-1)\sin\phi \qquad\qquad (4.21)$$

and

$$R = \frac{EFL}{2d_{pix}}(n-1)\tan\phi \qquad\qquad (4.22)$$

where λ_c is the central wavelength, T $(=1/d)$ is the number of lines per millimetre of the grating, n is the refractive index of the prism material and ϕ is the prism apex angle. EFL is the effective focal length of the camera system (see Eq. (4.3)) and d_{pix} is the pixel size. The factor of 2 assumes that two pixels are matched to the slit width. In practice, the number of free parameters is constrained by available materials and grating rulings, and given conditions within the camera system. Resolving powers (two pixels) of $R \sim 500$ are practical.

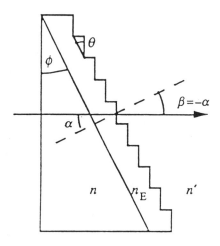

Fig. 4.4. A simplified schematic giving the basic geometry of a grism—a transmission diffraction grating deposited on the hypotenuse face of a right-angled prism.

4.5 POLARIMETERS—THE STOKES PARAMETERS

Polarization is notoriously difficult to measure in astronomy applications. With the exception of a few important cases (reflection nebulae, synchrotron emission from supernova remnants and cyclotron emission from magnetic white dwarf systems), the levels of polarization are quite small. Typically only a few per cent of starlight is linearly polarized by passage through the interstellar medium and much less is circularly polarized. Nevertheless, polarization is a powerful tool because it provides unique "geometric" information which measurement of intensity alone cannot do. To measure the polarization properties of light (fraction polarized, direction of vibration and handedness of rotation) all polarimeters "convert" the polarization information into brightness modulations which are directly measurable with an electronic detector. Polarimeters therefore benefit from photometer, camera and spectrometer designs. In essence, a photopolarimeter, an imaging polarimeter and a spectropolarimeter are all created by adding a "polarization modulator" to the design.

The modulator can be a rotatable plate of a special optical material (or composite material such as quartz and magnesium fluoride) which exhibits a crystal structure property called **birefringence** which makes the material sensitive to the orientation or plane of vibration of the incident light wave. Essentially, the crystal will distinguish between electromagnetic waves which are vibrating at right angles to one another, and will "slow-down" one of them. Alternatively, the modulator can be made of a material in which birefringence can be introduced by external means. For example, the application of an alternating electric field to certain crystals is one method—this is called a Pockels cell. Another way of inducing birefringence is by a variable mechanical strain—this is the photoelastic or Kemp modulator. In each case the modulator is followed by a polarizer (like Polaroid but usually a glass component rather than a plastic component to improve transmission). Operation of the modulator/polarizer combination results in a controlled, periodic variation of brightness of the transmitted light provided the incident light was polarized. These brightness variations are recorded by the detector. Conceptually, all polarimeters fall into this basic design structure; the light from the telescope must first pass through the polarimeter section comprising the modulator and polarizer, then through the camera or spectrometer part to arrive at the detector.

Although the simplest approach is to rotate a polarizer (such as Polaroid) in the beam, this is rarely done unless the remainder of the optical train is completely insensitive to polarization. A more robust approach is to construct the polarization modulator in two parts, a phase retardation device which introduces a known and controllable phase shift into the beam, and a fixed polarizer (or analyser) which only allows one plane of polarization to pass unhindered and reduces others by the factor $\cos^2 \theta$, θ is the angle between the polarizer's axis and the plane of polarization in the beam. The intensity transmitted by the analyser is therefore modulated by the action of the phase retardation device. Unlike the Pockels cell and Kemp modulator, in which perfect half-wave or quarter-wave retardation can only occur at a single wavelength, retardation plates can be made achromatic by combining two birefringent materials of different refractive indices. In general, the **retarder** is the first optical component in the system (apart from the telescope mirrors) to minimize spurious or instrumental sources of polarization. The analyser must also be carefully placed so as to feed a constant direction of vibration to the detector and not be "crossed" inadvertently with any partial polarizer in the system (such as a diffraction grating) which would reduce throughput.

A typical polarimeter layout for a camera or a photometer is shown in Fig. 4.5. The polarizing prism can also be a "double-beam" device which produces two polarized images on the detector known as the o (ordinary) and the e (extraordinary) images. A spectropolarimeter is constructed in a similar way, with the analyser in front of the dispersing element. To overcome brightness modulations caused by the atmosphere (seeing, transparency), polarimetry must either be done rapidly, or by a ratio method. Most photopolarimeters using PMTs operate using the rapid modulation approach, but this is not suitable for CCDs and array cameras because of readout noise. A partial solution is to "charge-shift" back-and-forth on the CCD as described in the Chapter 3 survey of instru-

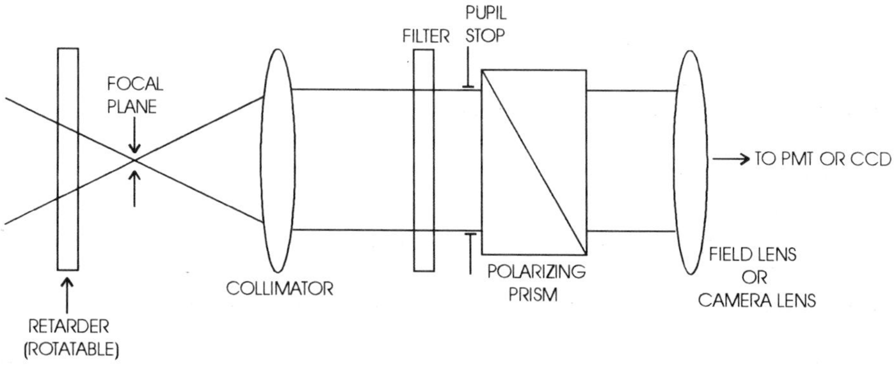

Fig. 4.5. A typical polarimeter layout for a camera or a photometer.

ments, but the best solution is to use a double-beam instrument and take ratios of images and/or spectra in the e and o channels.

Linear polarization is described by three parameters: intensity (I), degree (or fraction) of linear polarization (p), and the direction of the (fixed) plane of vibration projected on the sky (θ). Circular polarization is similarly described by three parameters: intensity (I), degree of circular polarization (q) and handedness of the rotation of the electric vector (+ or −). A more convenient way to express polarization information is to use the four **Stokes parameters** (I,Q,U,V). These quantities are phenomenological, that is they are more directly related to actual measurements. The Stokes parameters are easily related to the amplitudes (E_x, E_y) of the electric vector in two orthogonal directions and to the phase difference (δ) between the two components. The degree of linear and circular polarization is given by

$$p = \frac{\left[Q^2 + U^2\right]^{\frac{1}{2}}}{I}, \quad q = \pm\frac{V}{I} \tag{4.23}$$

and the direction of vibration of the linearly polarized part is given by

$$\tan 2\theta = \frac{U}{Q} \tag{4.24}$$

and it follows that

$$Q = Ip\cos 2\theta$$
$$U = Ip\sin 2\theta \tag{4.25}$$
$$V = Iq$$

The intensity of light transmitted by a retarder of retardance τ at angle ψ followed by a perfect polarizer with principal plane at $\phi = 0°$ or $\phi = 90°$ (upper/lower signs respectively) is given by

$$I' = \frac{1}{2}\left[I \pm Q(G + H\cos 4\psi) \pm UH\sin 4\psi \mp V\sin\tau\sin 2\psi\right] \tag{4.26}$$

where

$$G = \frac{1}{2}(1 + \cos\tau), \; H = \frac{1}{2}(1 - \cos\tau), \; \tau = \frac{2\pi}{\lambda}\delta \qquad (4.27)$$

There are two special cases of particular interest.

(1) *The quarter-wave retarder:* $\delta = \lambda/4$, $\tau = 90°$, $G = H = \frac{1}{2}$ which gives

$$I' = \frac{1}{2}\left[I \pm \frac{1}{2}Q\cos 4\psi \pm \frac{1}{2}U\sin 4\psi \mp V\sin 2\psi \right] \qquad (4.28)$$

(2) *The half-wave retarder:* $\delta = \lambda/2$, $\tau = 180°$, $G = 0$, $H = 1$ which gives

$$I' = \frac{1}{2}\left[I \pm Q\cos 4\psi \pm U\sin 4\psi \right] \qquad (4.29)$$

Note that the modulation is at four times the rotation angle ψ. Method (2) does not allow the circular component (V) to be determined, but it is more efficient in modulating the intensity to derive Q and U, and is the method most often used for stellar polarimetry. Solar magnetographs on the other hand must determine the circular component also, and therefore Method (1) is the basis for those instruments.

There are many ways to solve these equations for the Stokes parameters. As an example, consider the case of linear polarization (Method 2). The simplest solution, which also emphasizes the direct relation between the Stokes parameters and measured quantities, is to set the angle ψ to four discrete values ($0°$, $22.5°$, $45°$, and $67.5°$) which yields

$$I'(0°) = \frac{1}{2}(I + Q) \quad I'(45°) = \frac{1}{2}(I - Q)$$
$$I'(22.5°) = \frac{1}{2}(I + U) \quad I'(67.5°) = \frac{1}{2}(I - U) \qquad (4.30)$$

and solving for I, Q and U gives

$$Q = I'(0°) - I'(45°) \quad U = I'(22.5°) - I'(67.5°)$$
$$I = I'(0°) + I'(45°) \quad I = I'(22.5°) + I'(67.5°) \qquad (4.31)$$

Note that I is redundantly determined, and that Q and U have the same units as I. It is also common practice to form the normalized Stokes parameters by taking the ratio of Q/I and U/I and V/I; these ratios are sometimes referred to as the Stokes parameters in the literature and even given the same symbol.

4.6 INTERFEROMETERS

There are several types of interferometers in use, including the Fabry–Perot interferometer, the Michelson interferometer and the Fourier transform spectrometer (FTS) which is a scanning Michelson interferometer. Here we summarize the principles of just two of these instruments.

4.6.1 The Fabry–Perot etalon

The Fabry–Perot interferometer is an imaging spectrometer which is formed by placing a device called an **etalon** in the collimated beam of a typical camera system. A typical ar-rangement is shown in Fig. 4.6. The etalon consists of two plane parallel plates with thin,

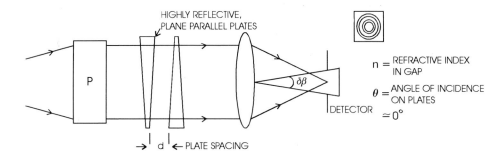

Fig. 4.6. A typical arrangement for a Fabry–Perot interferometer is shown.

highly reflective coatings on their inner faces. The plates are in near contact but separated by a distance d. Assuming that the refractive index of the medium in the gap is n (usually $n = 1$) and θ is the angle of incidence of a ray on the etalon (usually very small), then multiple reflections and destructive interference within the gap occurs and the wave-lengths transmitted with maximum intensity obey the relation

$$m\lambda = 2nd \cos\theta \tag{4.32}$$

For monochromatic light, the image is a set of concentric rings. To ensure that a suffi-ciently narrow band of light passes through the system, it is necessary to "pre-filter" the light. This can be done with a very narrow band interference filter. Usually a circular aperture isolates the central order which has an angular diameter $\delta\beta = \sqrt{(8/R)}$ and the free spectral range is given by

$$\Delta\lambda_{FSP} = \frac{\lambda}{m} = \frac{\lambda^2}{2nd} \tag{4.33}$$

and the resolving power (R) is

$$R = \frac{2\mathcal{F}nd}{\lambda} \tag{4.34}$$

where $\mathcal{F}$ is called the **finesse** of the etalon, which is a measure of the plate quality and the reflectance of the coatings; typical values are 30–50.

4.6.2 Fourier transform spectrometer (FTS)

The FTS is a scanning Michelson interferometer with collimated light as an input. A typi-cal scheme is shown in Fig. 4.7. For a collimated monochromatic beam, the intensity at the detector is determined by the "path difference" $\Delta x = 2(x_b - x_a)$, where x_a refers to the arm containing the fixed mirror A and x_b is the distance to the scanning mirror B. The **phase**

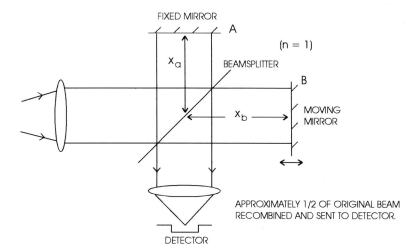

Fig. 4.7. The principle of the scanning Michelson interferometer is shown. As the mirror is scanned the intensity recorded by the detector is modulated to produce an interferogram. The spectrum can be extracted by an inverse Fourier transform.

difference is given by $k\Delta x$ where $k = 2\pi/\lambda$. The fraction of the incident beam in the output is given by

$$T(k,\Delta x) = \frac{1}{2}\left[1 + \cos(2k\Delta x)\right] \qquad (4.35)$$

from which it follows that $T = 1$ when the combining beams are in phase and $T = 0$ when they are 180° out of phase. Given an incident beam whose spectrum is $I(k)$, the signal F measured in the output is

$$F(\Delta x) = c \int I(k)T(k,\Delta x)dk = constant + \frac{c}{2}\int I(k)\cos(2k\Delta x)\,dk \qquad (4.36)$$

where c is a constant. The measured signal $F(\Delta x)$ is called the **interferogram** and the last integral is the Fourier cosine transform of the spectrum. Therefore, the transform of the interferogram is $I(k)$.

An FTS can have a very large resolving power. For example, since R is given by $4\Delta x_{max}/\lambda$ and with $\Delta x_{max} = 10$ cm we get $R = 400\,000$ at 1 μm wavelength. Moreover, *all* the light falls on the detector so, in principle, the signal-to-noise ratio is high. The primary disadvantage for astronomical work is the fact that the measurements require a time sequence to determine the spectrum, during which atmospheric conditions may vary.

SUMMARY

Consideration of the basic layout of an imaging system reveals the importance of matching the physical pixel size of a detector, such as a CCD, to the appropriate angular scale on the sky. The basic relationships for spectrometers are the fundamental starting point for selecting a suitable design, and it was shown that as telescopes get larger, gratings and spectrographs must get larger too. Also, the concept of the grism provides an extremely compact way to convert a camera into a spectrometer. Polarimeter modules can be added to cameras and spectrographs to convert vibrational information into intensity modulations which can be measured. Wave interference effects provide the basis of instruments such as the Fabry–Perot interferometer and the FTS, both of which are used in astronomy applications.

EXERCISES

1. Describe a basic photoelectric photometer. What precautions would you take to ensure that the signal remained constant even if the star drifted off-centre in the aperture? How would your design change if this were an "imaging" system? How would you extract a measurement of the magnitude of the star in this case?

2. Calculate the f/number of a camera lens system required to match 24 μm pixels to 0.2" on the sky for a 10-m telescope. Comment on whether or not this would be challenging. What is the field of view for a 1024 × 1024 pixel detector?

3. Consider the design of a diffraction grating spectrometer for a 10-m telescope. The two-pixel resolution element is 0.5" and the required resolving power is $R = 20\,000$. Assume that the configuration is Littrow. Two gratings are available, a first order grating blazed at 17.5° (2 tan $\theta_B = 0.63$) and an echelle grating blazed at 63.5° (2 tan $\theta_B = 4$). Determine the collimator size D_{coll} in both cases. Which is more practical? Assuming the telescope has an f/ratio of 15, what is the focal length of these two collimators?

4. Describe what is meant by a grism. Design a grism with an index of refraction of 2.4 and an apex angle of 30° which will have a central wavelength of 2.2 μm in the near infrared and a resolving power of $R = 500$ for 2 pixels. Assume the pixel size is 27 μm.

5. Explain the terms linear and circular polarization. How can a camera or a spectrometer be converted to measure polarization? Describe three kinds of polarization modulator and state one advantage and one disadvantage of each.

6. A polarization experiment provides counts at four angular settings each 22.5° apart of a halfwave plate $N(0) = 2000$, $N(22.5) = 1800$, $N(45) = 1000$, $N(67.5) = 1200$. Determine the normalized Stokes parameters, the degree of polarization and the position angle of the direction of vibration. Assuming that the errors of measurement follow a Poisson distribution so that $\sigma(N) = \sqrt{N}$, estimate the error in the normalized Stokes parameters.

7. Describe the basic design of a Fabry–Perot interferometer. For a resolving power of $R = 20\,000$ at $\lambda = 0.5$ μm with an air-spaced etalon of finesse 40, what is the gap d and the free spectral range $\Delta\lambda_{FSP}$?

8. What is the required scan length of an FTS working at a wavelength of 10 μm in the mid-infrared if the required resolving power is $R = 100\,000$?

REFERENCES AND SUGGESTED FURTHER READING

Born, M. and Wolf, E. (1987) *Principles of Optics*, 6th edn, Pergamon Press, Oxford. [Classic text.]

Clarke, D. and Grainger, J.F. (1971) *Polarized Light and Optical Measurement*, Pergamon Press, Oxford. [Basic introduction to polarimetry.]

Léna, P. (1988) *Observational Astrophysics*, Springer-Verlag, Berlin. [Very comprehensive generalized approach.]

Schroeder, D.J. (1987) *Astronomical Optics*, Academic Press, San Diego, CA. [Excellent treatment of telescopes and spectrographs.]

Wall, J.V. (Ed.) (1993) *Optics in Astronomy*, Cambridge University Press, Cambridge, UK.

5

Designing astronomical instruments

There are many important technical issues and design constraints to be aware of when developing new astronomical instrumentation. Of course, complete engineering details are beyond the scope of this book, or any one book, but the following sections will at least provide a basis for newcomers to instrumentation design.

5.1 BASIC REQUIREMENTS

The very first step in instrument design is to fully understand the application. What are the science goals? Sometimes the goals can be fairly general, such as "provide the most sensitive camera with the widest possible field of view consistent with the median seeing conditions". This approach is reasonable, on the grounds that the uses of such an instrument are so numerous. On the other hand, the science goals may be quite specific, such as "provide an instrument to search for planets around other stars via Doppler velocities in the 10 m/s range", or "provide an instrument to carry out a survey of redshifts of a very large sample of faint galaxies". In these two cases, the spectrographs involved would be quite different from each other and from a conventional "workhorse" spectrograph. For the planetary search the spectral resolution needs to be very high and the instrument must provide long-term stability, whereas for the faint galaxy survey the resolution is lower, but slit-masks and/or optical fibres are needed to provide a multi-object approach, otherwise this faint-object program would be much too slow. Clearly, the choice of instrument and the details of the design will depend on the kind of science to be done. If it is imaging science, what field of view is required? What is the angular resolution and the wavelength range? Is time resolution an issue? Are the measurements going to be read-noise limited or background-limited? Is the goal photometry, astrometry, acquisition and guiding, morphology? The basic requirements must come from the science goals, but beware of creating a monster with many heads!Too many scientific options in one instrument can be disastrous. It is wise to identify one or more ways in which the project can be reduced in scope (often called "de-scoped") if necessary. In turn, the science requirements are used to generate a specification for the instrument. Candidate designs can then be analysed in a design study phase and the best or most appropriate design can be selected. There is generally always more than one way of achieving the desired goals, and changes in technology result in fresh approaches and new ways to improve older methods.

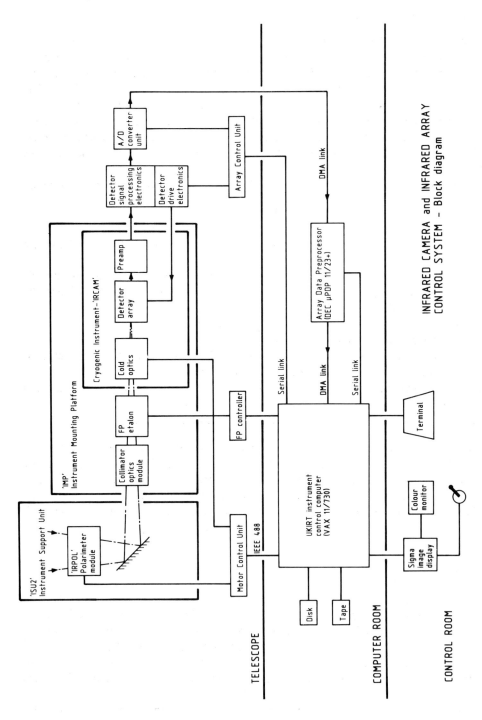

Fig. 5.1. A block diagram layout of an entire system is an essential starting point for any design. This one is for IRCAM, the first common–user infrared camera system developed for the 3.8-m UK Infrared Telescope (1984–1986).

5.2 OVERALL SYSTEM LAYOUT

The second step in the design process is to lay out in a pictorial form all the essential building "blocks" and their interconnections. Invariably, a modular approach results in the best design. As an example, the **block diagram** for the IRCAM infrared imaging system developed by the author's team at the Royal Observatory, Edinburgh (1984–1986) is shown in Fig. 5.1; the technology is out of date, but the principle is not. Basically, this near-infrared camera (see Chapter 8) is similar to an optical CCD camera. When examined at a level more detailed than the block diagram, each sub-system could use one of several different methods for detailed implementation, yet the overall concept is essentially the same.

Although it is hard to generalize, certain building blocks are almost always present. These are:

(1) the detector (the photon sensor itself and circuitry packaged close by)
(2) an opto-mechanical system (lenses, mirrors, filters, gratings, fibres and suitable mounts)
(3) an enclosure and cooling system (for the detector and other parts of the instrument)
(4) signal-processing hardware (e.g. amplifiers, noise suppression circuits) and the analogue-to-digital converter (ADC or A/D)
(5) detector "drive" electronics (pulsed and dc bias circuits)
(6) timing logic and synchronization circuits
(7) a "motion control" system and "housekeeping" system (motorized mechanisms, temperature control, status switches, other monitoring devices)
(8) an electronic interface to a computer (direct memory access, ethernet, telemetry etc.)
(9) a host computer plus peripherals
(10) an image display system and image processing/restoration software

These ten items form the basis of a great many astronomical instruments employing some form of electronic imaging device for photon detection. In fact, the above list could apply to almost any form of detector system used in astronomy if the items are understood in their most general sense. At the heart of all instruments is the detector. Usually, it is the performance of the detector system which determines whether the instrument is "state-of-the-art" or not.

5.3 OPTICAL DESIGN

After detectors, the optical design is the next major component of any astronomical instrument. Over the years, many astronomers have played a significant role in optical technology innovations. I strongly recommend more than a basic physics training in optical design for all young observational astronomers, and for anyone interested in building astronomical instruments.[†] Many good courses are available which teach practical methods of ray tracing and optical tolerancing. Excellent classic texts on optics include Born and

---

[†]Optics was cited as the number one opportunity for physics majors in a university survey carried out in 1996 in the USA.

Wolf (1987), Smith (1990), Kingslake (1978) and Schroeder (1987). A good strategy is to break down the optical design into the following steps and stages:

(1)   "first order" requirements
(2)   constraints
(3)   performance specification
(4)   ray-tracing and optimization
(5)   tolerance analysis

By "first order" requirements is meant a simple design, using "thin lens" formulae, and known facts about the system, such as the *f*/number and plate scale of telescope, the object/image distances and the location of any pupils formed, the required field of view and the desired magnification or image scale at detector. A typical "first order" layout is shown in Fig. 5.2. For example, an image is formed at the telescope focal plane, and this object is then re-imaged by the instrument optics onto the detector. Depending on the ratio of the focal lengths of the transfer optics, there will be a change in magnification of the new image. The transfer optics (in combination with all optics upstream) will form an image of the primary mirror of the telescope. This image is called the **entrance pupil**.

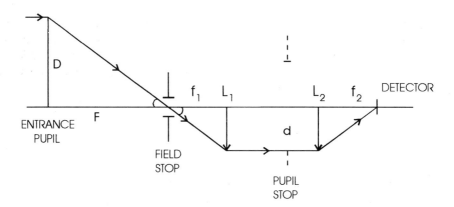

Fig. 5.2. A simple "first-order" layout of a camera system designed to collimate the beam from the telescope and then re-image the field onto a small detector. Pupil size and location is determined together with magnification and field of view constraints. Analysis of the f/numbers and focal lengths can be used to gauge what kind of "real" optics will have to be substituted.

At this stage, the usual simple equations of elementary optics can be employed.

The "thin lens" equation:

$$\frac{1}{f} = \frac{1}{s} + \frac{1}{s'}, \quad m = -\frac{s'}{s} = \frac{h'}{h}, \quad f/number = \frac{f}{D} \tag{5.1}$$

where *f* is the focal length, *s* and *s'* are the object and image distances respectively, *h* and *h'* are the object and image "heights", *m* is the lateral magnification and *D* is the clear aperture diameter of the lens. The same equation holds for spherical mirrors and *f* = *R*/2 where R is the radius of curvature of the mirror. The longitudinal magnification (along the optical axis) is m^2. With rays travelling left-to-right, the sign convention is that s is positive if on the left of the lens and s' is positive if on the right, and the negative sign indicates that the image is inverted when s and s' are both positive.

Snell's Law of Refraction:

$$n_1 \sin \theta_1 = n_2 \sin \theta_2 \tag{5.2}$$

where n_1 and n_2 are the indices of refraction on either side of the optical boundary and θ_1 and θ_2 are, respectively, the angle of incidence and angle of refraction with respect to the normal to the surface. Snell's Law and the Law of Reflection (*angle of incidence = angle of reflection*) are fundamental.

Optical "power" and the "lensmaker's formula":

$$\phi = \frac{1}{f} = (n-1)\left[\frac{1}{R_1} - \frac{1}{R_2} + \frac{t(n-1)}{nR_1R_2}\right] \tag{5.3}$$

where ϕ is the power of the lens, R_1 and R_2 are the radius of curvature of the front surface and the back surface respectively, and t is the central thickness of the lens. If t is less than 1/6 of the lens diameter, then the third term can be neglected. The sign conventions are that R is positive if the centre of curvature is behind the lens and negative if it is in front of the lens; light travels left-to-right.

Newton's equation:

$$xx' = f^2 \tag{5.4}$$

where the distances are now measured relative to the focal points; $x = s - f$ and $x' = s' - f$. This form is most useful for calculating the amount of re-focus required.

The Lagrange Invariant:

$$hnu = h'n'u' \tag{5.5}$$

where h and h' are the object and image heights, n and n' are the refractive indices in object and image space respectively (usually $n = n' = 1$) and u and u' are the (small) angles with respect to the optical axis of the same ray in object space and in image space; astronomers also use this in its area-solid angle ($A\Omega$) form, which is called the throughput or étendue. In general, throughput is defined as the area solid-angle product $A\Omega$, where A is the telescope aperture and Ω is the solid angle subtended by the entrance aperture (slit, seeing disk, diffraction disk). Multiplying by the actual transmission efficiency, e, gives the luminosité (L). The "information content" is then $LR = eA\Omega R$, where R is the resolving power. This is a "figure-of-merit" for a spectrograph.

Displacements and deviations:

$$\alpha \approx (n-1) A \qquad\qquad\qquad \textit{thin wedge}$$

$$z = \frac{(n-1)t}{n} \qquad\qquad \textit{parallel plate in converging beam} \qquad (5.6)$$

$$d = t \sin\theta \left(1 - \frac{\cos\theta}{n\cos\theta'} \right) \quad \textit{displacement by parallel plate}$$

where α is the angular displacement caused by a wedge of small angle A (angles in radians), z is the longitudinal (focus) displacement caused by a plane parallel plate of thickness t perpendicular to a converging (or diverging) beam and d is the lateral displacement caused by a plane parallel plate at angle θ in a parallel (collimated) beam. These formulae are useful when considering the effects of filters, entrance windows to dewars, dichroic beamsplitters and polarizing beamsplitters.

Of course, the grating equation and plate scale definitions given in the previous chapter will also be required.

Next, identify and list all the known constraints on the design. For example, the wavelength range (UV, visible, IR), the transmittance goals, restrictions or limits on scattered light—which are probably driven by the signal-to-noise calculations and the science goals, the desired back focal length and other opto-mechanical packaging issues (size, weight, thermal mass), polarization effects (due to birefringent crystals or boundary conditions), environmental concerns (thermal changes, shock and vibration), ability to test the optics and finally, the cost of fabrication.

Except in a few cases, it will not be possible to complete the design of the instrument by purely analytic means. The final step is therefore to enter the prescription into a "raytracing" program and develop a more sophisticated model. Many excellent programs are available, although cost and ease-of-use vary considerably. One of the larger packages, but an industry standard, is Code V ("code five") from Optical Research Associates. Other very popular packages include OSLO from Sinclair Optics, ZEMAX by Focus Software Inc., SIGMA by Kidger Optics (UK) and SYNOPSIS and ASAP from Breault Research Organization. A ray-tracing program cannot design a system for you, it can only trace what you enter, so the first-order analysis is very important and it is often helpful to begin with an existing design and modify it. Most ray-tracing programs will provide an algorithm which attempts to optimize a given design or search for different designs within constraints which you can control. In this way you can "explore" some options, but be prepared to use up a lot of computing time. A ray-tracing program can assist the designer in studying what the effect of these variations might be, and what compensation techniques (such as refocus) can be applied. It is also important to understand the limitations of a given optical design, so as to assess the impact on the astronomical goals, as well as the impact on cost and manufacturability.

Poor optics will degrade images irrespective of the number of pixels in the CCD array, and so an appreciation of the limitations is important. If the sine functions in Snell's Law are expanded in a Taylor series as

$$\sin\theta = \theta - \frac{\theta^3}{3!} + \frac{\theta^5}{5!} - \frac{\theta^7}{7!} + \frac{\theta^9}{9!} - \ldots \qquad (5.7)$$

and only the terms in theta to the first power are retained (i.e. sin θ is replaced by θ) then we get the familiar "first-order" or "paraxial" equations shown previously. A perfect optical system would obey the paraxial equations irrespective of the value of θ. Imperfect images caused by geometric factors are called **aberrations**. Including the third-order terms (i.e. replacing sin θ with $\theta - \theta^3/6$) leads to a useful set of equations for describing lens aberrations as departures from paraxial theory; these equations are called the **Seidel** or **third-order aberrations**. For monochromatic light, Seidel classified aberrations as spherical aberration, coma and astigmatism which all affect image quality, and distortion and field curvature which affect the image position. In multi-colour (or poly-chromatic) light there is also chromatic aberration and lateral colour. For completeness, each of these well-known effects is summarized here; details can be found in the references.

Spherical aberration is caused by the fact that a spherical surface, whether on a lens or a mirror, is geometrically speaking the wrong shape to ensure that all light rays converge to a focus at the same point. Spherical aberration is an "axial" aberration because rays at greater and greater radii from the centre of the lens (the marginal rays) focus closer and closer to the lens. A focal plane or detector placed on the axis will see a large blurry image instead of a point source. The circular image obtained has a minimum size called the "circle of least confusion". Spherical aberration in a lens can be minimized by varying the shape or "bending" of the pair of surfaces, since many different combinations of curvature can produce the same focal length. Alternatively, the lens power can be "split" between two or more "slower" lenses (larger f/number). Since the angular diameter of the blur circle is inversely proportional to the cube of the f/number, then splitting an f/2 lens into a pair of f/4 lenses reduces the spherical aberration of each by a factor of 8, and the combination has about 0.5 of the original spherical aberration.

If the curvature of the surface continuously departed from that of a sphere in such a way as to compensate for the difference between sin θ and θ, then spherical aberration would be eliminated and both marginal and paraxial rays would focus at the same point (for an on-axis object placed at infinity). A parabolic surface will achieve this ideal and this is the principle applied in modern reflecting telescopes whose primary mirror is a paraboloid (see the Exercises).

Coma is an off-axis aberration. The ray which passes through the centre of the entrance pupil from any field point is called the "chief" or "principal" ray. This ray defines the image height. Now consider parallel rays from the off-axis point passing through concentric points on the lens which lie on the circumference of a circle. This zone on the lens produces a different magnification from the chief ray, which results in a displaced image in the form of a ring whose diameter and offset increases with the distance of the transmitting zone from the optical axis. The final result is an overall image shaped like a comet with a conical flare. Coma scales linearly with off-axis angle and inversely as the square of the f/number.

Astigmatism is also an off-axis aberration and results from a cylindrical-shaped departure of the wave from its ideal spherical shape. All off-axis points lying in a plane which is skew with respect to the plane containing the optical axis form a focus at a point, which is displaced along the direction of travel of the ray, from the focus produced by rays lying along a line in the plane perpendicular to the first. A focal "line" is seen at one position and a similar line rotated through an angle of $90°$ is seen at the other point. A blurred elliptical or circular image is seen in between. Astigmatism scales as the square of the off-axis angle and inversely as the f/number. A tilted plate such as a dichroic beamsplitter or filter placed in a converging beam will produce astigmatism.

There is a natural tendency for optical systems to image better on curved rather than flat planes. This effect is called **field curvature**. In the absence of astigmatism, the image is formed on a curved surface known as the "Petzval surface". Positive elements usually have inward-curving Petzval surfaces (i.e. the surface bends back towards the lens) and negative elements have outward-curving fields, thus some measure of correction can be accomplished by combining positive and negative elements. Field curvature scales like astigmatism.

Distortion occurs when the image of an off-axis point, even though it is sharply defined, does not form at the position predicted by paraxial theory. The image may be well-corrected for coma and astigmatism, but is simply not at the correct location. Typical distortion patterns are "pincushion", "barrel" and "keystone" in which the displacements are toward the centre, away from centre and a different stretch from top to bottom of the image. Distortion does not lower the resolution of the imaging system and could be removed by computer processing if required; it is usually described as a percentage shift from the paraxial position.

Chromatic aberration is of course due to the fact that the refractive index of a material is a function of wavelength. Longitudinal chromatic aberration is the difference in focus as a function of colour and lateral colour is the difference in image height (or magnification) with colour. The blur circle diameter is inversely proportional to f/number and the Abbe V-factor or its equivalent ($V_{eq} = n - 1/\Delta n$) which is a measure of the dispersion of the material. Colour-corrected lenses can be designed by using two or more components, one positive and one negative, but materials with two different dispersion characteristics are also needed. Of course, there is no chromatic aberration with mirror systems.

In any optical system there will be a real aberrated wavefront travelling through the optics. At each location one can consider the difference between the real wavefront and the best fitting spherical surface. This difference is called the **optical path difference** or OPD. The OPD can be given in micrometres or in "waves" as a fraction of a reference wavelength such as 632.8 nm for a HeNe laser. The deviation of the wavefront from perfect, or the **wavefront error**, can be specified in terms of the "peak-to-valley" (P–V) optical path difference, which is the difference between the longest and shortest paths. This is readily observed with an interferogram image. If the OPD = $\lambda/4$ peak-to-valley, over the beam, then the system meets the Rayleigh criterion for diffraction-limited systems and is almost perfect. Alternatively, one can form the "root mean square" (rms) OPD summed over the entire wavefront, with the caveat that the rms wavefront error clearly depends on the distribution of the deviations or bumps over the wavefront. Two wavefronts could have the same P–V wavefront error, but very different rms values depending on the number of bumps and dips over the wavefront; as a rule-of-thumb, (P–V) OPD ~ 3 to 5 times the rms.

Any irregularity (of size δt) on the surface of a lens introduces a "bump" in the OPD of size $(n-1)\delta t$. For a mirror the effect on the OPD is $2\delta t$, i.e. it is doubled! This is why mirrors have to be particularly good to maintain the wavefront error; a $\lambda/8$ surface accuracy implies only $\lambda/4$ in the wavefront. If σ is the rms amplitude of the surface rough-ness, then the surface is "smooth" if $4\pi\sigma \ll \lambda$, and in this case the fraction of light within any angular radius θ compared to the light enclosed in the diffraction pattern of a perfect surface is the Strehl ratio $S = \exp[-(4\pi\sigma/\lambda)^2]$. For example, if $\sigma = \lambda/20$, then $S = 0.67$, and 33% of the light is scattered out of the diffraction spot. This is the total integrated scatter or TIS. For some design applications TIS is insufficient and it is necessary to measure the angular dependence of the scattering and determine a characteristic bi-directional re-flectance function (or BRDF).

Equally important is the area of the optical component over which it is important to maintain the desired OPD. This is called the "footprint" of the beam and is generally the same size as the optical component in the vicinity of a pupil image, but may significantly "underfill" a component near a focal plane. It is generally assumed that all the wavefront errors add in quadrature, i.e. square, add and take the square root. A tabulation of the contributions to the wavefront error for each optical component—usually called the de-sign residual—derived from the ray-tracing results, together with an allowance or "tolerance" for manufacturing, for changes due to temperature and for alignment errors is called a **wavefront error budget**.

Ultimately, performance of the optical design will have to be quantified. This can be done using encircled energy—or ensquared energy since pixels are usually square—(e.g. 80% within one pixel), modulation transfer function (MTF) across the field, a distortion map, limits on scattering and a wavefront error (WFE) budget. Note that the linear blur diameter due to diffraction by a circular aperture is $2.44\,\lambda\,(f/\#)$ and a diffraction-limited system would have 84% of the energy within this diameter; 50% of the energy falls within an angular diameter of λ/D.

Ray trace programs provide several tools for analysing and displaying the performance of an optical system.

(1) Spot diagrams are produced by tracing a large number of random rays through the system to produce a cluster of impact points on the focal plane to give a visual per-ception of the primary aberrations, such as coma or astigmatism. The rms spot sizes are geometric only, and do not include diffraction effects.

(2) Encircled energy plots give the total amount of energy within a circle of a given radius as a function of that radius and includes diffraction effects.

(3) Tangential ray fans show the variation of an aberration (i.e. where a ray lands in the focal plane) as a function of the axial height of the ray at the optical component (usually normalized to the diameter of the component). For example, the x-axis would correspond to $\pm d/2$ where d is the lens diameter and the y-axis would give the height in the image plane corresponding to a given x-value or height at the optic. The paraxial focus is at $(0,0)$. Spherical aberration gives a characteristic "s-shape" lying on its side. Coma shows a typical "u-shape".

(4) modulation transfer function (MTF) is essentially the modulation in the image as a fraction of the modulation in the original object, where by modulation we mean $[I(\text{max}) - I(\text{min})]/[I(\text{max}) + I(\text{min})]$. This "contrast" or visibility ratio is large for low spatial frequencies (many pixels across the bright and dark regions of a test pattern) but decreases quickly as the spatial frequency increases. Spatial frequency is either

given in cycles per millimetre or in line pairs per millimetre, with the highest possible frequency being determined by diffraction, $f = 1/(\lambda f/\#)$.

Many other options are available. A sample output is shown in Fig. 5.3. For further details see the references.

Of considerable importance in commercial refractive optics is the **anti-reflection coating**. Since most high-quality lenses are actually made from multiple simpler components, it becomes crucial to eliminate reflections between these components. The reflectance (R) of a surface is a function of the index of refraction (n) on both sides and the angle of incidence (θ). At normal incidence ($\theta = 0°$) in air ($n_1 = 1$) we have

$$R = \left(\frac{n-1}{n+1}\right)^2 \tag{5.8}$$

For glass with $n = 1.5$ the reflectance is 0.04 or 4%. Coating the lens with a very thin layer of material which has a refractive index intermediate between that of air and the original lens material can reduce the reflection losses by destructive interference at both boundaries. The basic relationship is given by

$$2nd = \left(m + \frac{1}{2}\right)\lambda \tag{5.9}$$

where m is an integer and d is the thickness of the film. Thus, the minimum thickness for an anti-reflection coating occurs when $m = 0$ and $d = \lambda/4n$. Note that since λ is the wavelength in air and λ/n is the wavelength inside the thin film (λ_n), then $d = \lambda_n/4$ which is why such simple coatings are called a "quarter wave" layer. The full solution of this interference problem using Maxwell's equations takes into account all of the other multiple reflections inside the thin film and shows that in addition to achieving quarter-wave thickness, the ideal value for n is $\sqrt{(n_g)}$ where n_g is the index of refraction of the substrate glass (and the surrounding medium is assumed to be air). A widely used material which almost meets this requirement for optical glass is magnesium fluoride (MgF_2) with $n = 1.38$. Clearly, however, the simple coating described above can only function perfectly at one wavelength, whereas modern astronomical instruments are designed to perform over a wide range of wavelengths (to take advantage of the response of a silicon CCD for instance). It is possible to develop multi-layer coatings using several materials which provide a reasonable average reduction in reflectance over a wider wavelength range. Such coatings are called "broad band" anti-reflection coatings. Because of the difficulties of securing durable broad band AR coatings, astronomers tend to favour mirror systems or split the wavelength interval by constructing a double-spectrograph or twin-colour camera. Nevertheless, AR coatings are very beneficial to astronomers, and have even been deposited on the surfaces of some CCDs; the large refractive index ($n = 4$) of silicon results in considerable reflection loses (about 36%) and therefore sets an upper limit of 64% on the quantum efficiency.

Similar principles are used to develop multi-layer interference filters for narrow band work, such as imaging nebular lines. These filters can be thought of as a "solid" Fabry–Perot etalon ($m\lambda = 2n^*d \cos\theta$) where n^* is an effective refractive index. Note that as θ increases from 0° the central wavelength is "scanned" to shorter wavelengths.

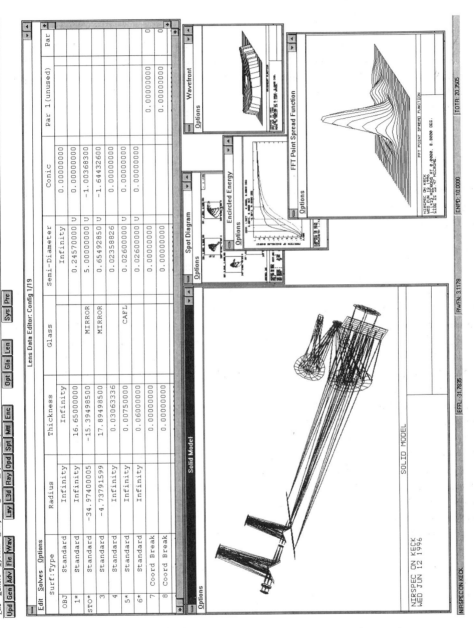

Fig. 5.3. The screen of a ray-tracing program (Zemax by FocusSoft) illustrating the input data, the layout and several diagnostics such as spot diagrams and encircled energy plots. Courtesy of Don Figer, UCLA.

In recent years optical designers have endeavoured to come up with complete solutions to image quality problems by appealing to specially designed corrector plates, thin lens-like pieces which "undo" some of the effects of other components in the optical stream. Undoubtedly, the most eminent proponent of this approach is the English designer Charles Wynne; most large telescopes with prime focus cameras employ a "Wynne corrector". An alternative approach is to make the surface(s) of the primary optical components depart from their normal spherical (or conic) shapes in such a way that they are actually compensating for the misbehaviour or aberrations which the normal spherical surface would cause. The "sag" or z-coordinate (with z along the optical axis) of the standard axially symmetric surface is

$$z = \frac{cr^2}{1+\sqrt{1-(1+k)c^2 r^2}} + \alpha_1 r^2 + \alpha_2 r^4 + \alpha_3 r^6 + \; ... \qquad (5.10)$$

where c $(=1/R)$ is the curvature, r is the radial distance from the axis and k is the **conic constant**. For spheres $k = 0$, and for a parabola $k = -1$, hyperbolas have $k < -1$, and k lies between -1 and 0 for ellipses; $k > 0$ corresponds to an oblate ellipsoid. The higher-order terms are the **aspheric** components which appear in systems like the "corrector plates" mentioned above. A different way of describing the higher-order terms which allows departures from rotational symmetry is to use the sum

$$\sum_{i=0}^{N} A_i Z_i (\rho \varphi) \qquad (5.11)$$

where the Z_i are called Zernike polynomials in the normalized radial coordinate ρ (rho) and the angular coordinate φ (phi). This is the terminology used to study a system of multiple mirrors like the Keck telescope.

Manufacture of aspherical optics is only practical with the aid of modern, precision computer-controlled milling and turning machines using diamond-tipped cutters. The final surface depends greatly on the choice of material and the nature of the asphere. Aspheric surfaces directly machined using diamond-tipped cutters are a particularly attractive technological innovation for infrared optical components which must be cooled to low temperatures. For example, it is possible to make an aspheric mirror by direct diamond-machining of a solid block of aluminium (aluminum) which has already been machined by normal lathes to provide its own mirror-support and attachments! The differential thermal contraction between metal supports and classical glass components is completely eliminated since mirror and support are made of the same material.

5.4 MECHANICAL DESIGN

Converting the optical design into an opto-mechanical layout which ensures that all the optical components are properly mounted, aligned and stable is often much harder than it seems, and can therefore lead to poor images. Several issues should be considered:

- choice of lens and mirror mounting schemes
- mechanical and thermal stress on the optics

- alignment of the optics
- flexure and stability under gravity if the instrument moves with the telescope
- method of attaching to the telescope
- stray light baffles and light-tight enclosures
- moving parts such as focus drives, filter wheel and shutter mechanisms
- ease of handling, assembly and dis-assembly
- integration of electrical wiring
- cooling systems, thermal paths and thermal mass

Since astronomical instruments exhibit an enormous range in size, weight and complexity it is hard to generalize. The mechanical design of a large and powerful CCD spectrograph like the LRIS on the Keck 10-m telescope, is quite different from that of a simple CCD camera head and filter wheel for a 10-inch telescope!

As with the optical design however, the first step is to draw a rough layout around the optical design in order to assign real thicknesses and dimensions to the support structure (lens barrels, mirror mounts, filter wheels etc.) and look for "first-order" problems such as collisions between components and vignetting. Try to estimate where the centre of mass of the whole instrument is. Look for poor design arrangements such as a heavy component supported by a horizontal rod which is either too thin or lacks a triangular support strut. Try to understand what might bend as the telescope carries the instrument around. Three-point (ball/groove/flat) "kinematic" mounting schemes may be required in various places to minimize stress and to enable units to be removed and put back in exactly the same place.

After this first look is complete a more detailed mechanical design can be developed based on a more careful assessment of the dimensions and properties of walls, struts, wheels, bearings and other support components required to meet specifications on strength, flexure, load-bearing, heat flow and so on.

A major asset in mechanical design is the availability of excellent computer-aided drawing tools such as AutoCad. This powerful drawing program runs on many platforms including PCs and workstations, and also allows one to import files from ray-tracing packages. For example, we routinely import our ray traces from Zemax into AutoCad and then build up the mechanical design around the real optical rays and surfaces. Fig. 5.4 shows an example of an AutoCad drawing.

Most instruments are constructed from aluminium (Al) alloys to save weight, although handling frames and large rotating bearing structures are always made from steel. Some use is made of copper, brass and stainless steel, fibreglass materials like G-10, and occasionally some plastics like Delrin. In general, there will be a total weight limit and a "moment" limit (such as, 227 kg (500 lb) at 1 m from the mounting flange) for instruments attached to a telescope. It is very important to know if and how the instrument will bend or flex as the telescope points to different parts of the sky. For instruments which are not fully enclosed in a vacuum chamber, a "space frame" structure of rods is often ideal for strength, weight and rigidity. A lightweight, but light-tight, enclosure can surround the space frame. Box-like structures with tongue-and-groove for light-tight fitting and triangular gussets and buttresses for additional stiffness are usually adequate for small, lightweight photometers and simple cameras. External surfaces are usually anodized and

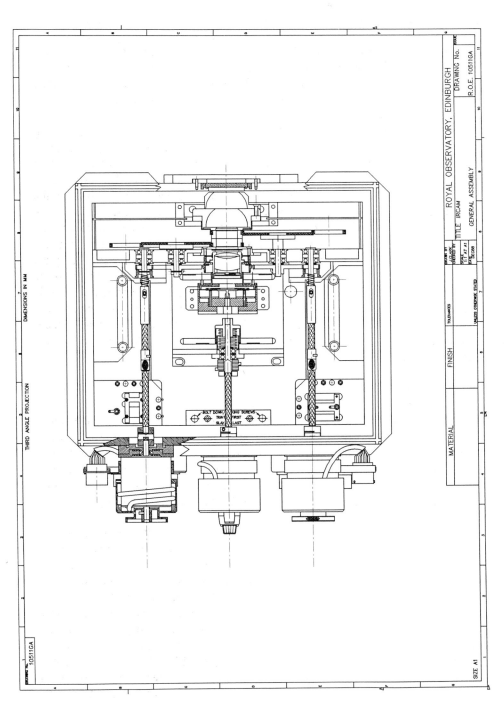

Fig. 5.4. An engineering drawing prepared using a computer-aided drawing package (Autocad from Autodesk).

internal surfaces are black anodized and/or painted with matt black paint. Black anodization is not really "black" at infrared wavelengths beyond 700 nm (0.7 μm) and special black paints are required (see Chapter 8). Infrared instruments pose additional mechanical problems because the entire optical system must be enclosed in a vacuum chamber and cooled to low temperatures. In this case, it is extremely important to evaluate the stress and deflections caused by atmospheric pressure on the (potentially) large surfaces of the chamber for a given thickness of walls.

All the basic properties of the materials to be used must be collected together, including the density, coefficient of thermal expansion, Young's modulus, yield strength, compression strength, shear strength, heat treatment, hardness, specific heat and emissivity. These quantities are tabulated in engineering handbooks. Many properties of materials are a function of temperature, so the range of applicability needs to be known or the "integrated" effect over the required temperature range should be used. Elementary physics can be used at this stage to estimate the tensile (stretching) and shear (tangential) stresses (force per unit area, F/A) on various rods, struts and plates in the instrument for comparison with tabulated limits on "yield" strength and "breaking" strength. The yield strength defines the stress beyond which the object will not return to its original shape when the forces are removed. Stress has the same units as pressure and is measured in newtons/m^2 (or pascals, Pa) in the SI system, but it is also commonly quoted in pounds per square inch (psi), especially in the United States. From a knowledge of the Young's modulus (E) and the shear modulus (S) of the material one can estimate the strain ($\Delta L/L$) and displacement since,

$$F = EA\frac{\Delta L}{L}, \quad F = SA\frac{\Delta x}{h} \tag{5.12}$$

where A is the area over which the force is acting, L is the original length, h is the separation between the two planes which have sheared by Δx. Advanced reference works on mechanical engineering provide equations for stress and strain on many complicated shapes, but ultimately it may be necessary to use a computer program to perform a finite element analysis (FEA) in which the structure is replaced by numerous much smaller identical mechanical blocks or elements whose loads and interface (or boundary) conditions are specified.

Aluminium is commonly used for small instruments. It is one-third as dense as steel and has 60% of the electrical conductivity of copper. Alloys are specified by four digits plus a suffix beginning with T (for heat treatment) or H (for work hardness); T0 is no heat treat (dead soft) and T10 is fully tempered (hard). For example, 1100-T0 is a very soft, nearly pure aluminium alloy, whereas 6061-T6 is a relatively hard yet easily machined alloy and readily available. Aluminium can be joined by a process called **heliarc welding** in which an electric arc is used as the heating (fusing) element in an inert (argon) atmosphere, or it can be electron-beam welded. The metal is weakened near the weld and this should be taken into account when estimating the strength of a joint. Like other alloys, aluminium can be heat treated in a variety of ways, such as **annealed, quenched** or **tempered**. Annealing is the process of heating a metal to above a transition temperature and then slowly cooling it. Quenching is similar, except the cooling is rapid. Tempering is an intermediate treatment in which previously hardened metals are reheated to below the transition point in order to relieve stresses and then cooled at a rate that preserves the desired property. Parts to be used at cryogenic temperatures in scientific instruments, especially metal optics, must be **thermally cycled** several times to release stresses and eliminate small changes in dimensions over time known as "micro-creep".

Table 5.1. Some useful parameters for common materials. The second entry in column 3 gives the percentage increase in the value at 77 K

Material	Thermal expansion %(77-295 K)	Young's modulus (psi)	Tensile strength (psi)
Aluminium (Al)	0.37	10×10^6 (add 10%)	45 000
Copper (Cu)	0.30	16×10^6 (add 10%)	45 000
Stainless steel	0.25	25×10^6 (add 10%)	85 000
Glass fibre (G-10)	~0.18	3×10^6 (add 15%)	40 000
Vespel	0.93	0.45×10^6 (add 25%)	10 000

Other well-used materials include the following. Stainless steel, which is an alloy of iron, chromium and nickel, is also used in astronomical instruments. The 300 series have a high nickel content and are all fairly non-magnetic, with type 304 being the best and most common. Invar is an iron–nickel alloy with about ten times smaller coefficient of thermal expansion than steel. Copper is soft, but provides the best electrical and thermal conductor and is often used as the thermal heat sink or cold block for the chip in cooled CCD cameras. Oxygen-free high conductivity (OFHC) copper is preferred for its purity, excellent conductivity and resistance to hydrogen embrittlement. Brass, an alloy of copper and zinc, was once very common in scientific instruments, but its use is more limited now. Fibreglass materials like G-10 are often used for thermal isolation in vacuum-cryogenic applications, and the Dupont plastic known as Vespel is very useful because it is impregnated with molybdenum disulphide (MoS_2).

For a large, complex instrument which has never been built before there is really no substitute for building a "space model"—a full-size replica constructed from the preliminary drawings but using light-weight materials such as foamboard, cardboard, thin metal sheeting, tubing and light woods (see Fig. 5.5). Although time-consuming to do well, the act of building the space model to match the preliminary drawings will almost certainly reveal design flaws and focus attention on areas of difficulty in the real instrument. Potential problem areas are (i) the order of assembly; (ii) optical alignment and verification; (iii) location and installation of baffles; (iv) electrical wiring paths; (v) cooling paths; (vi) general handling.

Mechanisms depend on the requirements of the instrument, but a common need is that of a filter wheel. Whenever possible, filters are located in a parallel beam close to the

Fig. 5.5. A "space" model of an infrared spectrograph constructed at the IR Lab at UCLA.

position of the pupil image. Doing so minimizes the size of the filter, reduces the influence of any dust or scratches on the filters and eliminates any focus shift when filters are changed. Filters placed in divergent beams may have to be larger and the optical path through each filter will need to be balanced to ensure that there is no change in focus when one filter is changed for another of different thickness and refractive index. If n filters are mounted in a wheel, the minimum diameter of the filter wheel is approximately nD_{pupil}/π, where D_{pupil} is the diameter of the pupil image. There are two common methods of driving a filter wheel; a direct drive shaft through the centre of the wheel, or an edge-drive using gear teeth around the perimeter. Edge-driven wheels usually employ a worm-gear and achieve very large mechanical advantage (or gear ratio). A direct drive may require intermediate gearing to yield some mechanical advantage. The rotating shaft can be driven by a DC servo motor or one of the many small stepper motors on the market. A DC servo motor runs continuously until stopped, whereas a stepper motor only moves when commanded to do so by receiving a pulse. Stepper motors are attractive because the mechanism can be operated "open loop" and the position of the wheel can be determined simply by counting pulses. Drives trains tend to suffer "backlash" which means that the amount of motion required is not the same in both directions of rotation. To counteract this effect the mechanism must be moved in one direction only or mechanically spring-loaded.

Fig. 5.6 shows a typical worm-driven wheel. The stepper motor shaft is attached to the worm shaft by a "flexible" wafer coupling which compensates for lack of precise alignment and any differential thermal contraction. Since the pitch of the worm threads is small,

Fig. 5.6. A typical worm-driven mechanism, in this case a simple filter wheel.

the worm cannot be rotated by turning the wheel and therefore it is safe to remove the holding torque by switching off the power to the motor. Typical stepper motors have 200 full steps per turn (1.8° per step) but are usually driven in "half-step" mode which gives 400 steps per turn (0.9° per step). If this motor was attached directly to the hub of the filter wheel then one step would displace the centre of the filter with respect to the optical axis by ~0.016R, where R is the radius of the wheel at the centre of the filter. For example, if $R = 50$ mm then the displacement is 0.8 mm or about 3% of the diameter of a 1-inch filter. When the wheel is worm-driven there is a huge gear ratio and resolution advantage. If there are 180 teeth around the circumference then it takes 72 000 steps (in half-step mode) to execute a complete turn, and one step now corresponds to a displacement of ~87 μrad/mm, or about 4 μm in the above example.

All of the parts of the filter wheel—the rotor, coupler, wormshaft, worm and the wheel—can be considered as disks with moment of inertia $I = \frac{1}{2}MR^2$ where M is the mass and R is the appropriate radius, except that the rotational inertia of the wheel is reduced by the square of the gear ratio. If the angular acceleration is α rad/ s^2 then the torque $\tau = I\alpha$.

Positional data can be obtained in one of two ways. In a "closed-loop" approach, the filter position is sensed using a continuous absolute position encoder, or at least a discrete encoding scheme (such as a switch) at each filter position. Various forms of absolute encoders are commercially available. The encoder is attached to the drive shaft or edge of the gear and gives a direct reading of its position. An "open loop" system has a single "datum" or reference position and relies on counting steps to know where the mechanism is and which filter is installed. The latter method is simpler to install and works surprisingly well. It is often used in infrared instruments because of the difficulty of getting an absolute encoder for cryogenic applications. If this approach is adopted then it is wise to include at least two "datum" or "home" positions. Miniature lever-type switches, Hall-effect switches and light-emitting diode (LED) switches are common in many instruments.

Once again, infrared instruments pose special problems because everything is cooled to cryogenic temperatures, not just the detector. Plastic mounted switches contract differently from their metal attachments, stainless steel ball and roller bearings must be degreased and a dry lubricant such as molybdenum disulphide (MoS_2) burnished on. Degreasing can be done by popping the seal and placing the bearing in a beaker of research grade ethyl alcohol for 5 minutes to loosen up the grease and then transferring the beaker to an ultrasound bath for 30 minutes to dissolve the grease. Finally, the bearings should be individually rinsed in alcohol, dried with a stream of dry nitrogen gas and stored in sealed plastic bags; cleanliness is essential.

Finally, bear in mind that you or someone else will have to physically handle the instrument and its sub-components, so think about safety issues and wherever possible build in handles and feet. Decide whether or not a lifting mechanism is required and think about storage containers.

5.5 CRYOGENICS AND VACUUM METHODS

As we shall see, almost all state-of-the-art detectors require cooling for optimum performance. There are several categories of cooling systems which might be required in cameras and spectrographs for ground-based astronomy depending on the application.

(1)   *Thermo-electric coolers and liquid circulation coolers:* These normally operate over the range −20 to −50 °C and are suitable for photomultiplier tubes, certain CCDs which have low dark currents, and high-speed applications such as telescope guiding cameras. This class of cooler is very popular in CCD cameras used on small telescopes because it is much easier and simpler to use than options below.

(2)   *Liquid and solid cryogens*:

Dry-ice (solid CO_2): cheap and readily available. Coming in the form of a "snow" it is most often used as the coolant for GaAs PMTs. It achieves temperatures around −76 °C (or 197 K).

Liquid nitrogen (LN_2): relatively cheap. Cools detectors (and other components if required) to −196 °C (or 77 K), which is the normal boiling point of liquid nitrogen. Almost all professional CCD cameras employ LN_2 cooling systems, and certain near-infrared devices such as HgCdTe arrays are also cooled with LN_2.

Liquid helium (LHe): more expensive, but will cool detectors to −269 °C or 4 K and is required for low band-gap semiconductor materials and bolometers used in infrared instruments. Even lower temperatures are obtained using ^{3}He systems in sub-millimetre and far-infrared bolometers. Liquid helium requires a special double-walled evacuated "transfer" tube and some experience in handling the transfer to recognize when liquid is being collected and transfer is complete. The LHe reservoir is usually pre-cooled with LN_2 which is then blown out with LN_2 gas and the LHe transfer begun immediately.

For liquid cryogens the cooling ability is expressed in terms of the product of the density (ρ) and the latent heat of vaporization (L_V):

$$\left(\rho L_V\right)_{LHe} = 0.74 \ \text{W h l}^{-1}$$

$$\left(\rho L_V\right)_{LN_2} = 44.7 \ \text{W h l}^{-1}$$

For example, a 10 W heat load boils away 1 litre (l) of LHe in 0.07 hours whereas 1 litre of LN_2 lasts 4.5 hours. By attaching a vacuum pump to the LN_2 vent and reducing the pressure above the liquid it is possible to solidify the nitrogen and achieves temperatures around 65 K.

Liquid neon (LNe) and liquid hydrogen (LH): provide an intermediate solution between LHe and LN_2, but are rarely used in astronomy applications.

3.   *Electrical heat engines or closed-cycle refrigerators*: Since LHe is fairly expensive compared to liquid nitrogen, and requires special care in handling, many infrared instruments and sub-millimetre receivers now employ powerful, multi-stage closed-cycle refrigerators. Typical two-stage Gifford–McMahon systems using 99.999% pure helium gas at about 300 psi as the working fluid (e.g. the 350 model from CTI Inc., USA, shown in Fig. 5.7) can provide two cold levels, typically 65 K and 10 K and extract heat at a rate of about 20 and 7 watts from each stage. Both single-stage and triple-stage versions are available, and some units provide about 100 watts of cooling power. If the mass to be cooled is very small, e.g. a single CCD, then much simpler systems such as small Sterling-cycle coolers can be used.

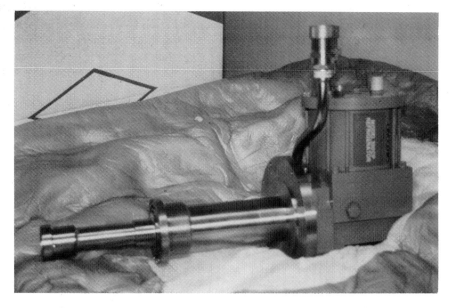

Fig. 5.7. The cold head of a two-stage closed cycle refrigerator. This unit is the CTI model 350.

The cooling system in an astronomical instrument can be a minor part, as in the case of a thermoelectrically cooled CCD camera head, or it can be the dominant issue as in the case of a large thermal infrared instrument in which *everything* must be cooled. Choosing the detector temperature will depend on factors like dark current and noise, whereas the temperature of the surrounding opto-mechanical structure may be determined by its thermal emission or simply by the environment. Optimum temperatures may not coincide with those readily available by means of liquid cryogens. Typically, steps in the thermal analysis of an instrument would be as follows:

- determine what is to be actively cooled, e.g. detector only, whole instrument
- tabulate the required operating temperatures, e.g. detector, optics
- calculate the thermal energy removed from the mass to be cooled
- determine the heat loads due to conduction and radiation
- select the appropriate cooling system
- estimate the cool down time
- estimate the hold time

Since the objects to be cooled are within an evacuated enclosure, convection is negligible. The heat H (joules) removed from a mass m (kg) which is cooled from a temperature T_h to T_c is given by

$$H = mC(T_h - T_c) \tag{5.13}$$

where C is the **specific heat capacity** of the material in joules per kilogram kelvin (J/kg K). The specific heat capacity of a substance usually changes with temperature and it depends on the conditions under which the heat is applied (i.e. constant volume C_v or constant pressure C_p), but for solids the difference is generally small. For aluminium, C is about 900 J/kg K, copper is 385 J/kg K, steel is about 450 J/kg K and water is 4190 J/kg K. Specific heat capacity is also tabulated in cal/g K; using the conversion 4.19 joules per calorie gives 1 cal/g K for water. As an example, to cool a mass of 1 kg (2.2 lb) of copper from 290 K to 80 K requires the removal of $1 \times 385 \times 210 = 80\,850$ joules of heat and 2.3 times that amount for aluminium. If this heat removal is to be accomplished in a time interval of $t = 1$ hour (3600 s) then the average power is $H/t = 22.5$ W.

The rate of transfer of heat Q_H (in watts) by conduction along a rod of uniform cross-sectional area (A) and temperature gradient dT/dx is given by

$$Q_H = -kA\frac{dT}{dx} \tag{5.14}$$

where k is called the **thermal conductivity** (W/m K) and is about 240 for Al, 400 for Cu and only 0.9 for glass. In steady-state conditions we can write dT/dx as $\Delta T/L$, where L is the length of the conductor. Moreover, if we think of heat flow as "current" and temperature difference as a "voltage" then by analogy with Ohm's law ($V = IR$) for electrical circuits, we can define a **thermal resistance** so that

$$\Delta T = Q_H R, \quad R = \frac{1}{k}\frac{1}{A/L} \tag{5.15}$$

which is a useful form, as shown below.

It is often the case that the optimum operating temperature of a detector does not coincide with that of the cooling stage. For example, liquid nitrogen is actually much colder than the typical CCD operating temperature of 135–150 K, and liquid helium is too cold for typical near-infrared detectors. What is needed is a temperature intermediate between the cryogenic bath and the warm (ambient temperature) body of the vacuum chamber. To achieve this it is usual to set up a thermal analogy to an electrical potentiometer circuit. One way to proceed is to attach a copper block (solidly) to the outer walls of the cryostat by means of pillars of fibreglass tube—a poor thermal conductor—and also to connect it by means of a copper wire braid to the copper face of the liquid nitrogen reservoir. A practical design is shown in Fig. 5.8. If R_{fg} and R_{Cu} are the thermal resistances of the fibreglass and copper braid respectively, then the intermediate temperature T_{ccd} is given by

$$\frac{T_{ccd} - T_c}{T_h - T_c} = \frac{R_{Cu}}{R_{fg} + R_{Cu}} \tag{5.16}$$

Alternatively, if the CCD temperature is chosen, then the ratio of R_{fg} to R_{Cu} can be found. The sum of the thermal resistances determines the total heat flow, which one tries to keep reasonably small. By placing limits on the heat flow then both resistances are fixed. For a given thermal conductivity (k), each thermal resistance can be achieved by the appropriate choice of A/L. Trimming the length and cross-section (number of wires) of the braid and the length and cross-section of the fibreglass tube enables a "balance" to be reached between the tendency of the LN_2 to cool down the copper block (and hence the CCD) and the tendency of the block to warm up to room temperature via the fibreglass pillars. In addition, a heavy-duty resistor can be attached to the copper block to enable the block to be heated (by passing a current through the resistor) at any time. A temperature sensor—usually a diode—is also threaded into the copper block, and a circuit controls the current to the heating resistor depending on the voltage measured across the diode which

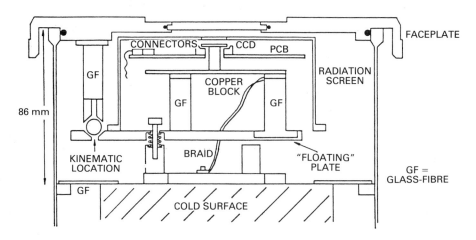

Fig. 5.8. A typical mounting scheme for a CCD detector in a liquid nitrogen cryostat. The arrangement allows the CCD to be operated at a temperature between 77 K and ambient. Courtesy Royal Greenwich Observatory.

has a constant current flowing through it. This allows some measure of control over the actual operating temperature. Temperature control units which handle these functions are available commercially (e.g. Lakeshore Cryotronic model 330).

Since k is a function of temperature, it is often more convenient to integrate over the required temperature range and give Q_H in the form

$$Q_H = \frac{A}{L}\left[I_{T_h} - I_{T_c}\right]$$ (5.17)

where L is the total length of the conductor and I is a tabulated property for many materials called the **thermal conductivity integral** which accounts for the variation of thermal conductivity (k) with temperature. In this expression, T_h and T_c represent the hot and cold temperatures between which the heat is flowing. If A and L are in square centimetres and centimetres respectively then I is in W/cm. Typical values are shown in the Table 5.2. Note how poor the conduction of G-10 is compared to aluminium. Optical glass is similar and therefore a large lens or mirror will cool much more slowly than the metal supporting it. Although only an approximation, this equation is extremely useful and can be used in a number of ways. For example, given A/L and the temperatures (T_h, T_c) we can derive the cooling rate Q_H. Alternatively, given the cooling rate, A/L and the initial temperature we can derive I_{T_c} and hence T_c.

Table 5.2. Values of the thermal conductivity integrals in W/cm for four materials

Temperature K	OFHC copper (W/cm)	6061 Aluminum (W/cm)	Stainless steel (W/cm)	G-10 fibreglass (W/cm)
300	1520	613	30.6	1.00
250	1320	513	23.4	0.78
200	1120	413	16.6	0.55
150	915	313	10.6	0.37
100	700	211	5.3	0.19
77	586	158	3.2	0.11
50	426	89.5	1.4	0.07
10	25	3.6	0.03	0.005

In large cryogenic infrared instruments, heat transfer by radiation is the dominant mechanism. The power radiated from a body of area A at an absolute temperature T is given by

$$Q_H = \varepsilon\sigma A T^4$$ (5.18)

where $\sigma = 5.67 \times 10^{-8}$ W/m^2 K^4 and ε is called the **emissivity** of the surface; $\varepsilon = 1$ for a perfectly black surface and is less than 0.1 for a shiny metallic surface. Polished aluminium foil yields an emissivity of about 0.05; anodizing increases the emissivity by a factor of 10 or more. The net rate of heat transfer by radiation from a body at temperature T_h onto a body at temperature T_c is given by

$$Q_H = \sigma A_c F_{hc}\left[T_h^4 - T_c^4\right] \tag{5.19}$$

where F_{hc} is an "effective" emissivity which also depends on the geometry of the dewar configuration, such as concentric cylinders or plane-parallel plates, both of which are quite realistic. In both cases, when the emissivities of the two surfaces are small ($< 5\%$) and equal, then $F_{hc} \sim \varepsilon/2$.

For a given temperature differential, radiation load is minimized by reducing the surface area and achieving the lowest emissivity (shiniest) surfaces. It is also possible to add n "floating" shields which reduce the radiated heat load on the innermost body by a factor of $(n+1)$; but careful application is critical. Various forms of floating shields are available. One form is called multiple layer insulation (MLI). Typical emissivities range from 0.03 for polished aluminium and gold foil to 0.32 for polished anodized aluminium to 0.95 for a matt black surface. A useful rule of thumb is that if $\varepsilon = 0.05$ and the hot and cold temperatures are 290 K and 80 K respectively, then the radiation load is ~ 10 W/m^2. Note that radiation load is very sensitive to T_h.

The cooling time for a mass m from initial temperature T_i to T_{i+1} ($T_i > T_{i+1}$) is found by dividing the heat removed by the effective rate of cooling,

$$\Delta t_{T_i, T_{i+1}} = \frac{mC_p(T_i)[T_i - T_{i+1}]}{(A/L)k(T_i)[T_i - T_{sink}] - Q(T_i)} \tag{5.20}$$

where T_{sink} is the temperature of the cold source (LN$_2$ or CCR stage) and the rate at which heat is removed by conduction to this reservoir is governed by the A/L and conductivity of the connecting path. The term $Q(T_i)$ is the rate at which heat comes back into the mass at temperature T_i (e.g. via radiation from the walls and conduction along shafts and other mounting structures). The total cooling time to an equilibrium temperature T_{eq}, given by $(A/L)k[T_{eq} - T_{sink}] = Q(T_{eq})$, is just the summation of equation (5.20) over all the individual cooling intervals. Cool down is a non-linear function of time. At first the cooling is very rapid and then it slows down as the heat coming into the system (by conduction and radiation) begins to balance the heat removed.

Dewars (or cryostats) come in various forms but are usually either single reservoirs for LN$_2$ or double reservoirs for LHe. Access to the interior is obtained by removing a plate which maintains a vacuum seal by using a rubber o-ring. The o-ring sits in a shallow groove but projects above the level of the surface to be sealed and is compressed by the opposing surface. In practice, it turns out that the cooling function is the most straightforward to accomplish and that it is the other aspects of cryostat design that cause the most headaches for designers. Among these other factors are mechanical flexure (bending), **hold-time**—how long the liquid cryogens last, out-gassing—the slow, natural escape of gases trapped on interior surfaces, electrical and optical access, and the ease with which the cryostat can be attached to other instruments such as spectrographs or polarimeters. For infrared work there is an additional problem, namely that of shielding the sensitive array detector from the flux of infrared photons emitted copiously by all warm objects such as the outside walls of the cryostat. This is accomplished by a double radiation shield, with the inner shield being as cold and as light-tight as possible.

Note also, that the larger the surface area the greater the deformation due to the atmospheric pressure of 14.7 psi (at sea level, only 61% of this value at the summit of Mauna Kea, Hawaii) on the walls of the dewar and the greater the stress on the joints. Metal vacuum chambers are most often cylindrical in shape because a cylinder is an ideal shape for resisting pressure. Flat plates are usually employed to seal the ends of the chamber. The deflection at the centre of a flat end plate clamped at its edge (e.g. by welding) is

$$\delta = \frac{3(1-\mu^2)R^4}{16Ed^3}P \tag{5.21}$$

and the maximum tensile stress, which occurs at the edge, is given by

$$S_{max} = \frac{3}{4}\left(\frac{R}{d}\right)^2 P \tag{5.22}$$

where μ is Poisson's ratio (typically 0.3 for metals), E is the elasticity or Young's modulus, R is the radius of the plate, d the thickness and P the external pressure. For 304 series stainless steel $E = 2.8 \times 10^7$ psi and for 6061-T6 aluminium $E = 10 \times 10^7$ psi. An acceptable value of R/d for steel is 30 which yields a relative deflection of $\delta/R = 0.002$. To achieve the same result an aluminium plate must be thicker with $R/d = 20$. In the case of aluminium, a 0.5 inch thick plate with a 10-inch radius ($R/d=20$) deflects 0.02 inch and the maximum tensile stress for $P = 14.7$ psi is 4.41×10^3 psi. Compare this to the yield strength of 6061-T6 material of 40×10^3 psi and 5×10^3 psi for almost pure aluminium. Since welding will weaken the strength of the material, the true yield strength is around 20 000 psi, which is still safe. The deflection of a circular plate with unclamped edges (e.g. an o-ring seal with relatively few bolts) is larger and requires up to a 50% increase in thickness for a given radius to achieve a comparable δ/R. Specifically,

$$\delta = \frac{3(1-\mu)(5+\mu)R^4}{16Ed^3}P \text{ and } S_{max} = \frac{3}{8}\left(\frac{R}{d}\right)^2 (3+\mu)P \tag{5.23}$$

where the maximum tensile stress now occurs at the centre.

Design objectives for a good cooling system, suitable for both laboratory and telescope usage, should include the following:

(1)   Minimum detector movement, either along the optical axis (defocus) or at right angles to the optical axis (image smear), to yield spatial stability of the CCD relative to the telescope focal plane. This means doing a careful mechanical and thermal design with detailed calculations and/or experiments to expose weaknesses.

(2)   Minimum effort to keep the system uniformly cold. When using liquid nitrogen as the coolant, this requirement simply means that the LN_2 should not run out too quickly, i.e it should have a good "hold-time". The hold-time of a cryostat depends on its thermal and mechanical design, the amount of power generated by the device being cooled and the quality of the vacuum. Typically, the hold time in hours is about 45 × volume of LN_2 in litres divided by the input power in watts. A hold-time of less than 12–14 hours is not acceptable for astronomy since a refill would be required during the night. For a 1-litre capacity cryostat to last 15 hours the heat input should be less than 3 W.

(3)   Good accessibility. When it is necessary to work on components inside the cryostat it is important to have enough "finger-room", good handling arrangements and adequate safety protection for delicate electronic parts including the CCD itself. Actual handles on the cryostat or on a large internal sub-assembly, are often overlooked.

(4)   Cost and manufacturing time. To achieve the above objectives usually means that a commercial cryostat purchased "off-the-shelf" will require some modifications or additions, which will cost time and money. Nevertheless, this can be the best approach when a good dialogue between the company and the observatory is established. The alternative approach is to "design your own" cryostat. This is a difficult route and may not save time and money. More often, researchers find it cost-effective to construct a "hybrid" cryostat by separating the "cooling function" from the function of supporting the detector and associated components. In the case of a CCD camera, a small "camera head" vacuum chamber can be attached to a proven, commercially available cryostat. Suitable LN_2 and LHe vessels are available from several companies including Oxford Instruments (UK), Infrared Labs (USA) and Janis Corp. (USA). A typical LN_2 cryostat used with a CCD camera is shown schematically in Fig. 5.9.

Moderately good vacuum techniques and methods of maintaining a good vacuum lead to excellent hold times. After initial evacuation the pressure in a cryostat may tend to rise slowly again due to "outgassing". A simple, widely used class of vacuum pumps which remove unwanted trace gases in the vacuum chamber by physically tying up molecules on a surface are called "sorption pumps". These can be constructed as small chambers about one inch in diameter filled with a sorbent material such as **activated charcoal** or one of the synthetic zeolite materials known as molecular sieves. These materials have a huge effective surface area of thousands of square metres per gram. At liquid nitrogen temperatures these materials will absorb air, and at very low temperatures of about 10 K **zeolite** can be used to partially absorb helium.

Vacuum technology accompanies the cryogenic techniques needed to support astronomical instruments and detectors. Air at 20 °C and 1 atmosphere of pressure (1 atmosphere = 760 Torr and 1 Torr = 132 Pa) contains about 2.7×10^{19} molecules per cubic centimetre and the average distance travelled between collisions, called the **mean free path**, is about 7×10^{-6} cm. In general,

$$\lambda_{mfp} = \frac{1}{\sqrt{2}n\pi d^2} \tag{5.24}$$

where n is the number density of molecules and d is the diameter of the molecule. A "rough" vacuum is about 10^{-3} Torr (mean free path = 5 cm) and a "high" vacuum would be 10^{-6} Torr (mean free path 5×10^3 cm). The capacity of a vacuum pump is specified in terms of the pumping speed at the inlet, which is just the volume rate of flow $S = dV/dt$ litre/s. The throughput of the flow is $Q_p = PS$ (Torr litre/s) and the throughput of the pumping line is given by $Q_p = C\Delta P$, where ΔP is the pressure gradient and C is called the conductance which depends on gas pressure and viscosity. Since this equation is also analogous to Ohm's law ($V = IR$) for electrical circuits, the net pumping speed of a pump and a system of pumping lines is given by

$$\frac{1}{S} = \frac{1}{S_{pump}} + \frac{1}{C_{lines}} \tag{5.25}$$

where the net conductance is found by adding the individual conductances like their electrical counterparts. Two equations are given for C (in litre/s). The first corresponds to

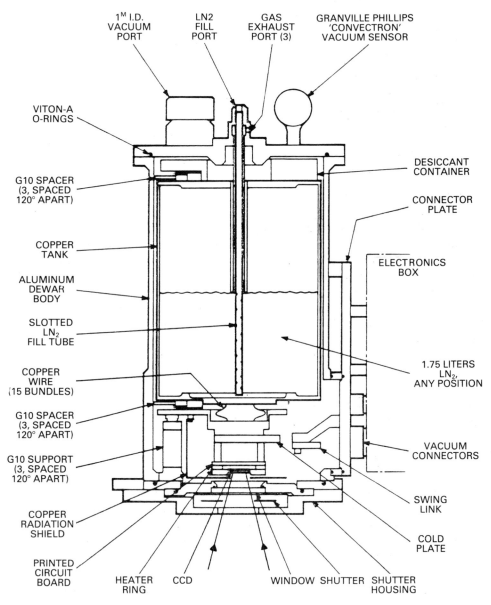

Fig. 5.9. A cross-sectional view of a typical liquid nitrogen (LN_2) cryostat illustrating all of the components needed in its construction. Courtesy NOAO.

viscous flow when the mean free path is small and the other to molecular flow when the mean free path is large compared to tube dimensions and C is independent of pressure. Both apply to air at 20 °C.

$$C = 180 \frac{D^4}{L} P_{av} \text{ or } C = 12 \frac{D^3}{L} \qquad (5.26)$$

It is assumed that the tube is circular with diameter D and length L in centimetres and the pressure is in torrs. Finally, the pump down time (in seconds) of a system with volume V from pressure P_0 to P assuming a constant net pumping speed S and no outgassing is

$$t = 2.3 \frac{V}{S} \ln\left(\frac{P_0}{P}\right) \qquad (5.27)$$

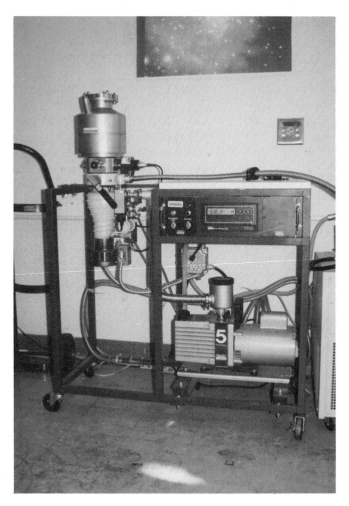

Fig. 5.10. A simple pumping station with a diffusion pump (vertical stack) and a roughing (rotary) pump in the lower right.

Typically, the chamber is rough pumped to about 5×10^{-2} Torr with a mechanical pump and then pumped to a lower pressure with a diffusion pump. Typical pump speeds are 100 litre/s at the inlet. It is important to know how long the roughing will require because the pressure in the diffusion pump section cannot be allowed to rise significantly. Outgassing from metals can be about 10^{-7} Torr litre s^{-1} cm^{-2} after pumping for 1 hour, but this can be reduced an order of magnitude after 24 hours of pumping. Plastics outgas at 100 times this rate and should be avoided. Pressure within the vacuum vessel can be monitored with several types of gauges. During rough pumping the most common gauge is a Pirani gauge in which the temperature of a filament depends on the rate of heat loss to the surrounding gas; heat loss decreases as the gas pressure falls. For lower pressures a Penning gauge is used. The principle of this device is the production of a small electrical current when molecules are ionized in a magnetically confined cold cathode discharge between two electrodes at room temperature; the lower the pressure the smaller the current. Fig. 5.10 shows a typical pumping station.

5.6 ELECTRONICS DESIGN

If the optics provide good images, but the electronics system is noisy, then you will end up with poor signal-to-noise ratios! Needs in this category generally split into analogue and digital electronics. As before, a block diagram layout using a "top-down" approach is best because it usually results in identification of much smaller units, some of which can be duplicated many times and others can be purchased commercially. The primary analogue circuits are amplifiers. Depending on the detector, there may be a pre-amplifier and a post-amplifier, either dc or ac coupling may be required, operational amplifiers (op-amps) may suffice or designs based on individual transistors (typically MOSFETs and JFETs) may be needed. Digital circuitry can be very compact and much effort is saved by using some form of microprocessor and programmable array logic and commercial data acquisition boards. One of the key components is the analogue-to-digital converter (ADC). A comparison of the voltage resolution of the ADC and the output signal strength from the detector will determine the amplifier gain. Careful attention to grounding schemes and isolation of the digital and analogue sections is very important for control of electronic noise. It is also wise to consider how the electronics will be packaged and integrated with the mechanical design.

Several circuit simulation packages are available for workstations and PCs to help the designer, of which the best known is probably SPICE (Simulation Program with Integrated Circuit Emphasis) which was originally developed at the University of California in 1975. Later commercial implementations (such as HSPICE and PSPICE) are very powerful and employ component libraries to store the characteristics of each device. Software such as PCAD enables designers to layout the circuit for manufacture on a printed circuit board (PCB) and the file can often be down-loaded by modem line directly to the company who will fabricate the board.

Basic knowledge about operational amplifiers, FETs, source followers, methods of isolation and noise reduction, ADCs, programmable logic arrays and microprocessors, power supplies and electromechanical devices such as stepper motors and electric shutters is very helpful.

Most solid-state imaging devices (CCDs, infrared arrays) require a pre-amplifier fairly close to the output pin before the signals can be sent down long lengths of cable. A typical design for a CCD preamp is shown in Fig. 5.11. Note the capacitor between the detector and preamp itself, indicating that this circuit is "ac-coupled". For higher-speed operation, as encountered with infrared detectors, it is necessary to eliminate this capacitor and directly or "dc-couple" the detector and preamp. Under these circumstances there will have to be an additional circuit component to "offset" the dc level prior to amplification. In some cases the preamp is placed inside the cryostat, as close as possible to the CCD. Other workers have had good success with preamps placed immediately outside of the cooled chamber provided the lines are well shielded and low-capacitance (e.g. twisted pair wires). This approach is obviously attractive from the point of view of accessibility. Good preamplifier performance is essential since it is at that point that small signals are being turned into large signals.

CCD VIDEO PRE-AMPLIFIER

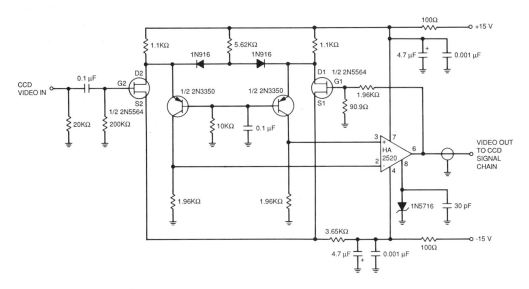

Fig. 5.11. A CCD preamp design based on an original concept used at the NASA Jet Propulsion Lab (JPL). Courtesy Lloyd Robinson.

Preamps can be designed with discrete components or with operational amplifiers (op-amps) which are constructed on a single integrated circuit chip and are "differential" amplifiers having two inputs labelled "+" and "−" for non-inverting and inverting respectively. A wide range of op-amps are available and therefore it is necessary to examine the data sheets carefully to select the best device for your application. Parameters to check are input impedance, output impedance, output voltage range, supply voltage range, common-mode rejection ratio which is the ability of the device to reject noise common to both inputs, input bias current, input offset voltage and its temperature-dependence, frequency response and noise. Manufacturers normally report the root mean square (rms) noise voltage divided by the square root of the frequency bandwidth of measurement (e.g. 10 nano-volts per root hertz), but watch out for the difference in voltage noise and current noise.

Field-effect transistors (FETs) are encountered frequently, both in detectors them-selves, such as CCDs and infrared arrays, and in external circuitry. FETs are three-pin devices in which the output current is determined by the voltage applied to the input termi-nal. The name derives from the fact that the electric field produced by the input voltage controls current flow. Two basic types are MOSFETs and JFETs (see Fig. 5.12). The MOS prefix indicates that the input terminal has a metal-oxide-semiconductor structure, whereas the J prefix indicates that the input gate of the device forms a pn junction with the substrate. In both cases, two n-type regions called the "drain" and the "source" are formed in a p-type substrate. In the MOSFET these two regions are linked by a thin channel of n-type silicon at the surface which forms the "S" part of the MOS input or "gate". The voltage applied to the gate controls the effective depth of the thin n-type channel and hence the amount of current that can flow from drain to source. If the gate is positive with respect to the source then the current increases. In the JFET, the MOS gate is replaced with a small region of p-type material diffused into the n-type channel to form a reverse-biased pn junction with a depletion region which effectively controls the flow of current from the drain to the source. As the gate voltage is made more negative the depletion region increases and conductivity drops.

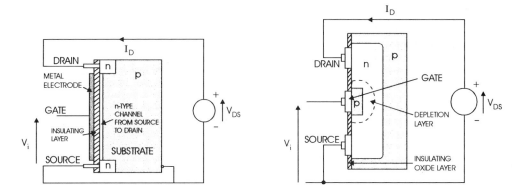

Fig. 5.12. Schematic representations of the basic structure of a JFET and a MOSFET. Both rely on controlling the width of a conducting channel by means of an internal electric field.

A common circuit found in CCD and IR array detectors is the "source follower" or common-drain amplifier in which the source voltage (the output) tends to follow the gate voltage (the input). When the product of the transconductance (g_m) and the source load resistance (R_s) is large then $1/g_m R_s \ll 1$ and $V_{out} \sim V_{in}$; in practice there is a loss of gain by about 20% but the advantage of these circuits is very high input impedance and fairly low output impedance. FETs are also used as simple switches. A sufficiently large negative gate voltage effectively cuts off current flow completely. Another use of FETs is as a constant current source when a constant voltage is applied to the gate. A simple constant current arrangement can be obtained by connecting the gate and source and placing the load between the drain and the drain supply voltage.

Digital circuits include timing pattern generators, analogue-to-digital converters, interface circuits or buffers and memory circuits. There is a lot more commercial hardware available for digital applications than for analogue applications. While it is certainly possible to use discrete TTL and CMOS logic chips for small "hardwired" systems, more flexibility is achieved by using one of the several classes of microprocessors together with some devices called **programmable logic arrays** (PLAs; also called PALs and PLDs). Two classes of device which go beyond the standard microprocessor chip are the **digital signal processor** (DSP) and **transputer**. Both of these devices can provide functions such as sequencing, timing patterns, data acquisition, input/output control.

Electrical interference and noise can be a serious problem in scientific apparatus with low signal levels. An excellent treatment of basic principles is given by Ott (1988). Ground-loops must be eliminated, signal wires must be screened or shielded, separate power supplies should be used for digital and analogue sections, switching-type power supplies should be avoided and either opto-isolation or capacitive isolation (e.g. the ISO150 from Burr-Brown) should be used. In the opto-isolator digital signals are turned into pulses of light from a light-emitting diode which are then detected by a photo-transistor and converted back into electrical pulses. A useful summary of good procedures is:

- establish a single point ground for each system and do not allow the analogue and digital parts of the system to share the same ground
- use large diameter multi-strand cable to reduce the impedance for ground connectors
- completely shield all signal cables, but ground the shield at the power source end only
- use optical fibres or shielded twisted pair cables where possible
- group cables by their function and sensitivity to noise, avoid long parallel runs, separate dissimilar types such as analogue signal cables, digital cables and power supplies
- use shielded cabinets

Electric motors fall into three categories: ac motors, dc motors and stepper motors. DC motors are used extensively in precision control applications requiring low power, whereas for high-power roles and less demanding precision ac motors are selected. In dc motors there is an almost linear relationship between speed and applied voltage, and between torque and current. A stepper motor consists of a central rotor surrounded by a number of coils (the stator) with diametrically opposite coils connected together so that energizing that pair of coils will cause the rotor to align itself with that pair. Applying power to each set of windings in turn causes the rotor to jump or step to the next position. To minimize the number of wires involved, the coils are grouped in sequence with, for example, the fifth coil connected to the first and similarly for the 9th, 13th, etc., so that when the rotor is at the fourth coil and the first and fifth are energized together, the rotor steps to the fifth because it is closest. The speed of rotation is controlled by the pulse repetition rate. Since the frequency can be changed easily by modifying the pulse waveform period then it is possible to build in acceleration and deceleration ramps.

For low-noise instrumentation it is generally best to use well-packaged linear power supplies and avoid switching power supplies despite the impact on size and weight, unless extreme care is taken to isolate and shield the switching supplies. It is also a good idea to employ an uninterruptable power supply (UPS) which takes over immediately if there is a power cut and provides power for long enough to ensure a safe and orderly shut-down of the instrument.

5.7 SOFTWARE

If the electronic imaging detector is the "heart" of an astronomical instrument, then software must be the "muscles"! Good software can make an immense difference to the ease with which an astronomer can learn to use an instrument efficiently and productively. As one of my colleagues has put it, the difference is like that between a "black-belt and a novice". Many levels of software may be required depending on the instrument. For example, software will be required for the following tasks:

(1)   Interface to the astronomer
  • Graphical User Interface (GUI—often called a GOO-EE)
  • Command Line Interface (CLI)
  • Scripting or Macro interface to allow sequences to be pre-programmed
(2)   Data acquisition
  • capture and storage of digitized data
  • manipulation of the data stream and attachment of "header" information
(3)   Data display and analysis
  • display of the data
  • "quick-look" facilities for analysing the data
(4)   Instrument control
  • conversion of high-level commands to hardware signals
  • handling of status signals from the hardware

Software is an area in which most astronomy graduate students can make a contribution, but it is very important to take the opportunity to develop good programming and documentation habits. More than one "level" of programming may be needed. For example, it is frequently the case that a "high-level" language is used for the user interface (C, FORTRAN) and "low level" coding of Digital Signal Processors (DSPs) and other microprocessors is done in the appropriate assembly language. Since this requires more training and more expertise than might be reasonably expected of an astronomer, try to stay with approaches which require only high-level languages.

Again, the top-down approach works best. Produce a flow chart of all the data and communication paths and develop an iterative approach to improve the model of the software as illustrated in Fig. 5.13. Keep the code modular rather than monolithic. Try to make it easy for the first time user by employing a Graphical User Interface, while also catering to the expert user by providing short-cuts and hot keys. Fig. 5.14 shows a graphical user interface developed at UCLA by Tim Liu for the NIRSPEC instrument for the Keck telescope.

Stick with a programming language which is widely accepted and supported. This probably means ANSI C, or perhaps FORTRAN or Pascal, and establish a set of documentation standards for coding the software and adhere to them, even if it takes more time: other people will be very grateful later.

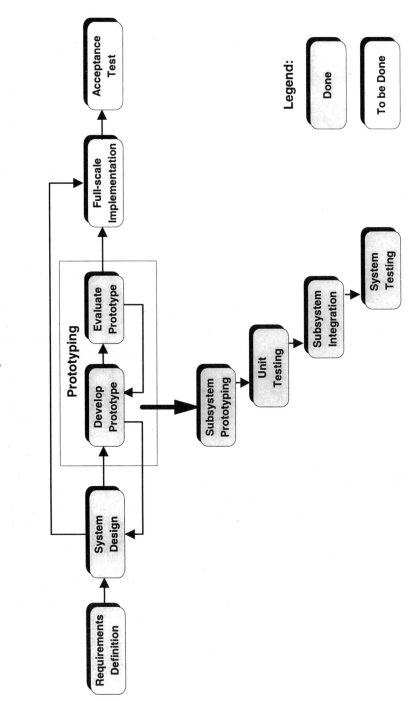

Fig. 5.13. A flow chart illustrating a good method of software development for astronomical instruments. Courtesy Tim Liu, UCLA.

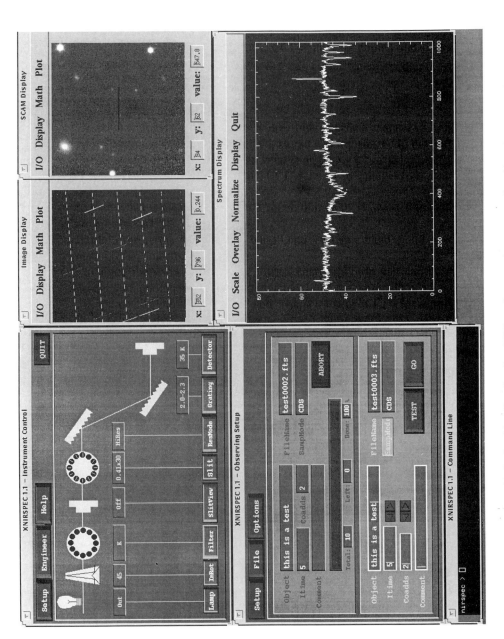

Fig. 5.14. An example of a Graphical User Interface (GUI) used to control a spectrograph and imaging system. This screen was developed for the NIRSPEC project using DataViews and IDL. Courtesy Tim Liu.

SUMMARY

The design and development of modern astronomical instruments embodies a broad range of basic physics and engineering principles. This chapter has touched on some of the background knowledge required to really appreciate what goes into the design of such instruments. It is important for observational astronomers to understand the technological basis in order to foster good communications with design engineers, and to recognize both the limits and opportunities of technology.

EXERCISES

1. Draw a block diagram layout of a cooled CCD camera system with a filter wheel which is to be used on a small telescope and remotely controlled from a "warm room" 50 ft away using a PC.
2. What is meant by an anti-reflection coating and why are they important in astronomy applications? What is the optimum refractive index of the coating to be applied to a silicon surface with $n = 4$? What thickness of coating is required at a wavelength of 2.2 µm?
3. Match the following three detectors to a 0.2-m telescope and then to an 8-m telescope: a Kodak KAF-4200 CCD with 9-µm pixels in a 2048 × 2048 format, a SITe CCD with 22-µm pixels in a 1024 × 1024 format and the Hughes-SBRC InSb array with 27 µm pixels in a 1024 × 1024 format. Which camera system will be the hardest to design?
4. Derive the relation for θ_{pix} which shows that, for a given telescope diameter and detector pixel size, the number of arcseconds subtended on the sky by each pixel is determined only by the f/number of the camera lens system.
5. Given the relation shown below for the angular blur size due to spherical aberration in a single lens of refractive index $n = 1.5$ and focal ratio $f/2$, determine the linear diameter of the image of a point source if the focal length of the camera is 50 mm.

$$\beta = \frac{n(4n-1)}{128(n+2)(n-1)^2} \frac{1}{(f/number)^3}$$

   How does this compare to the size of a typical CCD pixel? How can this limitation be overcome in practice?
6. The blur circle diameter for an on-axis point-source image in seconds of arc for a spherical mirror is given by

$$\beta = \frac{206\,265}{128(f/\#)^3}$$

   Determine the f/number which yields a diameter of 1 arcsecond.
7. Spherical aberration is zero for a paraboloidal mirror, but sagittal coma is given by

$$\beta = \frac{206\,265}{16(f/\#)^2}\theta$$

where β is the blur circle diameter in seconds of arc and θ is the off-axis field angle in radians. Determine the image blur due to coma $1'$ off axis for an $f/3$ mirror. How would this change if the primary was $f/1.5$?

8. (a) Calculate the diffraction limit for an $f/2$ lens with a focal length of 50 mm at a wavelength of 500 nm in the mid-visible. (b) The depth of focus is the amount of defocus which introduces $\pm \lambda/4$ wavefront error; $\Delta f = \pm 2\lambda(f/\#)^2$. Determine the depth of focus for the lens in part (a).

9. Calculate the thermal stress induced in an aluminium strut which is fixed at both ends if the temperature decreases by 210 K and compare this to the yield strength. Is there a problem? Ignore the temperature-dependence of the constants. Assume $\alpha = 24 \times 10^{-6}$ K^{-1}, Young's modulus $E = 10 \times 10^6$ psi and the yield strength is 40 000 psi.

    [Hint: $F/A = -\alpha E \Delta T$, because $\Delta L/L = \alpha \Delta T$]

10. An infrared cryostat has a surface area of 5 m^2. Assuming that the geometric factor is one-half the emissivity of 5%, calculate the radiation load on a 77 K interior from (a) laboratory temperature of 300 K and (b) mountain observatory temperature of 275 K. What could you do to reduce the load on the internal cold components?

11. Draw a sketch of a MOSFET and JFET and explain the basic principle of operation. How is it possible to use a MOSFET as a "switch"?

Advanced topic for discussion and research
Étendue: Also called luminosité or throughput. Consider a surface element of area dS and brightness B radiating into a solid angle dΩ. The flux (energy rate in J/s) dF is

$$dF = B\cos\theta\, dS\, d\Omega$$

dS is called the **entrance pupil** and could be the seeing disk, the Airy diffraction disk or the entrance slit of the spectrometer depending on the conditions. Integrating from θ to θ + dθ the value of d$\Omega = 2\pi\sin\theta d\theta$ yields

$$F = 2\pi B\, dS \int_0^{\theta_m} \cos\theta \sin\theta\, d\theta = BU$$

where θ_m is the maximum value of the half-angle of the cone of rays from element dS and the quantity $U = \pi \sin^2 \theta_m dS$ is called the étendue. When the étendue is evaluated for different systems (cameras and spectrometers) it is always the product of the area (A) of the entrance aperture and the solid angle (Ω) subtended by the entrance pupil; it is therefore also known as the $A\Omega$ (pronounced ay-omega) product. Assuming no losses due to absorption or reflection, then conservation of energy demands that $U (= A\Omega)$ is conserved.

12. Derive expressions for the étendue of a spectrometer with an entrance aperture defined by the slit, and derive a similar expression for a seeing-limited camera.

REFERENCES AND SUGGESTED FURTHER READING

Banmeister, T., Avallone, E.A., Banmeister III, T. (Eds) (1978) *Mark's Standard Handbook for Mechanical Engineers*, 8th edn, McGraw-Hill, New York. [Very extensive treatment of mechanics.]

Born, M. and Wolf, E. (1987) *Principles of Optics*, 6th edn, Pergamon Press. [Classic text.]

Kingslake, R. (1978) *Lens Design Fundamentals*, Academic Press, New York. [Classic text; mainly lenses.]

Moore, J.H., Davis, C.C. and Copland, M.A. (1983) *Building Scientific Apparatus*, Addison-Wesley. [Slightly dated, but outstanding introduction to laboratory practice.]

Ott, H.W. (1988) *Noise Reduction Techniques in Electronic Systems*, 2nd edn, John Wiley, New York. [Classic text; excellent.]

Photonics Handbook (1995), Laurin Publishing. [Published yearly.]

Rieke, G.H. (1994) *Detection of Light from the Submillimeter to the Ultraviolet*, Cambridge University Press, Cambridge, England. [Excellent, unified treatment of detectors.]

Schroeder, D.J. (1987) *Astronomical Optics*, Academic Press, San Diego, CA. [Covers telescopes and spectrographs; essential reading.]

Smith, W.J. (1990) *Modern Optical Engineering—The Design of Optical Systems*, 2nd edn, McGraw-Hill, New York. [Classic text; excellent.]

Storey, N. (1992) *Electronics: A Systems Approach*, Addison-Wesley, Wokingham, England. [Comprehensive and very readable; covers everything in electronics.]

Wolfe, W.L. and Zissis, G.J. (Eds) (1989), *The Infrared Handbook*, 3rd edn, published by the Information Analysis Center for the Office of Naval Research. [Classic reference.]

6

Charge-coupled devices

At the heart of all astronomical instruments is some form of detector. Having indicated repeatedly in earlier chapters that the detector of choice in modern instruments is the charge-coupled device (CCD), it is important to consider these remarkable devices in more detail. This chapter begins with the development history of CCDs from 1970 to the present and is an amazing story by itself. Subsequent sections cover all the basic principles of CCDs.

6.1 THE EARLY YEARS

6.1.1 Invention and development
The charge-coupling principle was invented in 1969 by Willard Boyle and George Smith and demonstrated in a simple one-line eight-pixel device by Gil Amelio, Mike Tompsett and George Smith at the Bell Laboratories in New Jersey, USA. Larger image-forming devices of 100×100 pixels were not introduced until 1973 and Boyle and Smith received the basic patent at the end of 1974. From the original small arrays available around 1973, CCDs have come a long way. Formats of 1024×1024 pixels are readily available for astronomy, with most observatories using one or more CCDs with 2048×2048 pixels. Fig. 6.1 shows a collection of CCDs. Smaller-sized CCDs are cheaply available for the amateur astronomy market and large "mosaics" of CCDs with 8192×8192 pixels[†] are in operation at many places.

6.1.2 The astronomical push
Many astronomy-related groups familiar with imaging technology—usually with vidicon-type systems—were alert to the potential of CCDs in the early seventies. Gerald M. Smith, Frederick P. Landauer and James R. Janesick of the Advanced Imaging Development Group at the NASA Jet Propulsion Laboratory operated by the California Institute of Technology (Caltech) in Pasadena, and Caltech scientist James Westphal were among the first to recognize the potential advantages of such an imaging device for astronomy

---

†These strange numbers are simply powers of 2; $2^{10} = 1024$, $2^{12} = 4096$. There is no fundamental reason to use these numbers. In fact, manufacturers of video CCD cameras use a different convention and rarely produce "square" arrays.

Fig. 6.1. A collection of CCDs including some very large buttable devices. Courtesy Gerry Luppino.

and space applications. In 1973, JPL joined with the National Aeronautics and Space Administration (NASA) and with Texas Instruments (TI) Incorporated (Dallas) to initiate a program for the development of large-area CCD imagers for space astronomy, in particular for the proposed "Galileo" mission to Jupiter. Originally scheduled for 1981, the Galileo spacecraft was finally launched in 1989 and arrived at Jupiter in December 1995.

During the period 1973 to 1979 Texas Instruments (TI) developed CCD arrays of 100×160 and 400×400 pixels, then 500×500 pixels and finally an 800×800 pixel array. Testing and evaluation of these devices was carried out at JPL by Fred Landauer and by a young engineer named Jim Janesick, who just happened to be an amateur astronomer. Having already approached one astronomer at a national institute about testing a CCD on a telescope, and having been turned down, Jim luckily met and teamed-up with Dr Bradford Smith a planetary scientist at the University of Arizona's Lunar and Planetary Laboratory. In early 1976 they obtained the first astronomical imagery with a charge-coupled device. Using the 61-inch telescope (designed for planetary imaging) on Mt Bigelow in the Santa Catalina mountains near Tucson, they obtained CCD images of Jupiter and Saturn using a special filter to pick out methane gas in the atmosphere of those giant planets. When they turned the telescope to Uranus (Fig. 6.2) and immediately thought something had gone wrong. It looked like a "donut"! After checking focus and everything else, Brad realized that it must be correct, they were observing "limb-brightening" of Uranus in the methane band for the first time. As Brad recalls vividly, "All who participated and who saw those images agreed that the potential of the CCD was superior to the other imaging equipment of the time".

At about the same time, NASA awarded contracts for the procurement of instruments for an earth-orbiting Space Telescope. One of the people awaiting NASA's decision was

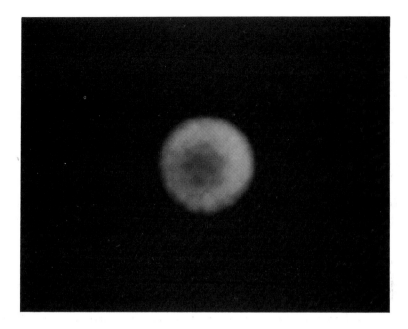

Fig. 6.2. The first astronomical CCD image. An image of the planet Uranus obtained in 1976 through a methane-band filter with a Texas Instruments CCD. Photo courtesy of Bradford Smith.

John Lowrance at Princeton University. John was working with SEC vidicon technology in the 1970's when CCDs were invented. SEC (or secondary electron conduction) vidicons seemed like the most appropriate imaging devices—at least in terms of their wavelength of response—for space ultraviolet/optical experiments such as the Orbiting Astronomical Observatory (OAO). While NASA and various advisory bodies deliberated, John continued to pursue the development of SEC vidicon systems. Finally, after many delays, NASA abandoned the OAO series in favour of the Space Telescope. By that time CCDs had been invented, and NASA JPL had begun their studies. Following a crucial meeting at which CCD results were demonstrated, the initial plan to utilize SEC vidicons for the Space Telescope was dropped and the concept of Principal Investigator instrument teams was introduced. A proposal from a team led by planetary scientist James Westphal of Caltech in collaboration with JPL was accepted for the inclusion of CCD cameras on the Space Telescope. Jim had heard about CCDs at a committee meeting a couple of years earlier. When he insisted on knowing more about the JPL results, the chairman (Bob O'Dell) sent him along to JPL to see for himself! There he met Fred Landauer and learned that CCDs were indeed very low-noise devices; 100 electrons (e) noise had been observed and 30-e was predicted. On his return to Caltech he mentioned these numbers to colleague, Jim Gunn of Princeton University, who was getting ready for a major project involving vidicon technology, and whose instant reaction was "that will revolutionize astronomy if it is true!"

John Lowrance, like Jim Westphal, moved away from vidicon technology and began working with CCDs. Luckily there was a key player in the game right on his doorstep—the

Electro-optics Division of the Radio Corporation of America (RCA) in nearby Lancaster, Pennsylvania. At the head of the RCA group working on CCDs was Dick Savoye, and Dick was enthusiastic about the astronomical applications, and moreover he believed that the technology possessed by RCA would yield devices extremely sensitive to blue light, as later demonstrated by the superb blue response of the thinned, backside-illuminated 512×320 RCA CCD. John Lowrance at Princeton and John Geary at Harvard each established good relations with RCA and began testing these devices in the late seventies. John Geary, having first tried an unthinned device on the 1.5 m and 60 cm telescopes on Mt Hopkins in April 1980, visited the RCA facility in Lancaster Pennsylvania shortly thereafter to show them the splendid results obtained so far and urge them to provide him with a thinned CCD. He received one of the very first thinned backside-illuminated CCDs manufactured by RCA; this device was considered a reject and was lying in the desk drawer of RCA engineer Don Battson. John put it on the telescope on Mount Hopkins in September 1980 and there it remained for almost a decade!

Meanwhile, the Texas Instruments (TI) chips evolved through a programme of systematic development toward the eventual goal of an 800×800 array. One of the key figures on that programme from the outset was Morley Blouke. Several approaches to the design and fabrication of CCDs were tried. A major constraint was that the device must be able to survive the harsh radiation environment around Jupiter and so two different constructions evolved—the buried-channel and the virtual phase CCD. Tens of thousands of CCDs were being manufactured under contract to NASA (the final number was 75 000) and, following the success of the collaboration with Brad Smith, JPL realized—as Jim Janesick stated in a proposal to the Director JPL in October 1976—that there was "... a need to expose and familiarize astronomers and scientists to the capabilities of the CCD for use in planetary observation and stellar studies".

Around this period (1974–77) other companies were also beginning to develop CCDs. One of the first companies to actually market a device was a division of Fairchild Semiconductor who produced a 100×100 CCD; Gil Amelio had moved from Bell Labs to Fairchild. At the Kitt Peak National Observatory in Tucson, Arizona (USA), Richard Aikens and Roger Lynds had been working on low-light-level imaging systems for astronomy for many years. Soon the Kitt Peak National Observatory began a programme of development of CCDs. With Steve Marcus, they began working on the Fairchild device. The Fairchild CCD201 and CCD202 image sensors were designed for TV applications and, although capable of high performance, they had a serious impediment for astronomical work due to the interline transfer construction (see below) which meant that they had columns of picture elements which were alternately light-sensitive and totally insensitive due to a cover by opaque metal strips; in terms of the image falling on the CCD these devices were half blind!

Not long after this initiative, Richard Aikens left the Kitt Peak National Observatory to set up his own company, called Photometrics, which has played an important part ever since in stimulating the manufacturing of CCDs and the development of scientific camera systems.

There was a time of great frustration in the late seventies about the lack of access to CCD technology by the mainstream astronomical community. Development of the Wide-Field/Planetary Camera, abbreviated WFPC but spoken as "wiff pick", was going

well and many people were now aware of the sensitivity and the scientific potential of
CCDs. Industry too was embracing the new technology, but commercially available products were scarce. During this interlude people were side-tracked into trying other forms of
less suitable solid-state imagers which were commercially available such as the Charge
Injection Device (or CID) from General Electric (America), or the interline transfer device
from Fairchild already mentioned. When 512 × 320 RCA CCDs appeared in the late seventies it was a welcome relief.

The first RCA CCDs were frontside-illuminated which meant they had a poor response
to blue light. Soon however, the thinned backside-illuminated CCDs appeared. Clearly,
RCA had "the secret" for treating or passivating the thinned backside surface, and these
CCDs displayed outstanding sensitivity over a huge spectral range—better even than the
TI chips. Unfortunately there was one weakness. The design of the on-chip output ampli-
fier was poor and so the CCD was 5–10 times "noisier" in electrical terms than the TI
CCD. Later RCA CCDs were much better. Sadly, in 1985, RCA withdrew from the CCD
market for commercial reasons. Detector development work has continued however, at the
David Sarnoff Labs.

In early 1980 a somewhat unexpected source of astronomical CCDs appeared. Craig
Mackay of the Institute for Astronomy in Cambridge, England, had been working on silicon vidicons. Progress was slow due to lack of funds. He had met Jim Westphal on Palomar Mountain in 1975 and was aware of the TI work on charge-coupled devices, but he
learned of a British source of CCDs by a curious coincidence. Silicon vidicons had good
spectral response, but they were "noisy". Craig needed a very low-noise amplifier design.
He contacted an eminent designer called Ken Kandiah at the British Atomic Energy Authority at Harwell and asked him to visit Cambridge. Kandiah offered a design based on a
Junction Field Effect Transistor (JFET) but recommended Craig to a David Burt at the
GEC Hirst Research Centre for a design based on the more readily available
Metal-Oxide-Semiconductor (MOS) transistors. When Craig met David Burt he learned
that GEC had a very advanced CCD programme. The following year, Craig and his PhD
student Jonathan Wright, put together a CCD drive system based on an existing vidicon
camera.

By March 1982 the noise associated with typical GEC CCDs was reported as seven
electrons while selected devices gave a mere three electrons, that is, the amount of charge
was uncertain by only three times the charge on a single electron. Remembering Jim
Gunn's excitement on hearing that devices with better than 100 electrons noise were possible, then three electrons was a truly amazing result. The GEC CCDs were also illuminated
from the front rather than the rear and so their blue response was poor. As discussed later,
chemical coatings are now applied to these devices to overcome this deficiency to a large
extent. Much later, both Craig and Jonathan formed independent commercial companies
to make CCD cameras. Craig's company is called Astrocam (formerly Astromed) and
Jonathan's company is called Wright Instruments. In a similar way, John Lowrance of
Princeton also branched out into making customized, very high-quality camera systems for
other institutes.

By June 1981, the date of the Harvard–Smithsonian conference on solid-state imagers,
the number of independent astronomy groups working on CCD systems had already grown
from 5 to 20. Devices in use came exclusively from TI, RCA and GEC (UK). Astronomers

were clearly pushing the technology as hard as they could, yet with only three manufacturers, one of which had low-noise devices (GEC), one of which had high quantum efficiency devices (RCA), and one of which should have had devices with both properties but was (a) having problems with blue sensitivity and (b) not available for sale anyway, it was understandable that people began to worry. When production of the GEC devices moved to EEV there was the inevitable hiccup in supply, and when RCA withdrew from the field, it seemed like the dream had become a nightmare. Eventually, TI CCDs "excess to NASA requirements" started to become available in the USA, exceptionally detailed studies of the TI chips by Jim Janesick and colleagues at JPL advanced the understanding of CCDs and their optimization, and new devices by companies such as Thomson-CSF (Europe) had been studied in detail by a team at the Royal Greenwich Observatory (RGO). In 1985 astronomers learned of a most exciting prospect. It was the formation of a team at Tektronix Inc. led by Morley Blouke to produce scientific grade CCDs with large formats and outstanding performance. The initial goal would be a 512×512 array with good-sized pixels (0.027 mm) leading to a chip with 2048×2048 pixels. Unfortunately, by mid-1986 it became clear that some sort of unexpected fabrication or processing problem, was resulting in large numbers of defects called "pockets" (see Chapter 7) thus rendering otherwise excellent low-noise devices unusable, and hopes were again dashed. This time the situation for many astronomical groups was serious because new instrument designs and funding for instrument developments had been tied to the expected Tektronix chips.

Morley and his team did not give up and, in collaboration with several interested parties, they valiantly followed every lead in an effort to understand, model and eliminate such problems. The research at Tektronix and at JPL led to an in-depth understanding of the solid-state physics of CCDs which, as Morley remarks, "ought to be much easier to understand than a transistor". Around mid-1988 Tektronix began to ship CCDs to customers with long-standing orders. Later, the Tektronix CCD group was spun off into a company called Silicon Imaging Technologies, Inc. (SITe).

Among the many people frustrated by the dry spell in CCD supplies during that era was Richard Aikens, founder and president of Photometrics Ltd. In an unprecedented move he contracted with a so-called "silicon foundry" (a division of Ford Aerospace later taken over by Loral Corporation) to produce a custom CCD with 516×516 pixels (marketed as 512×512 pixels) and this turned out to be an outstanding success. The advantage of this approach is that the silicon foundry, can quote for device production without having to consider the "end-use" product. In addition, Photometrics announced the availability of a chemical phosphor coating called Metachrome II which can be applied safely to any CCD by vacuum sublimation and thereby improve its response to blue light. In August of 1988 Lloyd Robinson reported excellent initial experimental results for another brand new device. As the result of a National Science Foundation grant to Lick Observatory, a contract was placed with E G & G Reticon Corporation to construct a large CCD suitable for spectroscopic applications; the format chosen was 400×1200. Finally, in the same year, new initiatives at EEV (UK) and at Thomson-CSF (France) were announced. A thinning programme and a mosaic construction programme had begun at EEV. Meanwhile, funded by the European Southern Observatory and the French agency INSU, Thomson-CSF in Grenoble had developed a "buttable" version of their excellent low-noise front-illuminated device and a 2×2 mosaic had been constructed at the European Southern Observatory near Munich.

These approaches set the trend that has continued until the present day. Astronomers work directly with a silicon foundry to obtain customized CCDs. Companies like Loral, SITe, Orbit, EEV and Thomson have provided these facilities. Other companies such as TI, Kodak and Dalsa cater to a mass market and provide CCDs for the many smaller businesses that now manufacture complete CCD camera systems (see Appendix 7 for a listing). In addition, some government-funded labs, such as the MIT Lincoln Labs in the USA, can provide special services. Most large observatories are using 2048×2048 pixel CCDs from either SITe or Loral. Smaller devices tend to have more specialized uses, such as in guide cameras or in small camera systems for the growing market of amateur astronomers.

Although devices with 4096×4096 pixels have been manufactured, most astronomical developments are concentrating on forming "mosaics" of high-yield formats like the 2048×2048 chips, and developing better response at shorter wavelengths. The largest single CCD manufactured so far is a $7k \times 9k$ device from Philips in the Netherlands.

6.2 BASIC PRINCIPLES

6.2.1 Absorption of light in silicon

The properties of any solid material depend on both the atomic structure of the atoms of the material and the way the atoms are arranged within the solid, that is, the crystal structure. Electrons can exist in stable orbits near the nucleus of an atom only for certain definite values of their energy. When individual atoms come close together to form a solid crystal, electrons in the outermost orbits, or upper **energy levels**, of adjacent atoms interact to bind the atoms together. Because of the very strong interaction between these outer or "valence" electrons, the outer orbits and therefore the upper energy levels are drastically altered. The result is that the outer electrons are shared between the different atomic nuclei. A diagram depicting the "energy levels" of the electrons for a combination of two atoms will therefore have two permitted levels near the core of each atom. A combination of three atoms will have three levels near the core because the outer electrons of all three atoms can be shared. The higher, unoccupied orbits are also split indicating that they too can in principle take two or three electrons. Even the tiniest sliver of a real crystal will contain many hundreds of millions of atoms, and so there is a huge number of split levels associated with each atom in the crystal because of the sharing of outer electrons. In other words, the energy levels or orbits are spread out into a "band". The lowest band of energies, corresponding to all the innermost orbits of the electrons, are all filled with electrons because there is one electron for each atom. This band of allowed, filled energy levels is called the **valence band**. Conversely, the upper energy band is empty of electrons because it is composed of the combined unoccupied higher energy levels or orbits of the individual atoms in the crystal. It is called the **conduction band** for reasons that will become apparent. Since the individual atoms have a gap between the permitted inner and outer orbits, that is, a gap in energy between the inner filled levels and the outer unoccupied levels, the energy region between the valence band and the conduction band in the crystal must be a **forbidden energy gap**. Fig. 6.3 summarizes this description. Note that the crystal must be

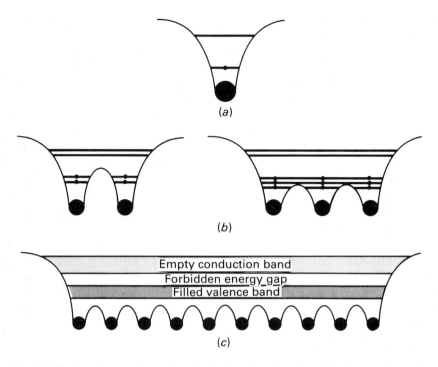

Fig. 6.3. The development of a conduction band and a valence band of electrons in a solid crystalline material such as silicon.

pure and contain atoms of only one kind, otherwise additional energy levels corresponding to the impurity atoms will be formed. More importantly, the periodic or repetitive crystalline structure must be unbroken to avoid distortions in the energy levels caused by abnormal sharing of electrons. Of course, in practice both of these conditions are violated in real crystals and this contributes to degraded performance of devices such as transistors and CCDs. In metals, the valence and conduction bands overlap and so any of the many valence electrons are free to roam throughout the solid and to move in response to the force of an electric field; an electric field could be produced by attaching a battery to both ends of the piece of metal. Therefore metals are excellent conductors of electricity (and heat). An insulating material on the other hand, has a highly ordered structure and a very wide forbidden energy gap. The conduction band is totally empty of electrons and so cannot contribute to an electrical current flow. Electrons in the completely filled valence band cannot move in response to an electric field because every nearby orbit is occupied.

In a semiconductor, a few electrons can be elevated from the valence band to the conduction band across the forbidden gap merely by absorbing heat energy from the random, microscopic, jostling motions of the crystal structure at normal "room" temperature. Electrons promoted to the conduction band in this way can then conduct electricity, that is, they are free to move under the influence of an electric force field. Interestingly, the corresponding vacancies or **holes** left in the valence band allow it to contribute to electrical conductivity as well, because there is now somewhere for the other electrons to go; descriptions of solid state devices therefore refer to **electron–hole pairs**.

Since the number of electrical charge carriers (electrons in the conduction band, holes in the valence band) is much fewer than in the case of a metal, semiconductors are poorer conductors than metals but better than insulators. The width of the forbidden energy gap in semiconductors is an important quantity. It is usually expressed in **electronvolts**. One electronvolt (eV) is the energy that one electron would acquire under the influence of a voltage of one volt. Most semiconductor crystals have a band gap energy around 1 eV. This energy is roughly 30 times larger than the thermal or heat energy in the crystal atoms at room temperature. Since the number of electrons which can be promoted to the conduction band by absorbing heat will vary with the temperature of the crystal, those semiconductors with larger band-gaps are preferred since transistors and other devices made from them will be less sensitive to environmental changes in temperature. For this reason silicon is preferred to germanium. If, however, the semiconductor is cooled to a low temperature and held there, then the random elevation of a valence electron to the conduction band can be virtually eliminated.

Table 6.1 is a section of the periodic table of the elements which shows that the primary semiconductors like silicon and germanium belong to the "fourth column" elements, which includes carbon. These elements have four valence electrons. Compounds of elements on either side of the fourth column can be formed and these alloys will also have semiconductor properties; gallium arsenide (GaAs) and indium antimonide (InSb) are III-IV (or "three-four") compounds and mercury-cadmium-telluride is a II-VI (or 2-6) compound. The column number corresponds to the number of valence electrons.

Table 6.1. Part of the periodic table of elements showing the location of semiconductors

IB	IIB	IIIA	IVA	VA	VIA	VIIA
		^{13}Al	^{14}Si	^{15}P	^{16}S	^{17}Cl
		Aluminium	Silicon	Phosphorus	Sulphur	Chlorine
^{29}Cu	^{30}Zn	^{31}Ga	^{32}Ge	^{33}As	^{34}Se	^{35}Br
Copper	Zinc	Gallium	Germanium	Arsenic	Selenium	Bromine
^{47}Ag	^{48}Cd	^{49}In	^{50}Sn	^{51}Sb	^{52}Te	^{53}I
Silver	Cadmium	Indium	Tin	Antimony	Tellurium	Iodine
^{79}Au	^{80}Hg	^{81}Tl	^{82}Pb	^{83}Bi		
Gold	Mercury	Thallium	Lead	Bismuth		

When a **photon** is absorbed in the crystalline structure of silicon, its energy is transferred to a negatively charged electron, the "photoelectron", which is then displaced from its normal location in the valence band into the conduction band. When the electron reaches the conduction band it can migrate through the crystal. Migration can be stimulated and controlled by applying an electric field to the silicon crystal by means of small metal plates called **electrodes** or **gates** connected to a voltage source. Absorption of photons in silicon is a function of the photon energy (and hence wavelength). The photon flux at depth z in the material is given by

$$I(z) = I(0) e^{-\alpha z}$$

where α is the absorption coefficient. At a temperature of 300 K (27 °C), $\alpha \sim 5\ \mu m^{-1}$ at $\lambda = 400$ nm in the blue, but only 0.1 μm^{-1} at 800 nm in the far red; the lower the value of α the greater the depth reached for the same absorption ratio. At 77 K (liquid nitrogen temperature) α reduces to 4.0, 0.25 and 0.005 μm^{-1} at 400, 600 and 800 nm respectively. Clearly, red (low-energy) photons pass deeper into the silicon before being absorbed. Eventually, for the reddest light, that is the lowest energy photons, there is simply not enough energy to elevate a valence electron to the conduction band. In other words, for each semiconductor there is a wavelength of light beyond (redder than) which the material is insensitive to light because the photons are not energetic enough to overcome the forbidden energy gap (E_G) in the crystal. The cut-off wavelength is given by

$$\lambda_c = \frac{hc}{E_G} \tag{6.1}$$

where h is Planck's constant and c is the speed of light. These wavelengths are given in Table 6.2. Some of these materials are sensitive well into the infrared region and will be discussed again in later chapters. Here we concentrate on silicon.

Table 6.2. Forbidden energy gaps and long-wavelength photo-absorption limits for some common semiconductors

Name	Symbol	T (K)	E_G (eV)	λ_c (μm)
Cadmium sulphide	CdS	295	2.4	0.5
Cadmium selenide	CdSe	295	1.8	0.7
Gallium arsenide	GaAs	295	1.35	0.92
Silicon	Si	295	1.12	1.11
Germanium	Ge	295	0.67	1.85
Lead sulphide	PbS	295	0.42	2.95
Indium antimonide	InSb	295	0.18	6.9
		77	0.23	5.4
Mercury cadmium	$Hg_xCd_{1-x}Te$	77	0.1 (x=0.8)	12.4
telluride			0.5 (x=0.554)	2.5

When silicon atoms in the crystal structure are deliberately replaced with other atoms the semiconductor is said to be **doped** (Fig. 6.4). If the impurity atom has more valence electrons than the semiconductor then it will donate these negative charges to the conduction band; such a material is called **n-type**. Conversely, if the impurity atom has fewer valence electrons than the semiconductor then a positively charged hole is left in the valence band ready to accept any available electrons; this material is a **p-type** semiconductor. In p-type material there is an excess of holes and so electrons are said to be the minority carriers of charge, whereas the opposite is true for n-type material. Junctions between p- and n-type regions are used many times in semiconductor structures to produce different devices.

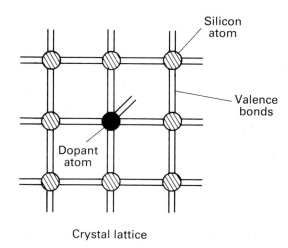

Fig. 6.4. An impurity atom or dopant in the silicon crystal lattice.

When a **pn junction** is formed, electrons from the n region tend to diffuse into the p region at the junction and fill up some of the positively ionized states or holes in the valence band thus making that region more negative than it was. Similarly, the diffusion of holes from the p to the n side leads to an increasingly more positive electrical potential. A narrow region forms on either side of the junction in which the majority charge carriers are "depleted" relative to their concentrations well away from the junction. Since the concentration of electrons in the n-type material is very much larger than in the p-type material, the flow of electrons would tend to be one way were it not for the fact that the diffusion process itself begins to build up an electrostatic potential barrier which restrains the flow of electrons from the n-type region; the build up of electrons on the p side starts to repel further diffusion. The magnitude of this potential barrier (V_0) depends on the impurity concentrations, that is, on the number of donor electrons at the junction that transfer into nearby acceptor levels and is just equal to the required shift of the energy bands needed to ensure that the Fermi level (E_F) remains constant throughout the crystal; the Fermi level is the energy at which there is a 50/50 chance of the corresponding electron energy state or orbit being occupied by an electron. For an intrinsic semiconductor, E_F lies halfway between the valence and conduction bands whereas for an n-type doped semiconductor the Fermi level moves up toward the conduction band and conversely p-type doping lowers the Fermi level.

The potential drop across the **depletion layer** is about 0.6 V for silicon and 0.3 V for germanium, half of the forbidden energy gap. Fig. 6.5 shows how this condition is represented by an energy band model of voltage or potential versus distance from the junction. The width of the junction region ($X_1 + X_2$) increases with increasing voltage (V_0) and decreasing number density of acceptor atoms.

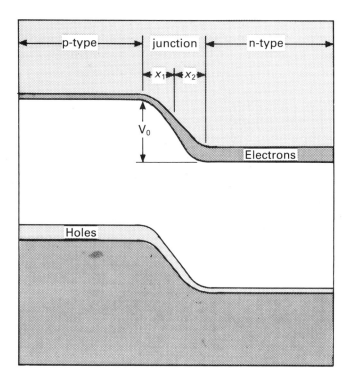

Fig. 6.5. An energy level diagram (voltage versus distance) of a pn junction. Two differently
doped layers of semiconductor silicon in contact form a "diode".

If a positive voltage is applied to the p side of the junction it will tend to counteract or
reduce the built-in potential barrier and attract more electrons across the junction, whereas
a negative voltage on the p side will enhance the internal barrier and increase the width of
the depletion region; these conditions are called "forward" and "reversed" bias respectively. Therefore, on one side of a pn junction there is a region which is more negative and
on the other side there is a region which is more positive than elsewhere in the crystal.
When light of the correct wavelength is absorbed near the junction an electron–hole pair
is created and the potential difference across the junction sweeps the pair apart before they
can recombine. Electrons are drawn toward the region of greatest positive potential buried
in the n-type layer which therefore behaves like a charge storage capacitor. Of course, as
more electrons accumulate the positive potential is progressively weakened.

6.2.2 Charge storage, charge-coupling and clocking
Basically, a CCD is an array or grid (Fig. 6.6) of numerous individual picture elements
(pixels) each one of which can absorb photons of light and utilize the energy to release an
electron within the semiconductor. If we are intent on making an imaging device, then we
do not want the photogenerated electrons to migrate away from the site of impact of the
original photons. To confine the electron within a pixel requires a special electrostatic
field to attract the charged electron to a specific spot. What happens to the next photon?

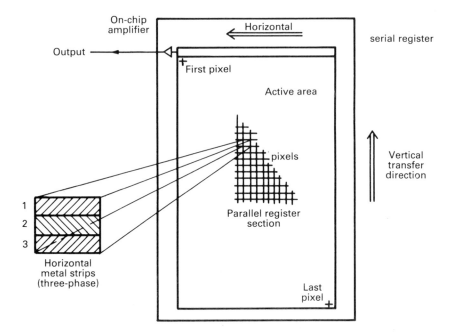

Fig. 6.6. The general layout of a CCD showing numerous square pixels laid out in a grid.

Clearly we need to create a storage region capable of holding many charges. This can be done by applying metal electrodes to the semiconductor silicon together with a thin (0.1 μm) separation layer made from silicon dioxide, which is an electrical insulator (see Fig. 6.7). The resulting structure behaves like a parallel plate capacitor which can therefore store electrical charge. It is called an **MOS** (metal-oxide-semiconductor) structure. An electric field is generated inside the silicon slab by the voltage applied to the metal electrode. If the material is p-type (the usual case) then a positive voltage on the gate will repel the holes which are in the majority and sweep out a region depleted of charge just as in the pn junction. When a photon is absorbed it produces an electron–hole pair but the hole is driven out of the depletion region and the electron is attracted towards the positively charged electrode. The MOS capacitor is the combination of two parallel plate capacitors namely, the oxide capacitor and the silicon depletion region capacitor, and therefore the capacitance is proportional to the area of the plates (electrodes) and inversely proportional to their separation. Since the voltage on the plate can be controlled the depletion width can be increased or decreased, and so the capacity to store charge can also be controlled. The depletion region shown in Fig. 6.7 is an electrostatic "potential well" or "bucket" into which many photogenerated charges can be collected. Typically, the number of electrons stored is just $Q = CV/e$, where e is the charge on the electron (1.6×10^{-19} C), V is the effective voltage and the capacitance C is given by the "parallel-plate" formula $C = A\kappa\varepsilon_0/d$ in which A is the area of the pixel, d is the thickness of the region and κ is the dielectric constant of the SiO$_2$ insulator (~4.5). As the voltage on the electrode increases, the "depth" of the well increases; other ways are needed to create side-walls to the well. Eventually, at a certain "threshold" voltage even the minority charge carriers due to impurities, electrons for a p-type semiconductor, will be drawn to the electrode.

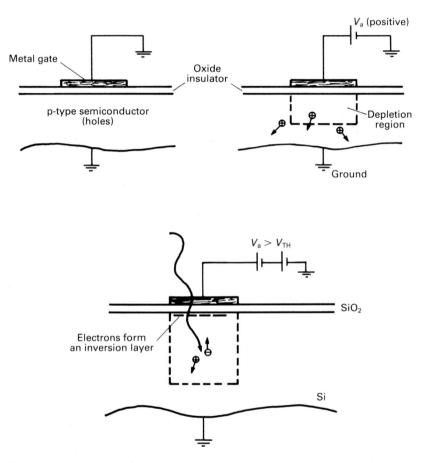

Fig. 6.7. A single metal-oxide-semiconductor (MOS) storage well, the basic element in a CCD.

It was Jerry Kristian and Morley Blouke who first pointed out that taking a picture with a CCD is a bit like measuring the rainfall over a rather large plantation! Suppose you distribute a large number of buckets in a rectangular pattern of rows and columns over a field. After it has stopped raining, measure the amount of water collected in each bucket by shifting the entire array of buckets towards a conveyor belt located at one end of the field, loading the buckets one whole row at a time onto the belt and then allowing the belt to convey each bucket from that row to a metering station where the amount of water in each is recorded. Then transfer another row onto the conveyor belt and keep repeating the sequence until all rows have been transferred to the metering station. By knowing the amount of water in each bucket the pattern of rainfall over the field can be generated. To visualize this pattern as a picture we convert the amount of rainfall in the bucket to a shade of grey—black for no rain, white for a full bucket—and mark this colour on our picture at

SECTION OF COD IMAGE SENSOR

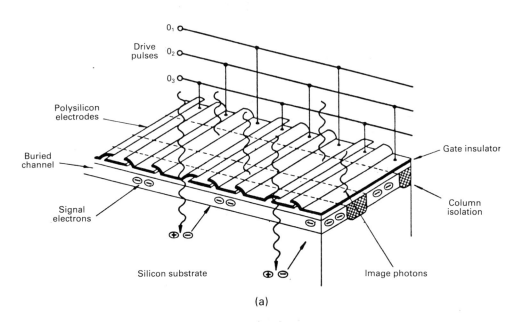

(a)

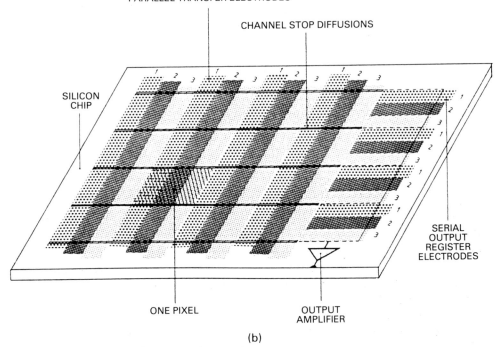

(b)

Fig. 6.8. Two schematic views from above showing the basic layout of a three-phase CCD.

the original location of the bucket. If instead of rain we think of photons of light, and our field of buckets is actually the pixels of the CCD detector in our camera, then we will have reconstructed the visual image of the scene.

Two views of a simple CCD construction called a **three-phase** structure are shown in Fig. 6.8. Semiconductor silicon (in this case p-type) is covered with a thin electrical insulating layer of silicon oxide on top of which are placed three sets of metal electrode strips. One of the three strips is set to a more positive voltage than the other two, and so it is under this one that the depletion region or bucket forms, and where the electrons accumu-

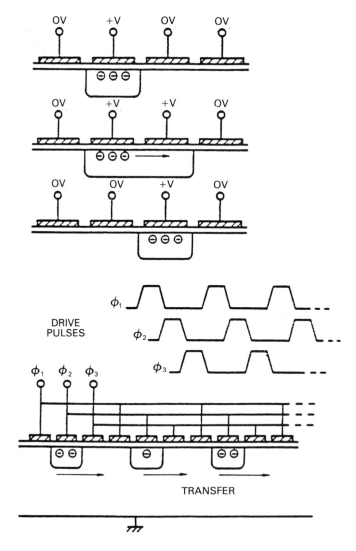

Fig. 6.9. Charge-coupling in a three-phase CCD and the associated timing waveform or clock pattern. In practice the degree of overlap between one electrode and the next depends on the CCD design.

late. We have created two walls of the well. By heavily doping the silicon crystal structure with a certain impurity it is possible to create a narrow "channel" which totally obstructs any movement of charge along the length of the electrode. These **channel stops**, and the triplet of electrodes define the pixel, with "photo-generated" charges being drawn to, and collected under the middle (most positive) electrode, like water in a well.

The unique feature of the CCD, which gives it its name, is the way in which the photo-generated charge, and hence the image of the scene, is extracted from the storage/detection site. It is called "**charge-coupling**".

To transfer charge from under one electrode to the area below an adjacent electrode, raise the voltage on the adjacent electrode to the same value as the first one. This is like lowering the floor of the well. The charges will now flow, like water, and be shared between both buckets (refer to Fig. 6.9). Transfer can be in either direction, and by connecting sets of electrodes together the entire charge stored on the two-dimensional imaging area can be moved simultaneously in that direction. When the voltage on the original electrode is reduced to zero volts the transfer is complete because the collapse of the storage well pushes any remaining charges across to the new electrode. Since it takes three electrodes to define one pixel, then three of the above transfers are required to move the two-dimensional charge pattern by one pixel step along the direction at right angles to the electrode strips. The process of raising and lowering the voltage can be repeated over and over and is known as **clocking**. These drive or clock pulses can be described in a diagram called a **timing waveform** (also shown in Fig. 6.9) and are not difficult to produce electronically.

6.3 CCD CONSTRUCTIONS

6.3.1 Interline and frame-transfer CCDs

Most CCDs are manufactured for use in video cameras and therefore the charge-coupling transfer must be done at very high speed (TV frame rates) and with light falling continuously onto the CCD. To achieve this, several manufacturers adopt an approach called **interline transfer**. In this scheme charges are moved sideways at high speed by one pixel to be relocated in a pixel which is shielded from light by an extra overlying strip of opaque metal. The charges are then coupled lengthways to transfer down the shielded column to the output row of the CCD. Another approach is that of **frame transfer**. In a frame transfer CCD there are no blind spots in the imaging area, instead a duplicate imaging area, contiguous with the first, is covered by an opaque mask. The entire charge distribution representing the scene is shifted very rapidly lengthways until it disappears under the mask. It is then transferred lengthways once more, but at a slower rate, to be finally read out. The transfer rate for readout is a little faster than the permitted time which one can dwell on the image scene, whereas the image-to-storage transfer rate is extremely high to minimize blurring due to the illumination which is still present. Frame transfer CCDs are best suited to astronomical applications, especially since the mask covering the storage area can be removed to yield a CCD of twice the original size! Blurring of the optical scene during readout is not a problem in astronomical applications because a light-tight electronically controlled shutter can be used to block off the incoming illumination.

6.3.2 CCD outputs

So far we have indicated that the charge distribution representing the image scene can be coupled or transferred lengthways by indefinitely sending clocking pulses to the three electrodes. But for a CCD of finite size then all the charges are simply going to pile up at one end!

In practice, of course, the transfer direction is terminated (as already shown in the previous diagrams) by a special row of CCD pixels called an **output register** whose electrodes are arranged at right angles to the main imaging area of the CCD—like the collection conveyor built in the rainfall measuring model—so that it can transfer charge horizontally rather than vertically. Since the output register is a single row it is usually called the **serial register** whereas the main area of the CCD is called the **parallel register**. At the end of the output register is a single output amplifier. Fig. 6.10 shows a photomicrograph of the output corner of a CCD. A complete clocking sequence therefore consists of the following:

(1)   A vertical shift of the entire image scene by one pixel. This delivers a row of charge to the output register.
(2)   A horizontal shift through all the pixels in the output register. This delivers each charge in that row to the output amplifier, one pixel at a time, where the charge can be detected and converted to a voltage which can be measured and recorded.
(3)   Another vertical transfer. This delivers the next row in the image.
(4)   Another horizontal transfer.

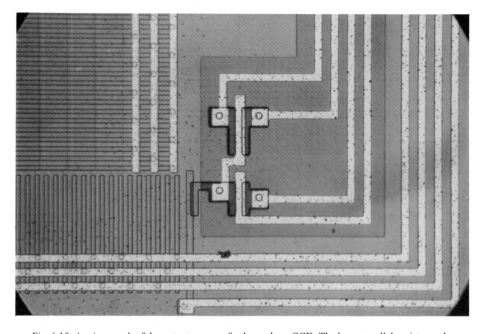

Fig. 6.10. A micrograph of the output corner of a three-phase CCD. The large parallel register and the serial register at right-angles to it, as well as the output amplifier can be seen. Photo courtesy of Pat Waddell.

The above process is repeated until all the rows in the CCD have been delivered to the output register and out to the output amplifier.

In general, the pixels (electrodes) associated with the output register are larger than those in the imaging area to ensure that they have more storage capacity and are therefore much less prone to saturation. We will discuss the effect of saturation later. The output amplifier of the CCD is shown schematically in Fig. 6.11. A packet of electrons with charge Q is allowed through the output gate onto an effective storage capacitance C (essentially a reverse-biased diode) which causes an instantaneous change $V = Q/C$ in the voltage of the input line of the on-chip transistor which in turn yields a voltage change at the output line.

Before the next pulse arrives the storage capacitor is "recharged". The readout is destructive, that is it can only be done once, but this is of little consequence. Unwanted electrical noise due to the periodic resetting of the output diode can feed through to the output transistor and degrade the sensitivity of the CCD to faint signals, but fortunately this type of "reset noise" can be totally eliminated by subsequent electronic signal processing (see Chapter 7).

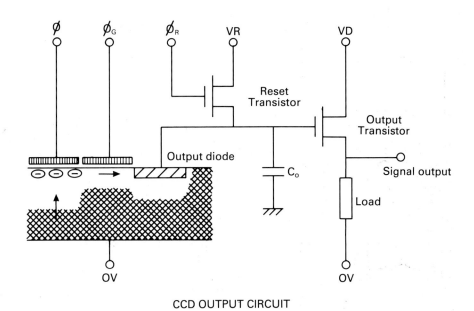

CCD OUTPUT CIRCUIT

Fig. 6.11. Output stage of a CCD showing how charge is extracted in pulses by reading and resetting the output node periodically. Courtesy EEV.

6.3.3 Buried-channel CCDs

While it is true that the simple CCD structures described so far can be constructed, that is metal electrode strips deposited directly onto an insulating layer on the top of a slice of uniformly doped silicon, they are far from ideal. In these CCDs the most positive potential lies at the silicon surface immediately under the insulating oxide layer; electrons are therefore stored and transferred at the surface of the silicon semiconductor and hence they are called **surface channel** CCDs. Unfortunately, the surface layer has many crystal irregularities and defects in the crystal lattice which can readily **trap** charge, but not so readily release it! The result is very poor charge-coupling and severe image smear.

To avoid the surface trapping phenomenon it is advantageous to "grow" another layer of silicon onto the existing p-type substrate to separate it from the insulating layer. If this is a more highly doped n-type layer a more complex depletion region is created with a potential minimum (or collecting layer) which is substantially buried inside the bulk silicon. This is the **buried-channel** CCD. All modern CCDs are of this construction.

The unit cell of a BCCD is shown schematically in Fig. 6.12 together with the distribution of potential. Recall the discussion earlier of the pn junction in which the most positive potential occurred on the n side but within the depletion region. All that has been added on the n-type layer is an insulating oxide and a metal control gate. The most positive potential still lies well away from any surface. If a voltage more positive than any of the gate voltages is applied to this n-type layer, a depletion region will form at the pn junction. An additional depletion under the gate is generated by a difference in gate potentials and the exact depth of the buried channel can therefore be controlled by the electrode potentials as before. Charge transfer within the bulk silicon is very efficient because the number of trapping sites are considerably fewer and much less noisy. The analogy with buckets of water still holds, and charges stored at the embedded collection layer can still be charge-coupled from one pixel to the next by pulsing or clocking the metal electrodes. A penalty is paid in terms of a somewhat reduced storage capacity, i.e. a buried-channel CCD will saturate before an equivalent surface-channel CCD. Incidentally, similar bene-

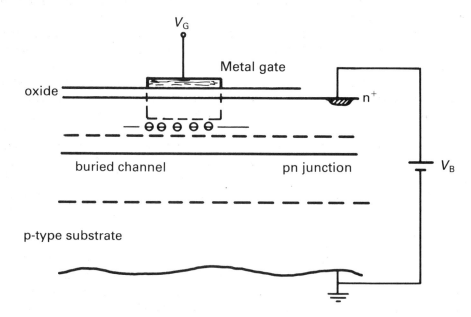

Fig. 6.12. (a) A single storage site in the buried-channel type of CCD. As the gate voltage increases the depleted zones finally meet.

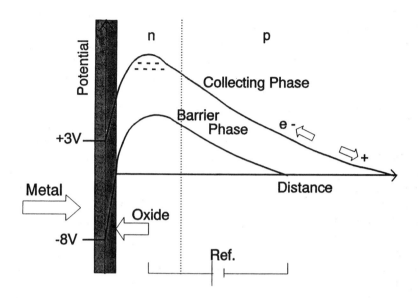

Fig. 6.12. (b) The collection layer lies well below the surface at the overlap between the gate depletion and the depletion of the pn junction. Courtesy Jim Janesick.

fits are accrued if the output amplifier is also constructed in buried-channel form. A full analysis of the buried channel construction using Poisson's equation ($dE_x/dx = -\rho/\kappa\varepsilon_0$) relating the charge density ρ and the electric field E_x can be used to derive the location of the potential minimum (see, for example, Janesick 1994, Rieke 1994).

In summary, therefore, to construct a practical high-performance CCD one starts with the buried channel, a region where electrons are confined in depth (z) below the gate but still free to move from side to side (in x and y); this is usually n-type on a p-type substrate. Next, narrow columns of heavily doped n-type material are diffused into the normal n-type region to produce channel stops which subdivide the x-direction and constrain electrons to move only in the y-direction along columns. A serial register is formed by leaving a gap at one end between the endpoints of all the channel stops and the edge of the CCD. Electrons emerging into this narrow strip could move in x once more. Finally a silicon dioxide insulation layer is grown on top and the basic three-phase electrode gate structure is added as rows or strips across the entire parallel area to define subdivisions (pixels) in the y-direction; the serial register receives similar treatment but the gates are at right-angles to the parallel section.

6.3.4 Two-phase, four-phase and virtual-phase CCDs

There are alternative structures to the three-phase CCD outlined above. If bidirectional charge motion is not required then the CCD can be doped along the direction of vertical transfer in such a way as to provide a barrier to backward-flowing charges. This is a two-phase CCD. The principle is illustrated in the upper panel of Fig. 6.13. An "implant" is diffused into one half of the substrate below each of the two electrodes. The presence of this layer affects the depth of the depletion region immediately beneath it in such a way that the depletion is always greater under the implant. With one electrode low and the other at some fixed voltage level, the result is a staircase-shaped well. When the low electrode is raised to a higher voltage than the fixed electrode, charges are forced to move left-to-right as shown in the figure.

Texas Instruments Corp. developed a single-clock CCD known as the **virtual-phase** CCD (Janesick *et al.* 1981). The principle of this device is very similar to a two-phase CCD as shown in the lower panel of Fig. 6.13. Consider again the two-phase CCD. One electrode was left at a constant voltage to produce an intermediate depletion region. There is really no necessity to do this with a semi-opaque overlying metal gate since the same effect can be achieved by a diffusion directly into the silicon substrate. Only one electrode

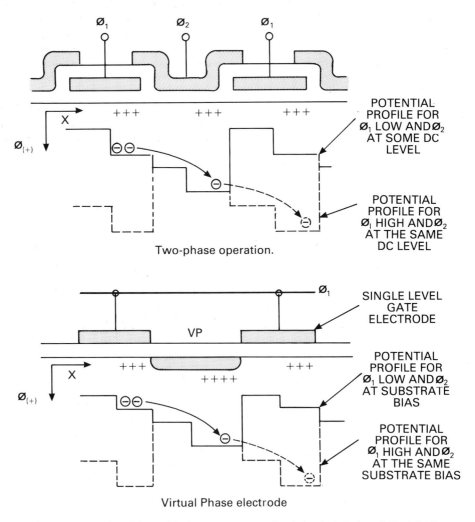

Fig. 6.13. A two-phase CCD and the Texas Instruments virtual- (or single-) phase CCD. Selective doping eliminates the need for the other electrodes by creating different "floor levels" within a pixel.

need be physically present and the other half of the pixel is left clear and uncovered except for its implant. This is the "virtual" electrode or virtual phase. As in the two-phase device, the roles of other electrodes are taken by a series of potential steps created by different levels of surface doping. These potential steps direct the flow of charge as the voltage on the single electrode is driven above and below the potentials of the uncovered or virtual electrode. Apart from a considerable simplification in the number of clocks, this kind of CCD structure was chosen to permit an improved blue response by minimizing the amount of absorption due to the polysilicon (conducting) electrodes when it is frontside illuminated. It is these CCDs, which are used on the Galileo Mission to Jupiter because with implants instead of polysilicon electrodes there is less likelihood of gate shorts caused by ionizing radiation. A few CCDs have also been made with four physical electrodes per pixel which would correspond to the four potential steps apparent in the two-phase and virtual-phase devices. This complexity gives complete control over all combinations of phases and can be used for special applications involving two alternating image scenes—so-called "chopped" signals—or to increase charge storage capacity.

6.3.5 Backside-illuminated CCDs
So far we have been describing CCDs in which the illumination or photon flux was compelled to pass through the overlying electrode structure in order to reach the depletion (or storage) region in the silicon. In practice, this simple approach results in severe absorption of blue light in the electrodes; in fact, the device has almost no blue response at all!

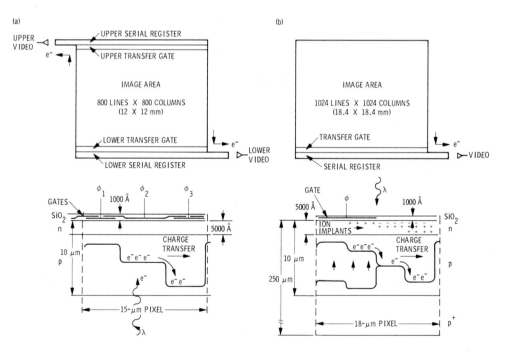

Fig. 6.14. Detailed comparison of the structure and dimensions of two CCDs from Texas Instruments. (a) A thinned and backside-illuminated three-phase CCD, and (b) the thick, front-illuminated virtual-phase device. Courtesy Jim Janesick.

Virtual-phase CCDs are one approach to solving this problem. Alternatively, the CCD can be turned over and illuminated from the back side! Before this becomes effective, however, the thick silicon substrate must be reduced in thickness (either mechanically or chemically) to only 10 μm or so. Fig. 6.14 compares both constructions. Such thinned, backside-illuminated CCDs have excellent response in the blue and violet. If thinned too much they lose their red response because the red photons need more absorption length and if this is not there, they will pass right through the silicon!

One disadvantage of the thinned CCDs is that they are more mechanically fragile and prone to warping. Some thinned CCDs are mounted to a supporting substrate. Also, inter-ference **fringing** can occur due to multiple reflections internal to the CCD substrate or between the silicon and the supporting substrate in a manner similar to fringes which can be seen when two flat glass plates are placed almost in contact at a small angle. For imaging, this problem can be eliminated by computer processing and calibration techniques discussed in more detail later, but it is a more serious drawback for spectroscopy.

6.3.6 MPP CCDs

Several years ago it was realized that the dominant contributor to the dark current in CCDs was thermal generation due to surface "states" at the Si–SiO$_2$ interface. In fact, surface dark current is 10^2–10^3 greater than dark current generated by the bulk of the CCD. Two factors control the dark current at the silicon–silicon dioxide "interface", the density of interface states and the density of free carriers (holes or electrons) that populate the inter-face. Electrons that thermally "hop" from the valence band to an interface state and then to the conduction band will produce an electron–hole pair that will be collected in the potential well of the CCD pixel. Free carriers could "fill" the interface states and inhibit the hopping mechanism and hence drastically reduce the dark current, but standard opera-tion of CCDs does just the opposite. In a CCD where the voltage on the gate electrode is such that the surface potential at the Si–SiO$_2$ interface is greater than the silicon substrate potential then the resulting depletion drives away all the free carriers and therefore *maxi-mizes* dark current from surface states. Dark current is now controlled solely by the density of interface states and is thus dependent on fabrication processes. If, however, the CCD can be operated in **inversion** mode then holes from the channel stop regions migrate to populate the interface states below the inverted gate and thereby eliminate surface dark current (by filling the hopping sites). If all three phases of a three-phase CCD are driven into inversion (i.e. all gate electrodes set very negative relative to the substrate) then there is no potential well in which to collect charges. Partial inversion can be accomplished by biasing a collecting phase higher than the substrate (say $\phi 1 = 3$ V) and biasing the two barrier phases into inversion ($\phi 2 = \phi 3 = -8$ V). Inverting two barrier phases reduces the dark current generation by 2/3. **Multi-phase-pinned (MPP)** CCDs are designed in a spe-cial way to allow operation in a totally inverted mode. Clearly, to obtain any charge stor-age capacity while totally inverted the potential of one or more phases must be offset from the others. For a three-phase CCD this can be done by doping the silicon beneath phase 3 with boron. In effect, the boron implant cancels the effect of the phosphorus doping used in the buried channel and reduces the potential in the region. When biasing all phases into inversion, charge will collect under phases 1 and 2 while phase 3 will now act as the barrier phase. As the clocks are driven negative, phase 3 will attain inversion before phases 1 and 2. Three-phase MPP CCDs could also be fabricated by implanting more phosphorus under phase 3 forcing charge to collect there and resulting in phases 1 and 2 becoming the barriers. Loral and SITe both use the boron implant approach to make MPP CCDs. By offsetting the positive clock level of phase 3 by the amount of the built in MPP potential, the CCD can be operated only partially inverted (to gain back some well depth). The well-depth of an MPP CCD is determined by the potential which is built in by the implant and can thus be 2–3 times lower than the partially inverted case.

One of the first demonstrations of the power of MPP technology was an 8-minute exposure at room temperature by a Loral 1024 × 1024 CCD with tiny 15-μm pixels before saturating on dark current. Any non-MPP CCD would saturate almost immediately (< 1 s) on dark current at room temperature. Needless to say, almost all CCD manufacturers now fabricate MPP CCDs and it is the availability of MPP technology that has really been responsible for opening up the CCD camera market to amateur astronomers and to a whole host of applications.

Care must be exercised with CCDs using MPP architecture if the ultimate in dark current performance is required. If the CCD becomes saturated because of very substantial light overload or because the power to the CCD was turned off while the device was cold, then trapped charges will raise the dark current (by a factor of 2–3 typically) for hours until it gradually decays back to its normal low state.

There are other pitfalls too. As explained above, when a CCD phase is inverted, holes from the channel-stops migrate and collect beneath the inverted gate thereby pinning the surface to substrate potential. Unfortunately, some of these holes become trapped at the $Si–SiO_2$ interface and when the clock is switched back to the non-inverted state to transfer charge, these trapped holes are accelerated out of the interface with sufficient energy in some cases to cause impact ionization which results in a "spurious charge". This is potentially quite serious in large arrays because spurious charge is produced during each pixel transfer and has a shot noise behaviour which can overwhelm the on-chip amplifier noise and dominate the observed readout noise of the chip. As an example, suppose on average that one electron of spurious charge is produced for 10 pixel transfers. After 1024 transfers this process would have resulted in about 102 electrons and an associated noise of 10 electrons rms, whereas the on-chip amplifier might be capable of achieving 3–7 electrons rms of read noise. To overcome the shot noise produced by spurious charge production, Jim Janesick and other investigators suggest three strategies. First, slow down the rise time of the drive clocks (falling edges are not important in this context) by adding an RC network at the output of the clock driver board to allow the holes to return to the channel stops slowly. Limit the clock voltage swing to the smallest value possible consistent with good charge transfer efficiency and hence reduce the driving electric fields and the acceleration of the charges; less spurious charge will be generated. Finally, a "tri-state" clocking scheme can be tried in which an intermediate clock level is established which is just above the inverted state but not enough for complete charge transfer. A slow transition from the inverted phase to the intermediate phase followed later by the complete swing allows the trapped holes to be released slowly from the oxide. This approach was employed by the JPL team on the 800 × 800 pixel virtual-phase CCDs on the Galileo probe to Jupiter.

The vertical registers of an MPP CCD are always inverted to produce low dark current, and the suppression of spurious charge in these registers is helped by the high capacitive load to the imaging pixels which slows down the clock edges. On the other hand, the horizontal register is clocked much more rapidly and would produce significant amounts of charge if it was being continuously brought in and out of inversion. Consequently, the horizontal registers of an MPP CCD do not receive the MPP implant. Unfortunately this is not the case for virtual phase (VP) CCDs which require that the horizontal register be switched in and out of inversion. Spurious charge produced by this effect actually limited the system noise on the Galileo chips to three times the intrinsic noise of the on-chip amplifier—even with tri-state clocking. Recall that the big advantage of VPCCDs, apart from simplicity of clocking, is the fact that they have significant blue response for a front-illuminated chip because of the absence of a polysilicon gate over the virtual phase.

MPP technology offers benefits other than dark current. For example, surface residual image charge and pixel non-uniformity are improved, and anti-blooming features can be employed. Pinning also increases the CCDs tolerance to high-energy radiation and allows the CCD to be erased rapidly because trapped charges are neutralized by the preponderance of holes in the inverted phase. The anti-blooming mode works as follows. During integration, suppose $\phi 3$ remains inverted at all times to form the barrier phase. Photogenerated charges therefore build up under $\phi 1$ and $\phi 2$. Phases 1 and 2, however, are slowly *switched* between the inverted state and just above the optimum full well voltage (typically -8 V to $+5$ V according to Jim Janesick for Loral, Lincoln Labs and SITe CCDs). The required switching rate obviously depends on the rate of charge build-up, but about 50 Hz is typical. What happens is that one of the two phases reaches saturation and charges begin to get trapped in the $Si-SiO_2$ interface. At this point the phases are switched. The holes generated by the phase going into inversion neutralizes the trapped electrons. Simultaneously, charge builds up under the other phase starting at surface full well (because of the transfer) and again electrons begin to enter the oxide and get trapped. Phases 1 and 2 are switched back, and the inversion eliminates the trapped charge under the second electrode. This process of back-and-forth switching continues during the entire integration period and inhibits the saturating pixel from blooming and bleeding charge up and down the column. In summary, the benefits provided by MPP operation outweigh the disadvantages of reduced full-well capacity, especially for applications where the exposure times are long compared to the readout times such as in low-light-level spectroscopy. This mode is also extremely attractive for amateur astronomers who have no desire whatsoever to keep a large canister of liquid nitrogen handy! Simple Peltier cooled cameras with MPP CCDs can provide excellent dark current performance.

6.4 ASTRONOMICAL REQUIREMENTS

6.4.1 Slow-scanning, cooling and optimization

For astronomy applications, CCDs cannot be used like normal TV cameras taking frames every 1/30th of a second. Instead, the CCD must be used as if it were a photographic emulsion in a camera; in other words we need to take long exposures to build up a charge image from a faint source. Even when the charge image is removed during the readout process, we cannot do that rapidly either because the charge transfer efficiency will be impaired and the electronic noise is greater at higher readout frequencies. In fact, compared to TV frame rates, astronomical CCDs must be read out very slowly and hence this mode is called **slow scan**. Typically, rates of about 50 000 pixels per second (50 kHz) are used which implies that it will take 20 seconds to read out an array with one million pixels. Clearly, the CCD must not be exposed to light during this time period otherwise there would be considerable smear. Although this readout time may seem long, it must be remembered that the exposures are typically much longer than this—1000 seconds or more.

With such long exposures, however, the second problem arises—dark current. To permit long exposures, astronomical CCDs must be cooled to temperatures well below the freezing point of water, which implies using a vacuum chamber to avoid frosting. The advent of MPP CCDs has helped considerably in reducing the cooling requirements and many applications can be met with simple thermoelectric (air- or water-chilled) coolers—which is great for amateur astronomers! For the most stringent applications in astronomical spectrographs, more cooling is required and most CCD cameras at professional observatories use modified liquid nitrogen cooling systems.

Finally, the performance of a CCD depends on how it is operated, and while modern CCDs are much more uniform in behaviour than earlier generations, there is still a certain amount of "optimization" that is required in terms of finding the very best clock voltages, bias levels and currents.

6.4.2 CCDs for amateur astronomers

Complete CCD systems packaged in small units ready for attaching to telescopes of apertures in the 10-inch range have flooded onto the market and opened up a whole new world of opportunity for amateur astronomers, and indeed for teaching uses at schools, colleges

Fig. 6.15. An image of the Horsehead Nebula in Orion obtained with a Photometrics CCD camera on a 6-inch Celestron telescope. Courtesy Richard Aikens, Photometrics.

and universities everywhere. A wide variety of cameras are available with small CCDs of only 192 × 165 such as the TC211, up to those with 2048 × 2048 pixels like the Kodak KAF4200. Some companies sell only the camera heads and electronics, others provide PC-based image processing software ready to support popular cameras, and a few companies provide complete turn-key packages containing camera, computer and software. Fig. 6.15 is an impressive 30-second exposure with a CCD camera attached to a 6-inch Celestron telescope. Prices are exceptionally good and likely to get better. Several suppliers are listed in Appendix 7.

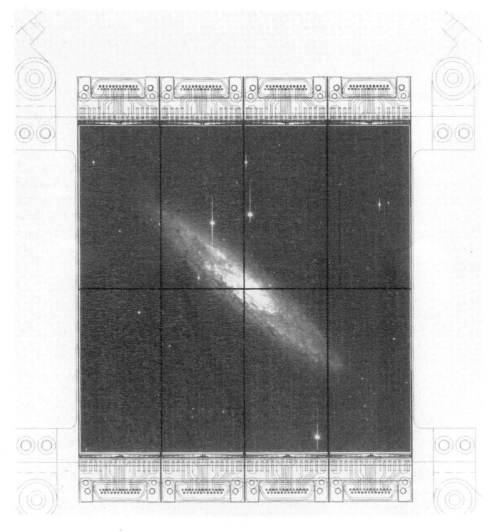

Fig. 6.16. A real image of the giant spiral galaxy NGC253 superimposed on a drawing of the 2 × 4 CCD mosaic (8192 × 8192 pixel) which obtained it. The camera was developed by the University of Hawaii and the image was obtained using the CFHT. Courtesy Gerry Luppino and Mark Metzer.

6.4.3 CCD mosaics

The search for larger and larger formats continues. While waiting on technology to catch up with astronomer's vision, the general approach has been to construct "mosaics" of CCDs (Fig. 6.16) which have been specially manufactured to ensure that one, two or even three sides have no connections and can therefore be butted very closely to one another. For example, Tony Tyson at Bell Labs has developed a 2×2 mosaic of thick Tek/SITe 2048×2048 CCDs with 24-μm pixels which yields $0.5° \times 0.5°$ on the 4-m telescope at Kitt Peak, the Kiso Mosaic Camera is 7×4 thick TI 1000×1018 CCDs with 12-μm pixels, and the University of Hawaii mosaic camera developed by Gerry Luppino's group is a 4×2 array of thick Loral 2048×4096 chips with 15-μm pixels. There are similar projects at ESO, NOAO, CFHT, RGO, Keck and a host of other places. At the time of writing, the largest mosaic is the one anticipated for the Sloan Digital Sky Survey which is a 5×6 mosaic of 30 CCDs, each with 2048×2048 pixels.

SUMMARY

Light can be absorbed and converted to electrical charge in a semiconductor using the internal photoelectric effect. The basic structure of a CCD is a two-dimensional grid of metal-oxide-semiconductor pixels controlled by overlying metallic electrodes arranged in strips. Charges accumulate where the light falls on the CCD and builds up an image. The pattern of charge can be "read out" by systematically pulsing or "clocking" the electrode strips to cause the charge pattern to "couple" from one pixel to the next and move along the columns of the CCD to an output register. Most modern CCDs are of the buried-channel construction. CCDs can be front-illuminated, or thinned and backside-illuminated. For astronomy applications, CCDs must be cooled and operated in "slow scan" readout mode.

EXERCISES

1. Describe with the aid of diagrams a three-phase surface channel CCD and show how photogenerated charges are collected, stored and transferred to the output.
2. Describe with the aid of a diagram what is meant by a "virtual-phase" CCD.
3. Draw a diagram of a buried-channel CCD. Cite its advantages over a surface-channel CCD.
4. Explain what is meant by "inverted" operation. Why is inverted operation of a CCD advantageous?
5. Explain the terms "conduction band" and "valence band", and describe what is meant by an "electron–hole" pair.
6. Calculate the wavelength of the lowest energy photon that can be detected by a germanium photodetector. Assume the band-gap for germanium is 0.67 eV.
7. Compare and contrast "interline" and "frame transfer" CCDs for astronomy applications. Why is the interline transfer approach attractive for standard TV and video rate applications?
8. What is meant by a "thinned" CCD? What are the advantages and disadvantages of thinned CCDs?

9.   What are the astronomical requirements for CCDs? Explain the terms "slow scan",
      "dark current" and "noise".
10.   What is meant by Multi-Phase-Pinned or MPP?

REFERENCES AND SUGGESTIONS FOR FURTHER READING

Amelio, G.F. (1974) Charge-coupled devices, *Scientific American*, February.

Benyon, J.D.E. and Lamb, D.R., (1980) *Charge-coupled Devices and Their Applications*,
    McGraw-Hill, New York.

Boyle, W.S. and Smith, G.E. (1971) Charge-coupled devices—a new approach to MIS
    device structures, *IEEE Spectrum,* **8**, No. 7, 18–27.

Janesick, J. and Blouke, M. (1987) Sky on a chip: the fabulous CCD, *Sky and Telescope*,
    September, **74**, 238–242.

Janesick, J. R., Hynecek, J. and Blouke, M. M. (1981) A virtual phase imager for Galileo,
    in Solid State Imagers for Astronomy, *Proc. Soc. Photo-Opt. Instr. Eng.* (SPIE), Vol.
    290, pp. 165–173.

Janesick, J.R. (1994) SPIE Lecture Series.

Kristian, J. and Blouke, M. (1982) Charge-coupled devices in astronomy, *Scientific American*,
    October, **247**, 66–74.

Mackay, C.D. (1986) Charge-coupled devices in astronomy, *Annual Reviews of Astronomy
    and Astrophysics*, **24**, 255–283.

Mackay, C.D. (1992) *The Role of Charge Coupled Devices in Low Light Level Imaging*,
    Published by AstroMed Press & Publications Department, Cambridge, England.

McLean, I.S. (1988) Infrared astronomy's new image, *Sky and Telescope*, March, **75**,
    254–258.

McLean, I.S. (1989) *Electronic and Computer-aided Astronomy; from Eyes to Electronic
    Sensors*, Ellis Horwood, Chichester, England.

Rieke, G.H. (1994) *Detection of Light from the Ultraviolet to the Submillimeter*, Cambridge University Press, Cambridge, UK.

Sequin, C.H. and Tompsett, M.F. (1975) *Charge Transfer Devices*, Academic Press, New
    York.

Troy, C.T. (1996) "How camera makers cope with a confused CCD market", *Photonics
    Spectra*, **30**, No. 2, Laurin Publishing Co., Pittsfield, MA, USA.

7

Practical operation of CCDs

Modern CCDs are fairly uniform and predictable in their operation and characteristics, but there are still many subtleties to successful operation, especially for the latest large-format devices customized for astronomy. Issues include maximizing the ratio of signal to noise, dealing with the peculiarities of individual chips, obtaining stability and repeatability in performance, and finding suitable methods of control and analysis. These problems are treated by recounting some of the developments which uncovered a host of effects and led to their correction. There is no intention here to provide a "constructors manual", only to alert the potential user to a host of practical issues.

7.1 CLOCK VOLTAGES AND BASIC ELECTRICAL FACTORS

CCD manufacturers provide a "data sheet" which gives the electrical pin connection diagram for the device (Fig. 7.1), the names and symbols for each pin, the voltages or range of voltages to be applied to each pin and the **timing diagram**, that is, a diagram showing the time sequence of the CCD drive signals and the relationship between them. Terminology varies, but certain basic functions are common to all. Voltages applied to CCDs are of two types. Fixed voltages, referred to as **dc bias** levels, which remain unchanged after switch-on, and pulsed or **clock** voltages which can be switched back and forth between two voltage levels known as the high and low levels. As described in Chapter 6, clock voltages are applied in a precise order and time sequence to charge-couple the electrons in one storage well to the next. Although the type of mounting package used (its size, number of pins and their names) differs from device to device, certain functions are required in all CCDs, and in particular in the frame transfer image sensors used in astronomy, namely:

(1) **Serial (horizontal) register clocks.** One pin for each phase or electrode used to define the pixels in the horizontal register is required, i.e. three for a three-phase CCD. In some larger CCDs such as the SITe (formerly Tektronix) devices, two completely separate serial output registers are provided.
(2) **Parallel (vertical) register clocks.** Again one voltage line is needed for each phase. In frame transfer CCDs the two-dimensional vertical or parallel register is split into

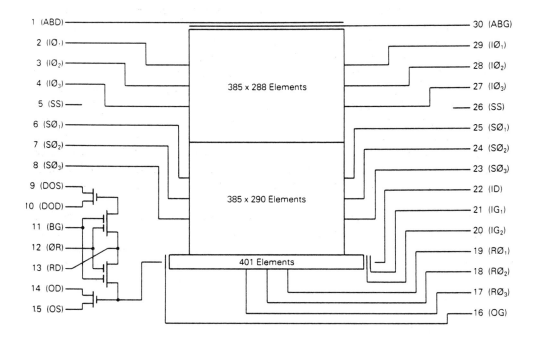

Fig. 7.1. The device schematic for a small frame transfer CCD to illustrate typical pin connections and terminology. Each manufacturer will provide information of this kind on a data sheet. Courtesy of EEV.

two identical sections—the "image" section and the "store" section—which can be controlled separately and therefore two sets of pins are needed.

(3)   **Reset transistor clock.** A single, periodically recurring voltage pulse is required to reset the CCD output amplifier or more accurately, the output charge collecting capacitor during the readout process. Each of these clock voltages will have a specified high and low level and, since the voltage difference or "swing"—usually in the range 5 to 10 volts—can have important consequences on performance, it is often arranged to select the levels on demand from the CCD controller. The total range of voltage required to operate a CCD is generally less than 20 volts.

The most important dc bias voltages are as follows:

(1)   **Substrate voltage** (V_{sub}). The reference for all other voltages. This voltage is usually, but not always, kept at ground or zero volts.

(2)   **Reset drain voltage** (V_{RD}). This voltage is applied to the "drain" terminal of the on-chip "reset" field-effect-transistor (FET) at the CCD output to establish the level to which the output node (capacitor) must return after each charge packet is read out.

(3)   **Output drain voltage** (V_{OD}). This voltage applied to the drain terminal of the on-chip output amplifier determines the operating point of that transistor.

(4)   **Output gate voltage** (V_{OG}). The output gate is essentially an extra (last) electrode in the serial output register.

Pulsed voltage signals, corresponding to individual charge packets, emerge from the source of the output transistor (OS)—also called the video output—which is connected to ground via another transistor external to the CCD to provide a constant current load. The output transistor source current (I_{OS}) is important for most CCDs.

Some CCDs have additional options. For example, the vertical register can be clocked up or down to independent serial registers. One serial register may terminate with an on-chip amplifier designed for low-noise slow-scan operation whereas the other register might have an output amplifier optimized for TV video rates (see Fig. 7.2 for an example). Some chips have a separately clocked gate known as a "summing well" which has the storage capacity of two serial pixels. Almost all CCDs have electrical input connections and test points; the manufacturer will specify whether these pins should be fixed at a high or low voltage. Remember also that voltages are relative to the substrate voltage which may not always be zero volts.

Normally, a CCD manufacturer will provide an initial set of operating voltages for a particular chip, but will not necessarily optimize these voltages for cooled slow-scan astronomical work. Builders of commercial astronomical CCD camera systems can provide this service. For some CCDs, small changes of one-tenth of a volt to clock swings or dc bias values can often yield substantial improvements in low-light-level behaviour. A compendium of useful voltages is included in Appendix 6.

7.1.1 The analogue signal chain

The analogue signal chain includes the preamplifier, post-amplifier, noise removal circuits and the analogue-to-digital converter. In addition, low noise dc power supplies are required to provide the bias voltages and a "level shifter" circuit is usually needed to convert simple 0–5 V TTL clock pulses to the levels required for the chip. Fig. 7.3 shows a typical signal chain after the preamplifier.

Mechanically, CCDs are quite robust unless heavily thinned. Jim Janesick recalled for me a horrific moment when during the course of handling one of the early 800×800 CCDs at JPL a 6/32 screw somehow rolled off a shelf above the workbench on a trajectory aimed squarely at the centre of the chip! He saw it out of the corner of his eye, and then all time seemed to stand still; an eternity passed in the blink of an eye as he saw the thinned surface bow under the impact and then give way. Thinned CCDs may not be able to survive a collision with a tiny screw but they can withstand substantial illumination overloads without permanent damage. However, they are integrated circuits of the CMOS type and their tiny gate connections can be short-circuited by static electricity discharges. Precautions must be taken when handling CCDs. For example, CCDs should be stored in electrically conducting containers, earthing-straps should be tied to the wrist during handling operations, no nylon clothing worn, and the work performed in a clean, ionized (electrically conducting) airflow. Finally, some kind of protection on the drive signals is essential if power supplies are used which have a rating even slightly above the maximum recommended voltages for the CCD. Usually this is achieved by the use of Zener diodes on the drive outputs. With these precautions, a CCD camera can last a long time. Most users of CCDs accept that individual devices may have imperfections that must be dealt with, but many are surprised by the significant effect that the CCD electronic hardware systems used at telescopes can have on data collection.

512 × 512 PIN-OUT

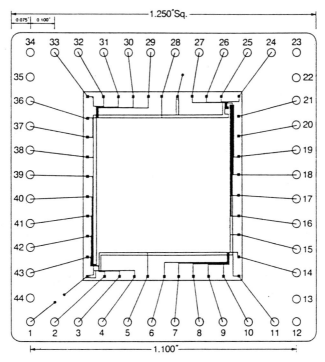

Note: Pin 1, backside contact & package body.

Functional Diagram

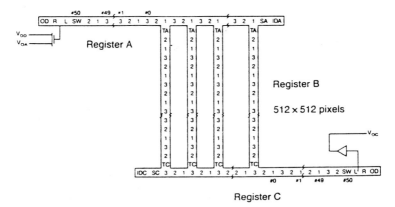

Register C

Fig. 7.2. The pin connection and functional diagram for a Tektronix/SITe 512 × 512 CCD showing additional features such as the double horizontal (serial) readout registers, one for TV rates the other for cooled slow-scan. Courtesy of Morley Blouke.

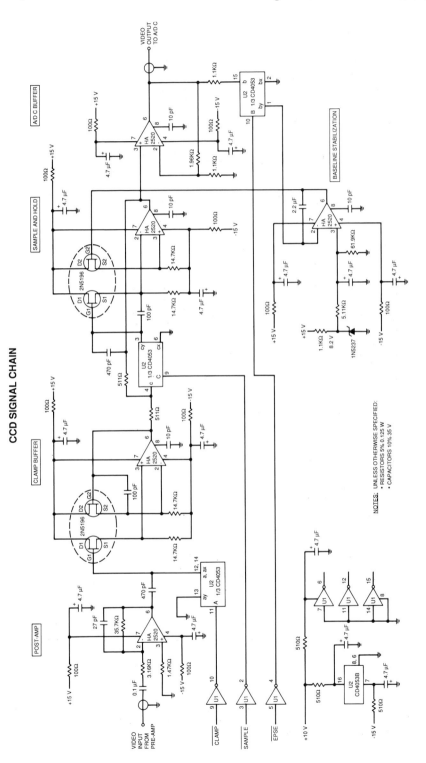

Fig. 7.3. Part of the analogue signal chain. A practical correlated double sampling circuit used by Jim Janesick at JPL. Courtesy Jim Janesick.

The heart of the electronic system is a signal processing unit designed to sample and filter noise, but proper grounding, power supply de-coupling and optimum control timing are also very important factors in achieving a low-noise system. Undoubtedly, "ground-loops" are the most common cause of noisy CCD systems. A ground-loop forms when two interconnecting parts of an electronic system are separately connected to ground via small but different impedance paths. As a result, a voltage difference can exist between "grounds", and currents can flow. Ground-loops between the telescope and camera body and driver electronics, within the driver electronics, or between the driver electronics and the computer system can cause interference with the readout electronics, giving clearly visible sloping, "stripe" patterns in the CCD images; these patterns are synchronized with the mains frequency. The solution to this problem is to have only a single ground point in the system, to which all the zero reference points and shields are connected; because many wires may radiate from such a single ground point it is often called a "star ground". All connections to the single system ground should be made as short as possible (< 1 m for frequencies up to 10 MHz), with the lowest-resistance electrical wires available. Wherever possible, circuit boards should use copper ground planes. Also, many designers carefully isolate the entire instrument from the telescope structure—even if the telescope is known to have an excellent earth ground—which means that extreme care is required when attaching any other piece of electrically-powered apparatus to the instrument in case a ground-loop is formed.

Similarly, electrical noise from motors, light dimmers, or computer parts can be "picked up"—by capacitance coupling, inductive coupling or radiative coupling—if inadequately shielded wires or components are used in the CCD system. Remember, the CCD system is capable of detecting signals of only a few millionths of 1 volt! Coaxial cables which have a surrounding braided copper shield are quite effective (90%), and "twisted-pair" wire is often used in preference to a single wire for carrying critical signals to the CCD. Signals transmitted over long distances, such as from telescope to control room, usually use optical fibres if possible. Most astronomical CCD systems have a low-noise preamplifier inside or very close to the cryostat to boost the CCD signal and most systems convert from the weak analogue output to healthy digital signals before transmitting the data over long cables. The physical environment at cold, high-altitude mountain-top observatories can cause "drifts" in the operating points of many electronic components; this effect can be overcome by careful component selection and packaging.

7.1.2 CCD sequencers and clock drivers

Much of the complexity in a CCD camera, or other electronic array detector system, revolves around the critically important timing diagram (Fig. 7.4). Numerous events must occur in strict time order and at precise time intervals. For example, in a typical CCD system the following sequence of events must occur in time sequence:

- check instruction status from operator
- prepare the CCD—unrecorded readouts to flush the CCD
- open the shutter for a timed period
- close the shutter
- read out the CCD according to a precise pattern

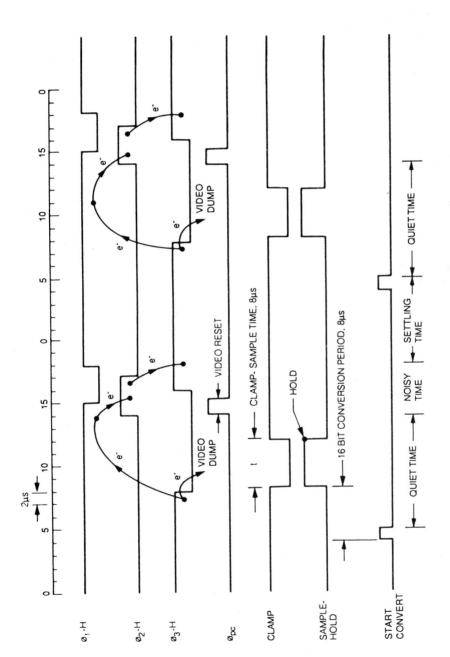

Fig. 7.4. A typical timing diagram (associated with the analogue chain shown in the previous figure). The waveforms are produced digitally (usually as 0–5 V TTL signals) and then converted to the required voltages by a "level shifter" circuit. Courtesy Jim Janesick.

- digitize the signal from each pixel
- store the data in a computer
- return the CCD to a "standby" mode, if appropriate

That part of the system responsible for such timing activities is often called the **sequencer**. Almost all CCD systems fall into one of the following design categories, either "hardwired" or "programmable".

Hardwired designs

By "hardwired" we mean that the electronic functions are carried out by circuitry, and cannot be altered by typing instructions on a computer keyboard. This approach results in a very compact and rugged system, and is favoured by those with travelling systems or with severe constraints on space and/or weight, such as small telescopes. Obviously, the hardwired approach is more restrictive in the sense that the circuitry has to be altered, perhaps by re-soldering wires, in order to effect a change, such as going to another type of CCD, or to an alternative clocking sequence.

In a typical hardwired CCD camera, the digital controller for the camera is built with commercially available logic chips such as TTL or low-power high performance CMOS. Readout of the CCD is in a fixed form, and all clocks and signal processing pulses are generated on a single board, usually called the "clock logic" board. This circuit board derives its time-base from a small device known as a quartz crystal oscillator. The clock logic board in turn feeds both the CCD clock "driver" board which has more current handling capacity than logic chips for pulsing the CCD electrodes, and the "analogue processor and digitization" board. A fourth board, containing several low-noise well-regulated dc power supplies for setting the fixed voltage levels on the CCD, as well as providing auxiliary functions such as temperature control, completes the so-called "camera head" electronics (Fig. 7.5). The whole four-board controller could fit into a small box of typically $13 \times 15 \times 20$ cm. Many of the pioneering astronomical camera systems were of this type and many amateur astronomers get lots of fun out of the do-it-yourself approach. Several small companies advertising in astronomy magazines can provide "kits" for enthusiasts to build hardwired systems for specific CCDs and small telescopes.

One example of a hardwired system is that developed originally by Jim Gunn of Princeton and Jim Westphal of CalTech. The clock logic is all CMOS except for the 50-ohm line drivers and receivers on the four coaxial cables providing communication between camera and data-recording system. A master clock running at 2.5 MHz is divided down to 1.25 MHz to give an 800 nanosecond control cycle period. CCD clock waveforms are generated by a counter-divider chain feeding a set of electronic latches which allows "wire-wrapped" programming. The clock generators are housed with the bias regulators, video signal chain, A/D, and parallel-to-serial converter and line driver circuitry in two "saddle-bags" clamped to the LN_2 cryostat containing the CCD. One external signal, a line start command, generates the following sequence: (a) the serial clocks (normally running) are interrupted, (b) a vertical transfer sequence is executed, and (c) the serial clocks started again. The first line start triggers a continuous frame signal for clock voltage control. Clocking signals go to a driver board inside the dewar.

Another variation on the "hardwired" approach is to allow the clock voltage levels to be configured in "digital" form with the values being selectable by tiny thumb wheel

Fig. 7.5. A typical camera "head" electronics box and CCD controller.

switches or from front-panel knobs attached to potentiometers. D/A converters—the op-posite of A/Ds—can be used to give an analogue output which goes to the chip through high power amplifiers that can cope with the rate-of-change through potentially large volt-age swings.

Programmable designs
By "programmable" we mean that the electronic functions can be changed by computer commands. This capability becomes very important when many different kinds of CCDs are to be used, tested or developed. There are a great many ways of providing pro-grammable systems, especially since the advent of digital signal processor (or DSP) chips and transputers (see Ch.5).

One of the earliest sequencers of this type was based on the 2901 "bit-slice" micropro-cessor, essentially a specialized microchip capable of accepting "instructions" in binary digit (bits) form and doing some arithmetic logic but without all of the memory manage-ment frills of a full microprocessor. The micro-instruction was 64 bits long, of which 32 bits were used to control the 2901 and the remaining 32 bits are available to control the CCD clocks, the signal processor and other units with which the camera must communicate. The key feature of the bit-slice approach was very precise timing which was not a function of the particular set of micro-instructions called. Timing and control signals for the CCD are generated by sequentially stepping through the micro-program instruction set, which itself is contained in eight Programmable Read Only Memories (PROMs) of 8 bits by 256 instructions. This approach, developed at the Kitt Peak National Observatory, became the basis of the early Photometrics camera systems when Richard Aikens left the observatory to start his own company.

If the microprogram is stored in PROM, the system can be operated without the main data-gathering computer and the program will always be available. There is the disadvantage that one must replace the PROMs with new ones to effect a timing or sequence change. This process is called "burning" PROMs and requires special, although very compact, equipment and a computer to develop the new PROM codes on. Good-quality Erasable PROMs (EPROMs) are available which at least save on the purchase of these microchips. PROMs provide a simple and reliable solution when the camera is fully commissioned and performing in a scientific instrument. For development engineers, however, it can be much faster and more convenient to replace the PROMs with Random Access Memory (RAM) chips. These devices must have the micro-program sent to them from the main computer, which has overall control, each time the sequencer is switched on; this process is known as "down-loading". Several different programs can be stored in the host computer.

A practical approach to CCD camera design requires room for expansion and change as new devices come along. It therefore makes sense to use a "modular" design. For example, one might design a CCD camera system to allow the following range of configurations

(a)   slow or fast readout
(b)   on-chip pixel binning (two axes)
(c)   digitization of only a subset of the image area
(d)   bidirectional charge shifting
(e)   overscanning

These modes can be built-up from a small set of primitive waveforms. The longest of these involves the digitization of a pixel value which typically takes 20–50 microseconds due to the integrating correlated double sampling method described later. Also, the operational modes require groups of consecutive sequences of the various waveforms. For instance, suppose the waveform sequencer is based on a Programmable Read-Only Memory (PROM) designed to accept 16-bit commands with two-parts from the microprocessor. The first part specifies which waveform is required while the second part specifies the number of waveforms to be generated, that is, a "repeat" count. To smooth the flow of commands and eliminate timing problems a FIFO (first-in-first-out) buffer can be placed in the 16-bit data path between the microprocessor and the waveform sequencer. This method provides a clean split between the hardware and the software with the microprocessor responsible for breaking down high-level commands, e.g. "take an image", into a stream of sequencer-compatible 16-bit commands. The PROM based nature of the sequencer, implies that it can be easily updated to provide any desired format of driving/readout waveform. Many controllers of this general design were developed by observatories all over the world when 8-bit microprocessors became available.

Many programmable controllers now incorporate powerful 32-bit processors, such as DSPs and transputers, which can provide significant intelligence and computing power within the controller itself, and yet are easily programmable with all the desired waveforms. The Motorola 56000 series DSPs are used in the Leach controller developed by Bob Leach (San Diego State University) and the Inmos T805 transputers are used in general-purpose systems like the ARCON developed at CTIO by Roger Smith, Alistair Walker and colleagues. These and similar systems are used at major observatories in Hawaii, La Palma, Arizona, Chile, South Africa and Australia. For more technical details the reader should consult papers in SPIE proceedings such as Crawford and Craine (1994) as well as the World Wide Web home pages for major observatories such as the AAT, CFHT, CTIO, ESO, IFA, NOAO, RGO and ROE.

7.2 DARK CURRENT AND COOLING

Random motions of atoms at normal room temperatures within the silicon lattice will release sufficient energy to give rise to a continuous stream of electron–hole pairs in the absence of light. This process for electron–hole pair production is called "thermal" because the energy source is heat. Depending on whether or not the CCD can be operated in inverted mode, as in devices with MPP implants, this **dark current** can be very substantial. At room temperature the dark current of a typical, non-inverted CCD is about 100 000 electrons per second per pixel (or equivalently, 1.8×10^{-9} A/cm^2 for 30 µm pixels) which means that the CCD storage wells will fill up, saturate and spill over on dark current alone in just a few seconds. If the CCD is read out very rapidly and continuously at a high rate (frequency) such as 5 MHz then the dark current is cleared after only a brief accumulation time (about 1/60th of a second) and so the CCD can be used for ordinary TV applications. Most astronomical sources are much too faint to yield a good image in such short snapshots especially since there is a small but definite "noise penalty" paid for every readout. Fortunately, the solution is straightforward—it is "cooling". Or at least that solution seems straightforward now, but when it was first suggested to Bell Labs engineers by Jim Westphal back in 1974 it was considered radical!

Since the thermal dark current undergoes a very rapid decay with decreasing temperature, a drastic reduction in dark current can be achieved by cooling the CCD to a temperature somewhere between those of "dry-ice" (solid carbon dioxide or CO_2) and liquid nitrogen (LN$_2$).

For charge-coupled devices there are three main sources of dark current. These are

- thermal generation and diffusion in the neutral bulk silicon
- thermal generation in the depletion region
- thermal generation due to regions called "surface states" at the junction between the silicon and the silicon dioxide insulation layer

Of these sources, the contribution from surface states is the dominant contributor for multi-phase CCDs. Dark current at the Si–SiO$_2$ interface depends on two factors, the density of interface states and the density of free carriers (electrons or holes) that populate the interface. Electrons can more easily thermally "hop" from the valence band to an interface

state (also called a mid-band state because the energy level associated with this disturbance in the crystal lattice lies in the forbidden energy gap of the normal semiconductor
crystal) and then to the conduction band, thus producing an electron–hole pair. The presence of free carriers will fill interface states and, if the states are completely filled, will
suppress thermal hopping (valence band electrons no longer have a stepping stone to the
conduction band) and therefore substantially reduce dark current. Classical CCD operation as described early in Chapter 6 actually depletes the potential well and the interface
of free carriers, thus maximizing dark current generation. Under depleted conditions, dark
current is determined by the quality of the silicon–silicon dioxide interface. Dark current
can be significantly reduced by operating the CCD in **inversion** mode as discussed in
Chapter 6. With this arrangement, holes are attracted up to the Si–SiO$_2$ interface at these
electrodes thus populating the troublesome states with holes. Not only does inverted operation improve dark current, but the neutralization of the interface states eliminates
"residual" images associated with saturation and problems of charge capture. The inversion phenomenon was noticed by Tony Tyson of Bell Labs while investigating RCA
CCDs, but its significance was not appreciated until a similar effect was uncovered later
by Jim Janesick and colleagues at JPL for the TI devices. Virtually all CCDs which are
capable of inverted operation are now run in this manner and new CCDs are manufactured
with a technology called **Multi-Phase-Pinned (MPP)** which allows the CCD to operate
totally inverted at all times while maintaining other performance characteristics. MPP
CCDs have achieved dark levels of less than 0.025 nA/cm^2 at room temperature, and
thinned backside-illuminated devices with a substantial part of the epitaxial layer etched
away to reduce bulk dark current contributions have attained 0.010 nA/cm^2. MPP technology also aids charge transfer efficiency by allowing higher operating temperatures to be
used.

Roughly speaking, in silicon the dark current $N_D(T)$ falls to about one third of its previous value each time the temperature (T) is reduced by about 10°C; typical observed behaviour is shown in Fig. 7.6. More precisely, the relation can be expressed as

$$N_D = 2.55 \times 10^{15} \ N_0 d_{pix}^2 T^{1.5} e^{-\frac{E_G}{2kT}} \quad \text{e / s / pixel} \tag{7.1}$$

where N_D is in electron/s/pixel, and N_0 is the dark current in nanoamps per square centimetre at room temperature (T_0), d_{pix} is the pixel size in centimetres, T is the operating temperature in K, E_G is the band gap energy in electron volts and k is the Boltzmann constant
(8.62×10^{-5} eV/K). The band gap energy (in electronvolts) varies with temperature and is
given by

$$E_G = 1.557 - \frac{7.021 \times 10^{-4} \ T^2}{1108 + T} \tag{7.2}$$

Typical operating temperatures for astronomical CCDs range from about 223 K to 150 K;
or −50°C to −123 °C, depending on the application and whether or not the CCD employs
MPP. It is possible to end up with a dark current of only a few electrons per pixel per *hour*
instead of 100 000 electrons per second!

Example

The room temperature (300 K) dark current of a MPP CCD with 12 μm pixels is found to be 0.01 nA/cm². Predict the dark current at −50°C.

Solution: First find the band gap for this temperature ($T = 223$ K) using Eq. (7.2), $E_G(223)$ = 1.1295 eV. The pixel area is 1.44×10^{-6} cm². Substitute in Eq. (7.1) to find that $N_D = 2.12 \times 10^{-2}$ e/s/pixel, or about 76 electrons per pixel per hour.

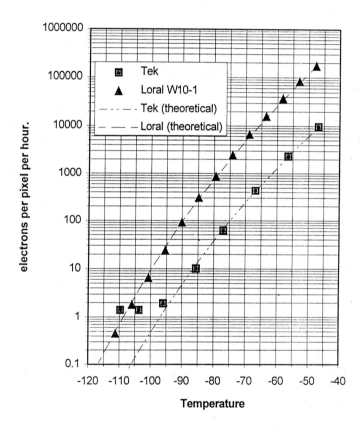

Fig. 7.6. Observations of dark current versus temperature (°C) for two CCDs compared to an "exponential" law $I = I_0 \exp[-eE_g/2kT]$. Both CCDs follow the model closely, although the Tek/SITe chip levels off at about 1.3 e−/pixel/hour. Courtesy Paul Jorden, RGO.

7.3 COSMIC RAYS AND LUMINESCENCE

With the elimination of dark current, cooled CCDs become capable of extremely long exposures. However, there is now another effect which comes into play. High-energy sub-atomic particles (protons) entering the Earth's atmosphere from outer space will generate a shower of secondary particles called **muons** which can be "stopped" by a sufficiently thick layer of silicon. The energy released can generate around 80 electrons per pixel per micrometre of thickness in the silicon. With a collection depth of 20 μm, a muon event is seen on a CCD image as a concentrated bright spot a few thousand electrons strong (Fig. 7.7(a)). Thinned backside-illuminated CCDs and those CCDs with **epitaxial** (i.e. thin) collection layers (like the EEV/GEC CCDs) are therefore much less prone to cosmic rays than the thicker frontside-illuminated arrays. High-altitude observatories may also encounter more problems than sea-level laboratories. Because these events are so conspicuous and concentrated to a few pixels it is generally straightforward to remove them from images of non-crowded fields and non-extended objects. Otherwise, the only approach is to take multiple frames of the same region and compare them, since there is very little chance of a cosmic ray event hitting the same pixel twice in quick succession.

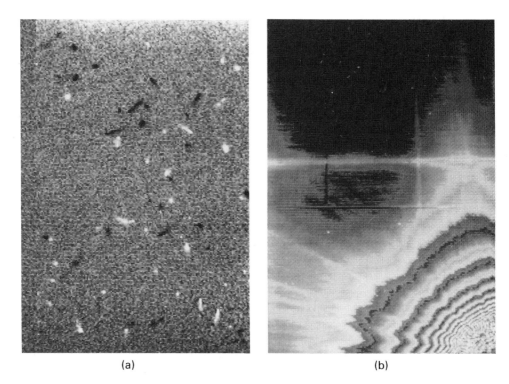

(a)   (b)

Fig. 7.7. (a) The difference of two 90-minute dark exposures with a CCD reveals numerous events even though there has been no illumination; these are ionizing particles or high energy photons. The difference-image shows the events in the second exposure in black. (b) Intense luminescence from the output amplifier of a CCD during an exposure. This effect can be avoided by turning the output amplifier off when it is not needed. Courtesy Pat Waddell.

Experiments conducted by Craig Mackay of the Institute of Astronomy in Cambridge (England) revealed that the events attributed to cosmic ray hits were actually a combination of cosmic rays and low-level X-ray emission from several forms of UV- and blue-transmitting glass (e.g. Schott UBK7 and GG385). In fact, the X-ray emission dominated by 5:1. This discovery alleviated the concern over the fact that the muon event rate was predicted by cosmic-ray physicists to be several times lower than the observed rate of events being recorded by CCDs. With fused-silica windows the actual event rate was only 1.5–2 events/cm^2/min, or 90–120 events/CCD frame/hour exposure.

Another source of unwanted signal in all CCDs and similar devices—such as solid-state infrared imagers—is **luminescence**. An electrical path with lower than normal resistance in the semiconductor caused by, for example, partial shorts between electrodes in the parallel or serial registers, can act like a light-emitting diode (LED). Incorrect voltages applied to certain parts of the device, in particular the output transistor amplifier, can cause similar problems (Fig. 7.7(b)). This effect can be a serious limitation in some CCDs because it will limit the integration time to only a few minutes. Device selection, or alternatively, active (computerized) electronic control of the applied voltages can circumvent this limitation.

The simplest examples of the latter approach is to electronically switch off the output transistor during a long integration and to reduce the clock level (or voltage potential) on the pixels. Just prior to the end of the exposure the transistor is switched on again, the clock swing increased to improve charge transfer efficiency and the horizontal register "purged" several times to clear it before the readout of the CCD frame is commenced.

One final concern expressed by some workers striving for the very lowest light levels is radioactivity from the heavy metals of which cryostats are made, such as copper, located near the CCD itself. Rumour has it, that some astronomers have actually discussed the feasibility of harvesting pre-nuclear age metals from sunken ships to fabricate "clean" CCD cryostats for astronomy!

7.4 BAD PIXELS AND CHARGE TRANSFER EFFICIENCY

Earlier sections described how charges are accumulated under the CCD electrodes until the end of the timed exposure and then moved out by charge-coupling of one pixel to the next through a repetitive clocking (or pulsing) voltage waveform applied to the electrodes. Unfortunately, as recognized immediately by the inventors Bill Boyle and George Smith, there are several possible impediments to the success of the transfer process.

Problems occur when the wells are (a) virtually empty or (b) almost full or (c) when a defective pixel gets in the way!

As the CCD wells fill up, the total stored charge can distort the electric potential of the applied clock voltage and thus charge transfer or coupling is degraded. Moreover, when the wells are completely full—a condition called "saturation"—there is nothing to stop the charge spilling over into the adjacent pair of pixels in that column; in general there is no serious sideways (row) spread because of the heavily doped channel stops. If the source of light causing the saturation is very strong then the adjacent pixels may also fill up and so the spreading continues and a streak appears up and down that column; this is also referred to as a charge bleeding. An example of **charge bleeding** is shown in the CCD picture in Fig. 7.8.

Global **charge transfer efficiency** (CTE) drops when a fixed amount of charge is "trapped" for *each* pixel transfer. Visually, the result is the appearance of a "tail" on point-like images; the tail will be parallel to the transfer direction. Trapped charge is not

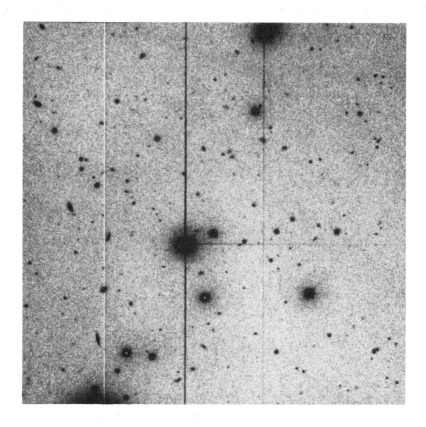

Fig. 7.8. An astronomical CCD image (displayed as a "negative") in which a bright star has
saturated in the exposure time needed to record the fainter objects. When the CCD pixel saturates
it spills over and "bleeds" up and down the column. The horizontal trail is evidence of poor charge
transfer efficiency in the serial register. The white line is a non-working or "dead" column.
Courtesy NOAO.

normally "lost" but more often it is "skimmed" to be released later, and is therefore related
to a general phenomenon called **deferred charge**, as illustrated in Fig. 7.9. Deferred
charge can have several origins including spurious potential pockets, design faults leading
to specific potential pockets at certain locations and charge **traps** associated with impurities. For most CCDs poor charge transfer efficiency occurs at a level of 10 to 50 elec-
trons/pixel and is mainly of concern only in very short exposures or in spectroscopic applications in which the dispersed background in the exposure is below 100 electrons. One
way around deferred charge effects is to **preflash** the CCD with a short flood of light from
an LED positioned so as to uniformly illuminate the array before each exposure. This
action adds a constant amount of charge, sometimes called a pedestal, to each pixel which
raises the base level to above 100 electrons (say), back into the acceptable region of good
transfer efficiency; note that a penalty of $\sqrt{100} = 10$ electrons noise is added to the readout
noise. The technique of introducing a substantial charge pedestal was termed (by someone

with a sense of humour) as adding a "fat zero"; inevitably someone nicknamed the smaller charge pedestal needed to offset the deferred charge problem the "skinny zero"! In astronomical circles the most common description is preflash.

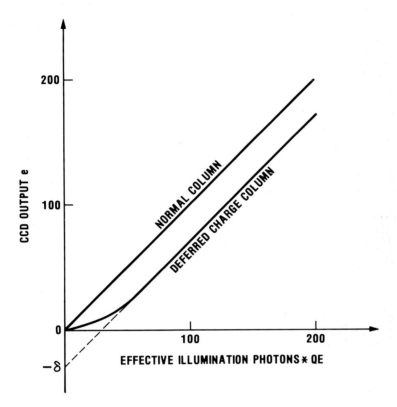

Fig. 7.9. A plot of the number of electrons detected at the output of a CCD versus the number of electrons present in the original pixel. If no charge is trapped anywhere during the readout then the graph should follow that for a normal column. It does not, indicating trapped charge whose readout has been "deferred". Courtesy Tony Tyson, Bell Labs.

Apart from these general problems of charge transfer inefficiency there are other difficulties related to specific faults on a given CCD. A break in one of the polysilicon electrode strips due to a fault in the manufacturing process can effectively stop or seriously delay charge from being transferred past that point in that column; this defect is called a "blocked column". Crystal defects in the bulk substrate silicon can sometimes spread into the depletion region during manufacture. Such defects disrupt the semiconductor nature of the silicon and, in general, behave like sponges which can soak up any charges in their area. Delayed release of charge can have a wide range of time constants (from milliseconds to hours) and columns with this sort of behaviour are usually irretrievable by image processing. Needless to say, a charge trap near the beginning of the horizontal (serial) register is a disaster! Even the design of the overlapping electrode structure can create pockets of resistance in the channel potential which trap charge. Trapping sites are particularly serious for surface-channel CCDs. Using our water-in-a-well analogy, traps would behave like small pot-holes at the base of the well; such pockets fill up easily but do not empty their contents merely by raising and lowering the floor of the well.

CTE is quoted in the form 0.999 99 per transfer or, in CCD jargon, people speak of CCDs with "five nines of CTE". What this means is that after one transfer 99.999% of the original charge will have made it successfully, after the second transfer 99.999% of that charge will be moved on so that 99.998% of the original charge is still intact and finally, after 1000 transfers, 99.00% (that is $0.999\,99^{1000}$) of the original charge on the initial pixel will have been transferred to its new location (the output, say), while 1% of the original signal will be spread over the intervening pixels. In practice most of that 1% is in the immediately adjacent (following) pixel. We have already mentioned that CTE is a function of temperature. It also depends on the clock frequency or "readout rate" and the rise-and-fall-times of the clock voltage pulses and their degree of overlap. The CTE in the parallel (column) direction is usually different from that in the serial (horizontal) register and there can be large variations from column to column.

Measuring the level of charge transfer efficiency is not as simple as it might seem at first sight. A relatively straightforward method, at least for CCDs which permit it, which is both qualitatively and quantitatively useful, is to flood the CCD with light, read out the device and **overscan** both registers to produce a resulting image which is several pixels larger, along both axes, than the real CCD. Most CCD controllers permit overscanning, but some CCDs have special additional pixels in the serial register which confuse the overscan experiment. This technique is also called "extended-pixel-edge-response" or EPER by some workers. When the signals occurring in the imaginary pixels following the last row and column of the real CCD are compared to the dark level and to the illumination level then any charge transfer problems leading to deferred charge will be obvious (Fig. 7.10). Some investigators lump all rows or columns together to give a mean measure of CTE, but careful examination of an overscanned uniformly illuminated image reveals big variations from column to column. Some averaging will be needed to reduce the effect of readout noise. The CTE is then given by

$$CTE\left(e^{-}\right) = 1 - \frac{Q_d}{NQ_0} \tag{7.3}$$

where Q_d is the net deferred charge in the overscan, Q_0 is the charge on the last real pixel and N is the number of pixels transferred.

The spurious potential pocket problem has plagued most manufacturers, but it is now well-enough understood. Early Texas Instruments and Thomson-CSF CCDs exhibited design-induced pockets in the transfer-gate region where the parallel (imaging) area mates with the horizontal (serial) register. By design, the charge-carrying or signal channel beneath the transfer-gate was constricted by channel-stops to properly herd signal charge from the CCD columns into the horizontal register. As charge moved from a wider to a narrower region there was a local decrease in potential effectively creating a step or pocket which trapped temporarily small quantities of charge; the effect can be overcome by a preflash or by adjusting the potential on phase one to produce a "fringing" field. Thomson-CSF corrected their problem with the design of a new set of manufacturing

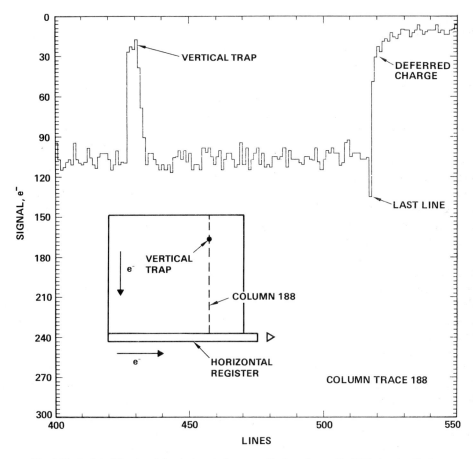

Fig. 7.10. A plot of the signal (in electrons) along a particular column of a CCD showing the loss of charge in a "trap" and the curved response following that trap and the last pixel indicating deferred charge. Courtesy Jim Janesick, JPL.

masks, but TI withdrew from the three-phase development before a correction could be made. Design-induced spurious potential pockets appears to be a thing of the past due to stringent scrutiny of the masks prior to fabrication. The most serious problem that the current generation of scientific CCDs has been facing is the isolated single-pixel-trap or "pocket" which affects CTE randomly on a local rather than global level. Some new CCDs were literally riddled with pockets. In striving for the perfect CCD the pocket phenomenon has taken its toll on CCD yields at several different manufacturers (Tektronix/SITe, Ford/Loral, Lincoln Laboratory) and has been studied with incredible intensity. Detailed tests have shown that these traps are localized to a single level of the polysilicon electrode within a pixel. When the effect first appeared with the Tektronix devices it threw everyone into a quandary. Why had this phenomenon not been observed before? It seemed that past RCA, TI, Fairchild, Reticon and Thomson CCDs never exhibited these single-pixel-traps to such an appreciable extent, even allowing for the improvements in other performance factors such as readout noise. This was the phenomenon which put the Tektronix devices "on hold". Two clues finally emerged. The TI family of 800 × 800 CCDs showed no large single-pixel traps and it was discovered that the impurity boron, used to dope the p-type substrate, had diffused through the thin (epitaxial) n-type layer of the buried channel construction during a high-temperature phase of manufacture. This effect was called the "p+ epitaxial diffusion tail". The Tek CCDs on the other hand exhibited very little diffusion of boron. The second clue was the realization that the change from the older practice of using silicon dioxide as the insulating layer on the semiconductor surface to an oxide–nitride double-layer could be relevant; the new material has significantly reduced short-circuits between the polysilicon electrodes and therefore improved the manufacturing yield. Pin-holes created by surface contamination during the oxide growing stage could allow nitride to diffuse through the oxide to the n-type semiconductor. At that tiny location the different insulator material would cause a slight variation in electric potential within the semiconductor. Ultimately, with all other problems solved, it will be traps in the bulk silicon crystal which limit charge transfer efficiency. Fortunately the quality of manufactured semiconductor silicon has greatly improved over the years and will probably continue to do so. Global CTE performance of selected Loral, Lincoln and Thomson-CSF CCDs are believed to be limited by "bulk traps" at the level of 0.999 999 5 (or "six nines five") per pixel transfer. In other words, only three electrons out of a total of 10 000 electrons are deferred for 512 pixel transfers.

7.5 NOISE SOURCES

A charge-coupled device is by its very nature a "digital" camera. The clocking procedure described earlier delivers a stream of charge "packets" from pixels in the image area all the way to the output amplifier; the charge (Q) in each packet is proportional to the amount of light in that part of the original image scene. As each charge packet arrives at the output field-effect transistor it causes a change in voltage to occur (of amount $V = Q/C$, where C is the capacitance at the output node); the smaller the node capacitance, the larger the voltage change for the same size of charge packet. For earlier CCDs, such as the RCA 512 × 320 CCD, the output capacitance was fairly high, e.g. $C \sim 0.6$ picofarads (pf) which yields about 0.25 microvolts (μV) per electron (in the charge packet), whereas for modern CCDs the node capacitance is < 0.1 pf which gives a healthy > 1.6 μV per electron. Much larger values are possible; some Kodak CCDs using an extremely small output MOSFET give 15 μV/electron yet the overall noise under slow-scan conditions is greater than 10 electrons because other noise sources become larger as the MOSFET gets smaller.

It is desirable that the noise performance of a CCD camera system be limited only by the output transistor of the CCD and not by any other part of the electronic system. To achieve this one must understand the noise sources associated with the CCD and take steps to get them to an irreducible minimum; this minimum is the ultimate **readout noise**. There are several potential sources of unwanted electronic noise in charge-coupled devices. These include,

- background charge associated with fat zero offsets
- transfer loss fluctuations
- reset or "kTC" noise

- MOSFET noise
- fast interface state noise

When a preflash is used to introduce a fat zero charge to aid transfer efficiency or elimi-nate charge skimming, then the consequence is a noise equal to the square-root of the total number of charges in a pixel.

During charge transfer a fraction of the charges are left behind. However, this fraction is not constant but may fluctuate and so an additional noise component is added to the signal noise. This "transfer noise" is given by

$$\sigma_{tr} = \sqrt{2\varepsilon n N_0} \tag{7.4}$$

where $\varepsilon = 1 - \text{CTE}$ is the fraction of charges not transferred, n is the number of transfer and N_0 is the original charge. The factor of two occurs because the Poisson-distributed noise happens twice, once for trapping and once for release. This effect can be of order 70 electrons for surface channel CCDs but is typically ten times smaller or better for buried-channel CCDs and astronomical light levels. For very large CCDs this effect implies that exceptionally good charge transfer efficiency must be achieved.

Noise associated with the re-charging of the output node is given by $\sqrt{(kTC)}/e$ or about $284\sqrt{(C)}$ at a CCD temperature of 150 K, where the capacitance, C, is in picofarads. This effect, called **reset noise**, is dominant and its removal by signal-processing is described in detail later.

Other noise sources associated with the output MOSFET, such as one-over-f ($1/f$) noise, can generally be made quite small by good manufacture, typically a few electrons. Traps which absorb and release charges on very short time scales, causing a fluctuation in the charge in any pixel, are called "fast interface states". The noise is given by

$$\sigma_{SS} = \sqrt{2kTnN_{SS}A} \tag{7.5}$$

where k is Boltzmann's constant, T is the absolute temperature, n is the total number of transfers (not the number of pixels), N_{SS} is the surface density of traps and A is the surface area. This effect is very serious for surface-channel CCDs but is normally quite small (of order five electrons or less) for good buried-channel devices. It is a remarkable tribute to the foresight of Boyle and Smith, and the skills of every scientist and engineer who has worked hard on CCD technology, to realize that devices are now routinely made with a readout noise of less than 10 electrons (see Fig. 7.11 a,b). The lowest noises that have been consistently measured are in the range 2–3 electrons.

7.6 SIGNAL PROCESSING AND DIGITIZATION

CCD outputs and the concept of readout noise was introduced in earlier sections. Here, we consider CCD signal processing in more detail, and in particular explain the crucially important technique known as **correlated double sampling** or **CDS**.

As each charge packet arrives at the output node it produces a voltage change which must first be amplified and then digitized by an analogue-to-digital converter. This cannot be done instantaneously, it requires a finite amount of time—hence the term

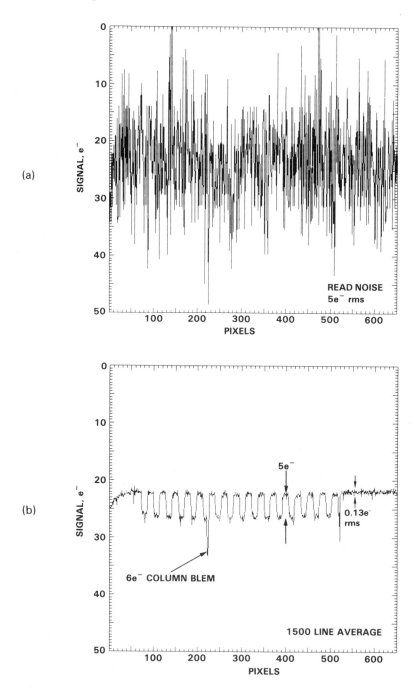

Fig. 7.11. (a) A single raw trace of the signal from a Loral CCD at the five-electron level. embedded in the noisy trace is an unseen 5e⁻ peak-to-peak square-wave pattern. (b) After 1500 lines have been averaged the random noise is only 0.13e⁻ as seen in the overscan or EPER region. A 6e⁻ column blemish has also emerged. Courtesy Jim Janesick.

slow-scan—if a high degree of accuracy is required such as can be achieved with a 16-bit A/D; a 16-bit A/D divides a specified voltage range, e.g. 10 volts, into 65 536 (2^{16}) parts and therefore each voltage interval is 152.5 μV in size. The A/D circuit matches up the actual voltage to the nearest number on the scale of 1:65 536.

To measure the voltage of each charge packet we need a "reference" voltage. We could use ground but, since it is important to reset the output capacitance back to some nominal value on each readout cycle—otherwise we would be forming the difference between one charge packet and the previous one while drifting away (in voltage) from the ideal operating point of the MOSFET—there is another way. The output capacitor can be recharged to a fixed voltage by briefly pulsing the gate of another transistor, called the **reset transistor** (see Fig. 7.12(a)), to briefly turn that transistor on (like closing a switch) so that current can flow from a power supply to charge-up the node to the supply level. When the reset pulse disappears the reset transistor is turned off (like opening a switch) and the output becomes isolated to await the next charge-dump from the horizontal register. As a capacitor is charged to a certain voltage level (V_{RESET}) it rises steeply at first and then levels off to approach its final value shown in Fig. 7.12(b). Again due to random thermal agitation of electrons in the material, there is "noise" or uncertainty on the mean value and so the final voltage can lie anywhere within a small range, given by

$$Reset\ noise = \sqrt{\frac{kT}{C}}\ volts\ or\ \frac{\sqrt{kTC}}{e}\ electrons \tag{7.6}$$

In this expression, k is Boltzmann's constant (1.38×10^{-23} joules/K), e is the charge on the electron, T is the absolute temperature of the output node in kelvins (K) and C is the node capacitance. If C is expressed in picofarads (i.e. 10^{-12} coulombs/volt) then this noise uncertainty (called the **reset noise** or, from the formula, **kTC** (kay-tee-cee) **noise**) is simply $400\sqrt{(C)}$ electrons at room temperature and about $250\sqrt{(C)}$ at 120 K (−153 °C); for a typical modern device this would yield < 80 electrons noise which greatly exceeds the readout noise of the MOSFET alone and so some means must be found to remove it.

7.6.1 Correlated double sampling (CDS)

Fortunately, removal of reset noise is quite straightforward due to the fact that whatever the final reset voltage actually is, and it must be in the range $V_{RESET} + \sqrt{(kT/C)}$ to $V_{RESET} + \sqrt{(kT/C)}$, it will get "frozen" at that value because the leakage of current through the switched-off reset transistor is exceedingly slow (its **RC time constant** is seconds compared to the microseconds between arrivals of discrete CCD charge packets) and hence, if this uncertain reset level is sampled by the A/D just prior to a charge packet being dumped at the output, and then again after the charge has been added, it will have *exactly* the same value in each sample. Forming the difference of these two signals therefore automatically eliminates this voltage level—without ever knowing exactly what it was! This technique is known as **correlated double sampling** (or CDS) because two samples of the CCD output voltage are taken per pixel and the offset due to the reset or kTC noise in each sample is the same, i.e. it is correlated.

In practice it is advantageous to integrate or sum the signals rather than merely "spot sample" them. Integration smooths out high-frequency noise and limits the effects of 1/f

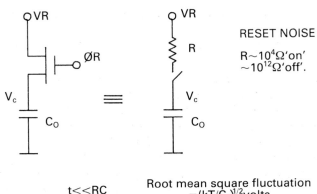

RESET NOISE

$R \sim 10^4 \Omega$ 'on'
$\sim 10^{12} \Omega$ 'off'.

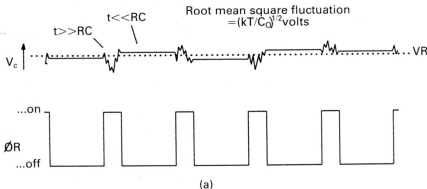

Root mean square fluctuation
$= (kT/C_0)^{1/2}$ volts

(a)

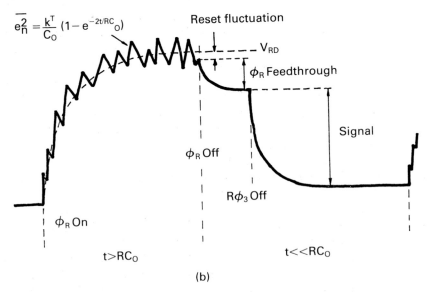

$$\overline{e_n^2} = \frac{kT}{C_0}(1 - e^{-2t/RC_0})$$

Reset fluctuation

V_{RD}

ϕ_R Feedthrough

ϕ_R Off

Signal

$R\phi_3$ Off

ϕ_R On

$t > RC_0$

$t \ll RC_0$

(b)

Fig. 7.12. (a) An equivalent "switch" circuit to explain the operation of the reset transistor. Note that the resistance, and therefore the RC-time constant, is very different between the off and on states. (b) The charging profile of the output of a CCD when reset. Notice that the curve is "noisy" but when the reset pulse disappears the last value of the signal becomes frozen.

noise for frequencies lower than the overall signal-processing time. A schematic version
of how a full, integrating CDS circuit works is depicted in Fig. 7.13.

Signals from the CCD output transistor are amplified by a low-noise solid-state operational amplifier (A1 in the diagram) and fed to the integrating amplifier A4 (with the capacitor in its "feedback loop" from the output to the input) by way of a three-position
switch. One of the three positions is ground, and the other two contain a *non-inverting* and
an *inverting* amplifier respectively, each with a gain of ×1 ; in other words in one case the
signal voltage is multiplied by 1 and in the other case by −1. Amplifier A1 is connected to
A4 via A3 (times −1 gain) for a precise period of time, for example twenty microseconds
(20 µs), during which the reset voltage level is being integrated (negatively) onto the capacitor on A4. Next, the switch is connected to ground for a few microseconds which
isolates the integrator and allows the signal charge packet to be transferred from the serial
horizontal register onto the CCD output node. Then the switch is thrown to the
non-inverting amplifier A2 (×1 gain) and the "reset plus voltage" level is integrated
(positively) onto the same integrating capacitor on A4 for exactly the same period of time.
Since A2 and A3 give identical but opposite signals in identical times, the resulting output
of A4 is the real *difference* signal, i.e. the signal due to the charge packet itself. Reset noise
is eliminated! The output of the integrator is fed to a Sample and Hold circuit which
"freezes" the value and then to a fast Analogue-to-Digital Converter (often in the same
commercial package) to convert the voltage to a number. The parallel output (typically 16
wires) of the A/D goes directly to the data input part of a computer.

Choosing the correct (pre-)amplifier gains can be important. To work at very low signal
levels, the total **system gain** should be such that the root mean square (rms) readout noise
corresponds to a few "bits" (also called Data Numbers or DN, and ADUs) of the A/D
converter. The maximum or saturation signal from a CCD is less than 500 000 electrons,
but some devices have a readout noise of only five electrons; the ratio of these two numbers is called the **dynamic range**. In this case the dynamic range is 100 000:1 whereas a
16-bit A/D can only give a range of 65 536:1. If we choose a gain to give one electron per
DN then the readout noise will be 5 DN and the A/D will saturate at 65 536 DN which is
only 13% of full well. If we use five electrons per DN then the readout noise will be 1 DN
(and not well determined due to digitization noise) but we can utilize 65% of the well. A
more realistic situation is that the full well capacity is less than 250 000 electrons, which
gives a maximum dynamic range of 50 000:1 which is within the 16-bit A/D range of
65 536:1. If the application is such that the photon flux nearly fills the CCD wells, then the
dynamic range will be limited by photon noise and not readout noise. For example, on a
signal of 250 000 photoelectrons the photon noise is $\sqrt{N}$ or 500 electrons and the dynamic
range is 250 000/500 = 500 and not 50 000. Since most CCD applications in astronomy
are for low light levels then it is better to use a high gain and work only in the lower part
of the wells, e.g. five electrons/DN in the last example yields a readout noise of 2 DN and
the A/D saturates at 65% of full well. A 10-volt 16-bit A/D has a resolution of 152.6 µV
per DN or 30.5 µV per electron if five electrons per DN are required. A typical CCD
output yields only 2 µV per electron and so an amplifier gain of 15 is needed to match
up to the A/D. It is possible to install switchable (or variable) amplifier gain settings,
but any method of changing the gain introduces another calibration step for astronomers.

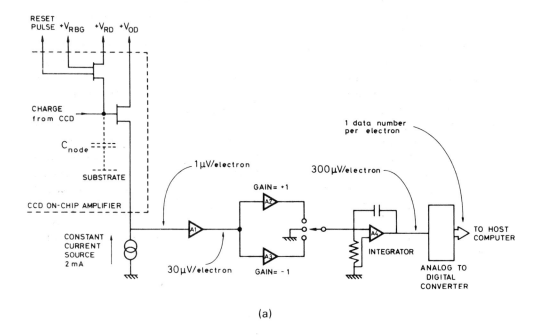

(a)

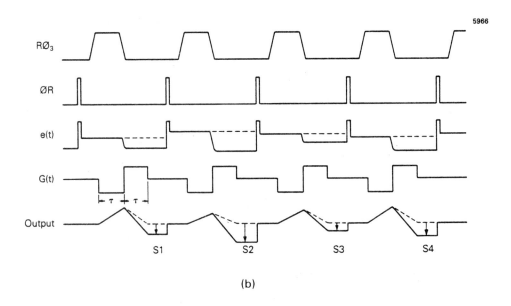

(b)

Fig. 7.13. (a) A block diagram of the principle of the correlated double sampling (CDS) method of removing reset noise before the signal is digitized and sent to a computer. (b) The associated timing diagram.

Digitization noise can be shown to be about $0.29g$, where g is the gain in electrons per DN. In the above example, with $g = 5$, this corresponds to a noise of 1.45 electrons to be added in quadrature with the readout noise.

The CDS integration, S/H delay and A/D time will determine the pixel readout rate, typically this is 10–50 kHz. In principle, the longer the processing time spent on each pixel the better the final noise performance. The CCD output noise from the CDS is inversely proportional to S_V, the sensitivity of the CCD output node ($\mu V/e^-$) and proportional to the CDS-filtered noise voltage spectrum $V_n(f)$ of the on-chip amplifier; $V_n(f)$ has two components, a white noise and a $1/f$ term. Readout noise will decrease approximately as the square root of the CDS integration period until correlation is lost by $1/f$ noise. A plot of readout noise versus CDS integration time for three CCD types studied at JPL is shown in Fig. 7.14; Thomson CCDs (not shown) lie between the Ford/Loral and Lincoln devices. With a CDS integration time of 8 μs the Lincoln Laboratory CCD gave two electrons noise. Typical pixel times range from about 20 μs for many CCDs to about 80 μs for EEV devices which have very low $1/f$ noise. For a 2048×2048 CCD with a 20-μs pixel period the total readout time is 80 s and for 50 μs it takes 3 minutes and 20 seconds to read the entire chip!

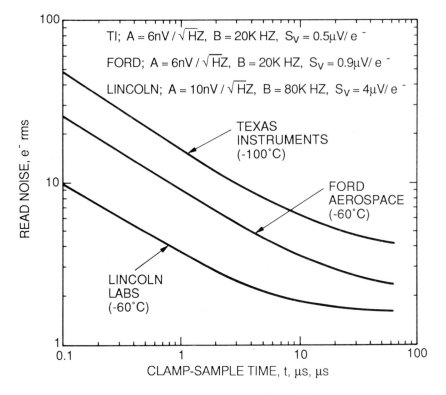

Fig. 7.14. The theoretical readout noise performance of three CCD types as a function of CDS integration or clamp-to-sample time. Although these plots were derived from an equation they were found to be in good agreement with experimental data obtained at JPL.

Low noise performance depends on the use of very low noise transistors such as the 2N5564 JFET in the JPL-style preamp and on good operational amplifiers in the signal processing chain, for example, Harris 2520 or National Semiconductor type LF357. The pixel clock timing should be synchronized with the crystal-controlled microprocessor clock and with a nonvarying (in time) instruction sequence in the microprocessor to avoid noise due to time "jitter" or changes in the amount of signal induced by accidental coupling from the microprocessor circuits. It is essential to maintain a precisely uniform readout rate since, due to the settling times of various components, temporal jitter can produce small variations in the steady or "baseline" voltage of the amplifiers which the CDS cannot cancel and so the noise will increase.

Janesick (1993) gives a very useful noise relationship for the read noise floor when off-chip amplifier noise is included:

$$\sigma\left(e^-\right) = \frac{\sqrt{\Delta f}}{S_V}\left(\sigma_{ccd}^2 + \sigma_{A_1}^2 + \frac{\sigma_{A_2}^2}{A_1}\right)^{1/2} \tag{7.7}$$

where σ is the "total read noise floor", A_1 is the pre-amplifier gain, Δf is the equivalent noise bandwidth (Hz), S_V is the node sensitivity (V/e^-), and σ_{ccd}, σ_{A1} and σ_{A2} are the frequency-dependent white noise figures for the CCD output amplifier, preamplifier and post-amplifier respectively (V/root-Hz); an on-chip amplifier gain of unity is assumed.

Example:
Take σ_{ccd}, σ_{A1} and σ_{A2} to be 10, 2.5 and 20 nV/root-Hz respectively and assume that the bandwidth is 250 kHz and the node sensitivity is 1 μV/electron. For a preamplifier gain of $A_1 = 22$, find the noise generated by each component and the total system.

Solution: Substituting the numbers in each term in the equation gives 5, 1.25 and 0.45 e^- respectively for the CCD, preamplifier and post-amplifier, and 5.17 e^- for the quadrature sum. Note that although the post-amplifier has the highest voltage noise, this is suppressed by the preamplifier gain.

7.6.2 On-chip binning
Although CCD architectures vary from one manufacturer to another, all frame transfer devices incorporate a structure which permits another type of signal processing called **on-chip binning**. On-chip binning of charges is the process of adding or combining charge packets from rectangular groups of pixels at a "summing point", which is the output node capacitor of the CCD. These groups of pixels effectively become single larger pixels, sometimes called "superpixels". On-chip binning is achieved as follows.

Any number of rows can be co-added together in the serial register simply by omitting the horizontal clocking sequence which usually occurs after each vertical transfer. Likewise, when the horizontal register is clocked, the charges from any number of pixels can be combined on the output capacitor by simply *omitting* the reset pulse which usually occurs between each horizontal transfer. By not running the horizontal (serial) clocks after a vertical transfer, the next vertical transfer will co-add the next row of charges into the horizontal register. Similarly, by not resetting the output transistor after a pixel has been read out, the charge associated with the following pixel will be co-added with the first. In this way rows and columns can be "binned" together (Fig. 7.15). The final picture has smaller dimensions and less resolution, but it may have higher signal-to-noise because more signal was combined "on-chip" before the noise of the readout process was added. Of course, care must be taken to avoid saturation. A 1024 × 1024 CCD using 4 × 4 pixel on-chip binning (omit four horizontal transfers and four reset pulses) will produce an image with only 256 × 256 pixels. This feature is most useful in high resolution echelle spectrometers where binning is usually slight (×2 say) or not at all along the spectrum so as not to lose spectral resolution, but substantial (×15 or more) at right-angles to the spectrum since it allows more signal per readout and so a much better signal-to-noise ratio is obtained in a faint spectrum. This technique is also used for optimizing the resolution in variable seeing conditions or to obtain a preliminary look at a field containing very faint sources. A clock sequence which has these properties can be "called up" when required if the sequencer is based on a programmable design.

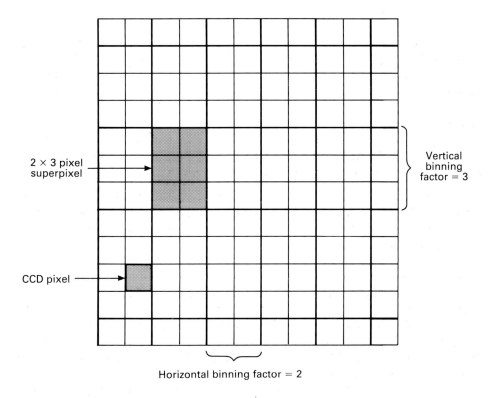

Fig. 7.15. The principle of on-chip charge binning.

7.6.3 Over-scanning

When a CCD is readout rapidly with essentially a zero exposure time, a picture appears on the screen which, although random, has an average value which is not zero. This value, called the **bias** value, must be subtracted from every image before any other arithmetic manipulation of the data can be performed. A bias frame can be obtained by closing the shutter of the CCD camera and setting the exposure time to zero. Alternatively, the bias level can be obtained by pretending that the CCD array is larger than it really is—say 10 pixels in each direction—so that the electronics continues to "clock" even though there is no longer any physical correspondence with points on the CCD; ten more vertical transfers are requested and ten more horizontal transfers are included each time the serial register is activated. Again, a programmable controller can deal with this request by simply calling-up a different "repeat count" for each type of clocking sequence. When the resulting image is displayed it will now have a "border" on two sides which contains no true signal, only the voltage corresponding to the bias level (Fig. 7.16). Many pixel values in this **overscan** area can be averaged to get a good estimate of the bias level. Thus, no extra frames are required. Bias frames will be discussed again when considering the detailed calibration of CCDs. Finally, just as you can send more pixels than are required, you can also opt to readout a subsection or "sub-array" of the CCD, usually by controlling the A/D unit with a (pulse) command to inhibit digitization except within the sub-area specified.

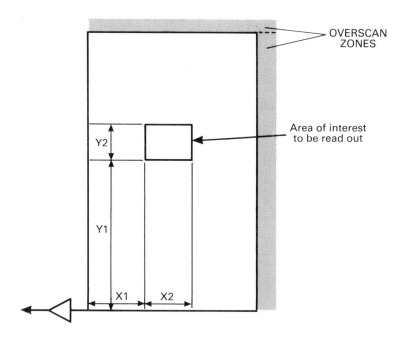

Fig. 7.16. The overscan region of an image is shown as well as an arbitrary sub-image.

7.7 UNIFORMITY OF RESPONSE

As Craig Mackay has aptly put it "the only uniform CCD is a dead CCD"! When a CCD is exposed to a scene in which the brightness is absolutely constant or uniform across the entire chip—this is called a **flat-field**—the actual output image from the CCD is generally not flat (Fig. 7.17). The image will be non-uniform in several ways. Some pixels will be

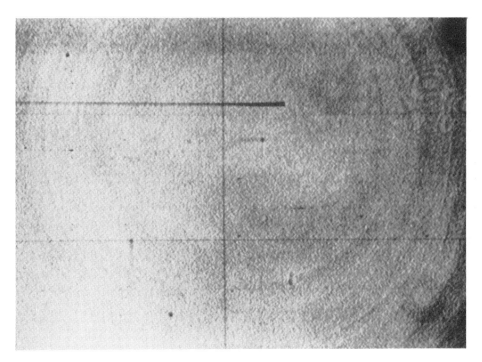

Fig. 7.17. A uniformly illuminated CCD. The apparently non-uniform image is called the response to a flat-field. Faint "tree-ring" structure is due to doping variations. A blocked column group can be seen and the fine vertical line is the join between the image and store regions of this frame transfer CCD. Courtesy CFHT.

more sensitive and some will be less sensitive than the average, and these pixels may be grouped in odd-shaped patterns or there may be a gross gradient in sensitivity from one corner of the CCD to the other. Bad pixels, blocked columns, out of focus dust specks on the window and even hairs or other debris on the chip will stand out dramatically. Some effects will even change with different light levels; this kind of behaviour is said to be "non-linear".

In addition, the flat-field response of a CCD is a strong function of colour, that is, of the wavelength and passband being used to illuminate the detector—mainly because photons of different wavelengths are absorbed at different depths in the silicon and the material is not homogeneous. At first sight, a newcomer to CCD technology might reasonably wonder if CCDs were all they are advertised to be! How is it possible that astronomers can tolerate such non-uniform detectors? First, it must be said that modern CCDs are significantly better than the one shown here, especially the smaller format devices which can now be produced with high uniformity. For large, thinned CCDs specially tailored for astronomy, there may still be problems. The important point is that these non-uniformities can be systematically eliminated by a pixel-by-pixel division of the observed region of sky by the corresponding pixels values in the flat-field frame. Suppose a_{ij} is the signal value on the pixel in the ith row and jth column of the image frame and b_{ij} is the corresponding value in the flat-field picture. For each pixel, the computer program forms the normalized ratio

$$r_{ij} = \langle b \rangle \frac{a_{ij}}{b_{ij}} \qquad (7.8)$$

where $\langle b \rangle$ is the mean of the entire flat-field frame.

This arithmetic task, although tedious for humans, is trivially accomplished by a computer. Pixel-to-adjacent-pixel variations in detector sensitivity of perhaps a few per cent of the average response is normal, and across the entire CCD variations as large as 20% can occur. Reducing such non-uniformities of response is a major pursuit, but there occurrence is not a serious limitation provided they are stable. One very big advantage of CCDs is that the pixels themselves are absolutely fixed in position, unlike in a vidicon-type TV tube where the pixel is defined by where the electron readout beam happens to land, and so stability in the uniformity of the CCD's response to light is a natural consequence. It is built in.

Charge-coupled devices and infrared arrays are manufactured in layers from a "mask set" produced by a CAD (or Computer-Aided Design) system. Small errors in the mask give rise to small but significant geometric variations in pixel size or in column widths. Stresses introduced by mechanical thinning can also cause variations in response. The silicon itself is not "pure", rather it is "doped" by impurities to form the required type of semiconductor, i.e., p- or n-type, but there are always patterns on a microscopic scale to this doping, often with an appearance which resembles "tree-rings", which show up in flat-fields. Sometimes there is also clear evidence of a cyclic or "step and repeat" pattern.

With a CCD selected to have a minimal number of bad pixels or columns one might expect to be able to correct the remaining non-uniformity in response by simply taking a ratio of images of astronomical sources to those of a uniformly illuminated white screen. But it is not quite so simple at the lowest light levels and the highest measurement accuracies. One problem is that the non-uniform response of a CCD is colour (or wavelength) dependent. Photons of different wavelengths are absorbed at different depths in the CCD on average. Frontside-illuminated CCDs, in which the electrode thickness may vary across the array, will therefore cause a variation in response to light which, because of the nature of the absorbing electrodes, will vary with wavelength. In thinned backside-illuminated devices, on the other hand, surface effects and multiple reflections from the back surface of the CCD cause colour-dependent non-uniformity. This phenomenon, known as **fringing**, is also discussed later in terms of calibration methods.

Obtaining an acceptable and appropriate flat-field is 90% of the solution to high-quality, very deep imaging with CCDs. Practical techniques for flat-fielding at the telescope are reviewed in Chapter 10 on characterization and calibration of instruments.

7.8 UV-FLASHING AND QE-PINNING

One of the earliest concerns was that the 800 × 800 CCDs to be used for the WF/PC instrument on the Hubble Space Telescope showed very poor response to blue and ultraviolet light, which was very bad news for a telescope expected to have great ultraviolet performance. Since tens of thousands of chips had been made, and since the WF/PC was a key instrument on this very expensive satellite, a solution *had* to be found. Of all the CCDs produced for astronomy only the TI and the RCA CCDs were thinned backside-illuminated devices. The RCA chip showed superior blue sensitivity and had useful sensitivity at ultraviolet wavelengths well below 0.40 µm (or 4000 A), but had a higher readout noise than was desirable. The thinned TI detectors had little or no response to blue light even though the exposed silicon surface was free of any electronic circuitry or other covering. Later, a second problem became apparent with the thinned TI CCDs. The effective quantum efficiency actually *increased* after exposure to light, so that the use of these CCDs as accurate, linear, brightness-measuring detectors was badly compromised. This effect was called **quantum efficiency hysteresis** or **QEH**.

Initially, chemical coatings were deposited on the TI CCDs to act as ultraviolet-to-optical converters. The idea for such a coating was due to Jim Westphal who, in an effort to learn more about ultraviolet behaviour, was browsing through the CalTech library and was surprised to find a 1968 volume by Samson which was highly relevant. He was even more surprised to learn that not only had knowledge of UV **fluorescence** been around for over 50 years and that a salicylate, like aspirin, was a phosphor which if smeared on a photocell would give it UV sensitivity, but the book also listed two little-used materials which fluoresced under UV illumination at the relatively long wavelength of 0.5 µm—ideal for CCDs. These substances were lumogen and coronene. Jim chose coronene because it could be obtained easily and he and Fred Landauer coated one half of a JPL/TI CCD and tested it at Palomar during an hour or so of twilight. [This use of a CCD on a telescope was shortly after the successful run by Brad Smith and Jim Janesick.] Incoming ultraviolet photons caused the coronene to glow in visible light, where the CCD was sensitive, and produced an effective efficiency to blue and ultraviolet light of perhaps 14%, i.e. one in seven photons at these wavelengths were actually detected. The coronene also seemed to act as an anti-reflection coating and so helped to improve the normal visible sensitivity.

Later, Martin Cullum and colleagues at the European Southern Observatory headquarters in Munich, Germany reported the use of a fluorescent laser dye coating painted onto front-illuminated EEV (GEC) CCDs to obtain quantum efficiencies as high as 25% at a wavelength of 0.35 µm (3500 Å) in the ultraviolet. The laser dye is reported as being more uniform in response in the blue (0.40 µm or 4000 Å) where coronene actually has a sharp dip in performance. The importance of achieving good UV, as well as good optical/red, response in CCDs can be judged by the international efforts to devise better **down-converters**. Many CCD vendors can now provide such coatings.

Workers at JPL also noticed a peculiar effect in a high-quality, thinned CCD which had been kept in a dry, sealed container. The device was found to have a much higher dark current than it had a few months earlier. Checking if it might be the coronene which had somehow affected the CCD, the JPL workers illuminated the device with an ultraviolet lamp to look for fluorescence. Soon after, the dark current was found to have fallen to normal! This clue led to a fix for both problems namely, poor UV response and QEH.

What was the explanation? It seems that imperfections in the thin "native-oxide", i.e. silicon dioxide (SiO_2), layer grown in air on the exposed backside surface of the CCD had created traps for electrons and holes. A net positive charge at the surface can develop, causing a shallow bucket or potential well which can trap electrons. This unwanted potential well must prevent photo-induced electrons from moving to the depletion region where the charge should have been collected. Since blue and UV photons are absorbed closest to the surface the result is a loss of sensitivity in the blue and UV. Traps can also absorb electrons from the silicon itself, electrons produced perhaps by photons; the electrons trapped in the unwanted potential well at the surface will affect the depth of that well and so the UV quantum efficiency will appear to change with illumination level. Time constants for the release of trapped charge can vary from seconds to days depending on the energy level and the CCD temperature. If the energy level of surface traps is midway between the valence and conduction levels, thermally induced dark current could be quite large.

The "fix"—exposure to UV light—works because energetic UV photons in the wavelength range 2000 Å to 2500 Å (which is not transmitted by the Earth's atmosphere) preferentially fill the most energetic or active trapping sites, creating a net *negative* charge on the CCD surface. Since like charges repel, the negatively charged surface generated by the "UV flood" eliminates the unwanted potential well and encourages electrons formed near the surface to move away from the surface toward the depletion region. Thus there is no loss of quantum efficiency and no QEH. The energy level of surface defects is forced away from the middle of the forbidden band and so the dark current is also reduced.

Fortunately, the negative charging procedure can be done near room temperature, and when the CCD is cooled to below $-100°C$ or 173 K, the charge is "immobilized", and provides a permanent field until the CCD is next warmed up. The quantum efficiency of the CCD is driven to its theoretical maximum and held or "pinned" there; this is the **QE-pinned** condition.

UV flooding requires wavelengths between 2000 Å and 2550 Å, and about 10^9 photons per pixel (at about 2139 Å) which can be provided by a zinc lamp. It also appears to require the presence of oxygen during the flood. At room temperature, the improvement disappears in a few hours, especially if humid air is present. In a constant vacuum at $-120°C$ the improved UV response can last for at least six months.

The decay of the UV sensitivity of the coronene-coated TI chip was noticed in an interesting series of communications between Jim Janesick at JPL and Bev Oke at the Palomar Observatory. Bev had UV sensitized his precious TI CCD from JPL and looked after it by keeping it dry and under vacuum in the CCD camera cryostat, whereas Jim had so many devices to test at JPL that cryogenic storage was impossible, so he kept them in a drawer! Initially each suspected that the other had made some error of measurement during testing. Eventually devices were exchanged, findings corroborated and the explanation found!

UV flooding is not a perfect solution. Apparently it works better for some CCDs than others, in particular, CCDs which are insufficiently thinned; thinning should be such as to reach all the way to the active layer where depletion occurs. For some CCDs the process will only work if oxygen (O_2) gas is present. The best results occur if the flooding is done warm and then the CCD cooled immediately afterward. Further it appears that if the UV radiation is too hard, i.e. too short a wavelength (e.g. 1850 Å) the process is reversed!

Negative surface charging can also be chemically induced. Exposure of a thinned CCD to nitric oxide gas (which is poisonous) for a few minutes dramatically improved the performance at 4000 Å in the blue. The effect was stronger than for UV flooding and somewhat more permanent. An alternative method for obtaining good blue response is the use of an arc or corona discharge near the CCD. It is hypothesized that the corona discharge or arc produces ionized oxygen, which charges the silicon surface by oxidation. Nitric oxide is also a strong oxidizing agent. Very little use has been made of corona discharge methods because of the possible danger to the CCD from the high voltage arc. However, as pointed out by Lloyd Robinson of Lick Observatory, high-voltage ion pumps have been used regularly on some CCD vacuum cryostats without destroying the CCDs. Lloyd even noticed an improved response in the TI 500 × 500 CCD in one of his cameras when the ion pump was switched on! Early users of the RCA CCD were concerned however, that ion bombardment of the silicon surface was causing some damage and most people try to avoid the use of ion pumps. The original success of the RCA chips in achieving superlative blue response was, in retrospect, not given sufficient attention. Having produced scanned electron-beam silicon vidicon targets for several years, RCA knew the importance of surface charging versus electron–hole pair recombination. Their 1976 patent application describes doping of the thinned backside layer to create a potential whose effect is to send photogenerated electrons inward toward the charge collection layer!

In 1986 the JPL workers once again came up with an even better method of obtaining control of the backside CCD surface of the thinned TI CCDs. Attempts were made to deposit a thin electrically conducting coating on the CCD that would allow an adjustable electrical voltage at the surface. Problems with good insulation between the conductor and

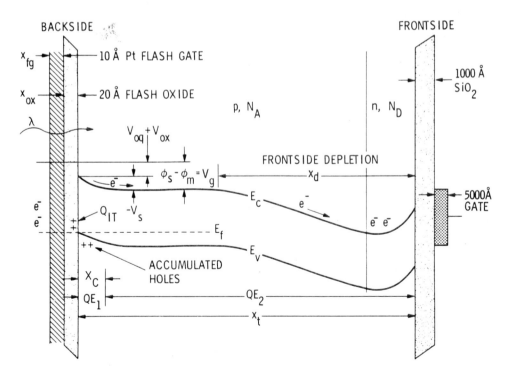

Fig. 7.18. A cross-sectional view of a thinned, buried-channel, backside-illuminated CCD with a ultra-thin coating of platinum called a "flash gate". The sketch is not to scale. The thickness of the flash gate is only 0.001 micrometres. Courtesy Jim Janesick, JPL.

the CCD limited the success of the proposed electrically controlled layer, but in the attempt it was discovered that simply the act of depositing an ultra-thin layer just 4 Å thick, of the right kind of metal, namely gold or platinum, produced an astonishing and permanent improvement in the UV, the extreme UV (XUV) and even the soft X-ray region. The new surface coating was termed the **flash gate** by Jim Janesick (Fig. 7.18).

Shortly after those events, while investigating oxide growth on the CCD surface, it was found that if the oxide was removed by an acid, the resulting quantum efficiency was very low initially, and moreover not one of the above surface treatments had any effect. Gradually, by exposure to air, the oxide grew back onto the surface and, lo and behold, the QE and the response to surface treatment returned to normal. Why? The explanation lies in the reduction in the number of surface traps as Nature smoothly and uniformly grows back the oxide layer. The most recent result has been the discovery that if the CCD is chemically etched (or "eaten away" with an acid), then oxidized by several days contact with de-ionized water (DI H_2O) or by a 1-hour exposure to steam at 100°C (Fig. 7.19), the CCD response with no other surface treatment is essentially perfect! Even devices that would not have responded well to UV flood are retrieved. Apparently, this new "good" oxide layer must give an overall negative charge to the CCD surface whereas the original "bad" oxide layer had so many traps that the surface was positively charged.

The highest possible quantum efficiency is achieved with a thinned backside-illuminated CCD. There are several possible techniques for thinning CCDs, including etching with solutions of hydrofluoric, nitric and acetic acids and mechanical lapping. For

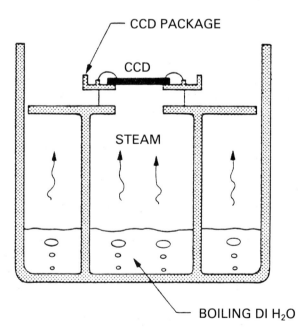

Fig. 7.19. A steam bath for a CCD! De-ionizing boiling water is used to assist in the preparation of the backside surface of a thinned CCDs. Courtesy Jim Janesick, JPL.

a considerable time the only excellent thinned CCDs which were not highly warped were those from RCA. Some thinned TI devices intended for the Hubble Space Telescope were so badly warped that custom corrective optics were required if the chip was to be used in a relatively fast camera system. For the most part, such difficulties have been eliminated today. SITe, Loral, EEV and Thomson have all developed thinning processes. Thinning and backside-passivation has also been developed by Mike Lesser at Steward Observatory. Mike is able to take thick devices and convert them to thinned devices or devices with UV-enhanced properties. See Lesser (1994) for an updated summary of all of the above techniques. The primary problem with thinned devices is still their fringing behaviour in the far red, which results in front-illuminated, phosphor-coated devices remaining very popular.

SUMMARY

There are many practical aspects to optimizing and using real CCDs in astronomy applications. This chapter has introduced all of the technical terms that astronomers are likely to encounter when trying to assess different CCD systems. Although it may seem that there are a great many problems in getting CCDs to excel, much of this has to do with the exceptionally stringent conditions required by astronomy applications. Over the years, CCDs have improved enormously, and astronomers have figured out how to get the most out of any given device. Excellent scientific CCD cameras can now be purchased commercially, but it is still necessary to "build your own" when you try something adventurous such as using a mosaic array of CCDs.

EXERCISES

1. Why must CCDs be cooled to low temperatures for astronomical use?
2. Explain what is meant by "inverted" operation and why is it advantageous?
3. What is meant by CTE or charge transfer efficiency? If a 2048 × 2048 CCD is quoted as having a CTE of "five nines", what fraction of the original charge in the last pixel will remain when it reaches the output?
4. What are the main sources of "noise" in a buried-channel CCD with a standard floating diffusion output?
5. Calculate the value of "kTC-noise", in electrons, for a CCD at 150 K if it has an output capacitance of 0.5 picofarads.
6. Explain the term "charge-trapping" and describe a method for overcoming the problem. What are the consequences of the solution?
7. A certain CCD requires a "fat zero" or "preflash" of 400 electrons. If it has a readout noise of 15 electrons what will be the final noise when the preflash is used?
8. A CCD camera seems to be showing a faint but regular pattern of sloping lines on bias frames, i.e. very short exposures with the shutter closed. What could be wrong?
9. What is meant by "over-scanning" a CCD? How could you use this technique to quantify the charge transfer efficiency (CTE) of a CCD?
10. When might you use on-chip charge "binning"? What do you need to watch out for when binning is used?

11. What causes "fringing" in CCD systems? Why is it worse at longer wavelengths?
12. What is meant by the term "down-converter"? Name two such materials and explain how they are used?

REFERENCES AND SUGGESTED FURTHER READING

CCD Imaging (1987) Booklet III, EEV Limited, Chelmsford, Essex.

Crawford, D.L. and Craine, E.R. (Eds) (1994) Instrumentation in Astronomy VIII, *Proc. SPIE* Vol. 2198, (in two parts); contains many articles on CCD systems, including controllers.

Gunn, J.E., Emory, E.B., Harris, F.H. and Oke, J.B. (1987) "The Palomar Observatory CCD camera", *Publ. Astron. Soc. Pacific*, **99**, 518–534.

Holst, G.C. (1996) *CCD Arrays, Cameras and Displays*, SPIE Press, Vol. PM30, Bellingham, WA, USA.

Howell, S.B. (Ed.) (1992) *Astronomical CCD Observing and Reduction Techniques*, Conf. Series, Vol. 23, Astronomical Society of the Pacific, San Francisco, CA.

Janesick, J.R. and Elliott, T. (1992) "History and advancement of large area array scientific CCD imagers", in *Astronomical CCD Observing and Reduction Techniques*, S.B. Howell (Ed.), Conf. Series, Vol. 23, pp 1–67, Astronomical Society of the Pacific, San Francisco, CA.

Janesick, J.R., Elliott, T., Daud, T. and Campbell, D. (1986) "The CCD flash gate", *Instrumentation in Astronomy VI*, David L. Crawford (Ed.), *Proc. SPIE*, **627**, 543–582.

Janesick, J.R. (1993) *CCD Camera Design and Application*, Lecture Notes, SPIE Meeting.

Jorden, P.R. (1990) "An EEV large-format CCD camera on the WHT ISIS spectrograph", *SPIE*, Vol. 1235, *Instrumentation in Astronomy VII*, D.L. Crawford (Ed.), pp 790–798.

Jorden, P.R. (1994) "ING CCDs—Properties of Operational Cameras", *ING La Palma Technical Note* No. 93, Royal Greenwich Observatory, UK.

Lesser, M.P. (1994) "Improving CCD quantum efficiency", in *Instrumentation in Astronomy VIII*, D.L. Crawford and E.R. Craine (Eds), *Proc. SPIE*, Vol. 2198, pp. 782–791.

Philip, A.G.D., Janes, K.A. and Upgren, A.R. (Eds) (1995) *New Developments in Array Technology and Applications*, IAU Symposium 167, Kluwer Academic Publishers, The Netherlands.

Rieke, G.H. (1994) *Detection of Light from the Ultraviolet to the Submillimeter*, Cambridge University Press, Cambridge, England.

Robinson, L. (Ed.) (1988) Instrumentation for ground-based optical astronomy: present and future, *Santa Cruz Summer Workshop in Astronomy and Astrophysics* (9th;1987), Springer-Verlag, New York.

White, M.H., Lampe, D.R., Blaha, F.C. and Mack, I.A. (1974) "Characterization of surface channel CCD image arrays at low light levels", *IEEE J. Solid State Circuits*, **SC-9**, 1–12.

8

Electronic imaging at infrared wavelengths

So great was the impact of the CCD that it is frequently said to have "revolutionized" optical astronomy. That same pronouncement, however, would seem like an understatement for the invention of the "infrared array" about a decade later. Infrared observations are extremely important in astrophysics for many reasons. For example, because of the Hubble expansion of the universe, the visible light from distant galaxies is stretched, effectively moving the spectrum to the infrared for the most distant objects.[†] Equally important, infrared wavelengths are much more penetrating than visible or UV light, and can therefore reveal the processes at work in star-forming regions which are typically enshrouded in clouds of gas and dust. Infrared arrays with one million pixels are now available, but infrared imaging systems are still much more expensive than the equivalent CCD cameras, and are therefore not easily obtainable for amateur astronomers—yet! Telescopes at relatively "bright" sites, where city lights have become a problem, can be revitalized by infrared imaging detectors because at these wavelengths the sky is already bright.

8.1 THE INFRARED WAVEBANDS

8.1.1 From Herschel to IRAS

For many years the infrared part of the spectrum was considered to be the region just beyond the red limit of sensitivity of the human eye, at a wavelength of about 700 nm (0.7 μm). With the advent of semiconductor silicon technology for imaging applications at very low light levels, optical astronomy extended its territorial claims to about 1.1 μm—the cut-off wavelength for detection of light imposed by the fundamental band-gap of silicon (see section 6.2). So where is the optical–IR boundary for ground-based astronomy? A reasonable response is that it occurs at a wavelength of around 2.2 to 2.4 μm, because at this wavelength there is a marked and fundamental change in the nature of the "background" light entering the telescope/detector system. For wavelengths shorter than ~2.4 μm the background light comes mainly from faint emission produced by

---

[†] The wavelength shift is measured by the scale factor $(1+z)$ where z is called the "redshift"; that is, $\lambda = \lambda_0(1+z)$. At a redshift of $z = 2.5$, when the universe was only $1/(1+2.5) = 0.29$ its present size, the familiar red H-alpha line of hot hydrogen gas at 656.3 nm moves to the infrared wavelength of 2297 nm (≈ 2.30 μm).

solar-induced photochemical reactions in the Earth's upper atmosphere, whereas at longer wavelengths the dominant source of background radiation is the thermal or heat emission from the atmosphere and telescope optical components.

Therefore, the domain of infrared astronomy extends to longer wavelengths from about 1.0 μm; the cut-off for silicon CCDs is 1.1 μm. **Near-infrared** is generally taken to be the interval from 1 to 5 μm, although the term **short wave infrared** (SWIR) is used for 1–2.5 μm, **mid-infrared** extends from 5 to 30 μm and **far-infrared** stretches from 30 to 200 μm. Wavelengths longer than about 350 micrometres are now referred to as the **sub-millimetre**, and although sub-millimetre astronomy is closely allied with infrared wavelengths in terms of the objects and regions of space which are studied, many of its techniques are more akin to those of radio astronomy.

Infrared astronomy had an early albeit somewhat accidental origin when, as early as 1800 in a series of papers, Sir William Herschel noted that a thermometer placed just beyond the reddest visible light in a spectrum of sunlight actually increased its temperature compared to two other thermometers set off to the side. Herschel called these unseen radiations "calorific rays" and proved that they were refracted and reflected just like ordinary light. The region of the spectrum discovered by Herschel was only the tip of the iceberg. Despite its early start, and some additional development of infrared detectors by Thomas Edison, and later by Golay, the major breakthroughs in infrared astronomy did not come until the 1950s—the era of the transistor—when simple photoelectric detectors made from semiconductor crystals became possible. The lead-sulphide (PbS) cell was used by Harold Johnson, to extend the "new" field of photoelectric photometry, and by Gerry Neugebauer and Bob Leighton at the California Institute of Technology in a pioneering, "two-micron survey" of the sky with an angular resolution of 4 arcminutes. For wavelengths beyond 3 μm however, something else was needed. One important step was the invention of the gallium-doped germanium (Ge:Ga) **bolometer** in 1961 by Frank Low who was at that time working for the Texas Instruments Corporation; Frank later moved to the University of Arizona in Tucson where he not only established a formidable infrared program, but he also set up a company, called Infrared Labs, which is very well known to everyone in the field. The gallium-doped germanium detector opened up much longer wavelengths to astronomers, especially with the aid of rocket, balloon and airplane surveys to get above more of the atmosphere. In the seventies, the lead sulphide cell was replaced by a photodiode made from indium antimonide or InSb (Hall *et al.*, 1975), typically about 0.25 mm in size, while improved bolometers and detectors using selectively doped silicon were also introduced for longer wavelengths.

Discoveries came fast and furious through the sixties and seventies. At Caltech, Gerry Neugebauer and his student Eric Becklin discovered many strong infrared sources including the centre of our own galaxy, and a very young optically invisible star in the Orion Nebula (now named the BN-object). This success rate stimulated the push for telescopes which were "optimized" for infrared work. By 1979, a new generation of 3–4 metre class telescopes dedicated to infrared astronomy were coming into operation including, for example, the United Kingdom 3.8-m Infrared Telescope (UKIRT)—the predecessor of which was the so-called "infrared flux collector" on Tenerife in the Canary Islands, a 60-inch telescope pioneered by Professor Jim Ring and colleagues at Imperial College London—and the NASA 3-m Infrared Telescope Facility (IRTF) whose first director was

Eric Becklin. Both of these telescopes were located on the 14 000 ft summit of Mauna Kea, Hawaii, which was rapidly becoming recognized as an exceptional site. Other tele-scopes optimized for infrared astronomy soon followed, and a great many "optical" tele-scopes began to provide optional "top-ends" with smaller secondary mirrors having much slower focal ratios (e.g. $f/35$) for infrared work; reasons for these slower f-ratios are given later. At wavelengths longer than 2.4 μm moonlight is almost undetectable and so infrared became the "bright time" or full Moon option at many optical telescopes where the new CCD instruments were best used for extremely faint work during dark of the Moon.

Thirty years later Frank Low and Gerry Neugebauer, together with astronomers in Eu-rope, played another key role in supporting the highly successful Anglo-American-Dutch **Infrared Astronomical Satellite** (IRAS) mission, which gave astronomers their first deep all-sky survey of the infrared sky. With the launch of IRAS in 1983, infrared astronomy took another quantum leap. This very successful mission mapped the entire sky at wave-lengths of 12, 25, 60 and 100 micrometres, produced a point source catalogue of over 245 000 sources (more than 100 times the number known previously), and made numerous unexpected discoveries, including a dust shell around the "standard star" α Lyrae (Vega). Unfortunately, IRAS had a lifetime of only 10 months in operation until the on-board supply of 475 litres of superfluid helium coolant, which held the detectors at 1.8 K and everything else at about 10 K, was finally exhausted. The 60-cm telescope and its detec-tors then warmed-up and lost their sensitivity. So successful was IRAS, that follow-up space missions involving "observatory class" satellites were planned by both ESA and NASA. The European project, called ISO (Infrared Space Observatory), was launched successfully in late 1995, while the American project, called SIRTF (Space Infrared Tele-scope Facility), has suffered delays due to budget cuts but is likely to be launched around 2002.

It is by no means easy to perform infrared astronomy from the surface of the Earth. To begin with, the atmosphere is not uniformly transparent at infrared wavelengths as it is in the visible part of the spectrum. Secondly, there is a lot of "background" emission at these wavelengths which reduces the contrast for detection of faint sources.

8.1.2 Atmospheric extinction and "windows"

Water vapour (H_2O) and carbon dioxide (CO_2) do an efficient job of blocking out a lot of infrared radiation. Fig. 8.1(a) shows a simplified transmission spectrum of the atmosphere over a broad wavelength range and Fig. 8.1(b) gives a more detailed spectrum in the near infrared. The water vapour content is particularly destructive for the longer wavelengths, but it is sensitive to the altitude of the observatory, and that is why most infrared tele-scopes are located at high, very dry sites such as Mauna Kea, Hawaii (14 000 ft). The absorption by these molecules occurs in certain wavelength intervals or bands, between which the atmosphere is remarkably transparent. These wavelength intervals are called atmospheric "**windows**" of transparency. Fortunately, there are many very important as-tronomical features to be seen at the wavelengths corresponding to the atmospheric win-dows. The standard windows are listed by central wavelength and the full width at half maximum intensity (FWHM) passband in Table 8.1. The J, H and K bands are sometimes called the Short Wave InfraRed (SWIR) and there are several variations on the definition

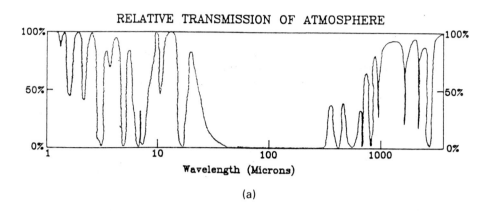

(a)

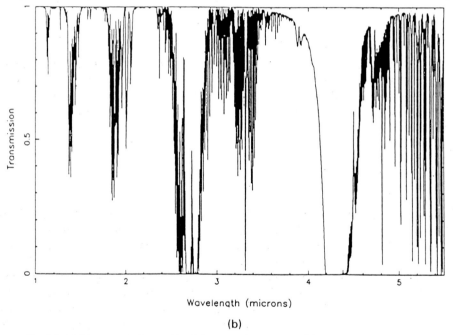

(b)

Fig. 8.1. (a) The transmission profile of the atmosphere over a wide wavelength interval from 1 μm to 1mm. Various atmospheric "windows" of transparency are apparent. (b) The near infrared (1-5 μm) transmission of the atmosphere above Mauna Kea (14 000 ft) for a typical water vapour level.

of these bands (such as K_{short}; 2.0–2.3 μm, K'; 1.95–2.30 μm and L'; 3.5–4.1 μm) which have been developed to improve performance especially at less dry sites. Interference filters can be manufactured to match these windows by several companies (e.g. OCLI, Barr Associates) and narrower passbands can also be produced. Several different filter sets are in use (e.g. UKIRT, Caltech) and therefore care must be taken when comparing observations with those of others.

Table 8.1. Infrared windows in the Earth's atmosphere

Centre Wavelength (μm)	Designation of the band	Width (FWHM) (μm)
1.25	J	0.3
1.65	H	0.35
2.2	K	0.4
3.5	L	1.0
4.8	M	0.6
10.6	N	5.0
21	Q	11.0

There is also a relatively poor window (designated X) from about 30–35 μm which is accessible from dry high-altitude sites such as Mauna Kea or Antarctica.

8.1.3 Atmospheric emissions

There are two major sources of solar-induced non-thermal emission that dominate the near infrared night sky from 1 to 2.5 μm. The first is the polar aurora, due mainly to emission from N_2 molecules, but which is negligible at mid-latitude sites such as Mauna Kea, the Canary Islands and Chile. The dominant problem is "airglow" which has three components:

- OH vibration-rotation bands
- O_2 IR atmospheric bands
- the near-infrared nightglow continuum

Of these, the strongest emission comes from the hydroxyl (OH) molecule which produces a dense "forest" of emission lines, especially in the 1–2.5 μm region. First identified astronomically in the optical red by Meinel, these emission bands are formed as OH molecules relax after absorption of UV photons from the Sun. The **OH emission** light comes from a thin layer of the atmosphere at an altitude of about 90 km and the strength of the emission can vary by a factor of 2 or more in half an hour due, it is believed, to large-scale wave motions in the mesopause. Since the line features are blended together at a spectral resolving power of $R < 600$, they are a significant drawback for low resolution spectroscopy in the near IR. OH emission is not confined to the near IR, but at longer wavelengths the OH emission is swamped by thermal emission.

8.1.4 Thermal emission

Since the atmosphere is at a finite temperature, it also emits thermal (blackbody) radiation with an **emissivity** (ε), which depends on how opaque the atmosphere is at that wavelength; the emissivity will be 1.0 (100%) if the atmosphere is totally opaque (i.e. a perfect blackbody), but may be less than 0.1 (10%) in good windows where absorption is low. Again, water vapour is the main problem, as it is responsible for much of the absorption from 3 to 5 μm, where the thermal emission of the atmosphere rises steeply. At wavelengths less than 13 μm, however, it is the thermal blackbody emission by the telescope and warm optics which dominate the background since both are at least 20 K warmer than the effective water vapour temperature.

To predict the thermal emission from the telescope optics and any other warm optics in the beam we need to know two quantities: the absolute temperature T (K) which determines the spectrum of blackbody radiation from the Planck function $B_\lambda(T)$, and the emissivity ε of each component which determines the fraction of blackbody radiation added to the beam. The Planck function is given by

$$B_\lambda(T) = \frac{2hc^2}{\lambda^5} \frac{1}{\left(e^{hc/\lambda kT} - 1\right)} \qquad (8.1)$$

where the units are $W \cdot m^{-2} \cdot m^{-1} \cdot sr^{-2}$ and h, c and k are the Planck constant, the speed of light and the Boltzmann constant. (Other forms of the Planck function are given in Appendix 3.) The peak in this function is given by Wien's displacement law:

$$\lambda_{max} T = 2898 \quad \mu m \cdot K \qquad (8.2)$$

For example, the peak occurs at $\lambda = 10$ μm for $T \approx 290$ K (17 °C), at 1 μm for T = 2898 K and at 0.1 μm (1000 Å) for a temperature of about 29 000 K.

Note that objects which "appear" black to our visual senses may not be black at longer wavelengths, i.e. they may reflect some infrared light; special infrared black paints are available. To estimate the emissivity of telescope mirrors (due to absorption by the coating or by a layer of dust on the surface) we can apply the law due to Kirchhoff and take one *minus* the measured spectral reflectivity. For example, if the reflectivity is measured to be 96% then the emissivity is 4% and is additive for the train of warm optics. There will be an additional component of emissivity from dust on the mirror surface.

When the thermal and non-thermal components of background are added together we get the total background flux entering the instrument. The typical resulting spectrum, shown in Fig. 8.2, is dominated by OH emission lines at the shorter wavelengths and then rises very steeply towards longer wavelengths due to thermal emission. The assumptions in this model include an atmospheric temperature of 253 K, an airmass of 1.5, a water vapour content of 1.2 mm, a warm optical train including primary, secondary and tertiary of the telescope plus two additional steering mirrors and a calcium fluoride window all at 275 K and with a total emissivity of 20%. Note the steep increase in background by many factors of ten towards the thermal infrared. With 20% emissivity, it is the warm optical train that dominates the thermal background. However, slight "bumps" can be seen in the smooth thermal background due to the superposition of the atmospheric thermal emission in those spectral intervals where the atmosphere is virtually opaque and the emissivity is unity. This difference in thermal contribution between the atmospheric bands is accentuated in an IR optimized telescope with a much smaller total emissivity. Most of the very large 8–10-m telescopes will try to achieve this goal.

To emphasize how bright the night sky is at infrared wavelengths we can compare the brightness in magnitudes of one square arcsecond in the blue ($\lambda = 0.43$ μm) $m \approx 24$ (no moonlight), with that at 2.2 μm in the near IR, $m \approx 13.5$, and at 10 μm the sky and telescope combined are brighter than $m \approx -3.0$ (depending on the emissivity and temperature)!

Background just inside dewar window

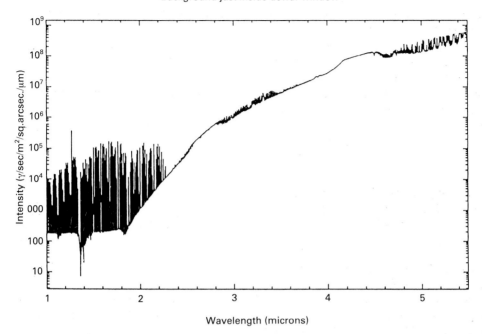

Fig. 8.2. The spectrum of total background radiation entering an infrared instrument as a function of wavelength. Note the logarithmic scale of the vertical axis which gives the number of photons/s per square metre of telescope aperture per square arcsecond on the sky per micrometre of wavelength interval. At the shortest wavelengths the dominant source is OH emission lines whereas at longer wavelengths the flux is due to thermal emission from the telescope and sky.

The most effective way of eliminating telescope background is to cool the entire telescope! On Mauna Kea at 14 000 ft above sea-level the temperature hangs around 1 °C. Temperatures in Antarctica near the South Pole are lower still, ranging from −13.6 °C to −82.8 °C and hence have stimulated the development of Antarctic astronomy. The Center for Astrophysical Research in Antarctica operates several instruments near the Amundsen–Scott South Pole Base, including an automated sub-millimetre telescope (AST/RO), a cosmic microwave background experiment (COBRA) and SPIREX (South Pole InfraRed Explorer) which is an infrared telescope and imaging system.

8.1.5 Chopping secondaries
Early infrared astronomers found a solution to the problem of a bright sky ... it is called "chopping". The infrared beam is rapidly switched between the source position on the sky and a nearby reference position, by the use of an oscillating or "wobbling" secondary mirror in the telescope itself. Typical wobbling secondaries on IR-optimized 3–4 metre telescopes such as the IRTF and UKIRT are small, 0.24–0.31 m in diameter, and have a slow f/ratio of f/35. On the larger 8–10 metre telescopes such as Gemini North and Keck I and II, these secondaries are bigger and more massive (1.0 m for Gemini and 0.5 m for Keck). Moreover, on alt-az telescopes the chop direction must be variable. Chopping typi-

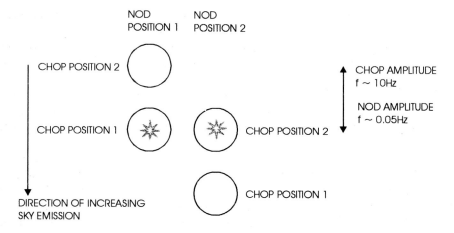

Fig. 8.3. Explanation of chopping and nodding to remove background flux and gradients in the background.

cally takes place at a frequency of about 10–20 times per second (10–20 Hz). In a photometer, this method involves isolating the astronomical object in a small aperture and first measuring the total brightness of "object plus sky" included in the aperture. Chopping changes the location of the image in the focal plane quickly so as to record the signal from a nearby piece of sky containing no objects in view. Once the difference is formed, the sky signal is eliminated, provided it has remained constant. In addition, it is usually necessary to move the entire telescope every minute or so to enable the sky on the "other side" of the object to be measured and thereby eliminate any systematic trend or gradient; this step is called "nodding" and the amount of the nod is usually the same as the "throw" of the chop for symmetry (Fig. 8.3). The difference between the pair of chopped signals for nod position 1 is given by

$$C_1(x) = S + B_{tel,1} - B_{tel,2} + \left(\frac{d}{dx} B_{sky} \right) \Delta x$$

where B_{tel} and B_{sky} are the telescope and sky backgrounds at the two chop positions separated by Δx. These terms are usually always much larger than the source flux, S. For the second nod position the signs are reversed and the difference signal is

$$C_2(x) = S - B_{tel,1} + B_{tel,2} - \left(\frac{d}{dx} B_{sky} \right) \Delta x$$

and adding these two results gives the required source signal

$$S = \frac{1}{2} \left(C_1(x) + C_2(x) \right)$$

Chopping and nodding are generally required at wavelengths longer than about 3.5 µm, and nodding is required for good background subtraction at shorter wavelengths too.

One reason for using a secondary mirror with a slow *f*/ratio is chopping, another is to significantly reduce the background on a given detector area by stretching the plate scale. For example, going from an *f*/9 secondary to an *f*/36 gives a smaller scale in arcseconds per millimetre by a factor of 4, and reduces the flux falling on each square millimetre by a factor of $4^2 = 16$.

The same strategy of chopping and nodding can be pursued with a panoramic detector too, either by chopping completely off the field of view of the array detector or by chopping a much smaller angle so that the source remains on the detector, and can be observed twice. This is common practice for point source objects.

An infrared secondary mirror is "undersized" to permit chopping and therefore it is over-filled by the beam from the primary mirror. Note that this means that the primary mirror no longer defines the entrance pupil of the system; it is now defined by the size of the secondary mirror. In general, the secondary mirror is not surrounded by a black baffle tube in the normal way because it is important to ensure that any image of the secondary formed inside the instrument is surrounded by sky, which produces a lower background than a warm, black baffle. Often, the secondary will be gold-coated for best infrared performance since gold is more reflective than aluminium in the IR. In addition, there will be either a small deflecting mirror or a hole at the centre of the secondary with access to the sky. Such precautions eliminate thermal photons from the central Cassegrain hole in the primary mirror. Telescopes built this way (such as UKIRT and the IRTF) are said to be infrared optimized.

8.2 INFRARED DETECTORS

Infrared detectors are classified as **photon** detectors or **thermal** detectors. In photon detectors, individual incident photons interact with electrons within the detector material. For example, if a photon frees an electron from the material the process is called photoemission; this is how a photomultiplier tube (PMT) works and it is also the underlying principle of the Schottky-barrier diode (e.g. platinum silicide on silicon) except that the free electron escapes from the semiconductor into a metal rather than into a vacuum. When absorption of photons increases the number of charge carriers in a material or changes their mobility, the process is called **photoconductivity**. Such devices are usually operated with an external voltage across them to separate the photogenerated electron–hole pairs before they can recombine. One carrier, e.g. the electrons, moves freely and the other carrier migrates very little until it recombines in the crystal lattice; this effect adds noise. If absorption of a photon leads to the production of a voltage difference across a junction between differently "doped" semiconductors the process is known as the **photovoltaic** effect. The interface is normally a simple pn junction which provides an internal electric field to separate electron–hole pairs; note that both the electron and the hole migrate and hence there is no recombination noise. More often the pn junction or diode is operated with an externally applied voltage to cause a reverse bias. The migration of the photon-generated carriers changes the electric field across the junction, effectively charging or discharging the diode as if it were a capacitor. Photovoltaic detectors are generally superior in noise performance to photoconductors. A modified construction of a photoconductor known as a blocked-impurity band or BIB detector contains an internal undoped "blocking layer" to give it the noise characteristic of a photovoltaic detector.

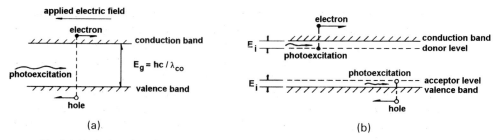

Fig. 8.4. Excitation of an electron due to absorption of a photon in (a) an intrinsic semiconductor, (b) an extrinsic semiconductor.

There are two classes of semiconductors used in infrared photon detection; **intrinsic** and **extrinsic** (see Fig. 8.4). In an intrinsic semiconductor crystal, a photon with sufficient energy creates an electron–hole pair when the electron is excited to the conduction band, leaving a hole in the valence band (see also Table 6.1 in Chapter 6); this is the same physical process that operates in a silicon CCD to detect photons. Common infrared intrinsic materials are germanium (Ge), lead sulphide (PbS), indium antimonide (InSb) and mercury cadmium telluride (HgCdTe), because these have band gap energies E_g less than silicon and therefore have a longer cut-off wavelength. Extrinsic semiconductors are intrinsic materials such as silicon and germanium doped with impurities so that a photon with insufficient energy to excite an electron–hole pair directly, can still cause an excitation from an energy level associated with the impurity atom. Donor atoms produce an energy level closer to the conduction band and acceptor atoms are those that produce a level near the valence band. Several extrinsic semiconductors are listed in Table 8.2 together with approximate values of their long wavelength cut-off points.

Table 8.2. Extrinsic semiconductors, doping material and long wavelength cut-off

Base	:Impurity	λ_c (μm)	Base	:Impurity	λ_c (μm)
Silicon (Si)	:In	8.0	Germanium (Ge)	:Au	8.27
	:Ga	17.1		:Hg	13.8
	:Bi	17.6		:Cd	20.7
	:Al	18.1		:Cu	30.2
	:As	23.1		:Zn	37.6
	:P	27.6		:Ga	115
	:B	28.2		:B	119.6
	:Sb	28.8		:Sb	129

In thermal detectors, the absorption of photons leads to a temperature change of the detector material, which may be observed as a change in the electrical resistance of the material as in the **bolometer**; the development of a potential (voltage) difference across a junction between two different conductors, as in the thermopile; or a change of internal dipole moment in a temperature-sensitive ferro-electric crystal, as in a pyroelectric detector. Of these, only the bolometer has been widely used in astronomy (see Fig. 8.5). The germanium bolometer is commonly used for wide band energy detection at wavelengths longer than 10 μm, but silicon bolometers have also been successfully introduced in recent years. In some instruments, several individual bolometers have been arranged in a grid pattern to produce an "array" of individual detectors. Bolometers absorb essentially all of the incident radiation (quantum efficiency = 100%) but are relatively noisy and are used mainly under very high background conditions.

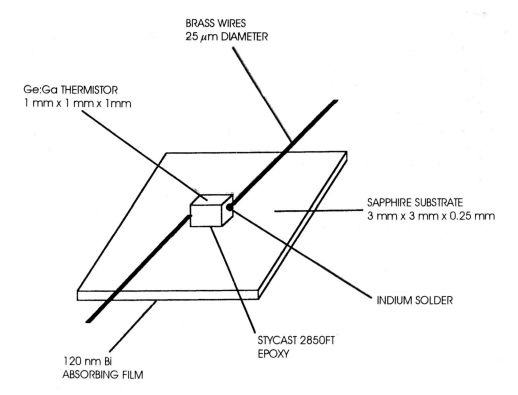

Fig. 8.5. The basic construction of a gallium-doped germanium bolometer.

By cooling these detectors and their local environment (see Chapter 6) with liquid cryogens such as liquid nitrogen (77 K) and liquid helium (4 K), or with electrically operated closed-cycle refrigerators, two very important results are achieved. Firstly, the heat or infrared emission from all the filters, lenses and metal support structure surrounding the infrared detector is completely eliminated. This is essential. If the infrared instrument itself, whether spectrometer or photometer, is not cold then the detector will be swamped instantly with thermal radiation. Secondly, operating the detector at a low temperature greatly reduces the detector's own internal, thermally generated background or "dark current" which results in a huge increase in sensitivity. Of course, in order to cool the detector and all surrounding metal and glass components, the entire unit must be placed inside a vacuum chamber with an infrared-transmitting window. This is similar to enclosing a CCD detector in a dewar or cryostat except that now the entire optical instrument, and all its mechanisms, must be inside as well! This is one reason why infrared instruments are harder to design and build than optical instruments.

The kinds of infrared detector systems available to astronomers up until about 1983, although quite sensitive, were not really ideal for making extensive infrared pictures of the sky with the same fine, seeing-limited detail of optical images. There were no infrared "photographic" plates or infrared TV cameras. All published infrared images of celestial objects prior to that time were actually maps made by scanning the infrared scene point-by-point, back and forth in a sweeping pattern, called a "raster", with a single sensitive detector having a small angular field of view on the sky. Usually, the field of view was a circle no smaller than 2 or 3 arcseconds. This is rather like using a photomultiplier tube, which is a single detector, to map out a scene which could be imaged in an instant with a snapshot by a CCD camera or a photographic plate! Alternatively, imagine using just *one* pixel in a CCD array of 500 × 500 pixels and then moving the image so that a different part of the scene fell on your pixel! Scanning was done by moving the entire telescope or some major optical component such as the secondary mirror. As a result, errors due to small miss-pointing of the telescope, imprecise registration of individual parts of the map, and slow changes in the background signal level—due to temperature changes, variations in

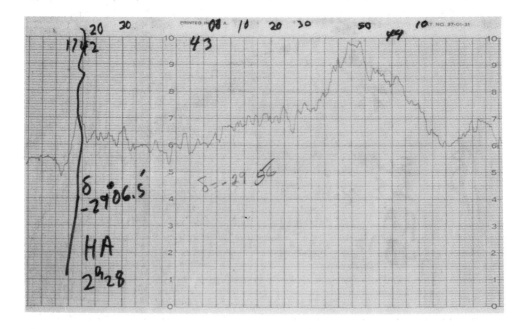

Fig. 8.6. A copy of the chart recording made by Becklin and Neugebauer to detect the infrared signal from the Galactic Centre.

the behaviour of the atmosphere and variations in the detector system—all introduce errors and artefacts. Moreover, even if the circular patches of sky are carefully overlapped in the mapping process it is impossible to record fine detail on the scale of the "seeing" (typically, 1 arcsecond or better).

Early infrared detectors were single-element devices ac-coupled to a phase-sensitive amplifier "locked" to the frequency of the chopper oscillations. Initially, the infrared signals were recorded only on a chart recorder as shown in Fig. 8.6 which illustrates the first detection of the bright infrared sources at the centre of our Galaxy. Compare this to the state of the art now by looking at the three-colour image in Plate 6.

In the mid-1980s, however, infrared astronomy received its greatest boost of all, with the introduction of two-dimensional **infrared arrays** tailored for astronomy applications. With these powerful electronic imaging devices it became feasible to pursue infrared astronomy much more vigorously from (very much larger) ground-based telescopes. Infrared array detectors have many similarities to silicon CCDs, but they do not employ the charge-coupling principle.

8.3 INFRARED ARRAY DETECTORS

8.3.1 The infrared "array" revolution—déjà vu

Lacking the long pre-CCD history of photographic imaging enjoyed by optical astronomy, it is easy to appreciate the staggering boost to infrared astronomy that occurred when the first infrared arrays were introduced.

Many forms of infrared array devices were constructed during the period 1974–1984 due mainly to the extreme importance of the infrared for military applications, such as heat-seeking weaponry and battlefield smoke conditions, but even by 1982 when I carried out a survey for the Royal Observatory Edinburgh (ROE) and UKIRT, formats were very small (1 × 8 or 2 × 16), very few array devices were actually for sale and virtually none of them had the requisite performance expectations for low-background astronomical applications. A handful of researchers had reported technical evaluations of a few devices and some obtained astronomical images, but prospects were bleak.

An early champion of astronomical array devices was Craig McCreight of the NASA Ames Research Center who led in-house tests and coordinated a major NASA-funded programme involving a number of other groups. As early as 1979, John (Eric) Arens and co-workers at the Goddard Space Flight Center tested a 32 × 32 pixel bismuth-doped silicon array made by the Hughes corporation at a wavelength of 10 μm, although only a 2 × 10 pixel area was functioning. A 32 × 64 platinum silicide Schottky barrier array was used by researchers at the Kitt Peak National Observatory (Al Fowler) and at the NASA Infrared Telescope Facility (Rich Capps) in collaboration with the US Airforce. A great deal of military funding was going in to the development of mercury–cadmium–telluride devices for mid-IR work, but it was difficult to obtain information on that technology. While those early results were encouraging, it was clear that none of the devices yet produced were ideally suitable for astronomy. Infrared charge-injection devices (CIDs) for various wavelengths and long linear photodiode arrays using individual, switchable MOS-FET multiplexers were tested and described in the technical literature; one of the best of these arrays was a 32-element linear array of indium antimonide (InSb) developed by Jim

Wimmers, Dave Smith and Kurt Niblack at Cincinnati Electronics Corporation which was used successfully by astronomers in near-infrared spectrographs. Each of these devices had some drawback for astronomy, such as poor quantum efficiency or high readout noise (~1000 electrons), or just not enough pixels! The most hopeful sign came from tests of a 32 × 32 array of indium antimonide detectors reported in 1983 at a NASA Ames workshop by Judith Pipher and Bill Forrest of the University of Rochester (USA). This device was a "reject" loaned to them by Alan Hoffman, a former colleague who was now employed by Santa Barbara Research Center (SBRC) in California. This kind of situation occurred quite often! Experimental arrays left behind in the development of new military devices were "loaned" by manufacturers to a few astronomers with good contacts.

From 1982 to 1984 UKIRT and NOAO staff worked closely with senior SBRC engineers (Alan Hoffman and Jim West) and marketing personnel (Dick Brodie and Carol Oania) to define the technical specification and costs of a new infrared array device suitable for astronomy. The original readout design suggested by SBRC had to be abandoned as too complex and subject to amplifier "glow". Luckily, an alternative readout device was suggested, and tested, by Al Fowler at the Kitt Peak National Observatory (now called NOAO), although with platinum silicide rather than InSb as the detector material. We all agreed that the alternative device was a bit noisier but should work, and fabrication began on the well-known 62 × 58 InSb array. In addition, Judy Pipher, Bill Forrest, Giovanni Fazio and others working towards the instrument definition for SIRTF were also negotiating for detector development work at SBRC and thus there was a significant incentive and an opportunity for parallel development.

At about the same time, through contracts being developed for second-generation Space Telescope instrumentation, several groups of US astronomers (in Arizona, Chicago and Hawaii) had obtained access to new arrays made from mercury–cadmium–telluride (HgCdTe or MCT) from Jon Rode of Rockwell International; subsequently, Jon moved up in the company and his role was taken over by Kadri Vural. Both Jon and Kadri have been strong supporters of the astronomy detector programmes and, over the years, Kadri has played a very significant role in encouraging and promoting the development of (MCT) arrays for very wide use in astronomy. [Note: mer-cad-tel (MCT) is also known as cad-mer-tel or CMT, especially in Europe.] This development would prove to be very important, because these array devices can be customized to the shortest IR wavelengths and could be run at 77 K using liquid nitrogen and operated by existing CCD controllers. Much of the stimulus for the mer-cad-tel array development was generated by funding for a proposed new instrument for the Hubble Space Telescope—which was still a long way from launch at that time. The new instrument was called NICMOS and the principal investigator was Rodger Thompson at the University of Arizona (see Chapter 11). Marcia Rieke and Rodger Thompson at University of Arizona, and Mark Herald at University of Chicago were the first to successfully demonstrate these arrays at the telescope. Several European sources of infrared array technology, for space applications in particular, were also recognized, and French astronomers were already using an InSb array with the charge-injection principle. Preparations for both ISO and SIRTF had stimulated work on longer wavelength devices too, so excitement was high!

The first project to develop a general-purpose, user-friendly infrared array camera based on the new 62 × 58 InSb array began at ROE in June 1984, and two years later in

September 1986 we delivered IRCAM to the 3.8 m UKIRT in Hawaii (Fig. 8.7). Al Fowler received his device at NOAO at about the same time, and so he and I were frequently in touch across continents, trying to compare results as we went along. Indeed, the development was not trouble-free. For example, the batch of InSb used to make the astronomy devices turned out to suffer a loss of quantum efficiency when it was cooled to operating temperature. This was a trying time for SBRC manager Dave Randall and scientists like Alan Hoffman and Geoff Orias. Fortunately, a boule with completely different doping had been developed simultaneously for the SIRTF project, and so the ground-based pro-

Fig. 8.7. IRCAM: the first common-user camera system on the UKIRT 3.8-m infrared telescope to employ the 62 × 58 InSb arrays from SBRC which were the first arrays specifically designed for ground-based astronomy from 1–5 μm. The author is on the right.

gramme within SBRC was able to acquire some of that material which performed very well. So little was known about how these devices would work, that both Al and I, as well as Alan's team at SBRC, wanted to be very cautious until we had "first light" on the telescope. In my case that event occurred at 8 a.m. on the morning of October 23 1986, in broad daylight. Together with my long-time colleague Colin Aspin—who had written much of the software for the camera—we obtained the first infrared image of a cosmic source with IRCAM, it was the Orion nebula (Fig. 8.8 and colour plates). Although the SBRC detector had an array of only 3596 pixels—that was 3595 more than before! In the following year IRCAM was used from January to August on a "trial" basis by staff and visitors, and then in September 1987 it became available to the entire scientific community and quickly absorbed over 80% of all telescope time. Over the next 3 years, two more cameras were delivered. Of course, these instruments have long since been upgraded with new detectors, but IRCAM 1 was an example of what could be achieved by a small team of astronomers and professional engineers (the engineering group was led by Timothy Chuter) working closely together to cover a wide range of disciplines, including electronics, mechanics, cryogenics, optics and software. IRCAM was also the first of a new generation of infrared instruments in which much more emphasis than ever before was placed on automation, computer-control, engineering reliability, and especially on ease-of-use through "user-friendly" software.

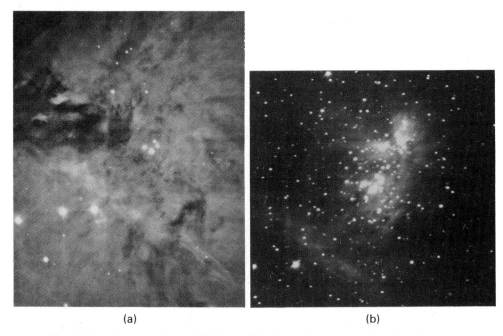

(a)   (b)

Fig. 8.8. (a) A photograph in visible light of the Trapezium region of the Orion Nebula is dominated by nebular emission. (b) An infrared image of approximately the same region as (a) obtained with the UKIRT IRCAM in 1986 at a wavelength of 2.2 μm. Numerous stars are revealed and the bright source above the Trapezium is known as the Becklin–Neugebauer object.

By March 1987, the first astronomical results from several of the new infrared arrays had begun to pour in. Short-wavelength mer-cad-tel was emerging as a powerful means for optical telescopes to extend their capability to 2.5 µm, and longer wavelength arrays were proving better than anyone dared hope. Formats were only 64 × 64 pixels of course, but plans for larger arrays were being laid. A key moment in infrared astronomy was a "workshop" on infrared array detectors in Hilo, Hawaii, in March 1987. Organized largely by Eric Becklin and Gareth Wynn-Williams for the University of Hawaii, Honolulu, with local support from David Beattie and I on behalf of UKIRT, since the UKIRT base was on the Big Island near the University of Hawaii, Hilo campus. That workshop attracted over 200 participants from all over the world, including CCD experts like Jim Westphal. To those of us who had straddled the apparent divide between optical and infrared astronomy, it was like history repeating itself. The euphoria was similar to the Harvard–Smithsonian meeting on optical CCDs in 1981. Six years later, in July 1993, Eric Becklin and I hosted a similar meeting at UCLA entitled "Infrared Astronomy with Arrays: the Next Generation". By then, everyone was already using 256 × 256 detectors for near-infrared work and 128 × 128 devices at mid-IR wavelengths, and plans for 1024 × 1024 NIR arrays were very advanced. Moreover, it was clear from the papers presented by 300 participants from all over the world that the new detectors had been assimilated into the subject and that a wide range of new astrophysics was being produced.

Just like CCDs, the advantages of the new array detectors are fairly obvious, but remember that the comparison is with a single-element detector and not with vidicons, image tubes and photographic plates. The main advantages can be summarized as follows:

- very time efficient because of large numbers of elements compared to the old single-element detectors; fainter detection limits
- high spatial resolution; usually seeing-limited, but new techniques permitting diffraction-limited resolution have also been successful
- no scanning; therefore much less prone to errors of measurement due to misregistration of scans
- very high sensitivity; partly the result of the small detector size
- "sky on the frame" gives the possibility of eliminating chopping, at least for NIR
- infrared spectroscopy is now viable

At the time of writing (1996), several kinds of arrays are in widespread use including 256 × 256 pixel arrays for near-infrared work, and 128 × 128 arrays for mid-infrared wavelengths, and "first-light" observations with NIR arrays having formats of 1024 × 1024 pixels were reported in 1995 by Klaus Hodapp for the University of Hawaii using a Rockwell mer-cad-tel array (Fig. 8.9(a)) and Al Fowler for NOAO using an SBRC InSb array. Unfortunately, infrared arrays are much more expensive than CCDs, typically $75 000–$150 000 per detector. Fig. 8.9(b) shows the 1024^2 InSb array from SBRC, which has an active area of 27.65 mm. From one to one million pixels in such a short period of time is both amazing and exciting.

Fig. 8.9. (a) The Rockwell 1024 × 1024 HgCdTe array has only 18.5 μm pixels and can fit into an 84-pin package.

Fig. 8.9. (b) The 1024 × 1024 InSb array from SBRC, which has 27μm pixels, is shown packaged in a 124-pin leadless carrier.

8.4 INFRARED ARRAYS—BASIC PRINCIPLES

To generate an infrared image an IR array detector must convert radiation into electrical charge by one of the means already mentioned—the photovoltaic effect, photoconduction or the Schottky barrier effect. Next, it must perform each of the following steps:

- store the electrical charge at the site of generation, i.e. in a pixel
- transfer the charge on each pixel to a single (or a small number of) outlets; this is the multiplexing task
- enable the charges to be removed sequentially as a voltage which can then be converted or digitized into a number for storage in a computer

Since each of the steps and stages in infrared detection are so similar to those employed in an optical silicon CCD it was natural to attempt to manufacture an entire CCD from some of the other known semiconductor materials. By selecting a material with a smaller energy band-gap between the valence and conduction bands one could, in principle at least, make a CCD. Texas Instruments were among the first to try. An entire CCD was manufactured out of the material mer-cad-tel or (HgCdTe) in the USA as early as 1974. Other companies tried indium antimonide (InSb), and still others tried silicon which had been specially treated or doped, all with varying degrees of success. The problems of manufacture turned out to be very difficult due to the limited experience in processing and purifying these materials compared to the extensive technology base which had already grown up around silicon. Finally, a different approach was taken—the **hybrid**.

In the hybrid infrared array, the functions of detecting infrared radiation and then multi-plexing the resulting electrical signal are separated, with the latter task going, of course, to silicon. At present then, infrared arrays are like "sandwiches" (as shown in Fig. 8.10) in which the upper slab is the IR sensor (e.g. InSb, HgCdTe; Si:XX, Ge:XX or PtSi), and the lower slab is a silicon multiplexer of some sort. The infrared part is really a tightly packed grid of individual IR pixels with minimum dead space between them. Initially, pixels sizes were quite large, 76 μm in the case of the 62 × 58 InSb array, but the current 1024 × 1024 arrays have pixels in the range 18–27 μm. Both slabs are provided with a grid of electrical connections in the form of tiny raised sections—referred to as "bumps"—of a soft electri-cal conductor called indium; indium remains soft at low temperatures. The slabs are liter-ally pressed together to enable the indium bumps to mate. Gaps between the bumps are usually back-filled with an epoxy to help maintain the integrity of the sandwich, especially through thermal cycles to low temperatures. What is the best choice for the multiplexer? CCDs would be acceptable since the potential wells provide the required charge storage, except for concern over their low-temperature operation. Charge injection devices (CIDs) can also be used (and a French-made InSb CID is one of the detectors on the ISOCAM instrument aboard ISO). Alternatively, a special array of "switches" made from metal-oxide-semiconductor field effect transistors (MOSFETs) in microscopic form can be used to access the signal from each IR detector (whether photodiode or photoconduc-tor). Charge storage may occur on the junction capacitance of the IR sensor itself or on a separate storage capacitor associated with the silicon circuitry. This turns out to be the best approach so far. The entire hybrid structure is often called a focal plane array (FPA) or a sensor chip assembly (SCA).

There are basically two approaches used in multiplexing the signal outputs from each unit cell. Either each pixel is read sequentially by connecting its signal to an output bus or each pixel can be accessed randomly for connection to the output amplifier. The latter method, called direct read out (DRO) is very attractive but requires more circuitry; the SBRC 62 × 58 InSb array had this feature. Sequential readout is easily implemented in the multiplexer using CMOS shift registers and this is the more common form.

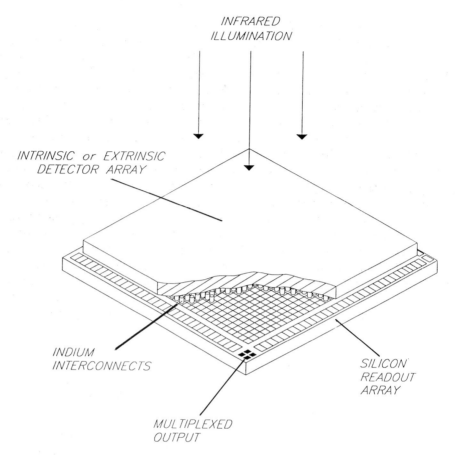

Fig. 8.10. Diagram illustrating the basic "hybrid" structure of infrared array devices. The two slabs are separated by tiny indium bumps.

8.4.1 How the "hybrid" structure works

Infrared array detectors are *not* based on the charge-coupling principle of the silicon CCD. This is an important distinction which has some practical implications. For example, IR arrays do not "bleed" along columns when a pixel saturates and bad pixels do not block off others in the same column. Also, since the pixel charge does not move, **non-destructive readout** schemes are possible and very effective. On the other hand, on-chip charge binning, charge-shifting and drift-scanning are not possible. Actually, a form of drift-scanning *can* be implemented but the charges are read out rather than shifted.

In the most common and successful multiplexer designs, the "unit cell" of an infrared array contains a silicon field-effect transistor (FET) used as a source follower amplifier, essentially providing a buffer for the accumulated charge in the infrared pixel. The term used to describe this structure is SFD—source follower per detector. Either the upper slab is constructed on an IR-transparent substrate, such as the sapphire substrate used for the NICMOS arrays, or the bulk semiconductor is physically thinned to enable photons to penetrate to the pixel locations on the underside. Note that other unit cell designs are possible. For example, instead of a simple SFD one could use a capacitive feedback trans-impedance amplifier (CTIA) which would allow charge integration to occur off the in-frared pixel. Charge injection can also be used. So far, the SFD approach has worked best.

When the absorption of a photon with a wavelength shorter than the cut-off wavelength generates an electron–hole pair in the semiconductor, at the location of the pixel, the pair is immediately separated by an electric field which can be applied externally or internally or both. For example, in the SBRC and Rockwell near-infrared arrays of InSb and HgCdTe, the electric field results from a reverse-biased pn junction. The depletion region produced by the reversed-bias junction acts like a capacitor which is discharged from its initial state (the full reverse bias voltage) by the migration of the electrons and holes; in effect, photogenerated charges are being "stored" locally at each pixel. Of course, there must be a small perimeter or boundary around each pixel to define each individual photo-diode, and this region cannot be too large otherwise there is a serious light loss, nor can it be too small because then there will be crosstalk between pixels, i.e. the charge from one pixel can be collected by an adjacent pixel. The voltage change across the detector capaci-tance is applied directly to the input gate of the silicon FET (across the indium bump interconnect), which in turns relays the voltage change to one or more output lines when it is switched on or "addressed" to do so. Again, there is no charge-coupling between pixels since each pixel has its own FET, and there is no overflow or charge bleeding since

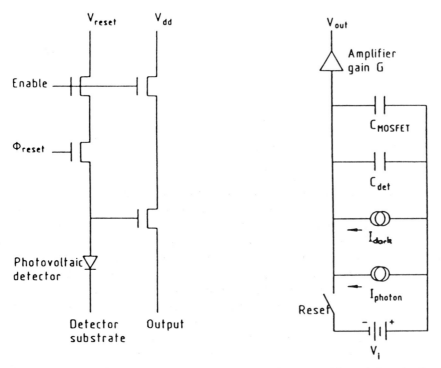

Fig. 8.11. The "unit cell" of a typical near-infrared photodiode array and its equivalent circuit. This is a four-transistor unit cell. In addition to the output source follower on each detector and its reset switch, there are two other FETs used for addressing. Note that the junction capacitance is not the only capacitor in the system.

the worst that can happen is that the diode becomes completely de-biased and no further integration occurs, in other words the pixel is saturated. In general, there is always a trade-off to be made in setting the effective pixel capacitance; the larger the capacitance the greater the charge storage efficiency or "full-well" capacity ($Q = CV/e$), but the greater the noise equivalent charge.

An equivalent circuit for a typical IR array with photodiode pixel elements is shown in Fig. 8.11 and the details of the process are summarized below:

- internal photoelectric effect produces electron–hole pairs
- an electric field separates the electrons and holes
- migration of electrons across the junction decreases the reverse bias, like discharging a capacitor
- the amount of charge is $Q = CV/e$ electrons where $e = 1.6 \times 10^{-19}$ C, V is the voltage across the detector and C is the effective capacitance (which is a function of pixel geometry, doping and bias conditions)
- each detector is connected to a source follower (SF) amplifier whose output voltage follows the input voltage with a small loss of gain; $V_{out} = A_{SF}V_{in} \sim 0.7V_{in}$
- the output voltage of the source follower can be sampled or "read" (with an analogue-to-digital converter) without affecting the input
- after sampling, the voltage across the diode can be RESET to the full reverse bias in readiness for the next integration

The reset action is accomplished with another FET acting as a simple on/off switch in response to a pulse applied to its gate.

8.4.2 Linearity

The equivalent circuit of the unit cell reveals that there are at least two sources of capacitance, the pn detector junction (C_{det}) and C_{FET}, the FET (plus other stray capacitance). In earlier InSb devices there was also a guard ring or gate around the pixel; this construction was eliminated with the 256×256 devices. Also, there are two sources of current to drive the discharge of the reverse bias, namely photoelectrons and "dark current" electrons. So far, many of the terms we have used are applicable to CCDs too. There is an obvious potential for non-linearity between photon flux and output voltage in these detectors which is absent in CCDs simply because the detector capacitance is *not* fixed, but does in fact depend on the width of the depletion region, which in turn depends on the value of the reverse bias voltage. Unfortunately, this voltage changes continuously as the unit cell integrates, irrespective of whether it is storing photogenerated charge or dark current charges. Using $Q = CV$ and taking both C and V as functions of V then

$$dQ = \left(C + \frac{\partial C}{\partial V} V \right) dV \equiv I_{det} \, dt \tag{8.3}$$

The rate of change of voltage with time (dV/dt) is not linear with detector current I_{det} because of the term $\partial C/\partial V$. Representing all fixed capacitances as C_{fix} then the denominator can be shown to be given by

$$C + \frac{\partial C}{\partial V} V = C_{fix} + C_0 \left((1 - \frac{V}{V_{bi}})^{-1/2} + \frac{1}{2} \frac{V}{V_{bi}} (1 - \frac{V}{V_{bi}})^{-3/2} \right) \tag{8.4}$$

where C_0 is the junction capacitance at zero bias and A_{det} is the detector area, N_A and N_D are the acceptor and donor doping concentrations (atoms/cm^3), with $N_A \gg N_D$ to form an abrupt junction, ε_s is the dielectric constant of the material in the junction (e.g. $17.7\varepsilon_0$ for InSb) and V_{bi} is the built-in potential of the diode given by

$$C_0 = A_{det} \left[\frac{e\varepsilon_s}{2V_{bi}\left(\dfrac{1}{N_A} + \dfrac{1}{N_D}\right)} \right]^{\frac{1}{2}} \tag{8.5}$$

$$V_{bi} = \frac{kT}{e} \ln\left(\frac{N_A N_D}{n_i^2} \right) \tag{8.6}$$

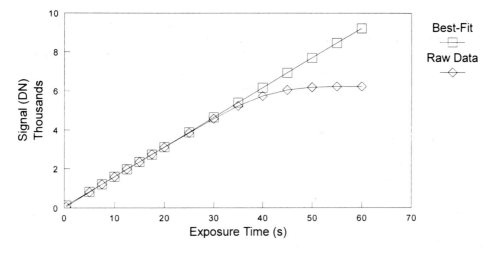

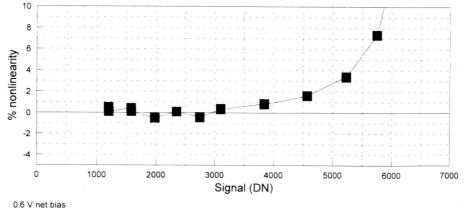

0.6 V net bias

Fig. 8.12. The intrinsic non-linearity of a reverse-biased photovoltaic array. The upper curve levels off as the device saturates. Linearity is better than 1% over almost 80% of the full-well depth.

where n_i is the intrinsic carrier concentration, T is the absolute temperature and the constants k and e have their usual meaning. A more detailed analysis is given by McCaughrean (1988) and references therein.

Values of the parameters can be obtained from manufacturers. As shown in Fig. 8.12, the effect is quite weak in the 256 × 256 arrays (which have relatively smaller pixels and therefore the variable component of C is smaller), rather slowly varying (< 10% worst case) and easily calibrated to high precision with a series of flat-fields at different exposure levels.

8.5 ARRAY CONSTRUCTIONS

8.5.1 HgCdTe and InSb photovoltaic devices

Rockwell International Science Center (Thousand Oaks, CA) has led the development of large format HgCdTe arrays for astronomy (see Vural, 1994). A process known as PACE-I (Producible Alternative to CdTe for Epitaxy) is used to fabricate arrays on 3-inch diameter wafers. Briefly, metal-organic vapour phase epitaxy (MOVPE) CdTe is deposited onto a polished sapphire substrate and then HgCdTe is grown via liquid phase epitaxy (LPE) from a Te-rich solution to a thickness of about 13 μm. The detector junctions are formed by boron implantation and then passivated by ZnS. Light enters from the backside through the sapphire which can transmit out to 6.5 μm. The percentage of Hg determines the cut-off wavelength. Pixel sizes range from 18–40 μm depending on the array.

Indium antimonide arrays specifically designed for astronomy are available from Santa Barbara Research Center (Goleta, CA) and arrays made by Cincinnati Electronics Corp. and Amber/Raytheon in the USA have also been applied to astronomy. Unlike the

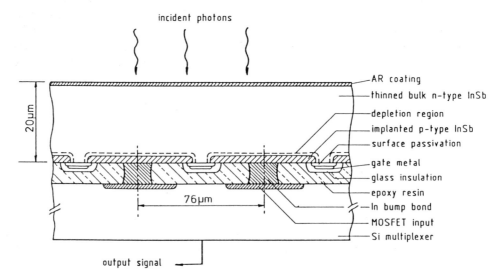

Fig. 8.13. The detailed construction of the original 62 x 58 InSb array from SBRC. A different passivation has allowed the gate electrode to be eliminated in later devices, but the basic bump interconnect construction is similar.

HgCdTe construction these arrays are produced with a substrate of InSb. In the SBRC process the pn junctions are diffused into the InSb while in the CE approach the diodes are grown as "mesas" on top of the substrate. The latter approach requires a small but finite physical gap between pixels. In the earlier SBRC arrays the photox front-side passivation was hygroscopic and required a gate electrode to control the front-surface potential (Fig. 8.13) and hence the dark current. With the 256 × 256 pixel array, a new gateless detector-side passivation was introduced which produced a marked improvement in performance. Since the indium bumps are on the surface of the InSb above the pn junctions, illumination must be from the backside. Therefore, InSb arrays must be thinned to about 10 μm by diamond machining or etching—compare the thinning of silicon CCDs—passivated and AR-coated.

A thick wafer of HgCdTe or InSb connected to a silicon readout using indium bumps and perhaps epoxy tends to pull apart as the device is cooled due to the difference in thermal expansion between the materials. If the layer is very thin, however, it behaves like a rubber sheet attached at many points to a rigid block. The contraction difference is much smaller locally. Because InSb arrays must be thinned for backside illumination anyway, they are inherently less sensitive to thermal mismatch. Special precautions and proprietary constructions are needed for HgCdTe arrays.

8.5.2 PtSi Schottky barrier devices

There is a fundamentally different infrared array device which is widely available and is frequently used in industrial surveillance applications, this is the platinum silicide (PtSi) array. A PtSi array is based on the **Schottky barrier** principle shown schematically in Fig. 8.14 and are very attractive for higher light-level applications because of their silicon-based fabrication process. When a metal such as platinum silicide is brought into contact with p-type silicon, electrons flow across the junction until stopped by the electric field created by the additional negative charges in the silicon; at this point the Fermi levels are equal. The result is an asymmetric potential barrier for conduction holes. Conduction holes in the metal must now overcome a slightly higher barrier to enter the valence band of the semiconductor. The barrier height ψ is determined by the contact potential and can

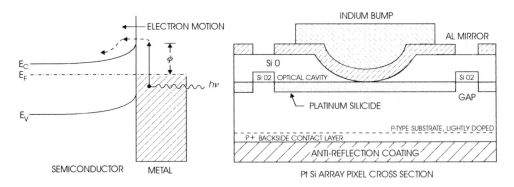

Fig. 8.14. The energy level diagram and physical construction of a typical platinum silicide Schottky barrier diode. These devices have low QE but can be produced in large formats more cheaply than InSb or HgCdTe.

be less than the semiconductor bandgap. For PtSi, $\psi = 0.22$ eV and therefore the cut-off wavelength is $\lambda_c = 5.6$ μm. The quantum efficiency is given by the product of the QE for absorption in the silicide layer ($< 10\%$) and the probability of producing a hole that will escape the barrier, $\frac{1}{2}[1 - \sqrt{(\psi/h\upsilon)}]^2$, so the QE is relatively low and falls steadily towards the long wavelength cut-off; the intrinsic absorption in silicon sets the short wavelength cut-off. Usually, the silicide layer is very thin and covered by a dielectric (insulating) layer of SiO_2 which has a metal (Al) surface layer. Together with an anti-reflection coating on the p-type silicon substrate, the entire device becomes an "optical cavity" which can be "tuned" in thickness so that interference effects increase the QE for certain wavelengths. Typical QEs are less than 2–3%. Nevertheless, these devices can be made very large and uniform, because it is an all-silicon process, and they are also not as expensive as the InSb and HgCdTe hybrid arrays. Ian Gatley at NOAO successfully demonstrated a JHKL four-channel camera called SQIID using four 256×256 PtSi arrays, which produced excellent images of relatively bright infrared sources such as M17, Orion and the Galactic Centre (see colour plates). McCaughrean and Angel used a Kodak 640×320 PtSi camera for imaging of solar system objects. Glass, Sekiguchi and Nakada (1995) describe a Japanese-made PtSi array of 1040×520 pixels with a quantum efficiency of 6% at J and 2% at K, operated at 60 K, in an infrared camera. Significant efforts (both in the USA and Japan) are being made to improve the performance of PtSi arrays.

8.5.3 Impurity band conduction devices

Initially, most long wavelength arrays used extrinsic silicon photoconductors, but these have been replaced by the technology called **impurity band conduction** (IBC). Although readily available, extrinsic silicon photoconductors doped with As, Ga or Sb suffer from many drawbacks which seriously degrade performance. For example, the quantum efficiency is dependent on the concentration of the selected impurity atoms, but there are also unwanted impurities such as boron which can nullify the effect of the given dopant. Usually the concentration of the impurity must be kept low to prevent tunneling or "hopping" so that the only way to increase QE is to make the infrared active layer very thick (several hundred micrometres) which leads to many operational problems. IBC detectors overcome these difficulties.

In an IBC device, a heavily doped infrared-active layer is placed in contact with a pure (undoped) epitaxial layer—the blocking layer—and the overall thickness of the device is greatly reduced (see Fig. 8.15). The blocking layer is isolated by an oxide layer from metal contact pads and the device is usually back-illuminated. Dark current due to "hopping" is prevented by the blocking layer which enables much higher doping levels to be used in the active layer. Although it is ten times thinner than the equivalent photoconductor, the device can achieve a high quantum efficiency. Because of the high donor density, the applied (bias) electric field causes a migration of holes toward the (negative) metal contact forming a depletion region similar to that in a photovoltaic device. As a result, IBC devices do not exhibit generation–recombination noise since the collected electrons are transported over a region devoid of holes. Under normal bias conditions IBC devices will not exhibit any "gain", but gain is possible at higher bias values. Photoconductive gain (G) is the ratio of the carrier lifetime to the carrier transit time in a photoconductor and $G > 1$ can occur because, on average, absorption of a charge carrier at the electrode triggers the release of

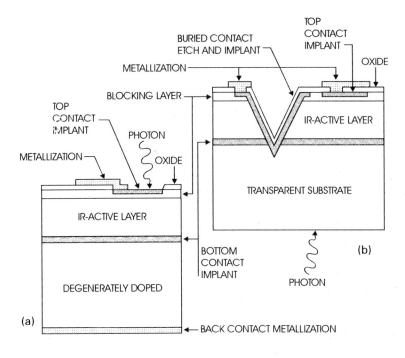

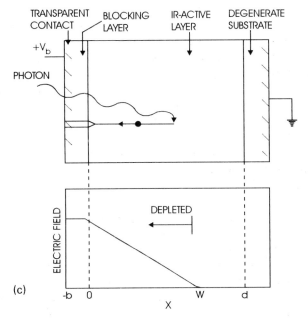

Fig. 8.15. (a, b) The structure of an IBC device. Both front-illuminated and back-illuminated constructions are shown. (c) Illustrating the detection principle in a typical blocked-impurity band device.

another carrier, either of opposite type from the same electrode or of the same type from the opposite electrode (see Rieke (1994) for a more extensive discussion). The original detectors in this class were invented in 1979 by Mike Petroff and Dutch Stapelbroek at Rockwell and were referred to as blocked-impurity-band (BIB) detectors, or BIBIBs for back-illuminated BIBs. Other manufacturers now make similar devices. IBCs are constructed in the following manner. The process starts by placing a (to be buried) contact on a relatively thick (300–500 μm) intrinsic Si substrate. The infrared active (doped) layer is then grown, followed by the blocking layer and the final contact is masked to define the individual pixels and establish the connection to the multiplexer via indium bumps. At the time of writing, arrays of 256 × 256 pixels and 320 × 240 pixels are available from Rockwell and Hughes/SBRC respectively.

For wavelengths longer than 40 μm there are no appropriate shallow dopants for silicon and therefore extrinsic germanium must be used. There are a number of problems with the use of germanium. For example, to control dark current the material must be relatively lightly doped and therefore absorption lengths become long (3–5 mm). Since the diffusion lengths are also large (250–300 μm) then pixel dimensions of 500–700 μm are required to minimize crosstalk. Large pixels imply higher hit rates for cosmic rays, especially for space applications, which in turn means that the readout device must be very low noise so that the background limit is reached in the shortest possible exposure time. But a large detector pixel means a large capacitance and more noise. Also, the photoconductive gain is inversely proportional to the inter-electrode spacing resulting in poor QE unless side-illuminated detectors with transverse contacts are used. Finally, because of the very small energy band gaps, these detectors must operate at liquid helium temperatures well below the silicon "freeze-out" range. Despite these challenges, small format arrays have been constructed and development is on-going.

Since very narrow band gap semiconductor crystals are harder to grow, it is tempting to seek a method of "engineering" small band gaps in wide-band-gap material such as gallium arsenide (GaAs). Using GaAs and AlGaAs junctions it is possible to create a "quantum well" or potential associated with either the conduction or the valence band. The quantum well is equivalent to the particle-in-a-box problem in quantum mechanics and so the well will contain energy levels or sub-bands. The energy difference between these sub-bands or levels ($h\nu$) is very small compared to the normal band gap. Transitions between these sub-bands provides the infrared detection process. See Gunapala et al. (1994) for a popular account and Rieke (1994) for more details.

8.6 DETECTOR PROPERTIES

8.6.1 Dark current and cooling

As with CCDs, an accumulation of electrons occurs in each pixel in the total absence of light. For near-infrared photodiode arrays there are at least three sources of dark current; diffusion, thermal generation–recombination (G–R) of charges within the semiconductor and leakage currents. The latter are determined mainly by manufacturing processes and applied voltages, but diffusion currents and G–R currents are both very strong (exponential) functions of temperature and can be dramatically reduced by cooling the detector. The sum of the diffusion, G–R and leakage currents is given by

$$I_{dark} = \frac{kT}{eR_{0_{diff}}}\left(\exp^{eV/kT} - 1\right) + \frac{2kT}{eR_{0_{GR}}}\left(1 - \frac{V}{V_{bi}}\right)^{\frac{1}{2}}\left(\exp^{eV/2kT} - 1\right) + I_{leak} \qquad (8.7)$$

where V is the voltage across the detector and R_{0diff} and R_{0GR} are the detector impedances at zero bias for diffusion and generation–recombination. Typical values for InSb detectors are about $10^5\,\Omega$ at 140 K and $10^{10}\,\Omega$ at 77 K respectively. Note that the "thermal voltage" $kT/e = 4.3$ mV at 50 K, and therefore $\exp[eV/kT] \gg 1$ for typical bias values ($V > 100$ mV) and temperatures ($T < 77$ K). In general, diffusion dominates at the higher temperatures, but below about 100 K the dark current is mainly G–R and falls almost exponentially until limited by small leakage currents.

Because the band gap is considerably smaller than for silicon, infrared arrays must be cooled to a lower temperature than CCDs. Dark currents below 1 electron/minute/pixel have been achieved in the most recent HgCdTe arrays at 77 K (LN$_2$) and about 0.1 electron/s/pixel in InSb arrays with low bias and temperatures below about 30 K. For BIB arrays with even smaller effective band gaps, the detector must be cooled to within a few degrees of the boiling point of liquid helium (4 K).

8.6.2 Noise sources

As with CCDs, the readout noise describes random fluctuations in voltage which are added to the true signal during readout. Similarly, readout noise is always converted from a voltage (V_{noise}) to an equivalent number of electrons (R) at the detector by using the effective capacitance (C), that is, $R = CV_{noise}/e$ electrons.

Noise in the multiplexer's output FET (similar to a CCD output) is a major noise contributor in IR arrays. When the 62×58 arrays achieved numbers of 300 electrons in the mid-1980s it was considered a breakthrough! Smaller output nodes have improved the number of microvolts (μV) per electron and have reduced this contribution by a factor of 10.

Reset or "kTC noise" occurs when the detector capacitance is recharged (i.e. the reverse bias is re-established). When the reset transistor is on, the voltage across the detector increases exponentially to the reset value V_{RD} with a time constant of $R_{on}C$, where R_{on} is the "on" resistance of the transistor; this time constant is very short. Random noise fluctuations in the reset charge at times very much longer than this time constant have a root mean square (rms) voltage noise of $\sqrt{(kT/C)}$ or an equivalent charge noise of $\sqrt{(kTC)}$, where k is the Boltzmann constant and T is the absolute temperature; typical values of C are 0.07 pF which yields about 54 electrons at 77 K. After the reset switch is closed, the time constant for the decay of the (uncertain) charge on the capacitor is $R_{off}C$, but $R_{off} \gg R_{on}$ and therefore the noise on the reset level is essentially unchanged through the pixel time. The unknown offset in voltage can be eliminated by taking the difference between the output voltages before and after reset. This procedure is, of course, just correlated double sampling. There are several possible strategies for reading out infrared array detectors and these are described in the next chapter which covers practical details.

Typical noise values for average arrays are $\sim$ 30–50 electrons in normal CDS mode, but $\sim$ 10–20 electrons have been obtained using "multiple reads" with the non-destructive readout feature of these arrays, but relatively long exposure times are required in order to tolerate the overhead in readout time required by these methods. Fifty electrons noise is equivalent to a photon-generated signal level of 2500 electrons (50^2) in the integration time, which is small compared to the maximum storage capacity of each pixel ($\sim 225\,000$ electrons for $C = 0.06$ pF and $V = 0.6$ V) or the linear range below 80% full well; note that the capacitance scales with pixel area. Typical background signals for broad band imaging can easily fill these wells in minutes to fractions of a second depending on wavelength. If the application is high-resolution spectroscopy then a read noise of ten electrons or better is needed. Longer wavelength detectors are typically a bit noisier, but are usually always background-limited by photons.

8.6.3 Quantum efficiency

The InSb and HgCdTe arrays used in the near IR have very good QE, typically 60% for HgCdTe and 80% or more for InSb because it is backside-illuminated and has an anti-reflection coating applied. Non-uniformities in thinning, the quality of the passivation applied to make the detector surface chemically inert, variations in doping density through a substrate or impurity centres in liquid phase epitaxy are all important in controlling the QE. Because the absorption cross-section changes with wavelength, the quantum efficiency also varies with wavelength, usually decreasing towards shorter wavelengths. Plat-

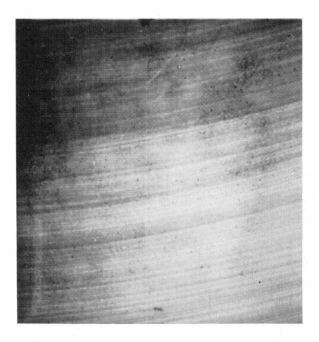

Fig. 8.16. A flat-field image for a 256 × 256 pixel InSb array showing characteristic "tree-ring" patterns. Flat-field variations in HgCdTe detector arrays are generally more random.

inum silicide arrays have QE of 2% or less. Mid-IR arrays have QE in the 30–40% range depending on operation; some of these devices experience "photoconductive gain". Uniformity is best for the PtSi arrays (~ 1%), very good for InSb at < 8% and somewhat poorer for HgCdTe. InSb arrays tend to show "tree-ring" patterns in the flat-field image (Fig. 8.16) whereas the HgCdTe arrays are more randomly structured.

8.7 THE IMPACT OF INFRARED ARRAYS

Figs 8.17 and 8.18, and the colour plates, show some of the remarkable results obtained with infrared arrays. Remember, when you look at these images, you are not looking at visible light. In many cases the objects are optically invisible! In the colour plate section, the blue/green/red colours are used to "translate" the infrared images into a representation of what your eyes might see if they were sensitive to wavelengths from 1–2.5 μm instead of 0.4–0.7 μm. With the addition of IR arrays to pick up where silicon CCDs end, it is now possible for astronomers to obtain seeing-limited images of comparable depth from 0.36–2.4 μm and images with thousands of times higher signal-to-noise ratios than before all the way out to 30 μm. Infrared array cameras can achieve diffraction-limited imaging and the readout noise and dark currents are sufficiently good to permit high-resolution infrared spectroscopy of faint galaxies. All of the data reduction and data-handling tech-

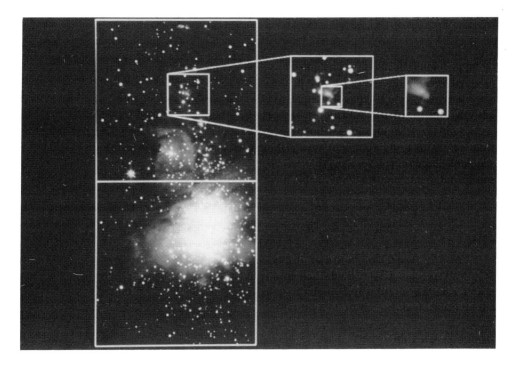

Fig. 8.17. The development of infrared array technology in the near infrared from 64^2 to 1024^2 pixels is illustrated in this JHK view of the Orion Nebula. Courtesy Ian Gatley, NOAO.

niques developed for CCDs are immediately applicable. Electronic drive systems and noise-reduction schemes are clearly similar, device physics is closely related, flat-fielding and other data reduction steps, image display and processing techniques are all identical. Infrared instruments are similar to optical instruments in most respects except that everything in the instrument, not merely the detector, must be cooled to cryogenic temperatures. Consequently, all major observatories now have infrared camera systems. The impact of infrared array technology has indeed revolutionized this field.

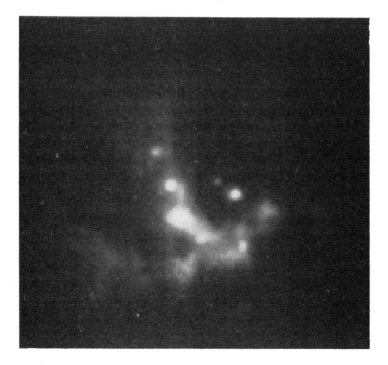

Fig. 8.18. An infrared image of the centre of our galaxy at a wavelength of 10 μm obtained with a 128 × 128 BIB detector showing the famous "mini-spiral". Courtesy Bill Hoffmann and the MIRAC team.

SUMMARY

Infrared array detectors are solid-state imaging devices made from low band-gap semiconductors. Although similar in concept to silicon CCDs in terms of the conversion of photons to electrons, they do not use an MOS unit cell to store charge and they do not use the charge-coupling principle to read out the electronic charge. Infrared arrays are constructed in two steps. First, a two-dimensional grid of infrared detectors is formed on a slab of IR-sensitive material and then a matching grid of field effect transistors is produced on a slab of silicon, together with additional circuits to "address" each pixel. The two pieces are mated by "bump-bonding" them together using tiny columns of indium. Thus the functions of IR detection and multiplexing the signals are separated. As for CCDs, readout noise, quantum efficiency, dark current and linearity are all important parameters. The introduction of infrared arrays has completely revolutionized infrared astronomy since there were previously no imaging devices of any kind.

EXERCISES

1.   Explain the terms "atmospheric window" and "thermal background" as they apply to ground-based infrared astronomy.
2.   Use Wien's displacement law to predict the wavelength of maximum emission from a blackbody at a temperature of 1500 K.
3.   Use the Planck function to calculate the monochromatic flux from a blackbody at a temperature of 300 K. Convert this to photons at a wavelength of 2.2 μm.
4.   Describe with the aid of diagrams how an infrared array is constructed and explain the basic principle of its operation. Assume that the detector is a photodiode of indium antimonide.
5.   Explain the charge storage method used in infrared arrays and how there can be a source of non-linearity between photon flux and output voltage.
6.   Why are CCDs not used as the silicon multiplexers for IR arrays?
7.   What is meant by non-destructive readout? Explain how this technique can be used to reduce noise.
8.   Describe the construction and physics of a platinum silicide Schotkky barrier array. What are the advantages and disadvantages of this type of infrared detector for astronomy?
9.   Describe the construction and principle of the blocked-impurity-band (BIB) detector. Why is this device better than a photoconductor?
10. What are the main sources of dark current in IR detector arrays? Why must infrared detectors be cooled to lower temperatures than CCDs?
11. Compare and contrast the construction, operation and performance of CCDs and IR arrays.
12. In what ways are infrared instruments more complex to construct than normal CCD cameras and spectrographs?

REFERENCES

Glass, I.S., Sekiguchi, K. and Nakada,Y. (1995) "An infrared camera based on a large PtSi array", in IAU Symposium 167, *New Developments in Array Technology and Applications*, 109–116, A.G.D. Philip, K.A. Janes and A.R. Upgren Eds, Kluwer Academic Publishers, Dordrecht, Netherlands.

Gunapala, S., Sarusi, G., Park, J., Lin, T.-L., and Levine B. (1994) "Infrared detectors reach new lengths", *Physics World*, December issue, 35–40.

Hall, D.N.B, Aikens, R.S., Joyce, R., and McCurnin, T.W. (1975) "Johnson noise limited operation of photovoltaic InSb detectors", *Applied Optics*, **14**, 450–453.

Low, F.J. (1961) "Low-temperature germanium bolometer", *Journal of the Optical Society of America*, **51**, 1300–1304.

McCaughrean, M.J. (1988) *The astronomical application of infrared array detectors*, PhD Thesis, University of Edinburgh, Scotland.

McLean, I.S. (1993) "Infrared instrumentation", in *Infrared Astronomy*, pp.337–378, A. Mampaso, M. Prieto, and F. Sánchez, Eds, Cambridge University Press, Cambridge, England.

Rieke, G.H. (1994) *Detection of Light from the Ultraviolet to the Submillimeter*, Cambridge University Press, Cambridge, England.

Vural, K (1994) "The future of large format HgCdTe arrays for astronomy", *Infrared Astronomy with Arrays: The Next Generation*, Ed. I.S. McLean, Kluwer Academic Publishers, Dordrecht, Netherlands.

SUGGESTIONS FOR ADDITIONAL READING

Allen, D.A. (1975) *INFRARED The New Astronomy*, Keith Reid Ltd., Shaldon, Devon, England.

Dereniak, E.L. and Crowe, D.G. (1984) *Optical Radiation Detectors*, John Wiley & Sons, New York.

Fowler, A. (1995) Ed. *Infrared Detectors and Instrumentation*, SPIE, Vol. 2475.

Gatley, I.S., Depoy, D.L. and Fowler, A.M. (1988) "Astronomical imaging with infrared arrays", *Science*, **242**, 1264–1270.

La Brecque, M. (1992) "IR Arrays in Space and on the Ground", *MOSAIC*, Vol. 23, No. 1, 12–23, National Science Foundation, Washington.

Lagage, P.O. and Pantin, E. (1994) "Dust depletion in the inner disk of Beta Pictoris as a possible indicator of planets", *Nature,* **369**, 628–630.

McLean, I.S. (1988) "Infrared Astronomy's new image", *Sky & Telescope*, March issue, 254–258.

McLean, I.S. (1995) "Infrared arrays: the next generation", *Sky & Telescope*, June, **89**, No. 6, 18–24.

McLean, I.S. (1994) Ed. *Infrared Astronomy with Arrays: The Next Generation*, Kluwer Academic Publishers, Dordrecht, Netherlands.

Norton, P.R., (1991) "Infrared image sensors", *Optical Engineering*, **30**, No. 11, 1649–1663.

Philip, A.G., Davis, Janes, K.A. and Upgren, A.R. (1995) Eds. IAU Symposium 167, *New Developments in Array Technology and Applications*, Kluwer Academic Publishers, Dordrecht, Netherlands.

Reidl, M.J. (1995) *Optical Design Fundamentals for Infrared Systems*, SPIE Press, Vol. TT20, Bellingham, WA, USA.

Rieke, G.H. (1994) *Detection of Light from the Ultraviolet to the Submillimeter*, Cambridge University Press, Cambridge, England.

Tokunaga, A.T. (1996) "Infrared Astronomy", in *Astrophysical Quantities*, 4th edition, Ed. A. Cox, AIP Press, submitted.

Vincent, J.D. (1990) *Fundamentals of Infrared Detector Operation and Testing*, John Wiley & Sons, New York.

Wolfe, W.L. and Zissis, G.J., Eds., 1989, *The Infrared Handbook*, 3rd edition, published by the Information Analysis Center for the Office of Naval Research.

Wynn-Williams, C.G. and Becklin, E.E. (1987) Eds *Infrared Astronomy with Arrays*, The Institute of Astronomy, University of Hawaii.

9

Practical operation of infrared arrays

It should now be clear from Chapter 8 that there are many similarities between CCDs and infrared arrays, both in terms of the physics involved and how the devices are operated. Almost everything that we have covered in previous chapters is therefore relevant. Although the unit cell is not an MOS capacitor and no charge-coupling is involved, infrared arrays still require a set of dc bias voltages and a set of clock voltages. The output signals (a few $\mu V/e^-$) must be amplified and digitized into Data Numbers, pixel-to-pixel variations must be corrected with a flat-field, kTC-reset noise must be eliminated with correlated double sampling, dark current must be minimized by cooling and all the same image processing methods can be used. This chapter illustrates the typical operation of IR arrays and describes some infrared instruments in order to emphasize the different issues that arise.

9.1 GENERAL OPERATIONAL REQUIREMENTS

Infrared array manufacturers will provide a data sheet which specifies the kind of packaging used (typically, 68, 84 or 124 pin leadless chip carriers), the pin connection diagram with the names or symbols and the function of each pin. Typical voltages and a timing diagram are also supplied. Voltages are of two basic types, low noise dc bias voltages and clock voltages which pulse between two fixed levels—a high and a low state. Note that unlike CCDs, charge storage capacity or well-depth is not controlled by the clock swing, but by an independent detector bias voltage. With some arrays there is a requirement to change the value of one or more dc bias lines between the integration state and the readout state. These lines are called "switched bias" lines. Certain IR arrays employ a multiplexer which allows any pixel to be addressed in any order (e.g. the original 62 × 58 InSb array from SBRC), but these Direct Read Out (DRO) devices require many clock lines and have given way to simpler multiplexers which employ CMOS shift registers. Hence, the readout of the array proceeds along columns and rows. Because of the higher backgrounds encountered, infrared arrays are generally provided with multiple outputs—at least four and often *many* more. Although the details differ from one class of device to another, certain features are common to all IR arrays.

1.   **Fast (column) register clocks**. When a given row is addressed, the fast register is clocked from column to column to enable every pixel in that row to be connected in turn to the output bus. Typical shift registers require one or two clock phases. A start pulse is also required in many cases.

2.   **Slow (row) register clocks**. Each row is addressed in sequence by pulsing the row register clocks. This is a slow clock because between each transition sequence the column shift register is operated to scan a row. Again, one- or two-phase clocks and a start pulse are common.

3.   **Reset clock**. A single voltage pulse is required to reset each pixel. Unlike a CCD, it is the charge collection node at each pixel site that is being reset rather than the output amplifier node. Some detectors cannot reset each pixel individually but must reset one row at a time, and some detectors have a "global" reset which can reset all the pixels at once. If the reset pulse is not sent after the pixel is addressed and the signal digitized then the IR detector simply keeps on integrating. This is a non-destructive readout. This feature is absent from most CCDs unless the output has been specially designed with a floating gate (such as the "skipper" CCDs).

Each of these clock voltages will have a specified high and low level. If the clocks are buffered in the on-chip multiplexer then the levels usually do not have to be adjusted from the manufacturers nominal values, because they only affect the shift register. If this is not the case, then some fine-tuning may be required to optimize performance of the readout chip.

In addition to power supply lines, the most important dc bias voltages are:

1.   The **substrate** voltage (usually ground) and the **detector substrate** voltage or detector "common" supply which may not be the same.

2.   The **output drain** voltage on the output transistor which should be minimized to reduce amplifier glow.

3.   The **unit cell drain** voltage. In some arrays the "bias" across the detector which establishes the well-depth is determined by the *difference* between the detector substrate (or common) and the unit cell drain voltage, i.e. $V_{detbias} = V_{detsub} - V_{dduc}$.

In some devices (e.g. the Aladdin 1024 × 1024 InSb arrays) bias lines which switch between two levels are required for the clamp lines (V_{ddCl} and V_{ggCl}) and pulse shaping with an RC time constant of ~1 μs is required on the reset lines (V_{rstG} and V_{rstR}). In general the detector outputs can be operated with either load resistors or a constant current source (~100-200 μA).

9.1.1 Array controllers

Arrays such as the HgCdTe devices developed by Rockwell, use very simple CMOS shift registers with 0–5 V clocks. The InSb arrays from SBRC on the other hand require clocks which operate from −3 to −7 V and hence a "level shifter" circuit is needed. Most arrays have multiple outputs and therefore parallel channels of digitization are required.

One very significant difference between CCDs and IR arrays is the readout rate and the operational speed of the electronics. Even at the shortest wavelengths (J band) the pixels can fill up in less than a few minutes in broad band applications and integration times drop to milliseconds in the thermal IR. Consequently, readout times for a full frame are much faster than with CCDs and there is no time to employ "integrating" dual slopes when doing correlated double sampling. The preamplifier must be dc-coupled to the output of the chip and higher speed electronics must be used in general, which has implications for system noise.

Infrared array controllers are nevertheless very similar to those for CCD systems. Some observatories employ identical architectures with minor modifications to handle speed and special requirements. For example, the ARCON (ARray CONtroller) developed by Roger Smith and Alistair Walker at CTIO for the CTIO and NOAO telescopes is a transputer-based controller which can run either CCDs or IR arrays. Many other simpler CCD controllers can also be used. It is very important, however, to look at the overall design from the perspective of the more challenging requirement, which is the infrared, and design the controller from that viewpoint (as in the ARCON system) otherwise it may fall short in some important ways. For example, can the controller make proper use of the non-destructive readout features of IR arrays? Is it possible to co-add many frames at high speed in "front-end" memory before sending a final frame back to the host computer? Can multiple outputs be handled?

Any infrared array controller must provide the following features:

- dc bias board
- level shifter board
- dc-coupled multichannel preamplifier board
- post-amplifier and ADC board
- clock generator board
- data acquisition board (including a co-adder facility)
- links to a host computer

One approach, developed at the UCLA Infrared Lab, is constructed around a front-end system based on the Inmos T805 transputer (Fig. 9.1(a)). This system consists of sixteen transputers and was designed to handle two independent arrays, each with four outputs. It was assumed that one of the arrays would work mainly in the thermal infrared and the other at shorter wavelengths, so one channel was designed to operate at twice the speed of the other. One transputer (the "root" processor) is situated on a PC-bus card in our host computer where it receives commands from the PC, passes them over the serial links to the front end subsystems: two data channels and a motor controller. Although this is a specific example, it contains all of the features of other controllers. Faster processors will almost certainly displace existing ones, but the required building blocks will be the same. Designs are often published in SPIE conference proceedings every few years (e.g. Fowler 1993, 1995) and many large observatories with CCD and IR controllers have home pages on the World Wide Web.

Each of the two channels consists of a "clock generator" transputer and a number of "data acquisition" transputers. The clock generator (Fig. 9.1(b)) is responsible for producing the waveforms required to drive the array and triggering the A/D converters at the appropriate points in the clock sequence. Acquisition transputers collect data from the A/D converters and perform operations such as co-adding of successive frames and correlated double sampling. The motor control transputer accepts commands from the root

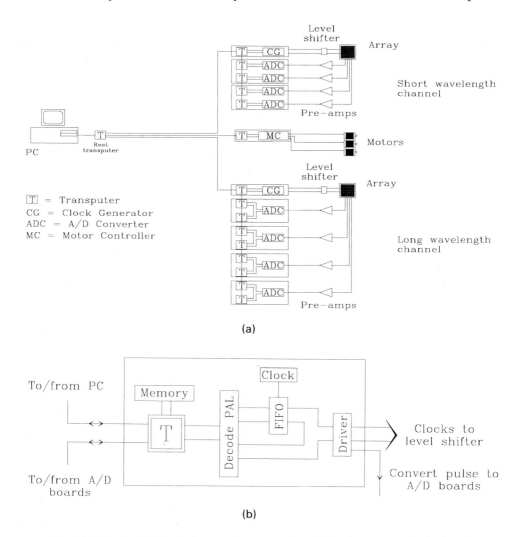

(a)

(b)

Fig. 9.1. (a) A simplified block diagram of a transputer-based infrared array controller developed at UCLA. (b) A schematic of the clock generator board.

transputer and feeds the required pulses to stepper motor driver chips to move the various mechanisms associated with the instrument.

Each A/D feeds its data to two identical first-in-first-out (FIFO) buffers, each 512 by 16 bits. A key feature of the design is that the output of each FIFO is "mapped" directly to the memory bus of a transputer. Control logic activates the input latches of the two FIFO buffers so that each buffer in turn receives and passes to its transputer alternate blocks (usually 256) of pixel values. The transputer has access, via control and status registers, to functions such as clearing the FIFO or setting the number of pixels per block, and to information such as whether the FIFO is empty, half full or full. Each channel of the camera has its own transputer-based subsystem to generate clock waveforms. The transputer has a FIFO buffer (4k by 9 bits) mapped to its memory bus. Waveforms generated as a series of words by the transputer program are fed via the FIFO to the level shifter board. Logic, under the control of the transputer, instructs the FIFO such that, once a sequence of numbers has been fed to it, the FIFO can be made to send out the sequence a preset number of times, by feeding the EMPTY output line of the FIFO to the RESET line. This output scheme takes advantage of the fact that the required waveforms consist mostly of a large number of repetitions of identical short sequences. Programmability of the transputer, which can generate arbitrarily complex waveforms, is therefore combined with the output speed of the FIFO buffer. A minimum time slice in the waveform of 50 ns is achieved, and the FIFO can hold 4096 waveform elements. The least significant bit of the FIFO output is connected to the data acquisition boards, where it triggers the conversions by the ADCs, and the remaining bits are used for the array clock signals and are therefore connected to a level shifter.

Dealing with cases such as the "start-of-frame", where a different sequence of pulses is needed, there is also a second memory-mapped output latch which feeds to the same output connector as the FIFO. A bit in a control register determines which output, either the FIFO or the "bypass" latch, is seen at the output connector.

The array output stages consist of amplifier FETs tied in a source-follower configuration, and they require current from a load resistor or current source. The preamp circuit shown in Fig. 9.2(a) contains both a load resistor and a current source, based on the low noise 2N4393 FET (Motorola, Inc.), for each array output; either of these can be engaged via a jumper. All four array output voltages are simultaneously transmitted through four identical preamp channels. Each channel has a gain op-amp, the CLC400, made by Comlinear Corporation, Fort Collins, CO, for the SBRC channel and the OPA620, made by Burr-Brown Corporation, Tucson, AZ, for the NICMOS 3 channel. These op-amps are arranged in a non-inverting configuration and are offset by an adjustable voltage level. The offset level is set by another OPA620 op-amp in a buffer configuration with a potentiometer at the input. In addition to the high current slew capability of the OPA620, this op-amp can also hold its output up to ± 4.2 V when using ± 5 V power supplies; this feature is needed to offset the rather large DC level of the NICMOS 3 array outputs. The CLC400 was chosen for its unique current feedback design which accounts for its very short settling time; this high-speed performance allows 2 MHz operation for the fast channel array. An active filter of the two-pole Bessel type, based upon the OPA620, can be engaged at the preamp outputs for bandwidth limiting, and thus, noise reduction. Figure 9.2 shows the typical preamp card, dc bias card and clock level shifter card layouts.

9.1.2 Comparison with CCDs

One very big practical distinction between IR arrays and CCDs is immediately apparent at the telescope, especially for the longer wavelengths. When an infrared image of a relatively faint object is obtained and displayed, almost nothing is visible except the flat field pattern of the device and any bad pixels! The reason is the huge infrared background flux (16 magnitudes per square arcsecond at 1.22 μm rising to negative magnitudes at 10 μm). To "see" anything you *must* subtract the background. This is most conveniently done

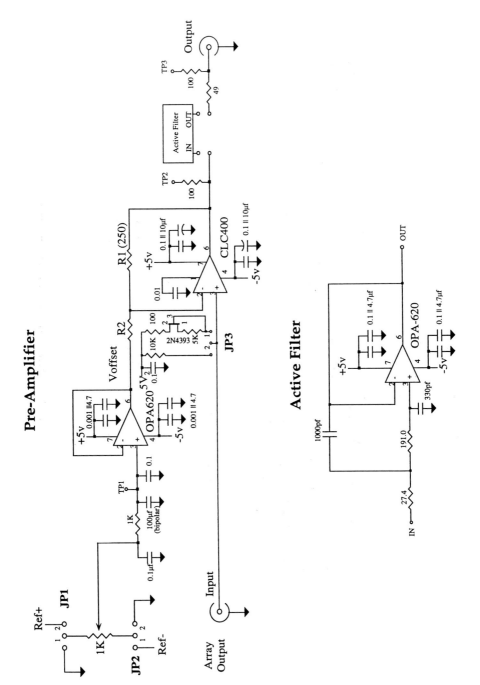

Fig. 9.2 (a) The dc-coupled preamplifier design used with the UCLA IR camera.

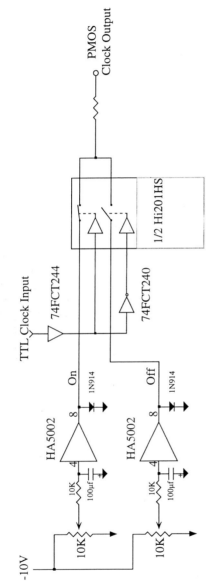

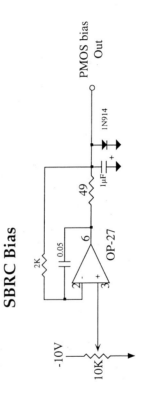

Fig. 9.2 (b) Typical clock level shifter and bias card layouts.

by moving off the target object to a relatively blank area of sky nearby and repeating the exposure. Now, if the difference of these two images is formed and displayed, then the background flux will be eliminated and the astronomical object will be visible; this is "nodding" if accomplished by moving the whole telescope and "chopping" if using the wobbling secondary. Infrared cameras must provide this feature automatically, or at least easily, otherwise it is very hard to know what is happening! Also, if you wait too long to observe the sky value then the difference will fail and remnants of the flat-field structure will reappear. Remember, because the background flux is so very large, even a variation in that background of ~1% may swamp your target star or galaxy! Sky subtractions do not correct for pixel-to-pixel variations in QE and so a flat field must still be carried out in order to perform photometric measurements.

9.1.3   Co-adding

If you are accustomed to CCDs but new to IR arrays you may be puzzled to see on the instrument control screen a request to set the "integration time" and select the number of "co-adds". This leads us to another practical difference between the use of IR arrays and CCDs. Despite a larger well-depth, the strong IR backgrounds result in very fast fill-up times and therefore IR arrays must be read out much more frequently (and faster). Exposure times of about 10 seconds or so may be all that is possible with a standard K-band filter (2.0–2.4 µm), whereas L-band exposures may be only milliseconds. It is extremely inconvenient to store all of these images independently, especially since ultimately all that will be done is to sum them. The images may as well be added together "on-line". Since each exposure is background-limited, i.e. the dominant noise source is the photon noise in the background flux, then there is absolutely no penalty in co-adding numerous exposures to achieve a longer total integration time.

There are two ways to implement the co-addition of frames:

- in memory in the host computer after the fact
- in a hardware "co-adder" in the front-end electronics

The advantage of the first approach is that all the original data is present and it is possible to perform post-processing "image-sharpening" algorithms such as "shift-and-add" provided the image sampling was adequate. Disadvantages of this method are that it requires much more disk memory and it significantly slows down operation of the system to allow each frame to be written to disk. Sometimes the disk write time is greater than the actual exposure time so that the telescope is being used inefficiently. In the second method, each frame is added into a hardware register or mapped to the on-board memory of the front-end processor in control of the digitization process; this processor is located near the instrument itself. Individual frames lose their identity as the sum builds up, but the process is very fast and efficient and essentially unnoticeable. That is, co-addition can be accomplished faster than readout. The apparent loss of the ability to perform shift-and-add can actually be reclaimed by implementing an automatic algorithm in the front-end processor, if it is fast enough. For instance, the program can find the brightest pixel in a defined region of interest and shift the entire frame by an integer amount before adding it into the sum. Several groups have successfully implemented this concept.

9.1.4 Data reduction

Data reduction of IR array frames can proceed in essentially the same way as CCD frames. With CCDs there is always an additive (dark + bias) term which needs to be subtracted before the data frame is divided pixel-by-pixel by the normalized flat field frame. Because of the large (additive) infrared background term with IR arrays it is generally best to subtract an average sky frame first, and then divide by the flat field; the flat field is usually normalized. Therefore,

$$Final\ Frame = \frac{Source\ Frame - Sky\ Frame}{Flat\ Frame - Dark\ Frame}$$

The sky *must* be close in time, with identical exposures and co-adds and should be an average of several frames before and after the target observation. Alternatively, the sky can be deduced from multiple exposures of the target field itself provided that each exposure corresponds to a small displacement of the telescope pointing so that sources never fall on the same pixel twice. Median filtering or pixel masking routines (see Chapters 10 and 11) can be used to remove point sources. The dark exposure (which includes the bias offset) should have the same exposure time and co-adds as the flat field frame and both should be averages of many measurements. Although it should be possible to scale from exposures of different durations, there may be subtle effects which don't scale and it is simply best to avoid these.

If linearization is required, as was the case for the 62 × 58 InSb array, this should be done on the raw data by subtracting the dark/bias frame, applying the correction on a pixel-by-pixel basis and then proceeding with the analysis.

9.2 DETECTOR OPERABILITY AND UNIFORMITY

Clearly, infrared array technology is still developing and even so-called "science grade" detectors may exhibit flaws. It is not surprising that some detectors show a variety of unusual (and unwanted) effects. Bad pixels in the IR-sensitive part of the array do not result in the loss of an entire column; the effect is localized. Problems in the silicon mux can lead to more widespread losses, but such failures are usually detected with a warm wafer probe test and the parts are never hybridized. HgCdTe arrays in general tend to show a random sprinkling of bad pixels, whereas the InSb arrays tend to have fewer bad pixels but more clustering. The NICMOS 3 arrays suffer from "residual" images due to charge persistence on subsequent readouts after saturation by a bright source. This effect, which varies significantly from detector to detector, can be reduced by using multiple resets. InSb arrays on the other hand exhibit many "hot" pixels with high dark current or "hot" edges producing a "picture frame" effect, especially after exposure to air. These effects can be eliminated by baking the detector in a vacuum oven at about 80 °C for about 14 days. Good anti-static precautions should be exercised when handling the chip. It is best to place the chip on an aluminium foil sheet and not leave it in its plastic socket. The process of "cleaning up" the array can be repeated. Keeping the detector under vacuum will delay the onset of these dark current effects, but outgassing and thermal cycling may still cause the picture frame and hot pixels to return eventually.

Both of the near infrared devices can show "glow" due to the readout amplifier, even when the drain voltage is reduced, but it is particularly noticeable in the NICMOS arrays when using multiple reads. The glow appears in the four corners and builds up with each non-destructive readout.

The SBRC 256 × 256 InSb arrays suffer from a capacitive coupling problem due to fluctuations of the column bus voltage which results in the loss of effective bias across the detector during integration. To achieve the desired well-depth more bias than expected must be applied which in turn leads to higher dark currents.

Thermal cycling can lead to cracking of the InSb substrates along crystal planes and lifting of the bump-bonds in both kinds of detectors. As with all of these devices, thermal cycling should be minimized. Defects can appear and grow in perfectly good areas of a chip after repeated thermal cycling. The integrity of the bump bonds is more problematical with the larger, mega-pixel arrays.

Arrays of extrinsic photoconductors exhibit many problems due to spiking and noise, which IBC devices don't show. Nevertheless, much is still to be learned about how to process and operate high-background arrays.

Temperature stability of infrared arrays may be more significant than for CCDs because of the higher dark currents and sometimes also because of the detector physics. While the HgCdTe devices will operate at 77 K, the InSb devices require a temperature intermediate between 4 K and 77 K, typically around 30 K. Thus a more complex dewar arrangement is needed for InSb. Similarly, mid-IR and far-IR devices all require near 4 K operation, but are quite sensitive to the exact temperature.

Uniformity of InSb detectors is better than for HgCdTe arrays, but both are less uniform than PtSi devices. Good science grade detectors typically exhibit variations in sensitivity from pixel to pixel which are less than ±10% of the mean value. This is still not good enough, of course, and a flat field scene must be observed with high signal-to-noise ratio to correct the pixel-to-pixel variations in quantum efficiency to a tiny fraction of the mean sky background. Basically, the same methods as used for silicon CCDs are applicable. A flat field source is obtained either by pointing the telescope at the inside of the diffusely illuminated dome or by using the twilight sky. At the shortest wavelengths, lamps can be used to illuminate the dome while at the longer wavelengths the dome's own thermal emission is sufficient. Actually, it is sometimes quite difficult to control these flux levels adequately since normal flat field lamps put out a lot of infrared radiation. When blank sky is used then it is essential to observe about 5–7 slightly displaced (by a few seeing disks) regions and form a median frame so as to eliminate any faint stars in the "blank" field. Sky flats often seem to give the best results for very deep imaging of faint sources whereas dome flats often work well for bright stars.

9.3 READOUT MODES

Unlike visible light CCDs, there is usually no shutter in an infrared camera system to determine the exposure time. This is partially due to the requirement that the shutter be cold (inside the dewar), and that it operate very rapidly to accommodate the short exposures demanded by the high backgrounds at IR wavelengths. Note, however, that this does not mean that the light path cannot be blocked for dark current measurements. All IR cameras contain a position in the filter wheel which is opaque, usually a metal blank or plug, at the appropriate cryogenic temperature to eliminate thermal emission. In general, however, IR arrays are continuously exposed to light and therefore the exposure time is controlled by the sequence of reset and readout pulses. A critically important feature of IR

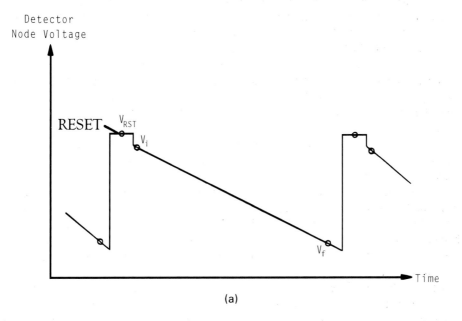

(a)

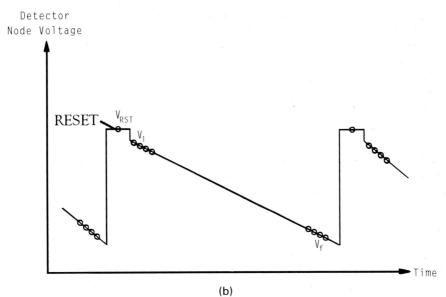

(b)

Fig. 9.3. Schematic variation of the output voltage as a function of time for a typical pixel in an infrared array detector. Associated readout modes are described in the text.

arrays is the ability to perform a non-destructive readout in which the charge on the detector photodiode junction is not altered by sampling its current value.

To prevent saturation at the longer wavelengths, IR arrays are read out much more rapidly than CCDs. On the one hand this means that the readout overhead is small, but the adverse consequence is that the electronic bandwidth is now much larger and noise is greater. In practice, the time required to read out an infrared array depends on the settling time of the on-chip circuits.

Suppose the device requires 4 μs for the output to settle when each new pixel is selected. The array is $256 \times 256 = 65\,536$ pixels, but there are four independent outputs operating in parallel so the effective number of pixels is only one-fourth as many, or 16 384. The readout time is therefore $16\,384 \times 4$ μs $= 65.5$ ms per frame. This is called the "frame time"—t_F—and the "frame rate" is the reciprocal of this number ($1/t_F$), or 15.3 frames per second, usually stated as 15.3 Hz. Similarly, the pixel time of 4 μs corresponds to a pixel rate of $1/(4 \times 10^{-6})$ pixels/s or 250 kHz. The SBRC arrays are capable of about 30 Hz frame rates, 500 kHz pixel rates or 2 μs pixel times.

Fig. 9.3 shows a schematic representation of the output voltage from an IR array. After reset, the signal shifts to the pedestal level and then the detector begins to discharge due to photocurrent and/or dark current. Note that the sign of this plot is arbitrary. Some detectors do discharge from more positive to less positive voltages, others discharge from negative voltages towards zero volts. The principle is the same. The primary readout modes are summarized below.

9.3.1 Single sampling

To clear the detector of charge and define the beginning of the integration period the array must be reset. Reset occurs pixel by pixel as each pixel is selected (addressed) in turn. The process is generally performed at the same rate as the normal readout, e.g. 65.5 ms in our example. Once a pixel has been reset it immediately begins to accumulate light. By the time the last pixel has been reset the first pixel has been integrating for almost 66 ms. Note that no digitization occurs during the reset cycle; no data is stored in memory. If the required exposure time is T_i, then readout must begin T_i milliseconds after the first pixel is reset or T_i–65.5 ms after the last pixel is reset. Relative timing between the "reset" waveforms and the "readout" waveforms must be precise and stable in order for the integration time to be well-defined and consistent. When the readout waveform is sent, each pixel address command is accompanied by a "command-to-convert" pulse rather than by a reset pulse. Thus, each pixel value is digitized but not reset.

This mode is incapable of removing kTC noise and drift because only one sample is taken, but it is an extremely important mode to have available because it directly measures the signal relative to the reset or bias level and unambiguously detects saturation. The bias/reset level can be determined by using the shortest possible exposure (the readout time) and no illumination (the cold dark slide position). The difference between these two frames will remove bias structure.

9.3.2 Correlated double sampling (CDS)

As in the output circuit of a silicon CCD, there is an imperfect reset condition in IR arrays. Two effects can occur. The reset voltage may drift due to temperature effects in the electronics and there are voltage fluctuations in the reset level due to the finite temperature of the charging circuit—this is the usual "kTC-noise". To remove these effects we can use a correlated double sample. There are several ways to do this. One possibility is to digitize the reset level in the sampling scheme already described and subtract the two results automatically. A second approach is to reset the pixel and then immediately digitize the level after the reset has been removed (called the "pedestal" level) but before moving on to address the next pixel. Both of these approaches were common with the first arrays to come out in the mid-eighties, but they turn out to be poor. The reason is that all of these arrays have a fairly strong time-dependent response to the reset action which requires milliseconds not microseconds to settle out. In addition, the act of de-addressing the current pixel in order to address the next pixel can add noise.

9.3.3 Reset–read–read or Fowler sampling

The best scheme is to reset the entire array pixel by pixel as before, taking perhaps $T_{ro} = 65.5$ ms, and then *immediately* read out the entire array again but non-destructively, digitizing the signal with no resets and save this frame. After waiting the required integration time (T_{int}), non-destructively read out and digitize the whole frame again (Fig. 9.4). This method is called **reset–read–read** mode or Fowler sampling (after Al Fowler at NOAO who was the first to promote its use in general) and is less noisy than other methods. The effective signal is the second read minus the first read, i.e. $S_2 - S_1$.

$$S_2 - S_1 = \left[(T_{ro} + T_{int}) \dot{N}_e + b + c \right] - \left[T_{ro} \dot{N}_e + b + c \right] = T_{int} \dot{N}_e$$

where the amplifier bias (b) and the unknown but correlated offset from the reset level (c) subtract out; $\dot{N}_e$ is the count rate in electrons per second.

This process is usually accomplished in a front-end co-adder by storing the first read as a negative number and *adding* the second number to the stored value (rather than over-writing the value). The unknown kTC-noise and actual reset level is eliminated. If the exposure is very long, there may still be drifts which are not removed by this method. One way to remove those effects is to take a "dark" frame in the same sampling mode with the same exposure time and subtract it. This of course also removes the dark signal.

There is a disadvantage to the reset–read–read mode, namely, that it makes it harder to know when the array has saturated. For instance, suppose the readout time is 66 ms, the pixel saturation level is 8000 DN but the flux rate from a certain star is 60 000 DN/s. In an exposure time of 0.1 s the flux is 6 000 DN in the pixel containing the star. Between reset and the first non-destructive read each pixel integrates for 66 ms, so the pixel with the bright star has a signal level of 3960 DN when it is recorded. When it is read again for the second sample, 0.1 s have elapsed and another 6000 DN of charge should have accumulated giving a total of 9960 DN. But the saturation level is 8000 DN and additional flux above this level will not significantly affect the detector which is now fully de-biased. So the actual value of the second sample will be 8000 DN and the difference—which is all that is saved—will be 4040 instead of 6000 DN. The detector is hopelessly saturated, but the displayed counts do not indicate this. The way to check is to use the single sampling mode and determine the true count rate on the star.

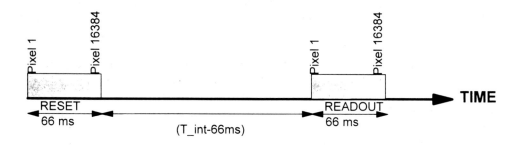

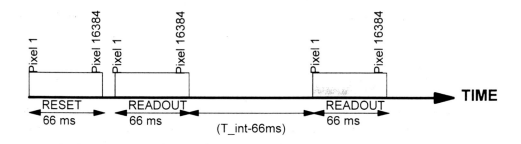

Fig. 9.4. Summary of various readout modes. Example is for a 256 × 256 array.

9.3.4 Multiple Fowler sampling

The technique is the same as the reset–read–read approach except that the initial (first) readout of the entire array following reset is repeated m times instead of just once. At the completion of the required integration time the final full frame readout is also done *m* times. Of course, the exposure time T_{int} must be long enough to accommodate *m* frame times. This is usually the case for narrow band imaging and near IR spectroscopy, where this technique is very effective.

The sum of the initial values is given by S_1 where

$$S_1 = \sum_{n=1}^{m} (nT_{ro}\dot{N}_e + b)$$

and T_{ro} is the readout time for one frame (e.g. 65.5 ms), $\dot{N}_e$ is the flux in photoelectron per second and b is the bias offset. The sum of all the final readouts S_2 is given by

$$S_2 = \sum_{n=1}^{m} \left[(nT_{r0} + T_{int})\dot{N}_e + b \right]$$

and the difference is

$$S_2 - S_1 = mT_{int}\dot{N}_e$$

so that by dividing by m we get the desired integrated flux. The signal $S_2 - S_1$ is m times as large but the readout noise adds in quadrature and is only $\sqrt{m}$ as large, so the effective readout noise in the final integrated flux is $\sqrt{m}/m$ or $1/\sqrt{m}$ smaller. For four reads ($m = 4$), the final readout noise is only half that obtained from a single pair of samples.

A complete mathematical analysis of multiple sampled schemes is given by Garnett and Forrest (1993). These authors show that both multiple Fowler-sampling and line-fitting (or "up-the-ramp" sampling—see below) are superior to correlated double sample in read-noise limited conditions (when the read-noise is itself dominated by white noise). Both sampling methods provide the expected $\sqrt{n}$ improvement, where n is the number of samples, unless or until $1/f$ noise dominates. Fowler sampling depends on duty cycle and is best when sampling the pedestal and signal levels each for 1/3 of the total observing time; a 2/3 duty cycle. Under these conditions Fowler sampling is approximately 6% inferior to line-fitting. For background limited performance, the difficulty is that for any non-destructive multiple sampling scheme, successive signal measurements are correlated in their noise. Theoretically, the best signal-to-noise comes from correlated double sampling.

9.3.5 Phased multiple CDS (PMCDS)
Low-frequency noise at 60 and 120 Hz from ground-loops and pickup can be eliminated with a variation of the multiple reset–read–read technique which we call phased multiple CDS. Instead of spacing the multiple reads as close together as possible, the spacing in time is adjusted slightly so that each read is evenly spaced in phase with a 60-Hz sine wave. For example, if the reads are spaced at an exact multiple of the 60-Hz frequency such as 83.3 ms then the 60-cycle noise would be maximized, but if four reads are spaced at 83.3 + (16.7/4) = 87.5 ms then they occur at phases 0, $\pi/2$, π and $3\pi/2$ along the sine wave. Adding the four signals will cancel the pick-up noise at this frequency.

9.3.6 Other schemes
(a) *Multiple convert per pixel sampling*: Since the signal processing electronics can perform faster than the array can settle, there may be circumstance in which is it worthwhile taking many A/D conversions after each pixel is selected and allowed to settle. These multiple conversions can be averaged together. This process is effective in smoothing high-frequency noise or in simply reducing the overhead from the multiple CDS approach.

(b) *Triple correlated sampling*: This is similar to the normal correlated double sample except that the reset level is digitized before it is released. Usually, the reset level following pixel reset at the end of the integrations is employed together with the normal non-destructive reads at the beginning and end of integration.

(c) *Ramp sampling*: In this approach the signal is sampled many times throughout the duration of the exposure rather than at the beginning and at the end. The signal therefore can be seen to "ramp" up (Chapman et al. 1990, Garnett and Forrest 1993). This approach is very useful if some pixels will saturate before the end of the exposure time. A best-fitting straight line can be applied to the data points to get the mean flux rate.

(d) *Drift scanning*: In this technique a segment or sub-array is read out and stored in memory as the sky is allowed to drift by at a rate which does not significantly smear the image in the short exposure. After each short exposure the memory address is "moved over" by one column to correspond to the movement of the image. This is similar to CCD drift scanning except that the array must be read out since there is no charge-coupling involved.

9.4 TYPICAL DEVICES

Since this field is changing rapidly, it is not clear how long any particular device will remain available or in use. Nevertheless, in order to explain the operation and behaviour of IR arrays we will consider two devices which typify near-infrared detectors, the Rockwell International HgCdTe 256 × 256 pixel NICMOS 3 array which has 40 μm pixels, and the SBRC InSb 256 × 256 pixel array which has 30 μm pixels. Typical operating conditions for both are given here. The 256 × 256 HgCdTe array from the Rockwell International Science Center is known as the NICMOS 3 array because it was initially developed under contract for the NASA/University of Arizona second generation Hubble Space Telescope instrument called NICMOS. At the July 1993 IR conference at UCLA, the principal investigator on this project, Rodger Thompson, had to remind the audience of the history of this development simply because of the flood of new, young scientists who had entered the field within the last few years!

Table 9.1. Operating conditions for the Rockwell and SBRC 256 × 256 arrays

		Rockwell HgCdTe		SBRC InSb	
Clock voltages	Pixel	0 to 5		SyncSlow	−3 to −6
(volts)	Reset	0 to 5		φ1 Slow, φ2S	−3 to −7
	Lsync	0 to 5		φ1 Fast	−3 to −7
	Line	0 to 5		φ2 Fast	−3 to −7.6
	Fsync	0 to 5		SyncFast	−3 to −6.2
	clear	0 to 5		φ Reset	−3 to −5
DC bias voltages	VDD	5 V		V_{dduc}	−3.2 V
	RHI	5 V		V_{gg}	−1.5 V
	AHI	5 V		$V3$	−2.0 V
	VF	5 V		V_{det}	−2.6 V
	MIRROR	100 kΩ to 0 V		—	—
	DETBIAS	0.6 V		V_{ddout}	−1.0 V

Applied detector bias	0.6 V (DETBIAS)	0.6 V (= $V_{det} - V_{dduc}$)
Output load	10 kΩ to +5 V	10 kΩ to +2 V
		or 250 μA current source
Temperature	70–77 K	28–35 K
Readout mode	Reset–read–read	Reset–read–read
Pixel rate (typ.)	250 kHz	250 kHz

9.4.1 SBRC 256 × 256 InSb array

The layout is shown in Fig. 9.5. This array has four separate outputs corresponding to groups of four columns. For example, signals from columns 1, 5, 9, 13, ... will appear on output 1, columns 2, 6, 10, 14, ... go to output 2, columns 3, 7, 11, 15, ... go to output 3 and finally, columns 4, 8, 12, 16, ... go to output four. The SBRC array uses a PMOS multiplexer which requires six separate DC bias voltages. The voltage V_{ddout} provides the current for the output amplifier. The amount of current depends on the load on the array output. Typically a 10 kΩ resistor to 0 V or +2 V is used, or a constant current source from an FET of around 250 µA, giving a total current of about 1 mA for the four outputs combined. It is important that V_{ddout} is always more negative than the raw output of the array. V_{dduc} and V_{det} are the unit cell bias voltages; V_{dduc} is the reset level (full reverse bias on the diode) and V_{det} is the voltage once the diode capacitance is fully discharged. Thus, the applied bias across the detector is the difference $V_{det} - V_{dduc}$, and this is the voltage that determines the full-well capacity of the pixel. In practice, V_{dduc} is kept at about −3.7 V and V_{det} is varied to adjust the well-depth; typical difference values are ~0.6 V, but up to 1 V is possible. Unfortunately, due to a phenomenon called "charge dumping" in these devices, the effective bias voltage can be 200–500 mV *less* than the applied bias. It can be advantageous to allow certain bias voltages such as V_{gg} to be changed or "switched" between integration and readout; making this voltage 0.5 V more positive during integration in our tests improved well-depth and eliminated several "hot" rows near the edge of the detector, but it also resulted in more individual "hot" pixels. As Al Fowler explains,

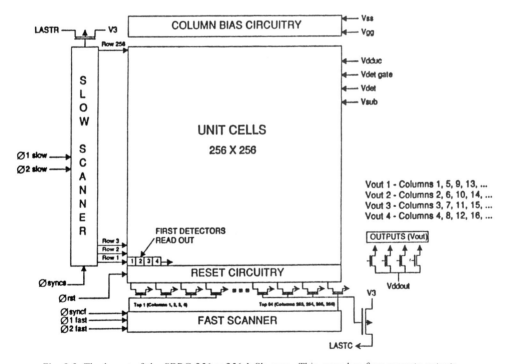

Fig. 9.5. The layout of the SBRC 256 × 256 InSb array. This array has four separate outputs corresponding to groups of four columns. Courtesy SBRC.

switching V_{gg} from -1.5 V to -1.0 V during the integration period puts the current source transistor in the sub-threshold region of operation and keeps the column buss voltage from collapsing and thereby further de-biasing the detector node. If V_{gg} is switched, it is important that this be done prior to de-addressing the last row or immediately after (< 100 ns) de-addressing the last row. Finally, $V3$ is a voltage used by the shift registers and draws about 1 µA per column or about 0.26 mA total.

The PMOS clock voltage levels swing between an "off" state (logic 0) around -2 or -3 V, and an "on" state (logic 1) of -5 to -7 V. If the off levels are too positive then the charge dumping effect is exaggerated. The clock waveform for the SBRC array is shown in Fig. 9.6. Parts of the waveform are repeated many times. The sequence for a single frame readout is:

- Part 1 (start of frame)
- Repeat 128 times: Part 2 (Start of odd row)
    - Repeat 32 times: Part 3 (Pixel pair)
    - Part 4: (End of odd row)
    - Part 5: (Start of even row)
    - Repeat 32 times: Part 6 (Pixel pair)
    - Part 7: (End of even row)

Note, of course, that the "Cnvrt" signal is sent to the ADC from the camera electronics and is not a clock level on the detector. The duration of each step in the waveform (except for the settling time pause indicated by the dashed line) is controlled by the electronics being used. The start-of-frame sequence begins by switching the four clock lines (two slow and two fast) on simultaneously to reset the shift register. The slow shift register which controls the rows is initialized by turning SyncS and $\phi2S$ "on" briefly, with the longer pulse applied to SyncS. In Part 2 the row is started by initializing the fast (column) shift register by switching "on" SyncF and $\phi2F$ together and then switching them off. Simultaneously the slow register is clocked to the first row by activating $\phi1S$, which then remains "on" for the entire row. Part 3 begins with switching "on" $\phi1F$ to clock the fast register and address the first pixel. A pause for pixel settling is required (e.g. 4 µs) and the exact position of the convert pulse to the ADC may depend on its operation (i.e. whether the ADC needs to sample before converting or whether it converts immediately). If the pixel is being "reset" then the convert pulse is not sent, and instead the Reset line is pulsed "on". The reset waveform should include a pause in the middle of the pixel time to ensure that the pixel is completely reset and to ensure that the read and reset waveforms are of the same duration. Next $\phi1F$ is turned "off" and $\phi2F$ is turned "on" to select the next pixel; note that the two clocks should not be on at the same time, since to do so will clear the shift register. After the second pixel has been digitized, $\phi2F$ goes "off" and $\phi1F$ goes back "on" again to select the third pixel. This process is repeated for 64 pixels (because the detector has four outputs) or 32 pixel pairs. At the end of the row $\phi1S$ is switched "off" (Part 4) and the next row initiated with SyncF and $\phi2F$ going high while $\phi2S$ clocks the slow register down one row. The process of clocking through the second row (Part 6) by pulsing the fast (column) shift register is the same as Part 3 except that $\phi2S$ remains in the "on" state for the row and $\phi1S$ is "off". Finally, in Part 7, switching off $\phi2S$ ends the row. The sequence described in Parts 2–7 is then repeated until all 256 rows have been clocked out.

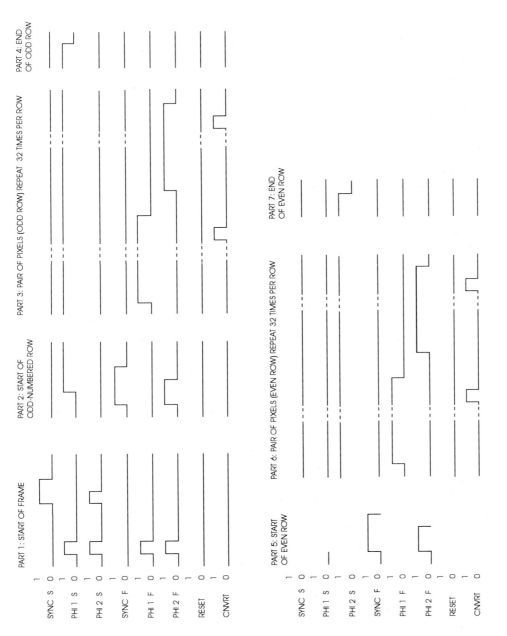

Fig. 9.6. The clock waveform for the SBRC array is shown subdivided into repeatable parts.

Sub-array readout is easy to implement. For example, suppose you wish to read out the central 64 × 64 pixels only. Start the waveform as in Part 1 and Part 2 above, but instead of holding ϕ1S "on" for the row and moving to Part 3, switch ϕ1S and ϕ2S "on" and "off" 48 times to skip down 96 rows, then enter Part 3. But as each of the next 64 rows are read out, Part 3 begins with a segment where ϕ1F and ϕ2F are clocked "on" and "off" to skip over 24 pixels. Skipping down or over can be done in 100 ns per row or column, consequently, these small frames can be read out at very high frame rates (e.g. 240 Hz).

The pixel readout rate is limited by the settling time of the array. As each pixel is addressed, the output voltage must slew and settle at the new level associated with that pixel. If slewing of the output from a very bright to a very dark pixel is incomplete at the moment the ADC is sent the command to convert, then the dark pixel will be recorded as "brighter" than its true value and a bright "tail" will appear. In tests carried out by Bruce Macintosh at the UCLA IR Lab, a settling time of about 4 μs was required for the pixel to settle completely. This corresponds to a frame rate of 15 frames per second or a pixel rate of 250 kHz; this assumes four separate outputs run in parallel.

Finally, Table 9.2 illustrates the loss of effective bias for one particular device.

Table 9.2. Well-depth versus applied bias and the charge-dumping effect

Applied bias ($V_{det} - V_{dduc}$, volts)	Full well (DN)	Full well (electrons)	Useful bias (volts)
0.60	6000	150 000	0.37
0.70	8000	200 000	0.48
1.00	11 000	280 000	0.68

9.4.2 Rockwell NICMOS 3 HgCdTe array

A schematic layout is shown in Fig. 9.7. The NICMOS multiplexer is divided into four independent quadrants with its own clock and bias lines and its own output. In practice, for a good chip with four excellent quadrants, it is usual to simplify wiring and tie together the corresponding bias and clock lines from different quadrants. The multiplexer is a CMOS design which is very simple to operate. Three bias voltages are required, namely, 0 V, +5 V and DETBIAS which is the reset level of the pixel. Since the other side of the HgCdTe photodiode is ground, then DETBIAS is the net reverse bias on the detector. As a pixel accumulates charge the signal voltage decreases. Typical values of DETBIAS are 0.5–0.7 V. There are six clock lines, each with a 5 V swing from 0 V (off—logic 0) to +5 V (on—logic 1), and the following functions. CLEAR resets the shift registers, FSYNC provides a start-of-frame, LINE is the slow (row) clock which triggers on the falling edge, LSYNC is the start-of-row clock, PIXEL is the fast (column) clock which triggers on both the rising and falling edges, and RESET is the pixel reset pulse. The NICMOS waveform is given in Fig. 9.8. Repetitions in the pattern are as follows:

- Part 1 (Start of frame)
- Repeat 128 times: Repeat 64 times: Part 2 (Pixel pair)

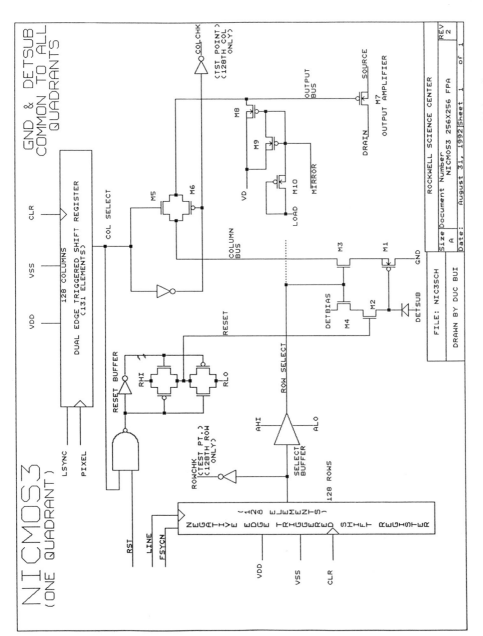

Fig. 9.7. A schematic layout of the NICMOS 3 256 × 256 HgCdTe array from Rockwell is shown. The multiplexer is divided into four separate quadrants of 128 × 128 pixels. Courtesy Rockwell.

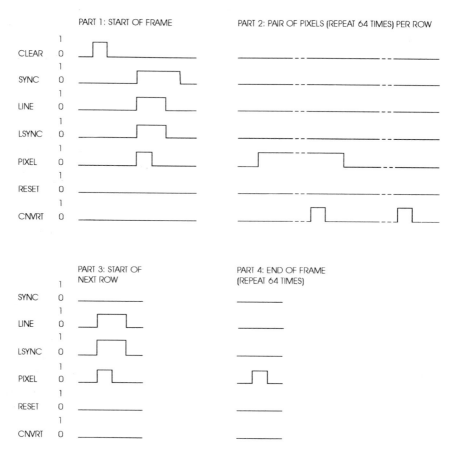

Fig. 9.8. The NICMOS waveform is shown subdivided into sections which can be repeated.

- Part 3 (Start of row)
- Repeat 64 times: Part 4 (End of frame/ Clear fast shift register)

The start-of-frame sequence begins with a pulse on the CLEAR line to reset the shift registers. FSYNC and LINE are set "on" and FSYNC is held high until LINE falls which initializes the vertical or row shift register and activates the first column. Simultaneously, LSYNC and PIXEL are pulsed "on" and LSYNC is held high until PIXEL falls which initializes the horizontal or fast register but does not yet activate the first column. Even after the horizontal register has been initialized, the first column is not activated because the shift register contains blank elements not connected to any pixels. Only after the next rising edge of PIXEL does the first pixel become valid; this begins Part 2. After a settling time of 3–4 μs the command to convert pulse is sent to the ADC and the falling edge of the PIXEL clock selects the next pixel. This step is repeated 64 times to clock across one row of the array. In a "reset" waveform the RESET line is clocked "on" and "off" instead of the convert line. In Part 3 of the waveform the falling edge of LINE clocks the vertical or slow register down to the next row. LSYNC and PIXEL initialize the horizontal or fast register, and again the LINE and LSYNC actions need not be simultaneous. After 128 rows have been clocked, the horizontal or fast register will have been initialized by the last repeat of Part 3. Part 4 is simply a waveform to rapidly clock through the fast shift register to clear it.

Dark current in the NICMOS arrays is extremely low on good pixels and the median dark current over the entire frame is actually dominated by a random distribution of isolated "hot" pixels. Corner "glow" is observed in the NICMOS arrays. The glow is independent of integration time and is associated with the array readout. Tests indicate that it is not simply related to the output amplifier being on, but may be related to "spiking" in the output amplifier when pixels are clocked or to shift register activity. The level is generally very low and of no concern except in multiple non-destructive readout applications with low flux levels.

Larger format HgCdTe and InSb arrays are now available. Both Rockwell and SBRC have developed devices with 1024 × 1024 pixels. A summary of their properties is given below.

9.4.3 1024 × 1024 near-IR arrays

The Hawaii 1024 × 1024 HgCdTe SWIR arrays (Rockwell)
This mer-cad-tel array has short wavelength cut-off detector material like the NICMOS 3 devices and is optimized for 1–2.5 µm. It was produced in a collaborative effort between the Rockwell Science Center in Thousand Oaks, California, and the University of Hawaii with support from the US Airforce. The array has a four quadrant architecture with only one output per 512 × 512 quadrant. Each pixel is 18.5 µm square, considerably smaller than the 40 µm pixels of the NICMOS 3 arrays. Typical full well capacity is about 60 000 electrons and readout noise is about 10 electrons with multiple sampling. The quantum efficiency at 2 µm is about 60% and very low dark currents of 4 electrons/minute/pixel have been observed at 77 K. Control is very similar to the NICMOS 3 detector.

The Aladdin arrays (SBRC)
The Aladdin program was a joint collaboration between the National Optical Astronomy Observatories (NOAO), the US Naval Observatory (USNO) and Santa Barbara Research Center (SBRC) to develop a large format infrared array for astronomy applications. The acronym ALADDIN stands for Advanced Large Area Detector Development in InSb. The Project was initiated in spring 1993 and SBRC began work in October 1993. Much of the detailed design of the new multiplexer was done jointly by Jim Wollaway at SBRC and Al Fowler at NOAO. A revision to the multiplexer design was carried out in spring 1996 to improve response speed, lower noise and add a facility to disable individual bad rows within a quadrant. Among the several goals that Al Fowler and Alan Hoffman had for the new detector were reduction in the "charge dumping" onto the detector node seen in the 256 × 256 arrays and the use of CMOS shift registers and transfer gates (T-gates) to eliminate interaction between logic voltage levels and the operation of the array.

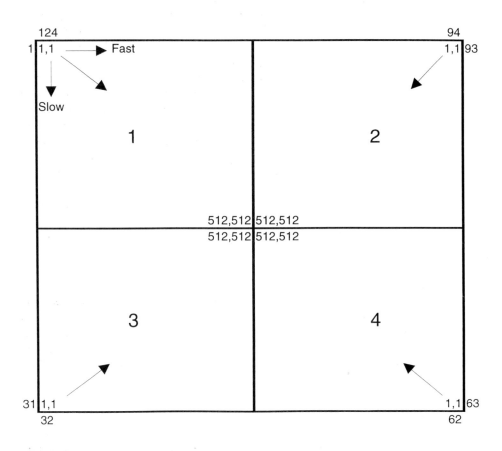

Fig. 9.9. The InSb 1024 × 1024 ALADDIN Sensor Chip Assembly (SCA) showing the edge-to-centre readout scheme.

A four-quadrant architecture in which the four 512 × 512 quadrants were completely electrically independent was adopted to ensure that any control logic defect would not ruin the whole chip. The design and operation of the CMOS shift registers incorporates several new ideas, one of which is that the array reads from each of the four corners towards the centre (see Fig. 9.9). This structure eliminates any anomalous behaviour where the four quadrants meet. Since the first row and column often exhibit artifacts which limit their

performance, reading from the corners towards the centre prevents such effects from producing a peculiar row or column in the centre. Remember that the InSb is continuous; it is
not subdivided into quadrants. All "quadrant" behaviour comes from the silicon multiplexer. In previous designs for InSb arrays the clock voltages were used directly for the
"enable" signals to all the unit cell FETs. Not only did that design allow a path for digital
noise to couple into the unit cell, it was also difficult to optimize both clock shift register
behaviour and array performance simultaneously. In the Aladdin design, the shift register
outputs are used to drive "transfer" gates, which in turn switch the enable voltages to the
unit cell. Since the enable voltages are now dc levels, they can be filtered to reduce noise.
Moreover, since one gate is always enabled, the unit cell enable signal is always returned
to a low impedance source which therefore guarantees positive control and reduced noise.
All input clocks are buffered as they enter the chip which eliminates the need to drive the
on-chip line capacitance and reduces noise coupling from the external clock lines. The
power to drive the on-chip capacitance is supplied by power supplies which can be filtered
to reduce external noise coupling to the chip.

A major difference compared to earlier arrays is that the Aladdin unit cell has only three
transistors and not four. While this does help the yield in manufacture of these huge arrays,
the adverse consequence is that you can no longer reset individual pixels, only rows at a
time. On the other hand, it is now possible to reset the whole array globally since the reset
enable line is controlled by a transfer gate which can be clocked.

The current in the unit cell is controlled by two current sources on each column bus. To
ensure stability, a minimum current I_{idle}, flows in each column at all times. In what amounts
to a "look-ahead" design, the column being read out AND THE NEXT column, both have
a higher current I_{slew}, which gives faster response and minimizes power dissipation. Column to column fixed pattern noise is also reduced. The detectors are essentially the same
as in the 256×256 devices and employ SBRC's "gateless" process. Each pixel is 27 μm
square which results in a rather large piece of InSb.

9.4.4 Mid-infrared arrays

Mid-IR arrays include the 62×58, 256×256 and 320×240 extrinsic silicon devices from
Hughes (USA) which were developed in parallel with the InSb arrays for astronomy,
64×64 extrinsic silicon arrays developed in France by LETI/LIR, a 128×128 pixel device
from Amber Engineering (USA) and 128×128 BIB detectors from Rockwell. Several astronomy groups have used the 128×128 arsenic-doped silicon BIB detectors from Rockwell which have a photon response out to nearly 30 μm. Low, moderate and high flux level
classes of device are available and a high-flux level version with 256×256 pixels is being
developed (1996) for astronomy. Rockwell also produces 128×128 antimony-doped silicon (Si:Sb) BIBs for wavelengths out to about 40 μm. Both of the high flux 128×128
arrays from Rockwell have 75 μm pixels and 16 outputs. Charge handling capacity is
about 2×10^7 electrons. Noise is relatively high at about 1000 electrons rms, but still
within photon shot noise on a full well. Operating temperature is in the 2–14 K range
depending upon the application and tolerable dark current, and high frame rates up
to 4 kHz are possible. The 256×256 Si:As BIB arrays from Rockwell have 50 μm detectors and a consequently smaller full well of about 17 million electrons. Readout noise is
about 1000 electrons rms and the device has 16 outputs.

9.5 INFRARED INSTRUMENTS

Every class of optical instrument has its infrared counterpart, but building infrared instruments is significantly more challenging for many reasons, not least of which is the fact that everything—not just the chip—must be reduced to cryogenic temperatures. Think of it. A large spectrograph with mirrors, slits, filters, diffraction gratings and mechanisms must be enclosed in a vacuum chamber and cooled to LN_2 temperatures to eliminate thermal emission. What are some of the issues?

9.5.1 General considerations

Obviously, optical materials which transmit infrared wavelengths are needed for vacuum windows and lens systems. Many of the more robust IR optical materials (e.g. zinc sulphide, zinc selenide) do not transmit well in the visible which hampers alignment and set up. On the other hand, crystalline materials like calcium fluoride and barium fluoride which do transmit both optical and IR light are very fragile and harder to work optically. In addition, refractive indices are less well-known than for the standard optical glasses. Of course, apart from the need for a vacuum window, one can try to use mirror systems throughout instead of refractive optics. Even here there are distinctions. Aluminium-coated mirrors actually have quite poor reflectivity in the IR compared to gold coatings, but gold does not reflect visible light well. In general, when reflective infrared optics designs are used the mirrors are gold-coated. Recently however, there has been a very serious resurgence of interest in the use of silver coatings which would be a good compromise for both wavelength regimes. It is conceivable that an entire 8-metre class telescope may be silver coated in the future.

Fig. 9.10. A lens holder design made in aluminium but with slits cut into the barrel to allow the tube to relax radially outwards as it contracts around a lens in a cryogenic camera.

Elimination of diffusely scattered light using blackened baffles requires care since anything truly black has almost 100% emissivity and will therefore be a strong infrared emitter unless very cold. There is a rule which says *"if it's black it must be cold, if it's white then make it gold"*. Black anodized aluminium is not adequate. Special infrared black paints such as Parsons black (Eppley Labs), Aeroglaze (Lord) and Nextel are required. Experimentation and care is required in applying these paints otherwise they will eventually flake off due to cryogenic cycling.

The necessity of cooling everything presents many problems. For example, the refractive index of any lens in the instrument will change with temperature. Hence, the rate of change must be known and the effect designed into the system so that the lens performs correctly after it is cold. All dimensions will change during the cool-down and worse, parts not made from the same materials will shrink by different amounts due to dissimilar coefficients of expansion. Lens holders could crush their optical components, optical separations will change and materials may experience stress. All these things must be calculated beforehand and each component must be constructed in such a way as to achieve the correct dimensions after it is cold. One way to avoid damaging a lens is to use a holder which is "springy". This can be achieved by cutting slits in the aluminium barrel as shown in Fig. 9.10.

In an infrared instrument it is essential to produce a real image of the entrance pupil. This image can be formed with a field lens or by collimating the beam and then re-imaging onto the detector. The "pupil" image is a perfect location to place a circular hole, called a "cold stop"; by matching the size of the cold stop to the size of the pupil image all unwanted, off-axis rays from warm structure in the telescope are eliminated. The best cold stop is a "Lyot stop" which has a central disk (of metal) with four tiny supports designed to mask scattered light and thermal emission from the secondary mirror. Cold filters to define the wavelength interval are placed close to the pupil image, preferably in a collimated beam. Again, precautions are needed to ensure that the metal holder does not crush the filter.

Dewar designs similar to those of LN_2-cooled CCD cameras can be "scaled up" in volume to accommodate the camera optics and filters, but at some point this becomes impractical and custom-designed cryostats are required. One approach is to build the instrument from the "inside-out" by making the vacuum chamber go around the optical design plus cooling system rather than folding the design to fit a given cryostat. Another way is to build the optics into a custom vacuum chamber and mate this to a standard cryostat. In any case, the optical train must be surrounded by a light-tight radiation shield to prevent thermal photons from the warm walls of the vacuum chamber reaching the IR detector. The "art" of designing an infrared instrument lies in minimizing the heat loading (in watts) and estimating that heat load correctly! Radiation loading is usually the dominant effect and can be minimized by using highly polished aluminium foil to line all the interior surfaces of the chamber and the radiation shield around the optics. The net heat radiated from a hotter body at temperature T_h onto a colder body at T_c and with a surface area of A is given in Chapter 5. Unlike in a CCD camera where little more than a 1 lb mass is to be cooled, large modern IR instruments may have to cool down hundreds of pounds. This implies a lot more cooling power and a much longer cool-down time (and warm-up time) than a CCD camera! Flexible copper braids must be connected from the cold reservoir to many places in the instrument to assist the uniform cooling of the components.

Since liquid helium boils away very rapidly, is harder to handle than liquid nitrogen and is much more expensive, many infrared instruments now incorporate closed-cycle refrigerators (CCRs) in preference to LHe. For small CCRs, vibration from the mechanical pump in the cold head is efficiently damped out using four supports of neoprene rubber and a

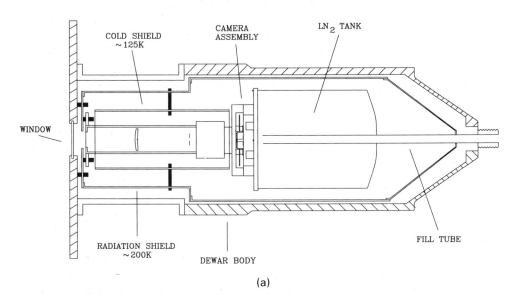

(a)

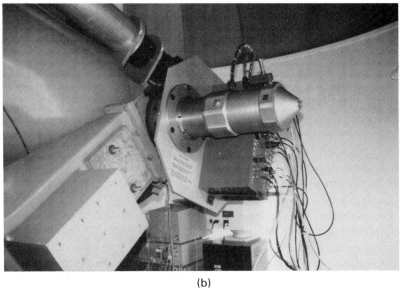

(b)

Fig. 9.11. (a) A schematic of a simple fixed-filter 2 μm infrared camera based on a HgCdTe array and a commercial liquid nitrogen dewar. (b) Showing the camera and electronics attached to the 24-inch roof-top telescope at UCLA.

stainless steel bellows. Larger heads may require more sophisticated damping systems. The compressor needed to supply the high pressure lines carrying the working fluid (high-purity helium) should not be attached to the telescope structure.

Many components must pass through the wall of the vacuum chamber, so there is a potential for leaks at many places e.g., the vacuum pumping port and pressure gauge fittings, electrical connections (vacuum feed through for wiring), drive shafts for filter wheels and other moving parts, LN_2/LHe filling ports or CCR cold fingers, the entrance window and of course the main O-ring-sealed cover plate on the chamber. Most designs try to ensure that all fittings come out through one face and that nothing comes through the main cover plate. Several instruments employ motors modified for cryogenic operation inside the dewar to eliminate mechanical shafts fed through the dewar walls.

9.5.2 Some infrared camera systems

Numerous IR camera systems are now available. Many large and small observatories alike, whether IR-optimized or not, have acquired cameras which at least cover the SWIR region from 1–2.5 μm using the Rockwell HgCdTe arrays. InSb cameras and IBC cameras for the 1–5 μm and 7–30 μm are generally found on telescopes and sites better-suited to infrared work, but are still surprisingly plentiful. Descriptions of many excellent systems can be found on the World Wide Web at such sites as IRTF, UKIRT, CFHT, ESO, NOAO, Calar Alto, UCO and many more. Useful technical details are contained in SPIE conference proceedings (see reference list).

A very simple cryostat design which illustrates the basic principles is shown in Fig. 9.11 (a) and (b). This is a fixed-filter 2-μm camera using a liquid-nitrogen-cooled HgCdTe array and it is used for student teaching and for research on our 24-inch telescope on the UCLA campus. In this case, a simple analysis shows that a single transfer lens of ZnSe will just be adequate to re-image the focal plane onto the detector and to produce a pupil image at which a cold stop is located. The lens, filter, cold stop and detector are all enclosed in multiple shields and cooled by conduction from the cold face of the liquid nitrogen reservoir. A system like this is relatively inexpensive. The cost is dominated by the detector ($60 000) and PC-controlled hardwired electronics designed by Ev Irwin of Rapax Systems and Engineering ($40 000). For a more versatile instrument on a site with much superior image quality, the design requirements are more stringent.

Short wavelength HgCdTe arrays, like the NICMOS 3 detector, have the big advantage that they can be cooled to LN_2 temperature using a single cryogenic vessel very much like a CCD camera. Indium antimonide and IBC systems, however, require colder temperatures and need a double-cryogenic vessel in which the colder, inner vessel (of LHe or the second stage of a CCR) is "protected" by the outer vessel at LN_2 temperatures. Fig. 9.12 shows one example of a mid-infrared camera system with this construction. The instrument is the MIRAC2 camera developed jointly by the University of Arizona, the Smithsonian Astrophysical Observatory and the Naval research Laboratory and operated by Bill Hoffmann, Giovanni Fazio, Lynne Deutsch, Joseph Hora and Aditya Dayal. MIRAC2 uses a Rockwell HF16 128 × 128 Si:As BIB detector. Reflective optics in the camera's liquid-helium cryostat provide achromatic, diffraction-limited imaging at a scale of 0.34 "/pixel (or at 0.25 "/pixel). Two cryogenic filter wheels contain a selection of filters including 15% bandwidth filters at 2.2 and 4.9 μm, a 5% bandwidth filter at 7.8 μm, 10%

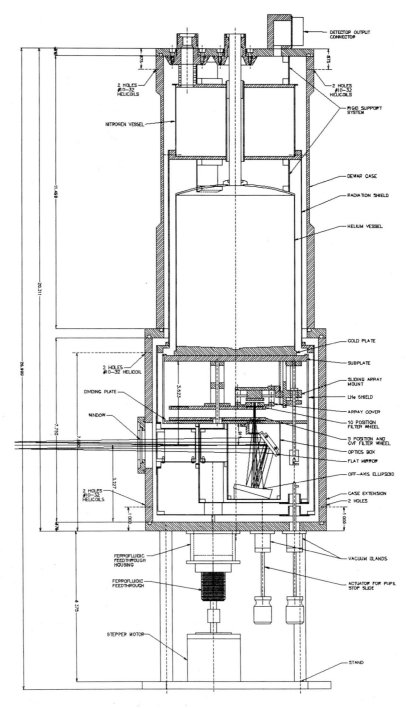

Fig. 9.12. The mid-infrared camera MIRAC2 illustrating the dual helium/nitrogen cryostat and side-looking configuration. Courtesy Bill Hoffman.

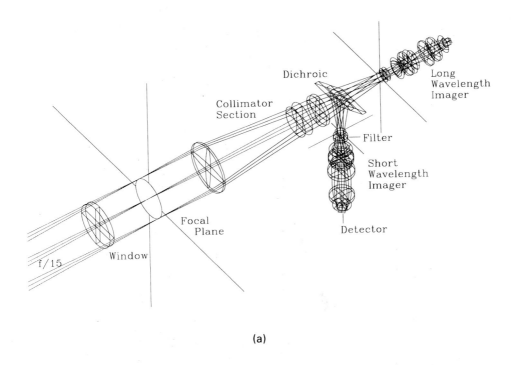

(a)

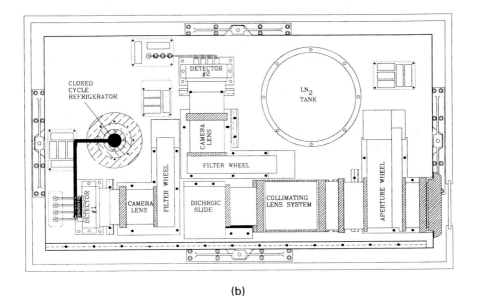

(b)

Fig. 9.13. (a) The optical design of a twin-channel infrared camera developed at UCLA. A dichroic beam-splitter provides simultaneous observations at two different wavelengths. (b) A plan view of the mechanical layout of the UCLA instrument.

bandwidth filters at 8.8, 9.8, 10.3, 11.7 and 12.5 µm, 4% bandwidth filters at 17.2 and 17.9 µm and a 9% bandwidth filter at 20.9 µm. A circular variable filter (CVF) together with standard N and Q broad band filters is also included. A Pentium-based PC with a DSP card controls the instrument and the chopping secondary of the telescope. Typical sensitivities are around 30 mJy/arcsec^2 at 11.7 µm with chop-nod and a total observing time of 1 minute.

An example of a more complex camera design is shown in Fig. 9.13(a,b). This unique camera, developed at UCLA, is equipped with a beam-splitter to provide simultaneous observations at two different wavelengths. A rotatable filter wheel is located in each beam and a selection of broad and narrow band filters is included together with grisms. At the heart of the UCLA camera is a pair of infrared array detectors; the SBRC 256 × 256 InSb array in the long wavelength channel and the Rockwell International 256 × 256 HgCdTe (NICMOS 3) array in the short-wavelength channel. The UCLA camera gives a pixel size of 0.68 "/pixel on the f/17 3-m Shane telescope at Lick Observatory and a size of 0.25 "/pixel on the *f*/15 10-m W.M. Keck telescope.

The results of an interesting calculation concerning the signal-to-noise ratios of simultaneous multi-wavelength observing in the near IR are shown in Fig. 9.14. Assuming that observations are always made in a background-limited mode, one can estimate the integration time which gives the same S/N ratio in any two different wavebands, such as J and K, and then form the *ratio* of these integration times as a function of the effective temperature

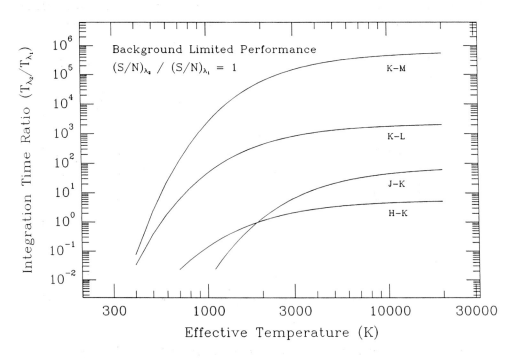

Fig. 9.14. A plot of the ratio of the integration times that yield the same S/N ratio in two different wavebands, such as J and K, as a function of the effective temperature of the source.

of the source. The figure reveals that simultaneous observations at J, H and K are only strictly compatible for effective temperatures around 2000 K, but the ratio remains within a factor of 10 of unity in either direction over quite a wide range. On the other hand, there is very poor compatibility with the L and M bands. The dependency on other factors was also examined including reddening and redshift. In most circumstances, the longest wavelength observation requires the most integration time except for very cool or heavily obscured objects, suggesting that either the J, H and K measurements are "nested" within the L measurement time, or that the J and H integrations are nested within the K integration.

9.5.3 Infrared spectrometers

It is relatively straightforward to design an infrared grism which mounts in the filter wheel of a camera to permit spectroscopy with a resolution of about $R = 500$ in the J, H and K parts of the spectrum. Fig. 9.15 shows an example of spectra obtained with the UCLA camera using grisms made from ZnSe. At longer wavelengths, the resin used in the grating replication process is absorbing. Some success has been achieved however, by a few groups who have ruled the grating directly into the prism material. For example, Zeiss has manufactured an L-band grism in KRS5 using this method.

To obtain an infrared spectrum with substantially higher spectral resolution than provided by a grism, requires an optimized design. Just like its optical counterpart, an infrared spectrometer relies on a diffraction grating, but unlike optical spectrographs, the grating and the entrance slit and all surrounding optics and metal must be cooled to cryogenic temperatures. The "image" on the IR array is now a spectrum of each spatial element along the length of the slit. In the absence of a source on the slit the spectrum will be

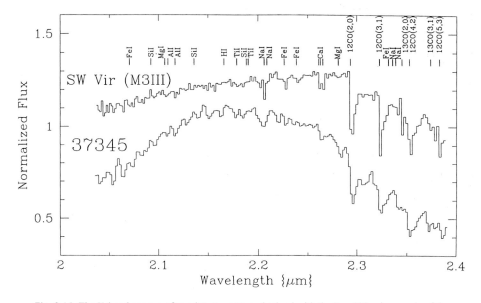

Fig. 9.15. The K-band spectra of two late-type stars obtained with the R = 500 grism mode of the UCLA twin-channel camera. Many absorption features are clearly seen including the CO bands at 2.3 μm. Courtesy Don Figer.

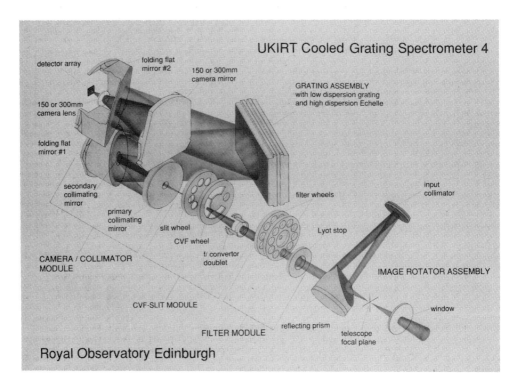

Fig. 9.16. The optical layout of one of the first infrared spectrometers to benefit from the introduction of IR arrays. It is the UKIRT CGS4 instrument.

dominated by the infrared OH emission lines from the night sky. Fig. 9.16 shows the optical layout of one of the first and best infrared spectrometers to benefit from the introduction of IR arrays. It is the UKIRT CGS4 instrument developed by the Royal Observatory Edinburgh team led by Matt Mountain (Mountain et al. 1990, Wright et al. 1993). Also shown (Fig. 9.17), is a spectrum of the suspected Brown Dwarf object Gl 229B obtained with CGS4 and compared with Saturn's moon, Titan.

The optics of CGS4 are almost all reflecting and this instrument was the first to fully pioneer the use of diamond-machined aluminium surfaces throughout to produce a design that is essentially athermal, eliminates adjustments after cool down and achieves much higher alignment tolerances.

Figure 9.18 shows the optical layout of a high resolution, cross-dispersed infrared echelle spectrograph for the Keck Telescope. This instrument is designed to use a 1024×1024 array, and is comparable in complexity to a number of large, cryogenic instruments being developed for the new generation of very large telescopes.

Since OH emission lines are a major problem for NIR spectroscopy, it is important to address this issue. One approach is to develop detectors with low noise and low dark current and go to higher spectral resolution. Another approach is "OH suppression" spec-

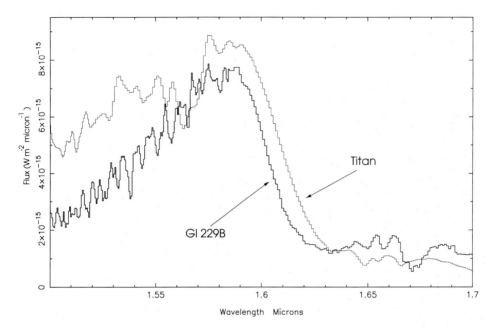

Fig. 9.17. CGS4 spectra of the suspected Brown Dwarf object Gl 229B and Saturn's moon Titan near the methane absorption edge at 1.6 μm. All the spectral features for Titan are believed to be due to methane. The narrow features short of 1.6 μm in Gl 229B are due to water vapour; those longward of 1.6 μm are not yet identified. Courtesy Tom Geballe, UKIRT.

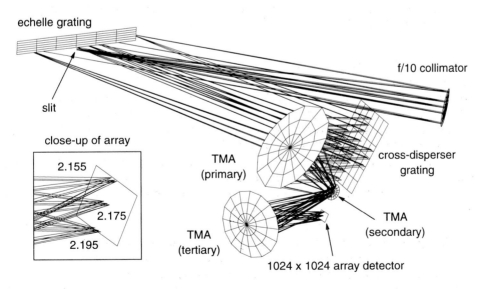

Fig. 9.18. The optical layout of a high resolution, cross-dispersed infrared echelle spectrograph (NIRSPEC) for the Keck Telescope being developed at UCLA. The advent of 1024×1024 arrays makes the echelle format attractive.

trometers. The goal of these instruments is to eliminate or mask out the OH lines from the final spectrum. Pioneering work in this field was done by Maihara and colleagues at the University of Hawaii (Maihara et al. 1993, Iwamuro et al. 1994). Briefly, a high dispersion spectrograph is used to generate a long (250 mm) spectrum from which the OH lines can be physically removed by a mask of blackened lines on a reflective surface at positions corresponding to the OH lines. The reflected light, which no longer contains any signal at the wavelengths of the OH lines, is recombined and then dispersed again at a much lower resolving power (~ few hundred) to yield a faint object spectrograph. The reduction in background can be a factor of 20 or more (Fig. 9.19). Another form of OH suppression instrument developed at Cambridge University by Ian Parry and colleagues provides $R = 500$ OH-suppressed spectra simultaneously for both J and H using an integral field fibre feed of about 100 infrared transmitting optical fibres. Fig. 9.20 shows the layout of another integral field instrument, the "3D" camera developed at the Max Planck Institute by Alfred Krabbe and colleagues.

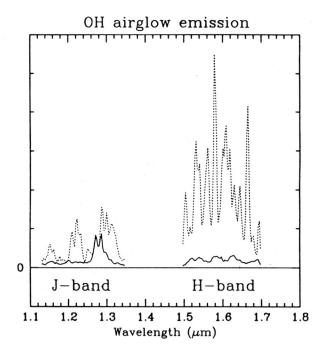

Suppressed (solid) and unsuppressed (dotted) sky emission spectra (resolving power~100) in the OHS slit mode. The remaining emission features are mostly due to O_2 band emissions (e.g., 1.27, 1.28 μm) and other unidentified weak lines.

Fig. 9.19. OH airglow emission before and after suppression in the J and H bands.

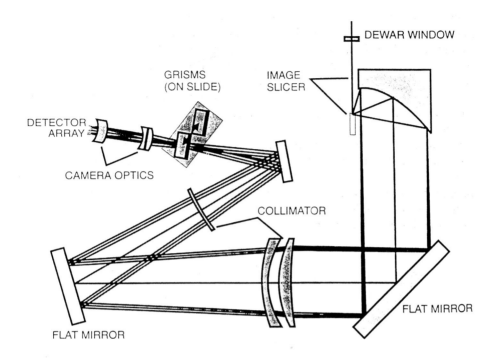

Fig. 9.20. The optical layout of the "3D" integral field imaging spectrograph developed at the Max Planck Institute. Courtesy Alfred Krabbe.

Mid-IR spectrographs follow similar designs and, as the low background of these (normally) high background arrays improve, spectroscopy in the 8–13 μm band will become very popular (see Jones (1996) for a discussion of mid-IR instruments). Finally, Fig. 9.21 shows a helium fill in progress for instruments in the forward Cassegrain module (black, centre) on the Keck I 10-m telescope which contains the Near-Infrared Camera (NIRC) developed by Keith Matthews and Tom Soifer at Caltech, and the Long Wavelength Spectrometer (LWS) developed by Barbara Jones and Rick Puetter at UC San Diego.

SUMMARY

We have seen that infrared arrays and infrared instruments, while sharing many similarities with their visible-light counterparts, also differ in several significant ways. Cryogenic operation requires much more attention to materials properties, whether optical or mechanical, and the cooling systems are more complex. Nevertheless, the new generation of large format infrared arrays that have emerged in the mid-1990s are incredibly powerful, and have had a significant impact on the way large observatories schedule their telescope time.

Fig. 9.21. A photograph showing the forward-Cass instrument module removed from its normal location in the Keck I telescope for cryogenic refilling. This module holds the near-infrared InSb array camera and the long wavelength Si:As array spectrometer.

EXERCISES

1.   Compare and contrast the operation of the Rockwell HgCdTe arrays and the SBRC InSb arrays.
2.   Why is it usually necessary to obtain a sky difference on a regular basis when imaging in the near-infrared at about 2.2 μm wavelength?
3.   Explain the necessity for a high-speed "co-adder" in infrared cameras.
4.   Draw a sketch of the output signal as a function of time for an infrared array and use the graph to illustrate what is meant by (a) single sampling, (b) correlated double sampling, (c) multiple sampling.
5.   Sketch the main features of an infrared camera. Indicate which components must be cold and what their temperatures would be for (a) a SWIR camera and (b) a mid-IR camera. What cooling apparatus would you use to achieve these temperatures?
6.   Describe the main features of an infrared spectrograph. What detector properties are particularly important for spectroscopy? Discuss the limitations imposed by OH emission lines in the near IR including ways to circumvent the problem.

REFERENCES AND SUGGESTED FURTHER READING

Chapman, R., Beard, S., Mountain, M., Pettie, D., and Pickup, A. (1990) "Implementation of a charge integration system in a low background application", *Instrumentation in Astronomy VII*, SPIE Vol. 1235, 34–42.

Fowler, A.M. (Ed.) (1993) *Infrared Detectors and Instrumentation*, SPIE Vol. 1946. [Many excellent papers. Highly recommended series for state-of-the-art reports.]

Fowler, A.M. (Ed.) (1995) *Infrared Detectors and Instrumentation*, SPIE Vol. 2475.

Fowler, A.M. and Gatley, I. (1990) "Demonstration of an algorithm for read-noise reduction in infrared arrays", *Ap.J. (Letters)*, **335**, L33–34.

Garnett, J.D. and Forrest, W.J. (1993) "Multiply sampled read limited and background limited noise performance", *Infrared Detectors and Instrumentation*, A.M. Fowler, (Ed.) SPIE Vol. 1946, 395–404.

Hoffman, W.F., Fazio, G.G., Shivanandan, K., Hora, J.L. and Deutsch, L.K. (1993) "MIRAC, a mid-infrared array camera for astronomy", *Infrared Detectors and Instrumentation*, A.M. Fowler, (Ed.) SPIE Vol. 1946, 449–460.

Iwamuro, F., Maihara, T., Oya, S., Tsukamoto, H., Hall, D.N.B., Cowie, L.L., Tokunaga, A.T., Pickles, A.J. (1994) *PASJ,* **46**, 515–521.

Jones, B. (1996) in *Instrumentation for Very Large Telescopes*, J.-M. Rodriguez-Espinoza, (Ed.) Cambridge University Press, Cambridge, England.

Maihara, T., Iwamuro, F., Yamashita, T., Hall, D.N.B., Cowie, L.L., Tokunaga, A.T., Pickles, A.J. (1993) *PASP,* **105**, 940.

McLean, I.S. (1994) (Ed.) *Infrared Astronomy with Arrays: The Next Generation*, Kluwer Academic Publishers, The Netherlands.

Mountain, C.M., Robertson, D.J., Lee, T.J. and Wade, R. (1990) "An advanced cooled grating spectrometer for UKIRT", *Instrumentation in Astronomy VII*, D.L. Crawford, (Ed.) SPIE Vol. 1235, 25–33.

Wright, G.S., Mountain, C.M., Bridger, A., Daly, P.N., Griffin, J.L. and Ramsay, S.K. (1993) "The CGS4 experience—two years later", *Infrared Detectors and Instrumentation*, A.M. Fowler, (Ed.) SPIE Vol. 1946, 547–557.

In November 1996 Rockwell International Science Center indicated that it would no longer produce the NICMOS 3 devices, and would replace them with an array called PICNIC. The PICNIC readout is a 256×256 pixel device with a similar structure to the NICMOS 3 array, but they are not pin-compatible. Common clock and bias lines are tied together on the multiplexer, greatly reducing the number of pads bonded to the chip carrier. Clocking requirements are also different. PICNIC arrays can only be reset by rows and the Read clock is introduced to enable the source follower outputs to be controlled. Details are available on Rockwell's Web page.

10

Characterization and calibration of array instruments

Electronic imaging devices such as CCDs and infrared arrays require calibration if they are to be used for quantitative work in photometry and spectroscopy. It is important to understand how the properties of the detector can be measured and how the behaviour of the detector affects photometric and spectroscopic analysis. This chapter describes several important steps in these calibrations and explains how to calculate signal-to-noise ratios.

10.1 FROM PHOTONS TO MICROVOLTS

The observed quantity in an experiment is the stream of photons, but the detected quantity is a small voltage (V_o) which is amplified and digitized. If N_p photons are absorbed in the integration time (t), then $\eta G N_p$ electrons will be detected. Here η (< 1) is the quantum efficiency and $G \sim 1$ is called the photoconductive gain and allows for intrinsic amplification within the detector. For a CCD and a photodiode infrared array, $G = 1$, but for an IBC detector or a photoconductor, G can differ from one. Multiplying by the charge on the electron (e) gives the total number of coulombs of charge detected, and the resulting voltage at the output pin of the array detector is

$$V_o = \frac{A_{SF}\, \eta G N_P\, e}{C} \tag{10.1}$$

In this expression, C is the capacitance of the output node of the detector (CCD or infrared array) and A_{SF} is the amplification or "gain" of the output amplifier which is usually a source follower (typically $A_{SF} \sim 0.8$); the suffix SF stands for "source follower". In practical terms, we need only know the combined quantity $A_{SF}\eta G/C$, but it is desirable to know these quantities individually too. Therefore, the first step is usually to determine the quantum efficiency or QE.

10.1.1 Quantum efficiency and DQE

In principle, quantum efficiency can be determined in the laboratory with a stable and well-designed calibration system constructed to properly illuminate the detector through a

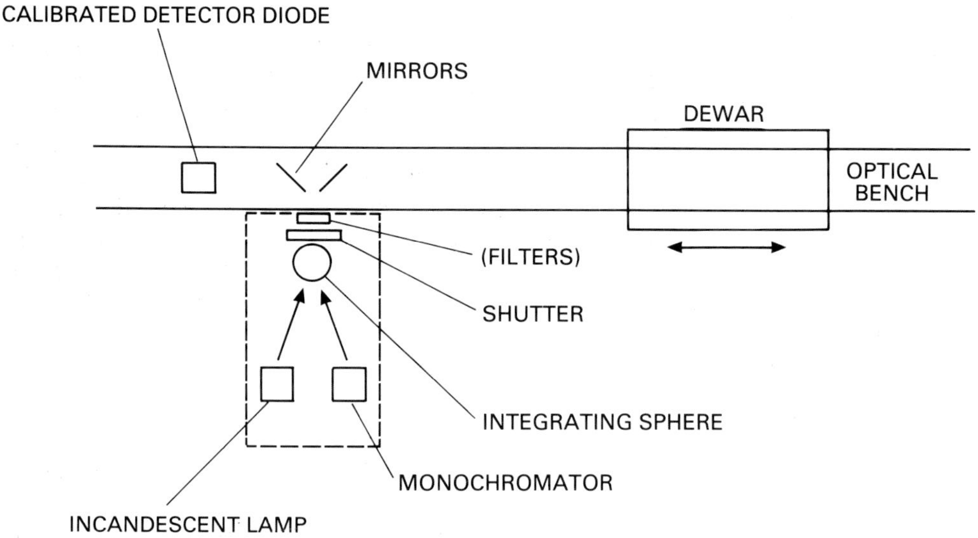

CALIBRATED DETECTOR DIODE

MIRRORS

DEWAR

OPTICAL
BENCH

(FILTERS)

SHUTTER

INTEGRATING SPHERE

MONOCHROMATOR

INCANDESCENT LAMP

Fig. 10.1. Laboratory arrangements for calibration and characterization of CCDs as used at
the National Optical Astronomy Observatories in Tucson.

known spectral passband with the minimum of other optics in the beam. One example of
an experimental set-up is shown in Fig. 10.1. Either an incandescent lamp or a grating
spectrometer can be used as a source of illumination. After passing through a device called
an "integrating sphere" which randomizes the light rays and produces a uniformly illuminated source, the light passes through a shutter and a filter holder. At longer wavelengths
a stable blackbody source (commercially available) can be used and the integrating sphere
is not needed. The light is then split by mirrors; part is directed towards the detector cryostat and part toward a calibrated photodiode. Exposures are taken at the desired wavelengths and recorded along with the signal from the calibrated photodiode. Since the camera unit is on an optical bench it can be moved closer or farther from the light source in a
controlled manner. This allows the experimenter to use the inverse square law for light as
a way of changing the illumination on the detector. For some wavelengths and passbands
it is also possible to use non-wavelength-dependent attenuating filters called "neutral density" filters since their attenuation can be determined fairly accurately with the calibrated
photodiode. With this set-up it is easy to obtain the relative QE as a function of wavelength, but to convert this to an absolute quantum efficiency at any given wavelength requires a precise calibration of the illumination, the exact transmission or "profile" of the
filter passband at each wavelength and an accurate determination of the solid angle on the
source subtended by a pixel. The latter is usually achieved by using a well-defined geometry controlled by baffles rather than optics since adding transmissive or reflective elements
to the set-up just adds other unknown quantities into the experiment. Filter profiles are
measured in commercial spectrophotometers. Note that in the case of an infrared filter the
scan needs to be done at the operating temperature (e.g. 77 K) since the passband broad-
ens and shifts to shorter wavelengths as temperature decreases. At infrared wavelengths it
is easier to be sure of the illumination level by using a blackbody source at a known tem-

perature because the energy spectrum is given by the Planck function $B_\lambda(T)$ which is determined only by the absolute temperature (T); several forms of the Planck function are given in Appendix 3. Good laboratory set-ups can yield both the relative quantum efficiency as a function of wavelength and the absolute QE. Electrical measurements can be used to determine A_{SF} independently. For A_{SF}, the simplest approach is to change the output drain voltage and observe the change in the output source voltage; the ratio will yield A_{SF}. To measure C a controlled charge Q can be injected and the voltage V measured, then $C = Q/V$. Alternatively, one can expose the detector to a substantial light level to yield a large output signal in which the dominant noise is shot noise. If N is the total number of charges collected then the measured voltage is $V = Ne/C$, and the noise is $\sigma_V = e\sqrt{N}/C$. By squaring the noise and forming the ratio we get

$$C = \frac{eV}{\sigma_V^2} \qquad\qquad (10.2)$$

allowing C to be determined from the mean signal V and the variance σ_V^2 of the voltage noise on the signal.

By observing the signal from a star of known brightness and energy distribution, one can take advantage of the fact that the solid angle (on the sky) of a telescope is very well-defined. Unfortunately, the product (τ) of all the unknown optical transmission is now included and so the derived quantity (assuming $G = 1$ and A_{SF} and C known from electrical measurements), is $\tau\eta$. While this is all that is needed for calibration, it is still very helpful to know where light is being lost so that improvements can be made. The observed quantum efficiencies of several CCDs are compared in Fig. 10.2.

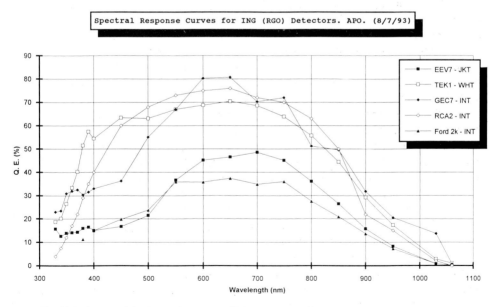

Fig. 10.2. Curves of the detected quantum efficiency (DQE) of a variety of CCDs. Courtesy Paul Jorden, Royal Greenwich Observatory.

When discussing systems which exhibit readout noise, as opposed to systems with pure photon counting detectors (PCDs), it is useful to introduce the concept of **Detected Quantum Efficiency** or **DQE**. The DQE is defined as the quantum efficiency of an idealized imaging system with no readout noise but which produces the same signal-to-noise ratio as the actual CCD system in question.

In the ideal case, the signal-to-noise ratio for a CCD pixel is given by

$$\frac{S}{N} = \frac{\eta N_p}{\sqrt{\left(\eta N_p + R^2\right)}} \tag{10.3}$$

where η is the quantum efficiency at the wavelength of concern, N_p is the total number of photons incident on the pixel in the exposure time, and R is the root mean square (or rms) value of the readout noise. An idealized detector with no readout noise would have QE equal to η' and a noise (N), given by Poisson statistics, of the square root of $\eta' N_p$. Equating the S/N ratios yields an expression for the DQE η' of

$$\eta' = \eta \frac{1}{\left(1 + \dfrac{R^2}{\eta N_p}\right)} \tag{10.4}$$

which shows that η' is less than η, and that the DQE of a CCD is dependent on the signal level (N_p) (Table 10.1). To keep the DQE within 10% of the QE then N_p must be greater than $10 \times R^2/\eta$.

Table 10.1. The detected quantum efficiency (DQE) as a function of readout noise R electrons rms and incident photon flux N_p photons/s for two values of the true QE (30% and 60%)

| Read noise | Incident photon flux (N_p) ph/s | | | | | |
R (e⁻)	1	10	100	1000	10 000	100 000
1	6.9	22.5	29.0	29.9	30.0	30.0
	(22.5)	(51.4)	(59.0)	(59.9)	(60.0)	(60.)
10	0.1	0.9	6.9	22.5	29.0	29.9
	(0.4)	(3.4)	(22.5)	(51.4)	(59.0)	(59.9)
100	0.001	0.009	0.1	0.9	6.9	22.5
	(0.004)	(0.215)	(0.4)	(3.4)	(22.5)	(51.4)

For a good CCD with five electrons noise and a QE of 0.5 (50%), N_p must be greater than 500 electrons, which corresponds to a star of visual (V) magnitude 22.0 in a 10-second integration time for direct imaging on a 4 m telescope. On the other hand, a spectrometer might require an integration time 1000× longer than this. Clearly, this equation puts stringent demands on CCDs for low noise and high quantum efficiency, and without both of these simultaneously there is still a need for intensified photon counting systems.

10.1.2 Photon transfer function

The digital signals actually recorded by the CCD system—usually called **Data Numbers** (or DN) or sometimes Analogue-to-Digital Units (ADU)—must be turned back into microvolts and then into electrons and finally to photons in order to calibrate the system. The relation between DN and microvolts at the CCD output depends on the "gain" of the amplifiers in the system, and the conversion between microvolts at the CCD output and an equivalent charge in electrons requires a knowledge of the capacitance (C) of the output node of the on-chip amplifier. The actual data counts or DN recorded in a given time by the CCD camera system are linearly related to the numbers of electrons in the charge packets by the following expression

$$S = \frac{\left(N_e + N_D\right)}{g} + b \tag{10.5}$$

where S is the recorded output signal in Data Numbers or counts, N_e is the number of electrons in the charge packet ($= \eta N_p$) and the system photon transfer gain factor is g electrons/DN, b is the (small) electronic offset or bias level (in DN) for an empty charge packet and N_D is the (small) residual dark current signal still present after cooling the device. The value of b can be controlled by the designer and is usually substantially above zero, by ten to one hundred times the readout noise (expressed in counts), to ensure that there are no problems with the A/D unit when it receives zero or slightly negative signals by chance. (Bipolar A/D units can be used which will properly record negative and positive signals.) Typical values of b would be in the range 50–500 DN depending on the particular CCD and its associated readout noise. [NOTE: Both the bias (b) and the dark current can be determined from measurements without illumination and can therefore be subtracted.] There are two ways to derive the transfer factor g in electrons/DN: (a) by calculation, knowing the overall amplifier gain and the capacitance of the CCD. (b) by a series of observations of a uniformly illuminated scene at different brightness levels. We will describe each method in turn.

Let V_{fs} be the full-scale voltage swing allowed on the A/D unit, and n be the number of bits to which the A/D can digitize. The full-scale range is therefore subdivided into 2^n parts, the smallest part—the **least significant bit** or LSB—is simply 1 DN. Thus the voltage corresponding to 1 DN at the A/D unit is $V_{fs}/2^n$; as an example, suppose the full-scale voltage is 10 V and the A/D is 16 bits then 2^n is 65 536 and so the ratio is 0.000 152 5 V, or 152.5 µV at the A/D is equivalent to 1 DN. Similarly, for a 14-bit A/D the range is 16 384 and 1 DN corresponds to 610 µV. Since we need the number of microvolts corresponding to 1 DN at the CCD rather than at the A/D, we must divide the number derived above by the total gain product A_g of all the amplifiers in the system; usually this means the on-chip amplifier (A_{SF}), the preamplifier (A_{pre}) and a postamplifier (A_{post}). To convert this number of microvolts to an equivalent charge of electrons we must then multiply by the CCD capacitance (C) and divide by the value of the charge on the electron (e). Therefore,

$$g = \frac{V_{fs} C}{2^n A_g e} \tag{10.6}$$

where $e = 1.6 \times 10^{-19}$ coulombs.

The value of C varies with the type of CCD, but is typically about 0.1 pF. In this case, we get $g = 95.3/A_g$. If the value of g is too small then large signals produced by the CCD will be above the maximum input level of the A/D unit. On the other hand, if g is too large then errors due to the process of conversion of analogue to digital signals—called **quantization errors**—will significantly add to the system noise. A rule of thumb is to arrange for g to be about 0.5 to 0.25 times the CCD readout noise expressed in electrons. So, for a 10-electron device we might chose g to be 5 e–/DN implying an external gain of about 27, if $A_{SF} = 0.7$. Notice that a full-scale reading of the A/D corresponds to 65 536 DN in this case which is equivalent to 327 680 electrons; this is smaller than the physical full well of some CCDs but larger than the well-depth of others. The transfer gain factor g can also be determined experimentally.

There are essentially two approaches. Obtain several exposures of a flat field and examine the mean signal and noise from each and every pixel independently before adjusting the illumination level, which must be absolutely constant during the measurements to avoid adding additional noise. Alternatively, take fewer exposures and examine the mean signal from a small array of pixels (in a good clean area of the CCD or IR array) which have first been flat-fielded to remove the dispersion (or noise) in the mean due to their individual variations in sensitivity. The latter method is the best and most widely used, but it will not work well unless the pixel-to-pixel variations in sensitivity are removed by good field flattening.

From either data stream X_i, one obtains a mean, dark/bias-subtracted signal (S_M) in counts and a **variance** (V_M), which is the square of the standard deviation of a single observation from that mean.

$$S_M = \frac{1}{n} \sum X_i \text{ and } V_M = \frac{\sum (X_i - S_M)^2}{(n-1)} \tag{10.7}$$

where the summation indicated by $\sum$ is over all n pixels in the data set. Now, if all is well, there should be only two noise sources, namely,

(1) photon noise on the signal photoelectrons (p)
(2) readout noise from the CCD output amplifier (R)

Since these two noise sources are independent and random they add together in a way known as **quadrature**, so that the total noise is given by

$$(noise)^2 = p^2 + R^2 \tag{10.8}$$

The noise-squared $(noise)^2$ is just the variance. It is important to realize that this expression applies to photoelectrons (e) and not to counts (DN). However, the measured quantities (S_M and V_M) are in DN. To convert from electrons to DN in equation (10.8) we need to divide each noise term by g (electrons/DN), to give

$$\left(\frac{noise}{g}\right)^2 = \left(\frac{P}{g}\right)^2 + \left(\frac{R}{g}\right)^2 \tag{10.9}$$

This equation is much simpler than it looks because the left-hand side is exactly V_M, the observed variance in DN. Also, the mean number of photoelectrons is $g(e^-/DN)S_M(DN)$

and the photoelectron noise (p) on this number is simply the square root of gS_M, so $p^2 = gS_M$ which gives

$$V_M = \frac{1}{g} S_M + \left(\frac{R}{g}\right)^2 \tag{10.10}$$

which is just the equation of a straight line in a signal-variance plot of $y = V_M$ and $x = S_M$. Plotting these "observed" quantities (noise-squared and signal) as the illumination changes will yield a straight-line of gradient (slope) $m = 1/g$ and the value of the intercept on the V_M axis when $S_M = 0$ gives $(R/g)^2$, and hence R since g is known from the slope (see Fig. 10.3(a)). Graphs of noise versus signal on a logarithmic scale (Fig. 10.3(b)) can also be used to display the same result. Each data point on the line should be determined several times to improve the straight-line fit. This method of deriving g and R is known as the "variance method" or the "photon transfer method" and is extremely powerful. Not only does it provide a completely independent way to derive R and g, it also shows where the CCD or IR array begins to become non-linear and saturate.

At the lowest signal levels the noise (variance) is dominated by the fixed readout noise whereas for larger signals it is dominated by photon noise, the statistical fluctuations in the photon detection rate itself. Finally, at the largest signals, the CCD (or the ADC unit) begins to saturate and the noise actually falls because many pixels have identical values. In principle, the photon transfer curve can be derived for every pixel on the detector so that the behaviour of each can be examined independently.

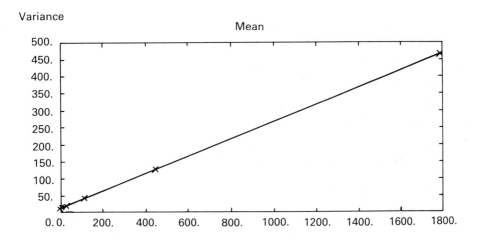

Least Square fit to variance/mean calibrationstatistics
gain calibration = 3.9611 electrons/adu
noise level = 14.0010 electrons rms
correlation coeficient = 1.000

(a)

Fig. 10.3. The photon transfer curves of noise versus signal for different CCDs. (a) a plot of variance versus signal; (b) a plot of noise versus signal on a logarithmic scale.

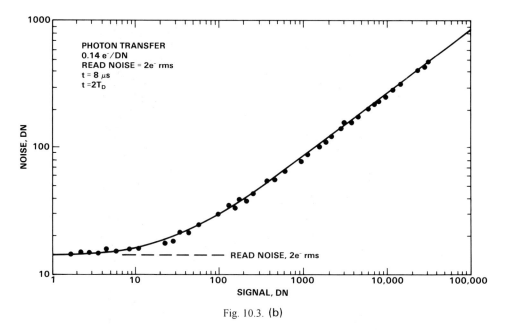

Fig. 10.3. (b)

In practice this technique requires some care. Several (perhaps five or seven) flat-field exposures are taken at each of many illumination levels, usually increasing by factors of two, and a dark frame is taken at each of the exposure levels. Sub-areas of the CCD or infrared array that are free from artifacts and show minimal non-uniformity are selected. For each illumination level the mean picture is computed and this is subtracted from the individual images to obtain pixel deviations. Alternatively, the mean, dark-subtracted image can be used to flat-field (divide) the individual, dark-subtracted images. The variance or square of the standard deviation of each pixel from the mean value in the clean sub-array is calculated and averaged over the several frames, and the mean bias-corrected signal value is derived from the difference between the averaged picture and an average of several dark frames of the same exposure. A least-squares straight-line fit can then be made to the data set to derive the slope and intercept.

10.2 NOISE, BIAS AND DARK CURRENT

In almost all CCD and infrared array systems it is necessary to purposely offset the video signal slightly positive so that the readout noise never drives the A/D input negative. Some systems employ bipolar A/Ds which allow negative-going signals to be properly handled but again, a "no-signal" condition will always correspond to some positive offset or bias signal. This electronic instrument signature is therefore known as the **bias level** (b) and can be determined easily by taking an unexposed frame (zero exposure time, shutter closed), also alluded to as an **erase** frame, as shown in Fig. 10.4. Several bias frames can be averaged to reduce the random readout noise while preserving any spatially coherent noise and the pixel-to-pixel bias levels. A mean bias frame must be subtracted from each exposed frame as the first step in data reduction. For some applications it may be sufficient

Fig. 10.4. A good bias frame showing no serious amplifier fixed-pattern noise or ground-loop interference.

to subtract a single, mean bias *number* from each pixel rather than a bias frame, but the normal practice in astronomy is to take many bias frames and use the **median** (middle) of that set. Infrared arrays show significant fixed-pattern noise on bias frames and therefore a single value for the bias is not appropriate.

Bias information can also be conveniently obtained in many CCD systems by using an **overscan**. That is, the sequencer is told to send more clock pulses than are actually required to vertically and horizontally read out the real CCD (e.g. 1034 × 1034 instead of 1024 × 1024). The entries in the image so obtained which lie outside the physical area of the CCD should only contain bias level signals, provided the CTE of the device is good. This latter point is important, since overscanning a uniformly illuminated CCD is one way of probing the charge transfer efficiency or CTE. Charges which are trapped during the primary transfer will slowly leak out and will therefore appear at the output as a result of the overscanning. Usually, the median bias frame is computed and subtracted from the object frame, then the overscan in the bias frame is compared to the overscan in the astronomical object frame to derive any small offset which is averaged and subtracted as a single number (scalar) from the object frame.

For CCDs requiring a preflash to alleviate global charge transfer problems, the preflash must also be applied to the bias frame. If the system is working perfectly then a bias frame should contain very little fixed-pattern structure; it should be dominated by random read-out noise variations. The standard deviation of a good-sized patch of the CCD therefore gives an immediate estimate of the readout noise in data numbers without recourse to the full "photon transfer curve" experiment described previously. If there is some unavoidable fixed pattern in the bias frame then it is straightforward to take the difference between two bias frames and divide the resultant noise by $\sqrt{2}$. The same applies to infrared arrays; the difference of two bias frames is obtained and the standard deviation of the pixel values in the difference frame is computed and divided by $\sqrt{2}$ to give the readout noise in DN.

In several CCDs it is necessary to turn off, or reduce to a very low level, the current through the output transistor and the clock voltage swing in the horizontal register during long exposures. If this is not done then an immense accumulation of charge can occur in the corner of the chip and along the rows nearest to the horizontal register. The horizontal register and output transistor need to be switched on again in sufficient time to allow them to be swept clean and re-stabilize before the integration time ends. Similarly, in infrared arrays the output drain voltage needs to be kept to a minimum value.

Dark current levels are determined by long exposures with the CCD shutter kept closed. It is advisable to perform these measurements at night or in a darkened room or in the actual telescope environment. Exposure times of 1 hour will be needed to accurately deter-mine the dark current levels. Several identical exposures of this duration will also enable cosmic ray and radioactivity events to be isolated and counted. Thinned CCDs or those with an epitaxial (or thin) layer are less prone to cosmic ray events. Sometimes "hot pix-els" cause the dark current to be non-linear, then for the highest accuracy many dark frames are required at the same exposure time as used for the object frame. Dark current is more significant in infrared arrays, and it may not be linear and scalable from different exposures. It is therefore prudent to either take a suite of exposures that encompass those actually used, or simply use the same exposure times and repeat the measurements to get a good mean value.

10.3 FLAT-FIELDING STRATEGIES

As already described in section 6.6, CCDs do not have a uniform response to light across their surfaces. Pixel-to-pixel variations in sensitivity (QE) arise due to physical differences between pixels as the result of fabrication processes and due to optical attenuation effects such as microscopic dust particles on the surface of the CCD. Although these variations can be as small as a few per cent of the mean sensitivity they must nevertheless be reduced much further because such variations result in a "noisy" image at a level corresponding to a few per cent of the sky brightness.

A common practice is to observe the inside of the telescope dome (if it is matt white) or a huge white card on the dome. In both cases the objects are so close that the telescope is actually completely out of focus, which helps to ensure that the field is uniformly illumi-nated (flat), but the beam passing through the instrument is filled in virtually the same way as when observing objects in the sky. Dome illumination is usually done with tungsten lamps which do not remotely mimic the spectrum of the night sky which is a complex composite of a thermal continuum, scattered light plus a number of practically monochro-matic (and variable) night sky emission lines. Optical filters often exhibit "red leaks", that is, an unwanted transmission to light at much longer wavelengths—typically double the original centre wavelength. Although the light leak is at a very low level (less than 1%), the sensitivity of the CCD is so good that even this is a problem. The solution is to "block" the red light with an additional filter. Since tungsten lamps are much redder than the sky, dome flat-fields with such filters would be a poor match. For faint objects it is the light of the sky which dominates and so it is better to try to use the sky itself as a flat-field, whereas for brighter objects it is their own intrinsic colour which matters and that neither matches sky or dome. Most workers have found it desirable to establish a set of narrow, rather than broad, passbands for imaging so as to limit the effect of colour-dependent non-uniformity.

Since, to first order, CCD detectors are quite linear, then only a simple arithmetic division pixel by pixel, by an image of a uniformly illuminated scene—the flat-field—of the appropriate colour, followed by re-scaling, is required.

Let I_{FF} be the uniform illumination of the flat-field source on a pixel in row i, column j whose quantum efficiency is η_{ij}. The observed signal from that pixel in DN is

$$\left(X_{ij}\right)_{FF} = \frac{1}{g} \eta_{ij} I_{FF} \tag{10.11}$$

where g is the conversion in electrons/DN.

The mean signal in the flat-field is obtained from averaging X_{ij} over all the rows and columns; call this S_{FF}.

$$S_{FF} = \frac{1}{g} \eta_M I_{FF} \tag{10.12}$$

where η_M is just the mean QE averaged over the entire array.

For the true image scene, which of course is not uniformly illuminated in general, we have a similar expression for the recorded signal on the pixel on the ith row and jth column.

$$X_{ij} = \frac{1}{g} \eta_{ij} I_{ij} \tag{10.13}$$

To eliminate the position-dependent QE response (η_{ij}) we form the *ratio* of image scene to flat-field

$$\frac{X_{ij}}{\left(X_{ij}\right)_{FF}} = \frac{I_{ij}}{I_{FF}} \tag{10.14}$$

and finally we rescale this ratio to the mean of the flat-field to give

$$\frac{X_{ij}}{\left(X_{ij}\right)_{FF}} S_{FF} = \left(\frac{\eta_M}{g}\right) I_{ij} \tag{10.15}$$

Thus the flat-fielded, re-scaled image constructed by the operations on the left-hand side of this equation differs from the true image scene I_{ij} by a single, constant scale factor η_M/g which can be determined by observations of a source of known brightness—a flux standard—if required.

Many flat-field exposures are averaged to increase the accuracy of the flat-field itself and to remove from the flat-field various artifacts such as cosmic ray events. Also, the colour of the flat-field should be as good a match as possible to that of the image scene since the quantum efficiency is typically a strong function of wavelength and the cancellation described above may not be perfect if the two scenes are significantly different in colour.

To detect the very weakest signals, particularly of extended images, it is necessary to correct for residual non-linearity by more complex means such as obtaining a series of flat-fields at different exposure levels and determining the response of each pixel individually by means of a polynomial-fitting routine.

As already mentioned, for many applications it is convenient and acceptable to use the inside of the telescope dome as a flat field, or better still a flat white screen which is uniformly illuminated with lamps. The lamps are usually attached to the secondary mirror support ring, or in some cases is it possible to use an independent projector. For very high accuracy work or for observations on objects fainter than the night sky itself, various sys-

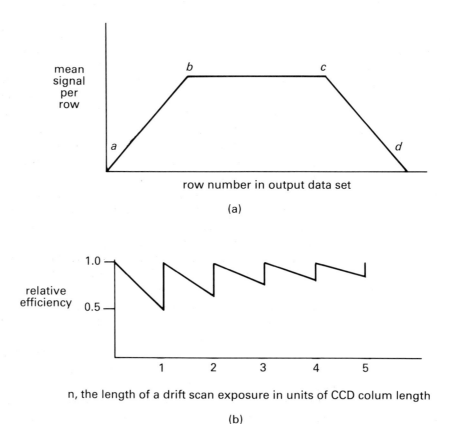

Fig. 10.5. The ramp-up and ramp-down of the mean signal from a given column with the drift scan technique for flat-field and the efficiency of the drift scan as the scanned area is increased.

tematic errors tend to dominate over the expected random errors from photon arrival statistics. The first advance in counteracting low-level systematics in the detector was the drift-scan technique developed by Craig Mackay and Jonathan Wright; this approach is based on the idea of time delay and integration (TDI) used in infrared detector systems. The CCD charge pattern is transferred along columns very slowly—i.e. the clocking rate is slow—while the image from the telescope is physically "scanned" along with it in precise synchronization, using a movable motor-driven stage. In this way the image keeps up with the charge pattern and every pixel along a column contributes to the final image of the object (Fig. 10.5). This technique averages over some of the fixed pattern noise on the CCD (at least in one dimension) and is effective in reducing systematics to below 1% of the night sky background level due mainly to the fact that the "colour match" in deriving the flat-field from the sky itself is much better than from an artificial lamp illuminating the inside of the telescope dome. There are also variations on this technique, such as the "short drift scan" in which the drift is over a much smaller number of pixels than in a whole column.

There are a number of steps which can be taken to reduce systematic errors to a level of at least 0.1% of the night sky background (assumed to be very uniform over the small areas imaged by CCDs) without the need for a mechanical stage. Tony Tyson and others have demonstrated that these methods enable a large 4-m-class telescope to reach its theoretical limit of 27th magnitude in a 6-hour exposure. Similar success in flat-fielding has been achieved with infrared arrays.

The dominant sources of error are the mismatch in colour between calibration flat fields and the actual night-sky background and, for some thinned CCDs, interference fringing due to unblocked night-sky emission lines. Generation of a master flat-field and sky frame from the object frames themselves has been shown to remove systematic effects to better than 0.03% of night sky. What this means is that after flat-fielding is complete, the apparent variations in brightness from pixel to pixel are, on average, only 3/10 000 of the mean sky brightness, assumed constant across the image. Similar results have been obtained using this approach with infrared array detectors. Essentially, this powerful technique involves numerous observations of a piece of relatively blank sky but with the telescope pointing to a slightly different position on the sky (displaced by, say, 5–10 arcseconds) for each CCD exposure. Positions can be chosen randomly or in a simple pattern, but it is best not to repeat the pattern exactly. When the sequence of disregistered exposures are examined later, and the frequency histogram or number of times a certain value or signal is found in a given pixel is plotted against those signal values, then one signal value (or a small range) will turn out to be most favoured. This value is just the signal corresponding to pure background night sky. An occasional faint object will fall on the given pixel very infrequently because of the semi-random displacement between exposures. The mathematical statistic used to determine the most likely sky value from such a data set is usually the **median** (or middle value) since it is extremely tolerant of a value discrepant from all the rest—including values due to bad pixels. If sufficient frames are taken, it is possible to calculate the **mode**, which literally is the most frequently occurring value. Clearly, the disadvantage of this technique is that it will not work on object frames which are too crowded, such as the image of a large galaxy or nebula, or a centrally condensed cluster of stars or galaxies unless much larger moves are made. A combination of dome flats and

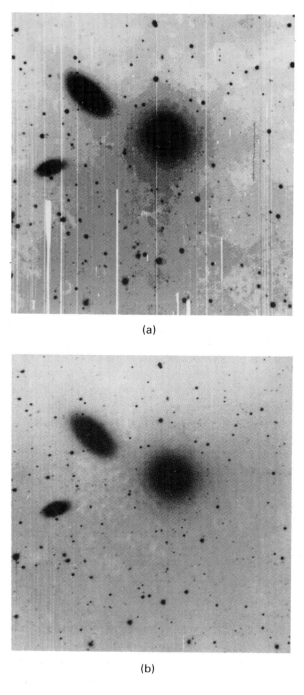

(a)

(b)

Fig. 10.6. (a) A raw CCD image with lots of defective pixels; (b) same image flattened by using "median sky flats" by shifting the images. Courtesy Harold Ables, US Naval Observatory.

sky flats is often recommended in that instance. However, with a telescope of the right scale and a 2048×2048 CCD, Harold Ables and colleagues at the US Naval Observatory in Flagstaff managed to apply this technique to the field shown in Fig. 10.6; the Tektronix chip was an early "set-up" device not intended for scientific use!

As a matter of practicality, it is essential to "normalize" the various flat-fields before applying a median filtering algorithm, since a drift in the mean level of the illumination, while not disastrous for the overall technique, will affect the calculation of the median value. It is very important that any additive effects which do not vary with the sky brightness should be removed before scaling. CCDs and infrared arrays exhibit both additive and multiplicative effects which require calibration. In the "additive" class we have

(1)   electronics pattern noise ("bias effects")
(2)   charge skimming or trapping
(3)   interference fringes
(4)   LED activity

and in the "multiplicative" class there are the effects of

(1)   quantum efficiency variations across the array
(2)   transmission of optics and coatings
(3)   thickness variations of thinned arrays and CCDs

Clearly, it is important to devise steps to remove these effects and to perform these corrections in an optimum sequence. In general, a good recipe for reduction and calibration of raw CCD images is the following pattern:

(1)   **Subtract Bias and Bias Structure:** A "bias structure" image, obtained by averaging many bias frames is subtracted and a further small scalar subtraction to eliminate any offset in bias between program and bias frames is also made using overscan data. For CCDs using preflash the bias must also use preflash.

(2)   **Subtract Dark:** Most CCDs exhibit some dark current or a low-level light-emitting diode (LED) activity, leading to "electronic pollution" during long exposures. To correct this, the "median image", rather than the "mean image", of many long (bias-subtracted) dark exposures, obtained as recently as possible, must be subtracted from the object frames. If these frames do not correspond exactly to the exposure time used on the illuminated frame, then careful experiments of dark current versus exposure time must be done to prove whether or not the dark current is sufficiently linear to be scaled from one exposure time to another.

(3)   **Divide by Flat Field:** Many exposures on a diffusely illuminated screen are normalized (adjusted to the same mean) and averaged, for each filter or spectrograph setting. Exposure levels in each flat-field (FF) and the illumination colours must be similar to the astronomical exposures; this means filtering the dome lights. For spectroscopic applications the colour of the FF is less important. The bias-corrected, dark-subtracted, normalized flat-field "master" averages are then divided into the dark-corrected object frames to calibrate for QE variations (and other effects such as out-of-focus dust spots) from pixel-to-pixel. Alternatively, the master flat-field is derived from a series of sparsely populated SKY frames each of which has been

displaced (semi-randomly) from the other. Since a star will not fall on the same pixel twice, then the median (middle value) or better still, the mode (the most frequently occurring value) of this set of dis-registered images will be a frame composed purely of uniformly illuminated sky. In practice, it is best to use a more limited number of sky frames (e.g. 7–13) closest in time to the frame being flat-fielded (e.g. three before and three after), and form a "running median" by stepping through the data set.

(4)   **Subtract Fringe Frame (Sky Subtraction):** Night-sky emission lines limited to a very narrow band of wavelengths can cause interference fringes on some back-illuminated CCDs bonded to glass, in sky-limited exposures. A "fringe frame" derived from an adaptive modal filter routine (see below) must be scaled and subtracted from the already flat-fielded object frame derived from Step (3). This step effectively performs a sky-subtraction.

(5)   **Interpolate over Bad Pixels:** Either a bad-pixel map for the particular CCD in use is supplied as input to a median interpolation routine or, if the two-dimensional displacement or disregistering technique (see above) is used during observing, then bad pixels at low levels will automatically get filtered out of the final image. Remaining bad pixels can be replaced by good data from other frames for that part of the scene.

(6)   **Remove Cosmic Ray Events:** A computer algorithm can be used to identify non-starlike point sources or the random offset technique can be used to eliminate (by median filtering) cosmic radiation events greater than several times the normal, random fluctuations in the sky background level.

(7)   **Registration of Frames and Median Filtering:** If the random offsetting technique is used, re-registration of data frames to a fraction of a pixel using interpolative software routines, followed by median-filtering and edge-trimming, is used to produce the final cleaned image.

NOTE: All of these basic steps also work for IR arrays, but extreme care is required because the backgrounds are larger and more sensitive to variations.

10.4 FRINGES AND SKY EMISSION

The occurrence of interference fringes with backside-illuminated thinned CCDs was described earlier (section 6.6). Sometimes in narrow band work, infrared arrays can also show significant circular fringe patterns caused by variable OH emission lines in the night sky. Because difference frames are routinely obtained with IR arrays, these "additive" effects are usually removed easily, except when the OH emission is extremely variable. Obviously enough bias, dark and flat-field frames must be taken to ensure sufficient accuracy in the median-processed images. Many astronomers unaccustomed to array methods balk at the thought of spending so much time on calibration data-taking. Although CCDs and infrared arrays are improving steadily, it is nevertheless absolutely essential to invest effort in calibrations if one is to extract the faintest signals and to use an instrument/telescope combination at the limit of its capability. There is usually sufficient of opportunity to take good calibration data during daylight or twilight hours.

Fringe removal can be performed by a technique called **adaptive modal filtering**. This routine computes the absolute difference between the mean and the median of values associated with a pixel over all the images in a set and rejects deviant values until this difference falls below a certain value or a maximum number of values have been rejected. A given pixel is then median-filtered over all the images. Of course, this technique fails with large, extended objects such as nearby galaxies and nebulae, but it is very suitable for relatively uncrowded fields in which two-thirds of the actual CCD area is occupied by sky. This fringe frame must then be scaled by trial-and-error and subtracted from the object frame until a patch of sky on the object frame is entirely flat.

Fig. 10.7 shows a 500-s CCD exposure on a 4-m telescope before and after fringe removal. The effect is remarkable!

10.5 LINEARITY

CCDs, if operated properly, are quite linear detectors over an immense dynamic range. That is, the output voltage signal from a CCD is exactly proportional to the amount of light falling on the CCD to very high accuracy, often better than 0.1% of the signal. The good linearity of the CCD makes it possible to calibrate observations of very faint objects by using shorter, but accurately timed, exposures on much brighter photoelectric standard stars. Linearity curves are usually derived by observing a constant source with various exposure times. This method assumes that exposure times can be very accurately controlled, which is generally a good assumption, and that $1/f$ noise sources are negligible. Simply put, if the exposure time is doubled the (dark + bias)-corrected signal should also double. Non-linear behaviour from CCDs can occur if incorrect voltages are applied. Care must be taken to ensure that the output transistor is operating in its normal linear regime and it is essential to use the correct clock voltages to ensure the CCD pixel is fully inverted, or use a CCD with MPP.

Infrared array devices commonly use reversed-biased photodiodes which exhibit an intrinsic non-linearity caused by the fact that detected photons (electron–hole pairs) actually discharge the reverse bias, thereby changing the capacitance and dark current levels associated with that detector pixel (see Chapters 8 and 9 for further discussion). Typically, these effects do not exceed about 10% and can be calibrated out to a small fraction of a per cent in the same way as for CCDs.

10.6 PHOTOMETRY

Photometry is the process of obtaining quantitative (numerical) values for the brightness of objects. Sometimes relative brightness is adequate and this may be in relation to other objects in the same field of view or nearby in the sky. For other studies it is imperative to derive the true or absolute amount of radiant energy reaching the Earth; this is much harder to do, but it is essential if we are to understand the distribution of mass and energy in the Universe. The term photometry of course derives from measurements of visual light, but it can be applied generally to the measurement of the energy transported by electromagnetic radiation. Some of the terminology of this field is confusing, so it may be useful to review a few basic concepts. An excellent introductory text on photoelectric photometry is the book by Henden and Kaitchuk (1982).

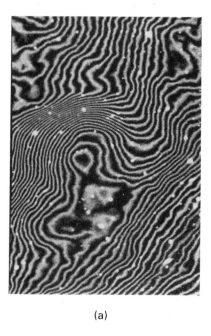

(a)

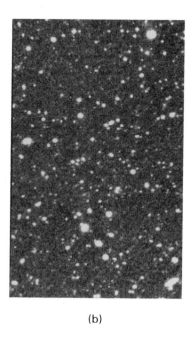

(b)

Fig. 10.7. A severe fringe pattern due to night sky emission lines on a deep 4-m telescope CCD image (a) before and (b) after processing.

Monochromatic flux: is the power received (in $W\ m^{-2}\ Hz^{-1}$) from integrating the specific intensity over the angular size of the source. Unfortunately, this use of the term flux is in conflict with its official radiometric definition in which the flux is the total power integrated over all frequencies.

Magnitudes: are relative measures of the monochromatic flux of a source. If F is the monochromatic flux due to a source, measured outside the Earth's atmosphere, and F_0 a reference monochromatic flux at the same wavelength, then Pogson's equation (see Appendix 3 if you have had no previous introduction to magnitudes) gives the corresponding magnitude (m) as

$$m = m_0 - 2.5 \log F + 2.5 \log F_0 \qquad (10.16)$$

If $m_0 = 0$, representing a reference star, then $2.5 \log F_0$ defines the reference or zeropoint of the magnitude scale. In practice, measurements are not monochromatic, but are made over a finite spectral band defined by a transmission filter. Therefore, there are different magnitude systems for different sets of spectral bands.

Bolometric magnitudes: this gives a magnitude corresponding to the total flux integrated over all wavelengths; the zeropoint is $F_b = 2.52 \times 10^{-8}\ W\ m^{-2}$.

Colour indices: this is the difference between magnitudes at two separate wavelengths, e.g. B–V, U–B in the conventional UBV system.

The original photometric system is the UBV system of Johnson and Morgan first introduced in 1953 and subsequently extended to the red ($R \sim 7000$ Å, $I \sim 9000$ Å). A modified UBVRI system known as the Kron–Cousins system is more frequently encountered in which R and I are re-defined; a version known as the Mould system is shown in Fig. 10.8(a). In addition, an important narrower band system was developed by Thuan and Gunn (the uvgri system) which avoids night sky lines and the serious overlap of the wide-band systems, and gives better flat fields for CCDs (see Table 10.2). Fig. 10.8(b) shows (on a logarithmic scale) the effect of a "red leak" in the B (blue) filter. The consequences of the leak depends on the "colour" of the illumination and the sensitivity of the detector at long wavelengths. Unfortunately, CCDs are sensitive at wavelengths beyond the original PMTs used to establish the UBVRI system.

Table 10.2. A summary of the major photometric systems

Kron–Cousins System		Thuan–Gunn system	
Wavelength (Å)	Width (Å)	Wavelength (Å)	Width (Å)
U 3600	700	u 3530	400
B 4400	1000	v 3980	400
V 5500	900	g 4930	700
R 6500	1000	r 6550	900
I 8000	1500	i 8200	1300

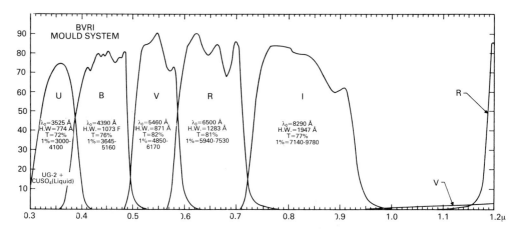

Fig. 10.8. (a) Standard filter bandpasses used with CCDs.

In general, the accuracy with which brightness can be measured on CCD frames is extremely high, and repeatability from one CCD image to another is excellent. Limitations on accuracy are really only introduced when comparing CCD photometry with the results of classical photoelectric photometers which employed photomultiplier tubes. Calibration of CCD systems against the old phototube systems to high accuracy (0.5% or 0.005 magnitudes or better) is very difficult due mainly to the gross mismatch between CCDs and classical photoelectric photometers in terms of the huge span in wavelengths to which CCDs are sensitive, and the effect which this has on the filter profile. To a lesser extent, small systematic errors can be introduced by the measuring method during CCD data reduction and analysis. In general, CCDs perform better, and can be calibrated more easily, if the spectral bandwidth is limited to a few hundred angstroms (i.e. a few hundredths of micrometres). Unfortunately, this is not "standard", although the Thuan–Gunn system approaches this ideal, and it does not easily permit a comparison with classical photomultiplier tube photometry. It has been said that classical photoelectric photometry was "a great idea at the time" but now that we have CCDs we should do everything over again! This may well happen by default when the Sloan Digital Sky Survey is complete.

Two basic procedures for obtaining photometric data from CCD and infrared images are:

(a)   aperture photometry
(b)   profile fitting

Each is discussed in turn and some additional detail is given in the next chapter on computers and image processing.

(a) **Aperture photometry:** Conceptually this is analogous to photoelectric photometry. After the image has been obtained, a computer program (e.g. *apphot* in IRAF) is used to reconstruct the signal which *would* have been obtained from an object in the field of view if the light had passed through a physical aperture (usually circular) of a certain diameter in arcseconds. The imaginary aperture is called a **software aperture**. Although this ap-

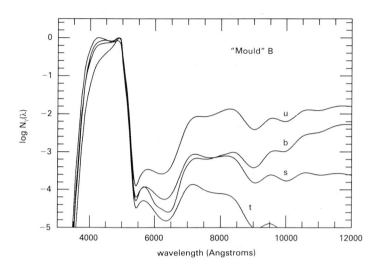

Fig. 10.8. (b) The effect of (accidental) imperfect blocking of the B(lue) filter is a "leak" of red photons to which the CCD is very sensitive. The consequence depends on the spectrum or colour of the source; u and b are balanced (filtered) and unbalanced artificial lamps illuminating the telescope dome, s is a typical solar spectrum and t is the twilight sky—which is quite blue.

proach sounds quite straightforward there are subtle practical difficulties which can influence the results. For example, the procedures used to estimate the background level, the centre of the optical image, and the shape of the image itself are important. Typically, the background sky brightness will be taken as the average in an annular ring with inner radius just beyond the limit of the object and extending a few pixels out. The algorithm should be capable of recognizing when one of the pixels in the sky annulus contains a star (higher signal than the rest) and it should eliminate this value from the average. The best procedure is to try a series of increasing values for the radius of the star aperture until the signal becomes reasonably constant. A plot of the derived **instrumental magnitude** (−2.5 log{counts/s within aperture}) versus the radius of the aperture yields a **curve of growth** from which "aperture corrections" can be deduced. Faint objects are best measured with a small aperture, but bright standard stars usually need a larger aperture.

(b) **Profile fitting:** Also called **point-spread-function** (PSF) fitting, where by PSF is meant the actual recorded shape on the detector of an unresolved, point source, this method relies on modelling the image rather than summing over the image. Mathematical curves are "fitted" to the real data using computer programs until a good match is obtained. This method is much more time-consuming for the astronomer, and for the computer! Various programs are available at major institutes for photometric reductions including such packages as DAOPHOT, ROMAFOT and STARMAN. Information on these software packages is available from IRAF(NOAO/Tucson), ESO(ESO/Munich) and STARLINK (RAL/UK) respectively, or from the original writers.

The stellar image is usually compared to a bell-shaped Gaussian intensity profile:

$$I(r) = I(0)e^{-r^2/\sigma^2} \tag{10.17}$$

where $e = 2.718$ is the base of natural logarithms, $I(0)$ is the peak intensity and r is the radial distance from the centre of the image. The quantity σ measures the width of the distribution; 68% of the light lies within $\pm 1\sigma$ and 98.7% of the light lies within $\pm 2.5\sigma$ Provided the PSF (i.e. σ) is constant across the image, programs like DAOPHOT will identify the bright stars, deduce their Gaussian profiles and subtract those profiles away, thereby revealing fainter stars.

It is often more convenient to describe the point-spread-function in terms of its full width at half of the maximum intensity; the FWHM (in pixels or arcseconds) is related to σ by the simple equation

$$FWHM = 1.665\sigma \tag{10.18}$$

The use of CCDs and infrared arrays for photometric measurements is founded on two basic assumptions, namely,

(1)   the response of each pixel is a well-defined function of exposure level, optical band-pass, and device architecture and control. Considerable effort may be required to fully optimize and stabilize detector control for precision photometry.
(2)   the incident signal from the astronomical source can be calibrated or transferred to the desired "standard" system.

For all techniques, whether profile-fitting or apertures, an understanding of the image profile and the centroid of the image is important since centring errors and inappropriate apertures or fitting parameters can lead to systematic effects. There are four other important issues; (1) passband mismatch of the filters, including narrow band filters; (2) red/infrared "leaks" in the filters which complicate flat-fielding; this can be very serious in the infrared where the backgrounds rise so steeply (3) the finite opening and closing times of electromechanical shutters; and (4) changes in the atmospheric attenuation or "airmass" for long on-chip integrations. Items (1) and (2) need to be eliminated by design. It is usually assumed that the basic CCD calibration procedures of bias subtraction, dark subtraction, flat-fielding and de-fringing have been applied correctly, that the effects are small or "infinitely well-known"! This is not always true. Extreme care and patience at the telescope is required to ensure that all necessary calibration data are obtained with very high accuracy so as not to limit the photometric determinations, and that nothing has happened "on the night" which might render reductions difficult, e.g. an accidental ground-loop, detector temperature instability, electronic failures, or inadvertent saturation which may result in latent images in some detectors.

Shutter-timing errors arise from the finite opening and closing time of electromechanical shutters. If T is the requested integration time, then $T + \delta$ will be the actual integration time and, since shutters are normally of the iris type, δ will be greatest near the centre of the CCD. It is straightforward to map δ. For example, take the sum of ten 1-s flat-field exposures and compare this with one 10-s exposure by forming their ratio $r(x,y)$ for each pixel which, provided the flat-field is constant, gives

$$r(x, y) = \frac{10(1 + \delta(x, y))}{10 + \delta(x, y)} \tag{10.19}$$

from which we get

$$\delta(x, y) = \frac{10(r(x, y) - 1)}{10 - r(x, y)} \tag{10.20}$$

The effect is typically about 7 to 11 ms with an 800×800 CCD (see Fig. 10.9), but could be quite serious for very large arrays. An alternative approach is to use "frame transfer". If the CCD is actually 1k $\times$ 2k with half of the long axis masked off permanently to act as a storage section, then instead of activating a shutter at the end of the integration, the charge pattern is rapidly transfer under the mask and then read out more slowly for low noise.

Once photometric values are obtained they must be compared or calibrated against well-measured "standard" sources so that different data sets can be intercompared, or the values related to theoretical predictions. With hindsight, the classical photometric systems such as the UBV system, may not be the most rational for use with CCDs, but such comparisons are inevitable. Photoelectric standard stars must be observed over a wide range of airmasses. The airmass (X) is given essentially by the sec ($= 1/\cos$) of the zenith angle of the object since this quantity is proportional to the thickness of atmosphere (treated as plane parallel slabs) through which the light passes. Straight overhead the airmass is 1.0 (because $\cos 0° = 1$) and at an altitude of 30 degrees the airmass is 2.0 (because $\cos 60° = 0.5$), thus the loss in magnitude from the top to the bottom of the atmosphere is doubled. In general then, if ζ is the zenith angle of a star and α is the difference between the true magnitude m and the magnitude that would be observed at the zenith, the true magnitude is related to the observed magnitude by

$$m = m(\zeta) - \alpha_\lambda \sec(\zeta) \tag{10.21}$$

where the suffix λ has been added to emphasize that α is a function of wavelength. Plotting the measured magnitudes obtained with the camera against the airmass ($X = \sec(\zeta)$) gives the slope of the line (α) and the intercept on the magnitude axis corresponding to $X = 0$ gives the true magnitude above the atmosphere (Fig. 10.10). Standard stars must also cover an appropriate range of spectral types (or temperatures) to obtain colour coefficients since the CCD magnitude derived will depend somewhat on how blue or how red the star is. Photoelectric standard stars have been tabulated extensively by Arlo Landolt for Selected Areas in the Equatorial Plane. The instrumental magnitudes are calculated from

$$I. M. = -2.5 \log \left(counts(e / s) \right) \tag{10.22}$$

and these can be compared on excellent nights with the standard stars to produce a fiducial set. The parameters to be determined are:

(1) the "zeropoint", i.e. the magnitude corresponding to 1 count/s for a star of zero colour-term above the Earth's atmosphere.

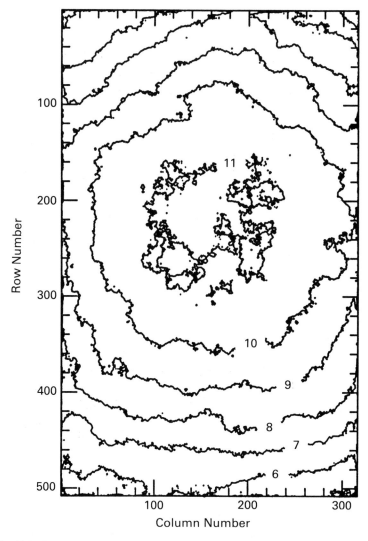

Fig. 10.9. Shutter exposure-time error map. An "iris" type shutter is open longest in the centre of the array.

(2)   the "colour equation" relating the CCD photometric system to the older photoelectric systems.

(3)   the "extinction" factor or light-loss through the Earth's atmosphere per unit airmass; at the zenith the airmass $(X) = 1$, whereas at an altitude of 30° above the horizon $X = 2$.

The resulting equation for the magnitude (m) of a star is given by:

$$m = -2.5 \log 10[\text{counts(e/s)}] - \alpha \times (\text{AIRMASS}) + \beta \times (\text{COLOUR}) + \text{ZP} \qquad (10.23)$$

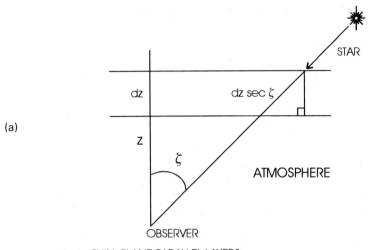

(a)

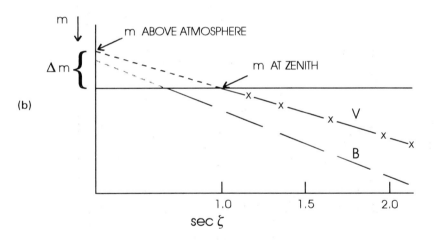

(b)

Fig. 10.10. A plot of observed magnitude against airmass to determine the extinction coefficients and zeropoints.

Peter Stetson, who developed the DAOPHOT package, has shown that during a long CCD exposure when the airmass has changed from X_0 at the beginning to X_1 at the end, that the best estimate is a "weighted" mean given by

$$\bar{X} = \frac{(X_0 + 4X_{1/2} + X_1)}{6} + O(e) \tag{10.24}$$

where $X_{1/2}$ is the airmass at the mid-point of the integration and $O(e)$ is a small error of about 1 part in 10 000. Colour correction, however, is still tricky since even two stars at the same airmass will undergo different extinctions if they are not identical in colour.

10.7 SPECTROSCOPY

Spectroscopic applications put the most stringent requirements on solid state detectors such as CCDs and infrared arrays and drive the quest for the lowest possible noise performance.

Typical CCD spectrographs were described in Chapter 3, the basic theory was given in Chapter 4 and infrared spectrographs were discussed in Chapter 9. Most spectrographs produce an image which consists of a long, narrow spectrum. The height of the spectrum is determined by the length of the slit in the focal plane, whereas the spectral coverage in the dispersion direction is determined by the grating properties and is usually matched to the number of pixels in the detector. In some cases, a large part of the detector format is underused. This is not the case for multi-object spectrographs, integral field mode spectroscopy or cross-dispersed echelle spectrographs.

A majority of large astronomical spectrographs employ diffraction gratings to disperse light, although prisms with transmission gratings applied to them are also frequently used. Spectroscopic calibrations proceed much as with imaging. A flat-field is required to remove optical interference effects caused by the near-monochromatic light and variations in the thickness of thinned backside-illuminated CCDs. Bias frames must proceed the derivation of the flat-field as before, but now, because of the much weaker signals and longer integrations, dark current may be more serious. Many dark frames may need to be averaged and subtracted from the object frame. The flat-fielded spectra must be "sky-subtracted". To do this it is normal to collapse or sum together all rows of the spectrum containing source flux and all rows containing sky spectra. After allowing for the fact that there will probably be more rows of sky spectrum than source spectrum, the pair of flat-fielded, summed spectra are subtracted. Next, the relationship between pixel number and wavelength is determined using arc lamp (emission-line) spectra containing numerous lines with accurately known wavelengths. Sometimes bright planetary nebulae like NGC7009 are used instead (Fig. 10.11)! Atmospheric extinction corrections are applied to the intensity and correction to absolute flux levels is accomplished by forming the ratio of the observed spectrum to that of a flux standard. Stars such as Alpha Lyrae, or fainter stars of the same temperature which are not too distant and therefore are not reddened by interstellar dust, are commonly used. Fainter stars, such as white dwarf stars which have almost featureless spectra, are also utilized as flux standards. A summary of the key steps are as follows:

(1)   Identify the direction of dispersion (is increasing wavelength the same as increasing pixel numbers?)
(2)   Interpolate over dead pixels or columns to prevent these extreme deviant values from ruining the subsequent steps, but keep a "map" of these locations so as not to forget that there was no real data in those pixels.
(3)   Sum up and normalize to unity the flat fields. Flat fields are usually taken with the spectrograph slit wide open and unrestricted by a decker. In this way, the orders overlap considerably giving a uniform illumination on the CCD when viewing a quartz lamp illuminating a white screen on the inside of the dome.
(4)   Divide the stellar spectra by the open-decker flat field to remove pixel-to-pixel sensitivity variations.

(5)   Some software packages (e.g. IRAF) require that you define the positions of the orders across the CCD. This may require observations of a brighter star if the program star is too faint. The program (e.g. "aptrace") will then know where to find spectra.

(6)   Extract the rectangular subsets of CCD pixels corresponding to the stellar spectrum. Do the same for the arc lamp and the narrow-decker (i.e. normal slit width) quartz-lamp exposures. In IRAF this is done with a program called "apsum".

(7)   Divide the flat-fielded stellar spectra by the "white-light" spectrum obtained with the narrow-decker (normal slit) and quartz lamp. This white light spectrum must itself be flat-fielded with the open-decker flat field before this division. The purpose of this step is to remove interference fringe effects rather than pixel-to-pixel variations in sensitivity.

(8)   Identify emission lines in the arc lamp used as a wavelength calibration device. The most commonly used lamp is a thorium-argon lamp spectrum. This is a painstaking step, but an important one if accurate transformations from pixels to wavelengths are required.

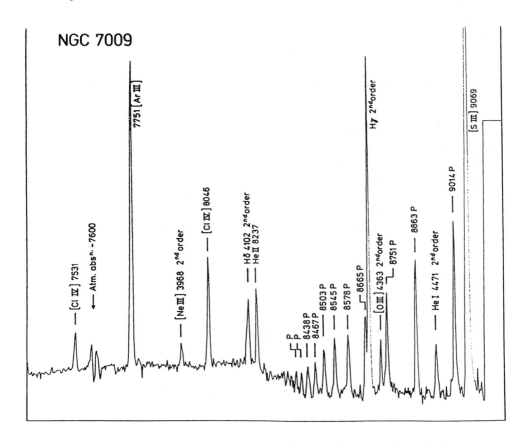

Fig. 10.11. The emission line spectrum of NGC7009 obtained with an EEV CCD attached to the AAT spectrograph.

One feature of CCDs of special use to spectroscopy is on-chip binning. In this mode the extension of the spectrum along the length of the entrance slit, and therefore at right angles to the dispersed spectrum itself, can be summed into one pixel while still on the CCD by a suitable clocking sequence (see section 6.3.2). The readout noise associated with this new binned pixel will be approximately the same as before even though the charges from several pixels have been summed together. Thus, apart from compressing the size of the image frame to be handled by the computer, a gain in signal-to-noise ratio is obtained. Care must be taken not to saturate the CCD by summing too many charges.

10.8 POLARIMETRY

Array devices used in polarimeters and spectropolarimeters are reduced in a similar way to achieve a basic set of clean frames before any other operations (e.g. determining Stokes parameters) are carried out. In the case of polarimeters, a set of flat-fields corresponding to different positions of the polarimeter waveplate are needed; since this element rotates then dust spots and other artifacts cannot be flat-fielded with just one waveplate orientation. After flat-fielding, the images must be very carefully registered to a tiny fraction of a pixel before forming the difference between pairs of frames corresponding to orthogonal polarization states. For example, if a halfwave plate is used to determine the linear polarization components $Q/I = p \cos 2\Theta$ and $U/I = p \sin 2\Theta$, where I is the total intensity, p is the percentage or fractional linear polarization and Θ is the direction of vibration of the electric vector of the linearly polarized component, then Q/I and U/I can be measured by the difference in counts at four waveplate rotations from an arbitrary starting point as described in Chapter 4.

Stars or objects of known polarization (e.g. the Crab Nebula) should be observed to verify the efficiency of the measurement and determine any scale factor. In addition, completely unpolarized sources should be observed to discover the "instrumental polarization" inherent in the system. The instrumental values of Q and U should be subtracted before p and Θ are calculated.

10.9 SIGNAL-TO-NOISE CALCULATIONS

A detailed calculation of the **signal-to-noise ratio** (S/N) of a CCD-type detector is quite complicated and depends on how closely one wishes to "model" the CCD behaviour. Here we will assume that the CCD electronics system is perfect and so there are no other sources of unwanted noise save the irreducible minimum readout noise. The simplest approach to analysing a CCD image is to construct a "final frame" by performing the following steps to subtract dark current and normalize with a flat-field:

$$FINAL\ FRAME = \frac{OBJECT\ FRAME - DARK\ FRAME}{FLAT\ FRAME - DARK\ FRAME}$$

The dark-subtracted flat-field frame in the denominator is usually "normalized" beforehand by dividing each pixel value with a constant equal to the mean or median of the entire frame; this step is included in the noise formulation below. For infrared arrays, the sky is often used as the flat-field source and so another version of this equation would substitute SKY FRAME for FLAT FRAME.

We assume that the only sources of "noise" are as follows:

(1)   readout noise, R electrons
(2)   photon noise on the signal (S) from the object
(3)   photon noise on the signal (B) from the sky background
(4)   noise on the dark current signal (D)

where the signals S, B and D in electrons per second are derived from observed counts in Data Numbers (DN) by multiplying by the transfer factor g electrons/DN, and we assume that the noise sources associated with the signals S, B and D all behave according to Poisson statistics so that the noise is equal to the square root of the number of electrons recorded. To find the original photon arrival rate for S or B, divide by the quantum efficiency (η), and to find the number of counts recorded by the A/D system (DN) divide by g electrons/DN.

For random independent noise sources, the noise terms can be added in quadrature (i.e. by square and add), and we must allow for the noise sources in each of the two calibration frames, the SKY/FLAT and DARK. We can further assume that the number of OBJECT, SKY/FLAT and DARK frames combined or "co-added" to form the FINAL FRAME are not necessarily equal, but for simplicity we assume that the normal practice pertains of keeping the same exposure time (t) for each image. This approach and terminology will work for both infrared arrays and CCDs. Suppose we have N_S OBJECT FRAMES, N_B SKY/FLAT FRAMES, and N_D DARK FRAMES.

Let,

$T = tN_S$ be the *total* integration time accumulated on the OBJECT FRAME

$f = S/B$ be the ratio of the source signal to that of the "background" signal per pixel

$\varepsilon_B = N_S/N_B$ is the ratio of the number of object and background frames

$\varepsilon_D = N_S/N_D$ is the ratio of the number of object and dark current frames

then, for a source covering n pixels on the CCD, we can estimate a total signal-to-noise ratio (S/N) of

$$\frac{S}{N} = S\sqrt{T}\left[u_r^2 + S + \sum_{i=1}^{n}\left\{\begin{array}{c}\left(B+D+\dfrac{R^2}{t}\right)+\varepsilon_D\left(D+\dfrac{R^2}{t}\right)\\ +(1+f)^2\,\varepsilon_B\left(B+D+\dfrac{R^2}{t}\right)+(1+f)^2\,\varepsilon_D\left(D+\dfrac{R^2}{t}\right)\end{array}\right\}\right]^{-\frac{1}{2}} \qquad (10.25)$$

where S is the *total* signal summed over n pixels, i.e. $S = \Sigma(S_i)$ and the term u_r represents the average, over n pixels, of any residual error due to failure in the flat field. The full derivation of this equation is given by McCaughrean (1988). Note that B, D and f are not the same from pixel to pixel. The terms in the denominator of this equation can be understood as follows. The first term is the residual non-uniformity and the second term is the shot noise in the source signal itself. The first term following the summation sign, ($B + D + R^2/t$), is the background and dark current shot noise and readout noise in the raw source frame and is always present. The next term contains only dark current and readout noise and is the error due to subtracting a dark frame from the object frame; the more dark frames that are co-added or averaged, the smaller is ε_D and the less significant this term. The third term comes from the application of the SKY/FLAT correction and the fourth term is the result of dark subtraction for the SKY/FLAT term. Again, the more SKY/FLATS that are used the smaller ε_B. Note that the additional scaling factor of $(1+f)^2$ arises when the object frame is divided by the flat field and the ratio renormalized by multiplying by the mean of the flat field. If the flat field signal is not large enough (ε_B is not $\ll 1$) then this term dominates for very bright sources and for very faint sources it contributes as much noise as the raw source frame.

We need to distinguish the cases of point sources and extended objects. A stellar object has a seeing disk of diameter θ_{FWHM} arcseconds, corresponding to the full width at half maximum intensity (FWHM), whereas each CCD pixel is a square θ_{pix} arcseconds on a side. The number of pixels covered by the star's image is approximately

$$n = \frac{\pi}{4}\left(\frac{\theta_{FWHM}}{\theta_{pix}}\right)^2 \tag{10.26}$$

and the summations (Σ) shown above must be taken over this number of pixels to estimate the S/N ratio. If the angular size of the object is much larger than the seeing diameter, then it is more convenient to deal with "surface brightness" in magnitudes per square arcsecond (as is done for the sky) and so $n = 1$ and each pixel is treated separately.

In the ideal case, with very accurate calibration data so that ε_B and ε_D become very small indeed, and if u_r—the residual non-flatness—is negligible then the equation is much simpler.

$$\frac{S}{N} = \frac{S\sqrt{T}}{\left[S + \sum_{i=1}^{n}\left\{\left(B + D + \frac{R^2}{t}\right)\right\}\right]^{\frac{1}{2}}} \tag{10.27}$$

This is the basic form used to estimate observing times and limiting magnitudes, and is sometimes called the "CCD equation". There are two further simplifying cases:

1. Background-limited or "sky-limited" case: B very much larger than $(D + R^2/t)$, and S is much weaker than B.

$$\frac{S}{N} = S\sqrt{T}\left[\sum_{i=1}^{n}(B_i)\right]^{-\frac{1}{2}} \tag{10.28}$$

Assuming the background is uniform ($\Sigma B_i = nB$), and using $S = f(nB)$ to relate the total signal (e/s) over n pixels to the corresponding background, and with $T = N_s t$ where the "on-chip" integration time is long enough to ensure that the term R^2/t is small relative to B, then

$$\frac{S}{N} = S\sqrt{\frac{N_s t}{nB}} = f\sqrt{nBT} \tag{10.29}$$

In this case, the accuracy of the measurement scales only as the square root of the integration time, an improvement in S/N by a factor of 5 implies 25 times longer in integration time. For a given background level B, the signal-to-noise ratio scales with the faint signal, but for a given signal level S, the S/N decreases as the square root of the increasing background. This is why it is much harder to get deep images at infrared wavelengths where the backgrounds are higher. Also, the signal-to-noise ratio scales only as the *diameter* (D_{tel}) of the telescope used because a larger telescope gives more background B in proportion to the area of the mirror $(\pi D_{TEL}^2/4)$ assuming seeing-limited conditions (as an exercise, show that S/N scales as D_{TEL}^2 in the diffraction-limited case). Similarly, the signal-to-noise ratio scales only as the square root of the detector quantum efficiency.

Example: Suppose that the source is only 0.1% of the brightness of the sky $(f= 0.001)$, to just barely detect it requires that $S/N = 1$ and therefore $nBT = 1\,000\,000$ photoelectrons. If the sky background (nB) gives about 1000 e/s, then this observation will take 1000 s. Note that for R^2/t to be negligible (a few per cent of B, say ~3 e/s), then R^2 must be less than $3t = 3000$, or $R < 55$ electrons if the entire observation is made in one exposure $(t = T)$. If, however, the exposure has to be made in five separate $t = 200$ s exposure blocks (because of well-depth limitations or to remove artifacts by median filtering or because of guiding problems) then R^2 must be less than $3t = 600$, which implies $R < 25$ electrons.

2. Detector noise limited case: R^2/t very much larger than $(B + D)$, and a weak signal S.

$$\frac{S}{N} = \frac{S\sqrt{T}}{\left[\sum_{i=1}^{n}\left(\frac{R^2}{t}\right)\right]^{\frac{1}{2}}} = \frac{St}{R}\sqrt{\frac{N_s}{n}}$$

(10.30)

which shows that the signal-to-noise ratio for a given source S, and detector noise R, scales linearly with the *on-chip* integration time and only as the square root of the number of repetitions. The signal-to-noise ratio also scales linearly with quantum efficiency and with telescope collecting area. One should try to select t to achieve background-limited operation, but if this is not possible then it is best to at least maximize t, reduce R and minimize over-sampling by reducing n.

Example: In a high resolution spectrograph the background plus dark current is 0.1 e/s/pixel and so for $t = 1000$ s we are formally readnoise-limited if $R^2 > 0.1 \times 1000 = 100$, i.e. if R is larger than 10 electrons.

Although simplified, the above treatment shows that the best detections and therefore the highest signal-to-noise ratios are obtained when

(a)   sufficiently accurate calibration frames are obtained
(b)   the readout noise and dark current are as small as possible
(c)   the on-chip integration time is as long as possible
(d)   quantum efficiency and telescope area as large as possible

The power (in watts) collected by a telescope of area A_{tel} (cm^2) in a wavelength interval of $\Delta\lambda$ (μm) from a source of apparent magnitude m (below the Earth's atmosphere), transmitted by an optical system of efficiency τ (<1) onto a CCD detector of quantum efficiency η (<1) can be simply estimated by

$$P(\lambda) = \tau(\lambda)\eta(\lambda)A_{tel}\Delta\lambda F_\lambda(0) \times 10^{-0.4m} \quad W \tag{10.31}$$

where $F_\lambda(0)$ is the flux in W cm^{-2} μm^{-1} (see Table 10.3) from a zeroth magnitude standard star above the atmosphere. The transmission factor τ is the product of all the transmission or reflectance factors in the system. For instance,

$$\tau = \tau_{tel}\tau_{optics}\tau_{filters}$$

Example: two telescope mirrors with a 95% reflectance each, six lenses with 96% transmission each and a filter with 80% transmission. The total transmission is $(0.95)^2(0.96)^6(0.8) = 0.57$.

Remember to subtract out the area of the central hole in a Cassegrain telescope when evaluating A_{tel} and allow for any stops in the system which reduces the effective aperture of the primary mirror; one example would be an undersized secondary on an infrared telescope. Note that this equation assumes that the filter which defines $\Delta\lambda$ has a "box-car" profile of exactly this width and that the spectrum of the source scales with that of the standard star, neither of which are true in practice. A more rigorous treatment requires that each profile be defined as a function of wavelength and the total power is then found by integration.

Since the energy of a single photon of wavelength λ is just hc/λ joules (or watt seconds) per photon, then the photoelectron detection rate is

$$S(\lambda) = (hc)^{-1}\tau(\lambda)\eta(\lambda)A_{tel}\lambda\Delta\lambda F_\lambda(0) \times 10^{-0.4m} \quad e/s \tag{10.32}$$

where h is Planck's constant, c is the speed of light and therefore $(hc)^{-1} = 5.03 \times 10^{18}$ J^{-1} μm^{-1}. Dividing by g electrons per DN gives the observed signal N in counts (or data numbers or ADUs). If we set $N = 1$ DN then we can derive the corresponding magnitude m_{zp} which is the "zeropoint" of the instrumental scale,

$$m_{ZP} = 2.5\log\left\{\frac{\tau\eta\lambda\Delta\lambda A_{tel}F_\lambda(0)}{hcg}\right\} \tag{10.33}$$

The zeropoint m_{ZP} can also be derived from observations of a standard star of known magnitude. Usually, one derives observed zeropoints $m_{true} - IM$, where $IM = -2.5\log$ {S (DN/s)} is the observed or "instrumental magnitude" and plots the numbers against airmass ($X = \sec\zeta$) and extrapolates the line to $X = 0$ (above the atmosphere). Note that the zeropoint is positive and numerically larger. Having obtained m_{ZP} by observations, we can then derive the product $\tau\eta$ which describes the system efficiency in the given passband.

$$2.5\log(\tau\eta) = m_{zp} - 2.5\log\left\{\frac{\lambda\Delta\lambda A_{tel}F_\lambda(0)}{hcg}\right\} \tag{10.34}$$

Example: The observed K′ band ($\lambda = 2.125$ μm, $\Delta\lambda = 0.35$ μm) zeropoint is observed to be 20.4 on a telescope with an area of 72 236 cm^2 with a camera using a gain of 25 electrons/DN. Taking $F_\lambda(0) = 4.34 \times 10^{-14}$ W cm^{-2} μm^{-1} gives

$$2.5\log(\tau\eta) = 20.4 - 2.5\log\left\{\frac{2.125 \times 0.35 \times 72{,}236 \times 4.34 \times 10^{-14}}{1.99 \times 10^{-19} \times 25}\right\} = -1.28$$

corresponding to $\tau\eta = 0.31$ (or 31%).

The signal from the sky or background (which includes thermal emission from the telescope at infrared wavelengths) depends on many things, including local conditions (temperature) and the amount of moonlight (phase of the moon and angular separation from the moon). A simple approach is to use the same form as before

$$B(\lambda) = (hc)^{-1}\tau(\lambda)\eta(\lambda)A_{tel}\lambda\Delta\lambda F_\lambda(0) \times 10^{-0.4m\,sky}\theta^2_{pix} \quad \text{e/s} \tag{10.35}$$

but with m_{sky} representing the sky brightness empirically as a magnitude per square arcsecond relative to a star of zero magnitude above the atmosphere, and introducing the pixel area on the sky θ_{pix}^2. In the thermal infrared F_λ can be replaced with the Planck function $(B_\lambda(T))$ and an emissivity factor (ε).

Since B is proportional to $A_{tel}\theta^2$ then

$$A_{tel}\theta^2_{pix} = \frac{\pi}{4}D^2_{tel}\left(206265\frac{d_{pix}}{D_{tel}(f/\#)_{cam}}\right)^2 \propto \frac{1}{(f/\#)^2_{cam}}$$

where d_{pix} is the pixel size of the detector and $(f/\#)_{cam}$ is the focal ratio of the camera system, which differs only by the magnification factor m from the telescope focal ratio. Many optical telescopes have focal ratios of about $f/8$ whereas infrared telescopes use $f/36$ or larger to reduce the background. For a given camera system, the background is independent of telescope diameter and depends only on focal ratio; two telescopes of the same focal ratio will provide the same background, although the plate scales will be different. At a good site with no moonlight, m_{sky} is about 22–23 in the blue, 21 in the visible and red, about 20 or brighter (and variable) near 1 μm and about 13 at 2.2 μm, assuming the standard passbands.

Example: Consider a CCD camera with $\tau = 0.5$ and $\eta = 0.5$ at a wavelength of 0.54 μm in the middle of the visible spectrum with a filter passband of $\Delta\lambda = 0.1$ μm on a 4-m class telescope, that is, a 4 m diameter primary mirror with a 1-m Cassegrain hole in its centre. With the appropriate value of $F_\lambda(0)$ from Table 10.3 we get $S = 3.136 \times 10^{10} \times 10^{-0.4m}$ e/s so that for $m = 17$ we obtain 5000 e/s whereas for the extremely faint sources seen on deep CCD images at $m = 27$ we detect only 0.5 electrons per second.

A useful expression for predicting the "limiting magnitude" of an array camera in the background- or sky-limited case is to solve for S in terms of S/N, convert to counts by dividing by g e/DN and substituting this equation into the definition of the instrumental magnitude to give

$$m = m_{ZP} - 2.5\log\left\{\frac{1}{g}\frac{S}{N}\sqrt{\frac{nB}{N_s t}}\right\} \tag{10.36}$$

The "limit" corresponds to $S/N = 1$.

Example: A camera with a gain of 10 e/DN and a zeropoint of $m_{ZP} = 18$ forms a star image across five pixels with an average background of 200 e/s/pixel what is the limiting magnitude for a 1 hour exposure?

Since we are assuming the background-limited case we are not concerned with the way the exposure has been broken up, all that matters is $T = N_s t = 3600$ s and $S/N = 1$.

$$m = 18 - 2.5 \log \{0.1 \times 1 \times \sqrt{(1000/3600}\} = 18 + 3.2 = 21.2$$

In reality, signal-to-noise calculations are often more complex than presented here because star images are not sharp-edged, filter transmissions are not simply represented by a wavelength interval, and the stellar energy spectrum is not the same as that of Vega.

Table 10.3. Absolute flux from a zero magnitude star like Vega

Symbol	$\lambda (\mu m)$	ν (Hz)	F_{λ} (W cm^{-2} μm^{-1})	F_{ν} (Jy)[a]
U	0.36	8.3×10^{14}	4.35×10^{-12}	1880
B	0.43	7.0×10^{14}	7.20×10^{-12}	4440
V	0.54	5.6×10^{14}	3.92×10^{-12}	3810
R	0.70	4.3×10^{14}	1.76×10^{-12}	2880
I	0.80	3.7×10^{14}	1.20×10^{-12}	2500
J	1.25	2.4×10^{14}	2.90×10^{-13}	1520
H	1.65	1.8×10^{14}	1.08×10^{-13}	980
K	2.2	1.36×10^{14}	3.8×10^{-14}	620
L	3.5	8.6×10^{13}	6.9×10^{-15}	280
M	4.8	6.3×10^{13}	2.0×10^{-15}	153
N	10.1	3.0×10^{13}	1.09×10^{-16}	37

[a] These units are called jansky. One Jy equals 10^{-26} W m^{-2} Hz^{-1} or $(3 \times 10^{-16}/\lambda^2)$ W cm^{-2} μm^{-1}, if λ is expressed in micrometres.

Example: 20 magnitudes is a factor of 10^8, therefore, $m = 20$ corresponds to 38 μJy at V, 6.2 μJy at K and 0.37 μJy at N.

SUMMARY

Calibration of array detectors such as CCDs and infrared arrays is not trivial and requires considerable care if these devices are to be used for precision brightness measurements rather than for morphological or more qualitative studies. When all the steps are followed, CCDs and infrared arrays are extremely powerful for quantitative astrophysics in photometers and spectrographs and other instruments.

EXERCISES

1. Explain the difference between QE and DQE. Produce additional entries in Table 10.1 for CCDs with 5 and 15 electrons of readout noise.
2. A sequence of flat-field observations is obtained at different exposure levels. The following signal and variance values were found: 200 DN, 75; 400 DN, 125; 800 DN, 200; 1800 DN, 475. Plot the data and derive the gain factor g (electrons/DN) and the readout noise R.
3. Derive the amplifier gain A_g for a detector with a capacitance of 0.1 pF and a source follower gain of 0.75 if a 16 bit A/D with a 10-volt swing is used and the value of $g = 25$ electrons/DN.
4. Explain the concept of drift scanning. Why does it produce a good flat field along the scan direction?
5. What is meant by "median filtering" and why is it a good technique for imaging very faint sources?
6. What is the difference between flat-fielding and fringe-correction? Why doesn't flat-fielding correct for fringing?
7. Explain the two approaches of extracting magnitudes from CCD and infrared images.
8. What is meant by the "zeropoint" of a magnitude scale?
9. Describe the major steps needed to calibrate a high-resolution spectrograph with a CCD or IR array detector.
10. What additional procedures are required to calibrate a CCD-based polarimeter camera?
11. Calculate the photon arrival rate for a 24th magnitude star in the V band on a 4-m telescope with a camera having an efficiency of 30%. Assuming that the pixels are 0.3 arcseconds, and the readout noise is 10 electrons and dark current is negligible, is the measurement background-limited? What integration time is required to achieve a signal-to-noise ratio of 10?

REFERENCES

Christian, C. (1987) *CFHT CCD user's manual*, Canada-France-Hawaii Telescope Corporation.

Green, R.M. (1985) *Spherical Astronomy*, Cambridge University Press, Cambridge, UK.

Hall, P. and Mackay, C.D. (1984) "Faint galaxy number-magnitude counts at high galactic latitude", *Mon. Not. R. astr. Soc.*, **210**, 979–992.

Henden, A. A. and Kaitchuk, R.H. (1982) *Astronomical Photometry*, Van Nostrand Reinhold, New York.

Howell, S.B. (Ed.) (1992) *Astronomical CCD Observing and Reduction Techniques*, Conf. Series, Vol. 23, Astronomical Society of the Pacific, San Francisco, CA.

Janesick, J.R., Hynecek, J. and Blouke, M.M. (1981) A virtual phase imager for Galileo, in *Solid State Imagers for Astronomy*, SPIE Vol. 290, 165–173.

Jaschek, C. and Murtagh, F. (Eds) (1989) *Errors, Bias and Uncertainties*, Cambridge University Press, Cambridge, UK.

Léna, P. (1988) *Observational Astrophysics*, Springer-Verlag, Berlin.

McCaughrean, M.J. (1988) Thesis, University of Edinburgh.

Stetson, P.B. (1987) "DAOPHOT: A computer program for crowded-field stellar photometry", *Publ. Astron. Soc. Pacific*, **99**, 191–222.

Plate 1. M20 (NGC6514), the Trifid Nebula, an ionized gas cloud in Sagittarius imaged with a 2048 × 2048 pixel CCD on the 0.9-m telescope on Kitt Peak. Three different colour filters (red, green and blue) were used to produce this image. Notice the lines coming from the brightest stars due to blooming of saturated pixels. Courtesy NOAO/Todd Boroson.

Plate 2. NGC253, a very dusty edge-on late-type spiral galaxy near the South Galactic Pole in Sculptor imaged with the same CCD camera and filters as Plate 1. Courtesy NOAO/Todd Boroson.

Plate 3. A 5 × 5 arcminute region of the Orion Nebula near the Trapezium cluster at near-infrared wavelengths. One of the images obtained with the first 62 × 58 InSb array developed by Hughes/SBRC for astronomy. This mosaic was made from 145 smaller images using the IRCAM array camera developed by the author for the 3.8-m UK Infrared Telescope in 1986. Three near-infrared filters (J, H and K) were used and colour-coded blue, green and red to produce a composite image with essentially 490 × 483 pixels. The reddish area to the upper right is the BN-KL region which contains optically invisible young stars. About four-fifths of the stars shown are invisible to CCD sensors. Courtesy Mark McCaughrean.

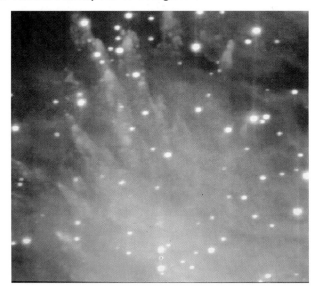

Plate 4. The optically invisible northern outflow from the Orion Molecular Cloud (OMC-1) near the BN object. An infrared image in three colours J (1.25 µm) = blue, iron emission [FeII] (1.65 µm) = green and molecular hydrogen emission H_2 (2.12 µm) = red. The field is 100 × 100 arcseconds. It reveals "bullets" of iron emission at the heads of hollow wakes of molecular hydrogen emission. Courtesy Michael Burton.

Plate 5. NGC2024, a star-forming region near the star ζ Orionis imaged in the near infrared using the UCLA twin-channel IR camera, with 256 × 256 pixels arrays, on the Lick Observatory 3-m telescope. This JHK-to-blue/green/red colour translation reveals a dense cluster of reddish stars embedded in the nebula's dark central dust lane. The field of view is about 8′ across. UCLA photo.

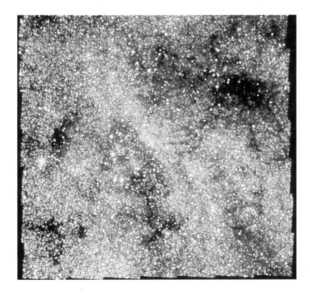

Plate 6. A mosaic of infrared images of the Galactic Centre spanning about 0.5° on the sky. Three filters (JHK) are colour-coded blue, green and red in this image obtained with 256 × 256 pixel PtSi arrays in a camera attached to the Kitt Peak 1.25-m telescope. Courtesy NOAO/Ian Gatley.

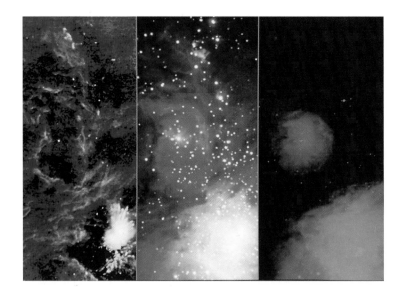

Plate 7. Another mosaic of the Orion Nebula (M42) including M43 to the north. In the centre is a JHK colour translation as shown before. On the right is an image in a narrow filter which isolates the Brackett γ line of ionized hydrogen, whereas the frame on the left is the same region seen through a filter which isolates emission from molecular hydrogen (H₂). The filamentary structure is remarkable and invisible at optical wavelengths. Courtesy NOAO/Ian Gatley.

Plate 8. CCD images of the planet Mars obtained near opposition (February 1995) by the Hubble Space Telescope's WFPC2 planetary camera. The north polar cap is clearly visible together with bright frosty clouds hanging around Elysium Mons (right) and the Hellas impact basin (extreme bottom). Courtesy NASA/Bob Williams.

Plate 9. The core of the Eagle Nebula (M16, NGC6611 in Serpens) imaged with the CCDs in the WFPC2 camera on the Hubble Space Telescope revealing intricate details, with ten times normal ground-based resolution, in the back-illuminated pillars of dusty gas of this star-forming region. Courtesy NASA/Bob Williams.

Plate 10. The Hubble Deep Field. Assembled from 342 frames obtained with the WFPC2 over ten consecutive days between 18 and 28 December, 1995. Colours represent blue, red and infrared passbands. The field of view is only about 1 arcminute. The faintest objects are near 30th magnitude. Courtesy NASA/Bob Williams.

Plate 11. A colour image based on VRI CCD exposures of Dwingeloo 1, a recently discovered nearby galaxy hidden by the Milky Way. The image was obtained with the Isaac Newton Telescope. The field is 6.4 × 8.2 arcminutes. RGO photo.

Plate 12. A "false colour" representation of Comet Halley displayed to exaggerate small changes in intensity; ROE photo.

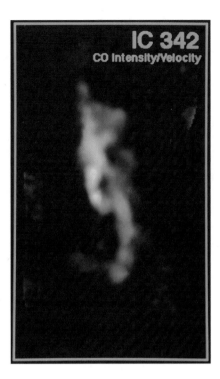

Plate 13. Imaging at radio wavelengths. A map of IC342 with colour coding to indicate the velocity of the gas as measured by the CO lines at radio frequencies. Courtesy Robert Hurt.

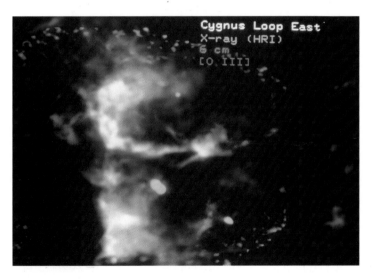

Plate 14. Images at very different wavelengths can be combined by colour-coding. Part of the Cygnus Loop supernova remnant is shown here. Green represents X-ray images from the Einstein satellite, blue is [OIII] optical emission from CCD images on the 1.52-m telescope on Mt Palomar and red represents radio emission at a wavelength of 6 cm observed using the VLA in New Mexico. Courtesy Jeff Hester/IPAC.

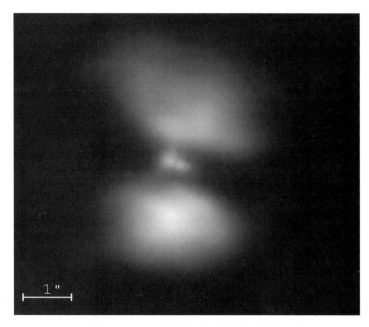

Plate 15. A portent of the future? A remarkable image of the Frosty Leo nebula (IRAS 09371+1212) in I, J and H bands obtained with the Canada-France-Hawaii Telescope IR Camera and the University of Hawaii adaptive optics system at the coudé focus of the CFHT. This adaptive optics image confirms the binary nature of this source (0.19″ separation) and clearly shows a bipolar jet emanating from just one of the stars. Courtesy François Roddier.

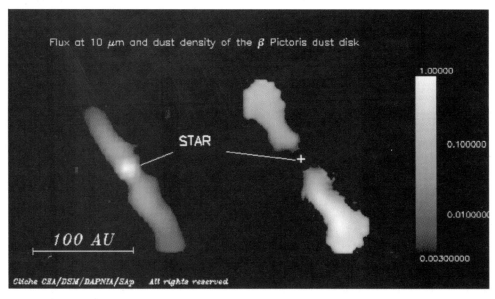

Plate 16. A mid-infrared 10-μm image of the dust disk around the star β Pictoris obtained with the ESO TIMMI camera using a 64 × 64 Si:Ga detector with a pixel scale of 0.3″. Emission (left) is seen extending over 100 astronomical units (AU). The dust density distribution is deduced and shown on the right. Courtesy Pierre-Olivier Lagage/SAp.

11

Computers and image processing

Computers are used in the control of telescopes and instruments, for acquisition of digital data from electronic detectors, for image analysis, numerical simulations and more. Over the years, astronomers have also played a significant role in developing algorithms for image enhancement and image restoration. Methods of analysis depend, of course, on the nature of the measurement and the properties of the instrument and detector, and requirements for computing power range from modest to mammoth, from PCs to Cray supercomputers. This chapter provides a very basic introduction for the novice to the kind of computer technologies which might be encountered by the users and builders of modern astronomical instruments, and it also gives a general introduction to the huge field of image processing and image restoration.

11.1 COMPUTERS

The growth rate of computer technology is enormous. Keeping up to date is best done by watching for reviews and surveys in both trade and popular magazines. The emphasis here is on those aspects of computer technology which are likely to be of most concern to astronomers, namely, acquisition of digital data from an instrument, image analysis facilities and data storage.

11.1.1 Data acquisition
The "data rate" is the number of digitized pixel values transferred per second to the host computer, or possibly an intermediate buffer or control computer, from the astronomical instrument. For an array detector, this rate is largely determined by the electronic configuration, although the actual application and device physics may also be factors. Recall that the integrating correlated double sampling (CDS) method used with CCDs requires an interval of time to digitize a pixel value, typically about 20 to 100 μs—corresponding to pixel readout rates of 50 kHz and 10 kHz respectively. For a 20 μs pixel time and digitization to 16 bits (65 536 voltage levels), the maximum output data rate is 800 000 bits per second. Note that this level of digitization is mandatory if one is to capitalize on the large dynamic range of CCDs. So this data rate is 100 kilobytes per second. Serial data rates are usually specified in bits per second (bps) and parallel rates in Mbytes per second.

Driven by diverse applications as far apart as digital audio reproduction and military radar, high-resolution analogue-to-digital converters (ADCs) are themselves getting faster and faster each year. There are literally dozens of different suppliers of ADCs, offering resolution from 12 to 24 bits and using a variety of techniques, from the standard successive-approximation method to new designs based on so-called "delta-sigma" and "multipass" architectures. Many of these devices are customized to their field of application and it is therefore important to look very carefully at their performance details. A critical parameter is the signal-to-noise ratio (S/N) in decibels (db)—more is better! Two commonly used ADCs in astronomy convert 16 bits in 2 μs and another performs 14 bit conversion in 0.5 μs; a 16-bit conversion in 0.5 μs is now available.

The transfer of data from the ADC to the computer can be accomplished in two ways, either by a high-speed **serial link** which is a single line along which signals are sent in a sequential pattern, or by a multi-way cable called a **parallel link** which sends all 16 bits at once. One of the most commonly encountered (and abused) serial standards is RS-232-C in which a "logic 1" level is represented by a voltage in the range −5 to −15 V, and "logic 0" by a voltage between +5 and +15 V; up to 25 lines are specified, although seldom used. In the RS-423-C standard, only two wires are used and the logic voltages are 0 and 5 V. The RS-422-C is similar but the lines are "balanced" to achieve much higher data rates. For parallel data transfers, many computers handle this flow by using **Direct Memory Access** (DMA). Parallel data highways within a computer system are called **buses** and there are different buses for data transfer to memory, for control and for addressing. The number of lines in the address bus determines the number of memory locations that can be specified. A 32-bit bus gives over 4 billion locations. Two commonly encountered systems are SCSI (pronounced "skuh-zee") and PCI. In other data acquisition systems the digital data are stored first in a large buffer memory called a "frame store" or "external memory", and in a few systems the data go directly onto magnetic tape. In general, digital data from an astronomical detector are saved initially on a magnetic disk drive (the "hard drive") and later transferred to magnetic tape or an optical disk for longer-term storage; this process is called "archiving".

For compatibility between two devices, several parameters must match besides the rate, in particular, the voltage levels, the timing, the format, the "code" by which numbers and characters are digitized, and the rules or "protocols" by which data transfers are acknowledged—called "hand-shaking"—and by which computer operations are tracked—called "housekeeping".

There is no truly universal standard for digital electronic data acquisition systems in astronomy, but there are several systems which are common. If you purchase a "packaged" CCD camera system you may not be particularly concerned about the approach used. On the other hand, if you are trying to ensure that the new system will work with some existing equipment, or meet the specification for a challenging new mode of observing, or just be supportable in the years ahead, then you do need to know more about the system involved.

One of the older methods still in astronomical service, is a modular system known as CAMAC. Functional circuit-board units such as counters, voltage-to-frequency converters, A/D boards etc., are constructed as "plug-in" modules which go into a central power supply unit called a "crate" which contains a set of connections or "data bus" which link each unit to a central crate "controller" module. With each of the units under digital control through an external host computer connected to the crate controller, the CAMAC system is capable of performing a wide range of control and data manipulation functions once it has been "programmed". A very basic instruction set known as "machine-language" for handling individual bits must be used. In practice, the program is written mainly in a well-established language, such as FORTRAN or C, with appropriate "calls" to machine-language sub-routines.

CAMAC technology came to astronomy from the field of high-energy physics many years ago. Interestingly, this link has continued. It was high-energy physics that brought us the World Wide Web, and a new instrument control philosophy/environment known as EPICS (Experimental Physics and Industrial Control System), developed at the Los Alamos labs in the USA, is now being used at several large astronomical centres. EPICS is based on a transport layer called Channel Access, a message system protocol over Ethernet hardware, which connects Unix host computers to smaller systems running VxWorks (Wind River Systems), typically in a VME crate. VME is a bus system used in mid-size computers such as the early Sun Microsystems machines. Many different CPUs are available to run in a VME bus system including Motorola MC68000 family, Sun Sparc and DEC Alpha. These run varieties of Unix, or an operating system called VxWorks which is particularly convenient for real-time systems. VME is capable of very high speeds, has extremely compact hardware and receives widespread commercial support. The VME bus is designed for input–output paths which are 32 bits wide.

Another type of data-acquisition system uses the IEEE-488 standard which is based on the Hewlett-Packard general-purpose interface bus (GPIB). Input–output data paths are 8 bits wide (parallel) and so transfer of a 16-bit word consists of two serial bytes. Each unit on the bus is independent and connections to the bus are by means of a standard 24-pin connector which allows other connectors to be added on to it to create a "stack" of several devices on the same bus. Instruments connected to the bus are designated talkers, listeners, and controllers. Again, a computer program is required to control the modules on the bus, but the support for this system is very widespread and many commercially available pieces of lab apparatus (such as oscilloscopes, temperature monitors, voltage-frequency converters) come with an IEEE-488 interface.

Eight-bit microprocessors from Motorola were extremely common in astronomical instruments in the nineteen-eighties, but these have largely been superseded by powerful devices called **Digital Signal Processors** (DSPs). DSP chips are made by several manufacturers (including Motorola and Texas Instruments) and are widely used in equipment of all kinds, from cellular phones to automobiles. Several observatory-built CCD and infrared camera systems employ DSP chips as the "intelligence" in a controller/ sequencer (e.g. the Leach CCD controller used at the Keck Observatory and elsewhere). Another advanced processor which has already found its way into many astronomy applications is called the **transputer**. The transputer was invented in the UK by Inmos (now part of SGS-Thomson) and employs a unique inter-processor linking system comprising four identical input/output lines; the structure is reminiscent of the covalent chemical bonds of carbon atoms, and in many respects, the transputer can be "combined" in linked chains and loops just like carbon atoms. Each transputer chip has its own bus, memory, external memory interface and serial links on a single piece of silicon. Transputers are programmed in a high-level language called **Occam** and are now widely used in CCD and infrared camera controllers.

A rather common method of data transfer is the use of a high-speed, single-cable serial link called **Ethernet** which is also used as a general means of inter-linking several computers in a Local Area Network or LAN. In the Ethernet approach (due originally to Xerox, DEC and Intel) messages and data are broadcast along a coaxial cable from the sender station to the receiver station. To send a message, the sending station senses the cable to see if it is free. If it is free the station transmits but it also continues to sense in case some other station on the net transmitted at the same time. If this were to happen a "collision" would have occurred and the transmission would be unsuccurred, so both stations stop sending, wait a random time interval and then try again. This is called "carrier sense multiple access with collision detection" or CSMA-CD. The hardware components of the Ethernet consist of the coaxial cable, a tap which connects the transceiver to the cable, and a twisted-pair interconnecting the transceiver and the controller. An individual Ethernet consists of a 500-m cable segment with up to 100 stations; the stations can be a wide range of devices from complete image processing workstations to terminals. Thin-wire ethernet is an extension of this concept that allows more computers to be added at shorter intervals along the line.

Of extreme importance has been the introduction of optical fibres. Optical transmission, which is insensitive to electrical interference, has rapidly emerged as the best high-speed, high-capacity, low-error communications link for data transmission. In an optical fibre, the electrical pulse which would have been sent down a copper wire is converted to a pulse of light by a transmitter at one end and then back into an electrical pulse by a receiver at the other end. Typically, the optical fibre itself has a very small diameter of only 50–75 μm.

The proliferation of personal computers (PCs) has also led to the development of numerous PC-bus cards and modules which can be "plugged-in" to free slots in the PC to provide all sorts of functions including "frame grabbing", "analogue-to-digital conversion", "motion control" for external hardware, Ethernet and so on. The mass market for PC-class computers means that they are cheap and well-supported with lots of software products, including image display and image processing packages. Many commercially available astronomical camera systems are based on PCs or Macs with processors which outperform earlier minicomputers by an order of magnitude!

11.1.2 Data file formats

There are numerous formats for image files such as bitmap (BMP), GIF, TIFF, JPEG and many more, but the recognized standard format among professional astronomers is the **Flexible Image Transport System (FITS)** developed by Don Wells, Eric Greisen and Ron Harten (1981). Sometimes instruments save data in a "native" form consistent with the software environment at the observatory. For example, the IRCAM cameras which we developed for UKIRT while I was at the ROE, worked within the ADAM/STARLINK environment and stored data in the HDS format. Nevertheless, the files were immediately converted to FITS for transport away from the observatory. On the other hand, the widely used IRAF data reduction facility expects data to be in its own internal format and therefore one needs to use a FITS conversion program. FITS is used by radio and optical astronomers and throughout astronomy. FITS files consist of three parts: a "header", the image "data" in binary form, and a "tailer".

A FITS header comprises an integer multiple of 36 lines of 80 bytes (the 80 bytes is a relic of 80-character punched cards) giving 2880 bytes, or 5760 bytes for 72 lines and so on. If less than 36 lines are used then the remainder must be filled out with the ASCII (American Standard Code for Information Interchange) character for a blank space (hexadecimal value of 20). Each line, also called a "card image", begins with a "keyword" in bytes 1 through 8, which identifies the information type for that line. The construction of the keywords is very specific. Each word must be left-justified and consist of only eight valid ASCII characters with no blank spaces except at the end, to pad out the keyword to eight characters if necessary. Longer keywords such as TELESCOPE are contracted to TELESCOP to remain within the 8-character limit. Uppercase letters, the digits 0 through 9, periods and hyphens are all that is allowed. Bytes 9 and 10 may contain an equal sign and a space if the keyword has an associated numerical or text value. Numerical values are always right-justified between bytes 11 through 30, whereas text strings begin with a single quote at byte 11 and must end (with a single quote) by byte 80. An optional "comment" can be added after the value if separated by a space followed by a slash (/). When a keyword has no associated value, then bytes 9 through 80 can contain any ASCII text characters. The following order of keywords is required: SIMPLE, BITPIX, NAXIS, NAXIS1, NAXIS2, ..., NAXISn, and END.

Table 11.1. FITS keywords and their meanings

SIMPLE	has the value in byte 30 of either T (true) of F (false): simply a statement of whether or not the file conforms to the FITS standard.
BITPIX	an integer describing the number of bits in the data values. Options are 8, 16, 32 for 8-bit, 16-bit and 32-bit unsigned integers. Floating-point data can be represented e.g. -32 and -64 for 32-bit and 64-bit respectively.
NAXIS	the dimension of the data array. If value is zero, no data follows. Value of 1 for 1–d data such as intensity values in a spectrum. For image data, NAXIS = 2 (e.g. rows and columns of CCD) and NAXIS = 3 would be used for a data cube of spatial coordinates versus velocity. The maximum value of NAXIS is 999.
NAXIS1, NAXIS2, NAXISn	Each specify number of elements along that axis, with convention that NAXIS1 is the axis whose index changes most rapidly and NAXISn is the axis whose index changes the slowest. For example, in a CCD image the number of columns would go in NAXIS1 and the number of rows in NAXIS2.

Completing the header is the keyword END which is located in bytes 1–3 and the remaining fields (to 80) are filled with ASCII blanks. Several optional keywords may be inserted after NAXISn and before END. For instance, BSCALE and BZERO relate the array values and the true values through the relation

True value = BSCALE × array value + BZERO

and can be used to convert signed 16-bit array values (−32,768 to +32,767) into unsigned 16-bit pixel values (0 to 65 535) by setting BZERO to 32 768.0 and BSCALE to 1.0. Standard self-explanatory additional keywords with associated character strings are OBJECT, TELESCOP, INSTRUME, and OBSERVER. DATE-OBS and DATE have character string values and are intended to record the date on which the observations were obtained and the date on which the header was written respectively. The usual format for the date is dd/mm/yy and Universal Time is preferred. The keyword ORIGIN is used with a character string to identify the institution creating the FITS file. COMMENT and HISTORY are two keywords which do not have associated values and any valid ASCII text can be inserted in bytes 9 through 80. Any number of COMMENT or HISTORY lines are allowed consistent with the 36-line header. If more than 36 lines are required then the keyword EXTEND should be inserted before line 36 and unused lines up to 72 will need to be padded with the ASCII blank character.

Immediately following the header (at byte 2881) begins the data in a continuous sequence according to the NAXIS parameters already declared. According to the FITS standard, 8-bit integer data must be represented by unsigned binary integers contained in one byte and 16-bit data values must be stored as signed binary integers in two bytes with the most significant byte first. This convention is not followed by many computers (especially PCs) and programs, consequently, "byte-swapping" may be needed to import FITS data into another system. Also, although 16-bit digitization is standard, many imaging systems use 12, 14 or 15 bits and some use 24 bits. The BZERO keyword can be helpful in offsetting the zero point of the stored values and the 24 bit numbers would require to be handled using the 32-bit convention of four bytes with the most significant byte first. Finally, the tailer of a FITS file is ASCII null (00) characters used to pad out the final 2880-byte record.

11.2 DATA REDUCTION AND ANALYSIS SYSTEMS

After the observations have been acquired and the data have been stored in digital form, the next step is "data reduction". You will almost certainly find yourself spending long hours in front of a powerful "workstation", like the one shown in Fig. 11.1, using one of many image-processing packages. Several major suites of computer programs have been developed specifically to support the reduction of astronomical data, and especially CCD-type imaging and spectroscopy. Most of these programs have been "packaged" within an environment which allows the users to select the appropriate task and even set up a sequence of tasks to be performed without writing and compiling computer code from scratch. The most well-known packages are:

- **IRAF**
- **STARLINK**
- **AIPS**
- **MIDAS**
- **IDL**

With the exception of IDL, which is a powerful commercial package not specifically designed for astronomy but nonetheless suitable, all the others were produced by as-

Fig. 11.1. An astronomer reducing data at a powerful workstation using an image processing and display package.

tronomers and are maintained by the astronomical community. Some of the historical development of these large and specialized software systems was described in Chapter 2.

IRAF, which is used extensively in the United States, goes beyond the already well-established AIPS (Astronomical Image Processing System) at the National Radio Astronomy Observatories (NRAO), in the sense that every effort has been made to make this software system as "portable" and as device-independent as possible. Clearly, the American system bears many similarities to the British STARLINK system and the European MIDAS system, but because of the importance placed on portability of the software much more emphasis has been given to the design of the command language. The IRAF also carries with it its own programming language for maximum portability and was developed more comfortably in the UNIX operating environment. IRAF is run in conjunction with a "windows" display system known as "SAO-image" on any typical workstation (Fig. 11.2). The main drawback to IRAF at the time of writing is that it lacks an easy-to-follow menu-style or **graphical user interface** (GUI, pronounced "goo-ee"), and so it takes a bit of getting used to. IRAF is available through STARLINK and MIDAS.

The ESO-MIDAS software system is built along the same lines as the STARLINK and IRAF systems and now runs on a wide variety of computers. The basic system consists of three parts, the monitor, the applications and the interfaces. MIDAS is a multi-process, command-driven system and the monitor part functions as a command interpreter, keeps logs, displays on-line help. Applications perform the actual data operations such as image display, graphics, general image processing (including filtering, interpolation, Fourier transforms), processing of tabular data, fitting software and data input/output management. The interfaces level binds the applications and the monitor level together. The MI-

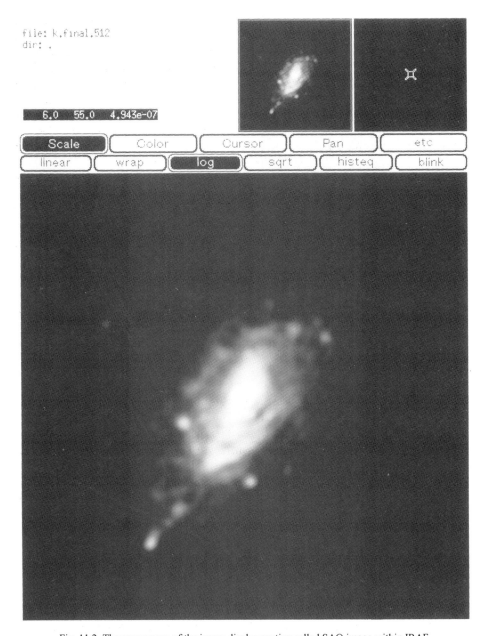

Fig. 11.2. The appearance of the image display routine called SAO image within IRAF.

DAS command language (MCL) provides tools to construct complex strings of commands together, each using results from the previous command. For more details see Warmels (1991) and the MIDAS World Wide Web page.

IDL (Interactive Data Language) is a complete visualization package for scientific, engineering and medical data which incorporates a powerful array-oriented language to enable users to create complex functions, procedures or applications without any conventional programming in FORTRAN or C. Operators and functions work on entire data arrays and there is immediate compilation and execution of the IDL commands. Image display and graphics capabilities are fully integrated with the computations and a superb tool called IDL Widgets, enables the user to create a graphical user interface (GUI) to IDL programs. This latter feature is particularly attractive to instrument builders. IDL is supported on Unix, VMS, Microsoft Windows and Macintosh systems. Many astronomers now use IDL for data simulation and modelling, and for the standard data reduction tasks (such as flat-fielding) normally handled by IRAF, but they still tend to return to packages such as DAOPHOT and DoPHOT for astronomical photometry.

Two other image/spectral processing systems developed by astronomers are worth mentioning and are still in fairly wide use, VISTA (developed by Lick Observatory, USA) and FIGARO (developed by Keith Shortridge at CalTech, after he left Alec Boksenberg's group at University College London, and further enhanced from his current position at the Anglo-Australian Observatory). Keith has supported the idea of treating data acquisition as a subset of data reduction since the earliest days of ADAM. FIGARO is the official STARLINK software package for spectral reductions.

There are also a growing number of very useful PC-based image processing systems which are much more attractive for the amateur astronomer with a CCD camera. These include the PC version of IDL, PMIS from Photometrics Ltd, Mira A/P and Mira Professional from Axiom Research, SkyPro and The Sky from Software Bisque, Hidden Image from Sehgal Corp. and Imagine 32 from CompuScope. These software packages all provide basic image arithmetic, image-enhancement and image-processing features.

Most data reduction packages (such as IRAF, STARLINK, MIDAS, IDL) provide hard-copy print-out of greyscale images, contour plots and other line graphics to a laser printer. The quality of such greyscale images is usually quite adequate for publication without further effort. Producing colour output is a little harder and more restrictive. Some institutes have suitable hardware for direct colour printing from screen images, or one can use the services of a commercial firm, or alternatively one can use the intermediate step of making a slide first.

By far the best way to make colour slides is to use a "slide-maker" unit. These systems have a camera attached to a special screen and software to import images for display. Many programs are available which can produce output ready for a slide-maker, e.g. Adobe Photoshop, Lotus Freelance, CorelDraw. The data must first be reduced using one of the many analysis packages and then saved in an appropriate format for transfer to one of these programs. Annotation or colour mixing or simple enhancements can be added and then the final image sent to the slide-maker.

If you don't have a slide-maker, but you are using SAOimage in IRAF or *kappa* in STARLINK or some similar program in which the image is displayed on a large monitor at high resolution, you can quite successfully photograph the screen to make a slide. When doing this, it is best to use a slow film (ISO 100 or slower), a relatively long focal length lens (70 mm say) to reduce distortion and a tripod-mounted camera with an automated exposure button to avoid camera shake. Make sure that the room is completely dark except for the screen being photographed, otherwise you will see reflections in the screen, and turn down the brightness of the monitor slightly to ensure that the exposure time is not too short, about 0.5 s or longer is fine.

Even temporary storage of digital images on computer hard drives requires considerable space. For example, with 16-bit digitization for a 1024 × 1024 CCD, a single frame occupies 2.1 Mbyte disk space which for a 1 gigabyte disk implies room for about 475 images—ignoring the extra space for informational data called "headers". It is by no means unreasonable to obtain about 50–100 images per night at the telescope, making 250–500 images in a single observing run, and during image processing and analysis the number of intermediate images will often be five times as many—at least temporarily. Many CCDs used in astronomy are larger than this! So computer disk space is crucial, and large observatories possessing an entire suite of CCD instruments will need hundreds of gigabytes of storage capacity. Magnetic tape data archives are no longer practicable, especially since the information on the tapes can become lost or unreadable after a period of 5–7 years unless the data are re-copied. The next generation of astronomical data archives will be based on optical laser disks. At the time of writing, a typical double-sided double-density disk the size of a familiar audio compact disk (CD) will hold 1.2 gigabytes = 1200 Mbytes, and the drive units are small and compact.

11.2.1 The IRAF package

Since the use of IRAF is very widespread it is worth providing a brief introduction to this system. The IRAF Command Language (CL) organizes the many system and application programs (called tasks) into a logical hierarchy of packages. A package is a collection of related tasks, and the major packages are

- system—system utilities
- language—the Command Language itself
- dataio—data input and output
- images—general image processing
- noao—astronomy packages from NOAO
- plot—general graphics utilities

The first two packages (system and language) include tasks that are available as soon as you enter IRAF. For tasks within the other packages you must first "enter" those packages by typing its name. Subpackages exist within the parent package and can only be reached through their parent package. Below is a "tree" of some of the more commonly used tasks. You will need to consult your local IRAF documentation for more options.

system	language	dataio	images	noao	plot
allocate	bye	rfits	hedit	*digiphot*	contour
deallocate	ehistory		imarith	*apphot*	
devstatus	eparam		imcopy	qphot	
directory	logout		imdelete	*imred*	
help	lparam		imheader	*ccdred*	
lprint	print		imhistogram	combine	
			imrename	*proto*	
			imstatistics	fixpix	
			tv		
			display		

Subpackages are italicized.

Assuming that you have logged in to a computer running IRAF and changed to the IRAF directory (probably by clicking on the x-term window and typing "cd iraf"), you enter IRAF's command language by typing "cl" (without the quotes) and hitting the return key. The computer should return with a prompt cl>. Typing a "?" (without the quotes: this will apply to all further typing instructions in this section) will list the IRAF packages and you can get help on any one of them by typing "help packagename" at the prompt. Similarly, typing "help taskname" will give information on the parameters that each task requires. If you execute a task unintentionally, then use the "Control C" combination (the Ctrl key and the C key together) to terminate the task. To leave and subpackage and return to the parent package, type "bye" or alternatively use "Control D".

Most likely, your first action will be to import your data into IRAF from a data tape or disk using the dataio package and rfits task. For instance,

    cl> dataio
    da> devstatus mycomputer!mta   [This checks whether or not the tape or disk drive is
                                     already allocated.]
    da> allocate mycomputer!mta
    da> help rfits   [Hit spacebar to advance a page; "q" to quit.]
    da> epar rfits   [epar or eparam is shorthand for edit parameter set-
                                     tings associated with this task.]

Simply edit the displayed list by typing in the values you want and press "Control D" to exit. For example, set

    fits_file = mycomputer!mta
    fits_list = 1-999   [Or whatever you want.]
    iraf_file = raw

to create IRAF image files for every FITS file from 1 to 999, or as many files as are on the tape if it is less than this number, and name the files raw001.imh, raw002.imh, and so on.

    da> rfits
    FITS data source (mycomputer!mta):<return>
    File list (1-999): <return>
    Iraf filename (raw): <return>   [Since you already set the parameters you can hit the
                                       "Return" key at each prompt.]

When all the files have been read, there should be an indication of this and the prompt will return. Be sure to deallocate the device for the next user

    da> deallocate mycomputer!mta
    da> bye
    cl>

and remove the data tape or disk from the device.

The next step is to display the raw images on a screen. In IRAF this is done with a facility called **SAOimage** which creates a display window separate from the IRAF command window. From outside the IRAF window pull down a menu and select SAOimage; a new display window will be created. You can drag and re-size this window in the usual way. Go back to the IRAF window, click in it and type the following in response to each prompt (remember to end with a carriage return):

cl> images
im> tv
tv> display imagename   [Where imagename might be raw012(.imh); try dev$pix for a built-in image of M51.]

Note that there are actually four frame buffers available and you can specify one of these by adding the numbers 1–4 after the imagename; no entry defaults to buffer 1. Also, it is not necessary to type the extension ".imh" after the imagename.

SAOimage provides mouse-controlled buttons to select grey or false colour (click the left mouse button on "colour"), then choose a "colour mapping" (click on "cmap"), chose any of the options "Gray", "BB", "HE", "18", "A", "B" by clicking on it and finally, change the appearance of the display (contrast and stretch) by holding down the left mouse button and dragging the mouse around. Watch the colour bar at the bottom of the frame and reset at any time by clicking on "Gray" or "BB" or one of the others. You can magnify or demagnify the display and you can "pan" around within the displayed image. All of these features are fairly typical of any image-processing package. Click back in the IRAF window when you wish to leave the display window.

IRAF can be used to perform all the steps necessary to reduce CCD images, such as dark subtraction, flat-fielding, registration and photometry. Suppose you want to average a few dark frames and then subtract them from each of your image frames. You need to enter the ccd reduction subpackage "ccdred" from the major package "noao" and then execute the task called "combine". For example,

cc> help combine   [Read the on-line help.]
cc> epar combine   [Edit the parameter file. You will be concerned with the parameters: images (the input data), output (name for final frame), combine (either an average or a median), exposure and scale. The last pair determine how to scale the frame—by exposure time or by the mode. Control D to exit.]
cc> combine

To subtract two images of equal exposure time, such as the average dark frame from the individual image frames, the steps would go something like:

cl> images   [Enter the major package. Read the help file and edit the epar file. The key parameters are the operand 1 (the first file), op (+,-,*,/), operand 2 (the second file), result (the output file).]
im> imarith file1 - file2 outputfile[You can also type the filenames directly on the command line.]

Be sure to think about a logical naming scheme before you start. The output file should be distinguishable from the raw data frame and it should be fairly obvious that it has been dark-subtracted e.g. f012d, for the dark-subtracted version of file number 12. The new data array, "outputfile" (or f012d in this case), can then be displayed. Do the same to a set of flat-field images to subtract the dark from each and then use "combine" again to put all the dark-subtracted flats together. This time you will probably chose the "median" option within combine and use the "scale" option to scale by the mode instead of exposure time. Eventually you will have a final flat-field frame, and you can use "imarith" again with "/" as the "op" and the flat-field frame as "operand 2" to divide each dark-subtracted data frame by the flat-field. A lot of the above procedures can be "scripted" using the IRAF command language to alleviate you of doing repetitive tasks.

IRAF also allows users to create their own tasks by writing scripts. An IRAF script is basically a series of tasks grouped together under a single umbrella name. A user-defined task will have a "param" file like other tasks, and it may have both required and optional parameters. Scripts are written to files with a .cl suffix. They are defined at the command line with

> cl> task mytask = filename.cl

where mytask is the name of the task defined in the script, and filename.cl is the file in which it is located. The scripting language in IRAF is very simple. A script consists a task declaration, followed by a definition of the parameters. The body of the script is started with the word "begin" and ends with the word "end". In between, IRAF commands are entered, one on each line. Variables may also be employed, but must be declared by type at the beginning of the body of the script. The usual programming statements are allowed as well (i.e. while loops, if-then statements). Comments start with the symbol "#".

EXAMPLE: As a simple example consider a script that will add two images together to make a third image, which will then be divided by a constant if that option is chosen. Further, assume the constant will most often be 1000 and will not usually be changed. The script to do this is provided below courtesy of Harry Teplitz. Explanatory comments begin with #.

```
#  IRAF SCRIPT addimages.cl
#  script to add two images and divide by a constant
#  first declare the task name and the required parameters
procedure addimages (image1, image2, image3)

#  next list all the parameters, and give the prompt for them
#  that will appear in the param file

   string image1          {prompt="first image"}
   string image2          {prompt="second image"}
   string image3          {prompt="output image"}
   real number=1000       {prompt = "number to divide by"}
   bool divbool           {prompt = "divide by number?"}
#  notice that the optional parameter is given a default value
#  start the main program
begin

#  add image1,image2 to make image3
#  notice that the required parameters are entered in
```

```
#  parenthesis and separated by commas
    imarith (image1, "+", image2, image3)

#  check to see if division should be done
#  notice that the commands inside {} are only
#  executed if divbool is "yes"
    if (divbool==yes) {
    imarith (image3, "/", number, image3)
    }
#  end program
end
```

This script could be executed at the command line as follows:

```
first define the task
    cl> task addimages=addimages.cl
then give command
    cl> addimages image1 image2 image3
```

For simple aperture photometry of isolated stars you can use the "qphot" task in the "apphot" subpackage of the "noao" package. The parameter file contains options to allow you to vary the size of the annulus around the star from which the sky background will be deduced; "annulus" is the inner radius in pixels and "dannulus" is the width of the annulus. For standards stars, specify a list of several "apertures" (e.g. 2, 3, 4, 5, 6 pixels) for the star image up to the value of "annulus" to provide a "curve of growth" from which you can later make aperture corrections for other stars if required. You can execute qphot and move about the SAOimage window placing the cursor on star after star. Hitting the f key produces a tabulation of centroid positions and magnitudes appear in the xterm window, while pressing the space bar sends the result to a printable file, with the same root name as the image but with the extension *.mag.1*.

For more sophisticated photometry and crowded-field photometry you will need to go to packages which invoke Point Spread Function (PSF) fitting such as DAOPHOT which can be run as part of IRAF. DAOPHOT is a very large program developed over many years by Peter Stetson of the Dominion Astrophysical Observatory in Canada. Another PSF-fitting program called DoPHOT was developed by Paul Schechter (MIT). Interestingly, a PC version of DAOPHOT has been developed with Stetson's permission by Smirnov, Ipatov and Samus of the Institute for Astronomy at the Russian Academy of Science. Most PSF-fitting programs start with a fit to a mathematical function and then modify the PSF with corrections based on the mean differences between the observed profiles and the functional form. In very crowded fields, an iterative process is used in which the best estimate of the PSF is scaled to subtract out the effects of nearest-neighbour stars on the PSF. A new PSF is derived from the cleaned-up frame and the process is repeated.

11.2.2 The IDL package
IDL handles images by loading them into its memory. Procedures involving very large numbers of images can also use the concept but only keep a few in memory and writing out to files often. While in memory, images are stored in two parts. First a string containing the

FITS header, and then a matrix of real numbers containing the pixel data. STSCI has provided IDL scripts for reading from FITS files, manipulating the header information, and writing to FITS files. Simple display routines are available as well (tv, tvscl). For example to read in a display a FITS file called "myfile.fits", one might enter the following at the command line.

```
IDL> image = readfits('myfile.fits', header)
IDL> tvscl, image
```

IDL commands are either procedures or functions. A procedure is a command that executes a set of operations without creating a new variable. For example, "tvscl" is a procedure that will plot a matrix of numbers with autoscaling. Procedure commands have two kinds of parameters, arguments and keywords. Arguments follow the procedure name and are separated by commas. They are identified by IDL according to their order on the command line. Arguments may be required or optional. Keyword parameters are also separated by commas, but they are identified by their names. For example to write the data contained in the matrix "image" to the file "outfile.fits" with the header information in "hdr_str" (including the header is an optional argument), with undefined pixels set to zero (NaN is the keyword for undefined pixels):

```
IDL> writefits, 'outfile.fits', image, hdr_str, NaNvalue=0
```

Functions are very similar except that they create new variables, as in the readfits example given above. Functions also have the form

```
IDL> new_variable = function(parameters)
```

IDL also allows users to create their own procedures and functions using the IDL programming language. This language is a full programming language like FORTRAN or C++, even though it is a higher-level interface for routines that are actually written in C. All IDL commands are available in programs, as are a variety of common programming statements. The language is constructed to make guessing the right syntax very easy. For example, the command to print to the screen is "print". IDL is not optimized specifically for astronomical uses. However, a large number of useful programs are already available from STSCI to perform complicated functions such as reading and writing FITS data.

Programs are usually written in files with ".pro" extensions. They can be loaded into IDL's memory with

```
IDL> .run myprogram
```

to load the program contained in the file "myprogram.pro". Once this is done, the program may be executed just as any other function or procedure. Variables may be declared explicitly in IDL (a good programming practice), but the language is very forgiving. Comments begin with ";". Large blocks of programs, such as the body of a loop, are started with "begin" and finished with "end". The main body of the program begins with the function or procedure declaration and ends with "end".

EXAMPLE: As a simple example, consider a script that will take a list of FITS files with a common rootname and subtract a bias image from each, and optionally divide by a flat field. The output will be written to a new series of fits files.

```
;   IDL program "subflat.pro"
;   This program will subtract the bias from a series of images that have a common root
;   their name followed by a number. It will then optionally divide by a flat field image.
;   Output will be written to fits files with a different root name

;   first define the procedure
;   the "$" allows the definition to continue on the next line
;   capitalization and indentation are stylistic, not required by the IDL language
PRO SUBFLAT, root, out, first, last, $
        BIASNAME=biasname, FLATNAME=flatname, DIVFLAT=divflat
;   first we check to see if bias and flat are defined. if not, they will have the default names
;   of "bias" and "flat"

    IF NOT(KEYWORD_SET(BIASNAME)) THEN biasname='bias'
    IF NOT(KEYWORD_SET(FLATNAME)) THEN flatname='flat'

;   now define the loop variable to start with first
    I = first

;   now read in the bias data
    biasdata = READFITS(biasname + '.fts')

;   if flat fielding will be necessary read in the flat field, notice that the conditional lines
;   are inside begin/end
    IF KEYWORD_SET(DIVFLAT) THEN BEGIN
        flatdata = READFITS(flatname + '.fts')
    END

;   next loop through the images.  Notice that the body of the loop begins with "begin" and
;   ends with "end"

    WHILE (I le last) DO BEGIN
        infits = root + STRTRIM(i,1) + '.fts'        ; make name for input fits file
        data = READFITS(infits, hdr)                 ; read that file into data and hdr
        outdata = data - biasdata                    ; subtract off the bias from data
        IF KEYWORD_SET(DIVFLAT) THEN BEGIN
            outdata = outdata / flatdata             ; divide by flat if requested
        END
        outfits = out + STRTRIM(i,1) + '.fts'        ; construct output filename
        WRITEFITS, outfits, outdata, hdr             ; write data, w/ same header
        I = I+1                                      ; increment the loop variable
    END

;   end the procedure
END
```

This program could then be loaded at the command line with

```
    IDL> .run subflat
```

(the .pro is assumed). Then it could be executed on im1.fts, im2.fts, im3.fts, with bias.fts and flat.fts, to make out1.fts, out2.fts, and out3.fts, with:

IDL> subflat, 'im', 'out', 1, 3, bi='bias', fl='flat', /div

Notice that only enough of the keyword names must be given to unambiguously identify them. Also notice that a keyword may be set using /keyword_name instead of explicitly giving it a value. This is very useful for yes/no types of keywords.

11.3 PCS IN ASTRONOMY

Scientific applications of personal computers (PCs) is widespread for many applications such as word-processing, mathematical calculations, spreadsheets and many other uses. PCs are also being used for instrument control (e.g. CCD cameras), small telescope automation (robotic observatories) and for CCD image data reduction (especially among amateur astronomers and in student teaching labs). In recent years, the rapid increase in the computing power and speed of personal computers has led to some impressive image processing being done on these platforms, including merging three images at different wavelengths to form a colour composite for transfer to a slide-maker unit. A real break-through was the introduction of **CD-ROM** (compact disk with read-only memory) technology. By purchasing a computer with "multi-media"—which means having a CD-ROM drive in addition to the normal magnetic disk drives (e.g. 3.5-inch disks)—you can gain vastly more information in one place than ever before. For example, the entire Hubble Space Telescope Guide Star Catalogue which contains 15 million stars is available on CD-ROM.

PCs are an area of common interest to professional and amateur astronomer. Numerous CCD manufacturers, such as Photometrics Ltd, First Magnitude Corp., and Axiom Research Inc. (USA), LE2IM (France) and AstroCam (formerly AstroMed) in the UK, provide complete PC-based camera systems with associated image-processing packages. AstroCam Imaging Systems CCD 2200 cameras are normally supplied with a host PC running under the UNIX operating system. Photometrics Ltd sell a range of "top-end" systems which can stand alone or be interfaced to PCs, Macs and Sun computers. Other companies provide CCD cameras ready to be interfaced to PC systems (e.g. Santa Barbara Instrument Group (SBIG), SpectraSource Instruments). Moreover, the number of amateur astronomers world-wide with CCDs on small telescopes has increased so dramatically that Sky Publishing Corporation has initiated a magazine called "CCD ASTRONOMY".

A wide range of software products of interest to both professionals and amateurs are also available, such as the "Voyager" star chart program which has been around for a while, or "Epoch 2000" with 45 000 stars, 13 000 faint deep-sky sources, 7700 asteroids and 650 comets listed. Among the better CD-ROM based programs is "Redshift" (Maris Multimedia Ltd) which, in addition to the star chart features with 250 000 stars and 40 000 deep-sky objects, includes a collection of 700 full-colour images of the planets, moons and deep space objects, 29 short (QuickTime) movies and the *Penguin Dictionary of Astronomy*. Programs such as Epoch 2000 will also allow you to control an 8-inch Schmidt-Cassegrain telescope by Meade Instruments Corp., which can be fitted with a CCD camera such as the ST-6 from Santa Barbara Instrument Group (SBIG). Both amateur and professional astronomers can now have access to very large database systems. For example, the CD-ROM entitled "Selected Astronomical Catalogues" which was released by NASA in 1991 contains 114 catalogues including, for example, the SAO Catalogue. Professionals, such as Rudy Schild at the Smithsonian Astrophysical Observatory, have used multi-wavelength CCD imagery to produce beautiful true-colour digital pictures which are now available on CD-ROM (Smithsonian Collection: Volume 4 in the series Voyage to the Stars by Astronomical Research Network).

The easy access to PC-controlled CCD cameras is also a big help in both undergraduate and graduate teaching in astronomy. For example, several years ago at UCLA we equipped our 24-inch $f/17$ Cassegrain telescope with two imaging systems, a Star 1 CCD camera system purchased from Photometrics and an infrared camera designed and built in-house. The IR camera, known as KCam, is based on a 256×256 pixel infrared array sensitive to wavelengths from 1 to 2.5 μm (these devices are described in Chapter 8). An optical device called a dichroic beamsplitter which consists of a glass substrate with a special coating, is placed in the beam just above the focal plane at an angle of 45 degrees. Infrared light is reflected through a right-angle to KCam and optical light is transmitted to the Star 1 camera. In this way we can observe in the optical and the infrared simultaneously! At a bright city site such as the UCLA campus, it is very advantageous to work at infrared wavelengths. Both the Photometrics camera and our own KCam system are operated independently with PC computers. The telescope is also controlled from a PC through a simple motor control chip called a CY500 which can operate the RA drive stepper motor. The "Voyager" software running on a MacII is used for star-finding, and image processing routines in IRAF (or PhotoShop) enable us to process and display images.

Axiom Research's Mira Professional and Mira A/P programs developed by Mike Newberry contain aperture photometry software with a graphical user interface, whereas Ron Haitchuck, Arne Henden and Ryland Truax (United Software Systems, Flagstaff, Arizona) provide a DOS-based program called CCDIR. Reviews of these and other excellent PC software products can be found in magazines such as *CCD Astronomy*.

11.4 PRINCIPLES OF IMAGE ANALYSIS AND PROCESSING

Raw CCD data may be quite uneven or peculiar in appearance until processed. Most processing steps must be carried out pixel-by-pixel. For example, we will need to remove the pattern of dark current which was accumulated during the CCD integration. To do this we subtract a data array containing the "dark" frame from the data array containing the "raw" frame using a vector arithmetic algorithm which moves from pixel to pixel calculating the difference and entering the answer in a new data array. Similarly, the frame can be divided by another frame (flat-fielding) by a pixel-by-pixel division of the number in the first data array with the corresponding pixel value in the second data array, and the quotient is entered into that pixel location in the output data array. Of course, simple scalar arithmetic is also possible. For example, each and every pixel value can be multiplied by a constant number and the product is entered into that pixel location in the output array.

For most astronomers, image processing means simply, "data reduction", and is largely associated with the simplest aspects of visualization and mathematical manipulation of the two-dimensional (x,y) array of numbers stored in the computer to represent, for example, the intensity on each pixel of a CCD camera. Some of the required manipulations are straightforward, for example, add, subtract, multiply or divide each pixel value by a constant or, by the value found at the same pixel location in another digitized image, such as a dark, sky or flat-field frame. Software is also needed for numerous other manipulations, including logarithms, statistics, smoothing, noise suppression, image enhancement, contouring and profile fitting.

11.4.1 Displaying images

Clearly this is the first step. We will always want to "see" the image, either in a raw form or in various stages of reduction to final form. Software is required to "map" the image onto the screen. Initially, this mapping is a linear mapping of true intensity to values in the display range, which might be 0–255 for instance to give 256 "levels". The conversion is stored in an **LUT** (or look-up table). If the weakest intensity is set to correspond to 0 and the brightest signal assigned 255, then all intermediate signals are binned into the intermediate levels. It is generally advantageous to have a cursor which can be moved over the displayed image and "read back" to the screen the (x,y) pixel coordinates and the true intensity at that point in the image (not the scaled value from 0 to 255). Images which have had vector or scalar arithmetic processing can be displayed as normal images or they can be displayed as

- pseudo three-dimensional image
- two-dimensional contours or isophotes
- one-dimensional cuts or cross-sections

In addition, the reduced/corrected image can also be further treated with one of several processes:

- low-pass filter, to remove the finest spatial details
- high-pass filter, to enhance fine spatial details
- block smoothing, replaces each pixel with the average of those in a $n \times n$ block
- Gaussian smoothing, replaces each pixel with a weighted average depending on the width of a bell-shaped "Gaussian" profile of width σ pixels.
- unsharp masking, enhances fine details by subtracting a low-pass-filtered (fuzzy) image from the original

Other processes which the computer must be able to supply include:

- Integer and non-integer pixel shifts in which the numbers in the original data array are mapped to new locations in a new data array and, if the shift is not an integral number of pixels, the new intensity value is "interpolated".
- Bad pixel corrections by interpolation across the incorrect pixel value in both x and y to estimate a replacement value.
- Deconvolution and image restoration.

These techniques, of which there are many, attempt to correct an image for deterioration due to atmospheric turbulence or known aberrations. The final (observed) image is a convolution of the initial image and a point spread function which is the instrumental response to a perfect point source. Deconvolution is achieved using a Fourier transform.

In the above discussion we have assumed that the 256 display levels represent shades of "grey", but it is also possible to match each interval to a particular shade of colour. When the distribution of brightness in an image is represented by arbitrary colours it is known as a **false-colour** representation (see Plate 12). False-colour display is really a simple form of image enhancement.

Since even visible light astronomical images are usually obtained with a combination of optical filters and detector sensitivities that doesn't remotely match that of the human eye, the term **true colour** must be used very cautiously. Many spectacular colour images are available to the general public and they are often represented implicitly as true or natural colour, but if you could really see these objects with your eyes they would simply not appear as colourful as shown. Rudy Schild at the Smithsonian Astrophysical Observatory is among the many people who have strived to get these issues right, and I follow his terminology here. In most circumstances three images are used to balance the colours. By adjusting the contrast of the three originals until they are equal and then changing their individual brightness levels, it is usually possible to balance the image so that the colour of an average star such as the Sun is white. Distinctions that result from using different filter combinations are as follows:

- spectrally augmented colour—e.g. replacing the "red" filter with an "infrared" filter which makes red stars look redder and blue stars look even bluer
- substitute-filter colour—e.g. the use of "narrow-band" filters to isolate specific features, such as the methane absorption band in Jupiter to enhance the cloud formations
- emission highlighting—e.g. the use of narrow-band filters to enhance nebular emission light (pink H-alpha, green oxygen) against starlight
- colour translates—e.g. images made at wavelengths completely invisible to the human eye (infrared, radio, X-ray)
- enhanced colour—e.g. increasing the colour saturation to improve colour contrast (such as Voyager images of Jupiter)
- two-filter colour—e.g. when images are only available at two wavelengths then the third must be interpolated to produce an image which looks natural but really has no information at the third wavelength

Again, see the colour plates section for examples.

11.4.2 Image enhancement

If the dynamic range in the image is large, $I_{max} >> I_{min}$, then the resulting linear mapping does not have good contrast. One way to bring up the faint end is to significantly reduce the intensity level assigned to 255 (white). For instance, if we set the white level to 10% of the peak signal, then the remaining range of signals is mapped into 255 resulting in a display in which all the brighter objects are white, but all the fainter signals are now visible. The display is said to have been "stretched" and it is clear that the "transfer function" is steeper. In fact, the 0 (black) level can also be moved and need not correspond to the weakest signal. Thus, any "window" of signal levels can be stretched from 0 to 255 display levels (see Fig. 11.3 (a,b)).

A variation on this linear stretching approach is to add a point in between the black and white levels and use a different steepness of transfer function in each part. For instance, a

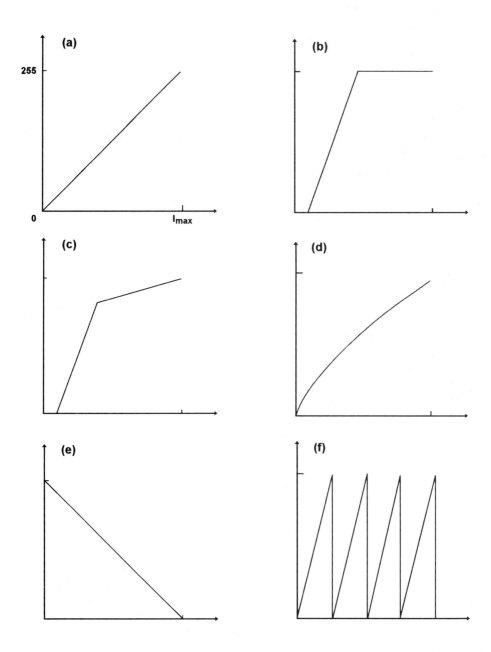

Fig. 11.3. Examples of look-up tables (LUTs) or display-stretching transformations. (a) linear; (b) steep linear—higher contrast; (c) stepped-linear; (d) logarithmic; (e) inverted; (f) saw-tooth or wrap-around.

steep transfer function could be applied to display data with signal levels of 0–10% of the peak value and these could be mapped into 0–200 levels, with the remaining signal levels (10–100% of peak) being mapped to the 201–255 levels as in Fig. 11.3 (c). Such a plot gives a good stretch to the faint end without grossly "over-exposing" the bright end.

Since the transfer function is now non-linear, although it is composed of two straight-line segments, then we might as well consider any non-linear mapping of signal to display levels, and there are several techniques which are useful in astronomy applications for bringing up faint sources near the noise level while reducing the inevitable saturation in the image from the bright regions.

Enhancing the contrast of faint objects near the sky brightness level can be done with a non-linear transformation such as a "logarithmic curve" (Fig. 11.3 (d)) which rises steeply at first to increase the contrast of faint objects, but levels off more slowly to compress the bright end of the map. Sometimes better contrast is achieved by "inverting" the colour table, so that black becomes white and vice versa. This is achieved with a LUT like the one in Fig. 11.3 (e). Repetitive, linear (saw-tooth) ramps can also be used to "wrap-around" all the grey levels several times (Fig. 11.3 (f)). Finally, one of the most powerful non-linear distortions is a transformation which "equalizes" the histogram of signal values versus the number of pixels with that signal (Fig. 11.4). Histogram equalization is very good at bringing out faint objects near the background level.

A **low-pass filter** is a "smoothing" process (rather than a stretching process) which reduces the contribution of high spatial frequencies, or fine detail, in an image while maintaining the lower, more slowly changing spatial patterns. Simply replacing the value associated with each pixel in an image with the average of its immediate neighbours constitutes an elementary low-pass filter. Since noise often contains high spatial frequencies (i.e. relatively large and random changes from pixel to pixel), then low-pass filtering can reduce noise. The appearance of a low-pass filtered image is "softer" or more blurry than the original, but it is useful for bringing up faint nebulosity in noisy images.

In a **high-pass filter** the process is one of "sharpening" which emphasizes fine details (high spatial frequencies) by suppressing the lower spatial frequencies or more slowly

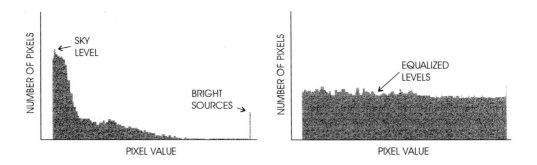

Fig. 11.4. Illustrating a look-up table for mapping true intensities to display values such that the final histogram of signal levels appears to have been "equalized". This is a useful way of bringing up the contrast to reveal faint objects.

changing spatial patterns. The image looks "sharper" but the noise pattern will also get enhanced unless the original image is just fuzzy rather than noisy.

Both of these processes are described by simple **convolutions**. A convolution is described by the following process:

$$G = F * H$$

where

$$G(x, y) = \int \int_{-\infty}^{+\infty} F(x - x', y - y') H(x', y') dx' dy' \qquad (11.1)$$

The way to visualize this action is to think of x, x', y, y' as pixel coordinates in an image $F(x,y)$, and $H(x',y')$ as a weighting function or "kernel". At the point $(x-x',y-y')$ the image has a certain value (the brightness at that point) and we multiply that brightness by the value of the weighting function at the point (x',y') and integrate (or in practice *sum*) over all the values of x' and y'. Usually, the weighting function is restricted in its range and is zero beyond a certain range of pixels. The next pixel in the image is selected and the whole process is repeated until the entire image has been covered. A simple way to implement these filters is to establish a small matrix (with an odd number of elements e.g.

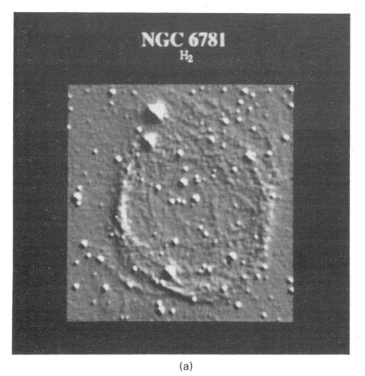

(a)

Fig. 11.5. (a) A method of enhancing sharp features and edges using bas-relief, and (b) an illustration of "stretching" the display with a steep linear look-up table as in Fig. 11.4. The object is a planetary nebula imaged in the light of molecular hydrogen emission at 2.122μm with the UCLA twin-channel camera.

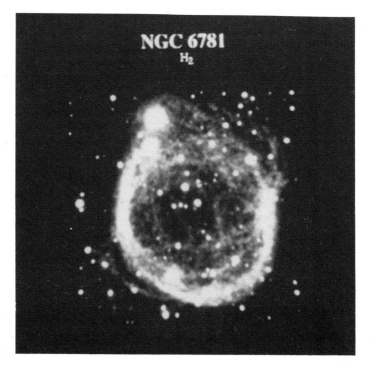

Fig. 11.5. (b)

3 × 3, 5 × 5, 7 × 7) and "scan" this matrix (or kernel) over the image. The new intensity corresponding to the current location of the centre of the matrix is found by summing all the products of each weighting value in the matrix with the underlying intensity at that point in the image; if parts of the matrix overhang the edge of the image then those elements contribute zero to the sum. Two simple examples of low-pass and a high-pass filters are shown below using only 3 × 3 matrices:

LOW PASS

1	1	1
1	1	1
1	1	1

Straight average of nearest
neighbours—strong blurring

0	1	0
1	10	1
0	1	0

More weight given to central
pixel—modest blurring

HIGH PASS

0	−1	0
−1	16	−1
0	−1	0

More sharpening

−1	−1	−1
−1	8	−1
−1	−1	−1

Less sharpening

In the low-pass filter, brightness values surrounding the central pixel are "added in" which smooths or blurs fine detail, whereas in the high-pass filter the adjacent signals are "subtracted out" to give emphasis to the central pixel.

Specific two-dimensional functions can be used to define H, such as a radially symmetric Gaussian profile of the form $\exp\{-r^2/2\sigma^2\}$, where σ is the radius at which the function has only 61% of its initial value.

Median filtering is a powerful yet simple example of non-linear filtering which provides a way of eliminating single pixel "spikes" with highly deviant brightness due to cosmic rays, for instance. In terms of the matrix approach mentioned above, each element is a 1, but the replacement value is not the sum of the products. Instead, the products are sorted into a "list" with increasing intensity (I_{min} to I_{max}) and the intensity which lies in the middle of this odd-numbered sequence (i.e. the median intensity) is selected to become the replacement value. Note that even when the filter is centred on the pixel with the spurious value, that value will be at one end of the ordered list and will never be selected as the replacement value. This method works on pixels that are too bright or pixels that are too dark. It is possible to design the algorithm so that it will only act on single-pixel events and/or exceptionally deviant values and leave well-sampled star images alone—otherwise the median will remove the stars too!

Median filtering is deliberately used to remove stars and faint compact sources in deep sky imaging so that the sky itself can be used as a flat-field. The ordered list of intensities is obtained by displacing or "dithering" the telescope numerous times in a random pattern between individual background-limited exposures so the faint point-like sources never fall on the same pixel twice.

There are several enhancement filters which are based on the "gradient" or rate of change of brightness in the image. For example, taking the difference between an image and a copy of itself in which the data values have been "shifted" over by a small number of pixels produces an estimate of the slope and enhances edges where the gradient is large. A more sophisticated form of this filter produces "side-illuminated" or "bas-relief" images which have a 3-D appearance (see Fig. 11.5(a)). Other filters in this category are the Laplacian and the Sobel filter.

One of the most well-known filtering methods is the **unsharp mask**. This technique has been used very successfully by astro-photographer David Malin to enhance prime focus images from the Anglo-Australian Telescope, and can certainly be used with electronic images from CCD cameras and infrared arrays. Basically, the image is first smoothed with a low-pass filter to create the blurry, "unsharp" mask and then this frame is subtracted from the original image. Features with low spatial frequencies are removed leaving an image dominated by high spatial frequencies. To control the harshness of the effect, the original image and the unsharp mask can be scaled differently to give more weight to the original image.

11.4.3 Image restoration

Image restoration is the recovery of images from raw, "image-like" data whereas image reconstruction implies the production of an image derived from data with a more complex form of encoding so that an image is not readily discernible. Image restoration is the common form of image analysis used in UV–visible–IR astronomy because these disciplines have direct imaging detectors such as CCDs, infrared arrays and microchannel plates. Radio astronomy on the other hand is often faced with reconstruction of images from interferometric (Fourier) data.

Although this field had been around for a long time, it received an immense boost following the discovery of the spherical aberration in the primary mirror of the Hubble Space Telescope which resulted in images with a large halo of light around a sharp core. Since the behaviour or point spread function (PSF) of the image was measurable, various techniques could be applied to retrieve the true image. Among these algorithms and methods were:

- **Richardson–Lucy** deconvolution
- **Maximum Entropy** deconvolution
- **Pixon**-based Image Reconstruction

Some techniques such as the Richardson–Lucy method are distributed with popular image processing packages such as the Space Telescope Science Data Analysis System (STSDAS) which runs under IRAF. This package also includes Maximum Entropy and Clean algorithms. Commercial packages such as MemSys5 from Maximum Entropy Data Consultants Ltd, England are also widely used. Nick Weir of Caltech has developed the MEM front-end program for MemSys5 and this combination is also available through the Space Telescope Science Institute. Pixon Image Reconstruction software is available by arrangement from Dr Richard C. Puetter at University of California, San Diego.

The mathematical basis for most of these methods is a theorem known as Bayes theorem and the methods are said to be Bayesian. Image reconstruction in its most general form is an "inverse" problem in which the data, $D(x)$, is related to the true signal, $I(x)$ through a relationship such as

$$D(x) = \int dV_y\, H(x, y) I(y) + N(x) \tag{11.2}$$

where x and y can be n-dimensional vectors and the quantity $H(x,y)$ is called a "kernel" function which describes how the act of measurement distorts or corrupts the true signal, and $N(x)$ is the "noise" or error associated with the measurement either due to instrumental effects or signal strength (counting statistics) or a combination. $I(x)$ is also called the "hypothesis" of the data or simply the "image". Usually, $H(x,y)$ is a simple "blurring" function known as the **point spread function** or PSF and the equation can be reduced to a convolution integral:

$$D(x) = \int dV_y\, H(x - y) I(y) + N(x) \tag{11.3}$$

In the more general case of image reconstruction, rather than image restoration then $H(x,y)$ is more complex.

Linear inversion methods, such as Fourier deconvolution, offer a simple closed-form solution for $I(x)$,

$$I(x) = F^{-1}\left[\frac{F(D)}{F(H)} \right] \tag{11.4}$$

where $F(f(x))$ is the Fourier transform of the function $f(x)$ and $F^{-1}(f(x))$ is the inverse Fourier transform of $f(x)$. Although computationally expedient, linear inversion methods tend to propagate noise and generate unwanted artifacts in the solution. Non-linear methods are of course more complex, but most can be interpreted in terms of a Bayesian Estimation scheme in which the hypothesis sought is judged to be the most probable. Bayesians use conditional probabilities to factor the joint probability distribution $p(D,I,M)$ where D, I and M are the data, the unblurred image and the "model" of the image. The mathematical expression of the model reflects knowledge of the physics of the imaging process, the instrumental properties and the method of modelling the data (e.g. as a discrete sum). Two Bayesian formulations are expressed below:

$$p(I|D, M) = \frac{p(D|I, M)p(I|M)}{p(D|M)} \qquad (11.5)$$

$$p(I, M|D) = \frac{p(D|I, M)p(I|M)p(M)}{p(D)} \qquad (11.6)$$

where $p(X|Y)$ is the probability of X given that Y is known. In the first equation, the model is fixed and this is the basis of Maximum Likelihood and Maximum Entropy methods, whereas in the second formulation the unblurred image (I) and the model (M) are taken together and the goal is to maximize the probability of the image/model pair given the data. This is the basis for the more recent pixon-based algorithms. Each term on the right-hand side of these equations has a physical significance. The term $p(D|I,M)$ measures the likelihood of obtaining the data (D) given a particular unblurred image (I) and model (M), that is, it is a "goodness-of-fit" (GOF) parameter. The term $p(I|M)$ is particularly interesting since it makes no reference to the data, and therefore this term can be assigned a value *prior* to making any measurements. Consequently, $p(I|M)$ is known as the image "prior". In maximum likelihood methods, the image prior is assumed to be a constant and the goodness-of-fit term would be the standard chi-square distribution $p(D|I,M) = \exp(-\chi^2/2)$. Typically, images restored by these methods show residual errors associated with the brightest sources and often produce spurious sources sprinkled across the field. In Maximum Entropy (ME) methods, the image prior is given by $p(I|M) = \exp(\alpha S)$ where S is called the **entropy** and is a measure of the degree to which the uncertainty in the information has been eliminated ($S = -\Sigma p_i \log p_i$ for $i = 1...n$, where p_i is the probability that the random variable X takes the value x_i) and α is "weighting" factor between the GOF and the image prior. Maximizing the entropy minimizes the information uncertainty. Since many of the spurious sources caused by pure goodness-of-fit methods are the result of "over-fitting" of the data (i.e. using too many free parameters), Maximum Entropy overcomes this by asserting that the best image prior is one in which the image is completely flat! This is both the strength and weakness of the MEM method. It is a strength because it tends to reduce the number of free parameters and the tendency to over-fit the data, but it is a weakness because a "flat" image is not the most likely image prior. A real (astronomical) image will have regions which are indeed very flat (e.g. the sky background), and regions which vary slowly in brightness (e.g. reflection nebulosity, elliptical galaxy starlight), and regions with significant structure on the finest spatial scales (e.g.

stars). In regions of the image devoid of any changing structure we could use much larger cell sizes (coarser resolution) to record our data because we are not losing any information, whereas we will need small cells to sample the finest structure in the image. Maximizing the image prior $p(I|M)$ could therefore be achieved by decreasing the total number of variable-sized cells used and packing as much signal as possible into each of these cells while still maintaining an adequate goodness of fit. This is the basis of the remarkably powerful new image reconstruction method developed by Rick Puetter and Robert Peña at UC San Diego known as "pixons"—the term they coined to describe the variable-sized cells. The pixon sizes follow the information density in the image—small when needed, large when not—and in fact each pixon represents a degree of freedom. Practical implications of the method require some choice of pixon shape, usually a radially symmetric cell with a "fuzzy" boundary, but the best shape should be related to the image properties. Pixons naturally eliminate signal-correlated residual errors and spurious sources because the technique eliminates all of the great many unconstrained degrees of freedom inherent in other approaches. The degrees of freedom represented by the pixons are in fact, the only ones constrained by the data. Excellent reviews of the Pixon technique with numerous examples are given in Puetter (1995, 1996). Fig. 11.6 shows examples of the Pixon

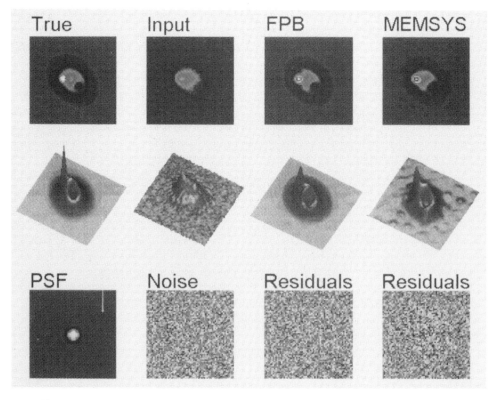

Fig. 11.6. An example of image restoration using the pixon method developed by Rick Puetter and Robert Peña at UCSD and a comparison with other approaches. Courtesy Rick Puetter.

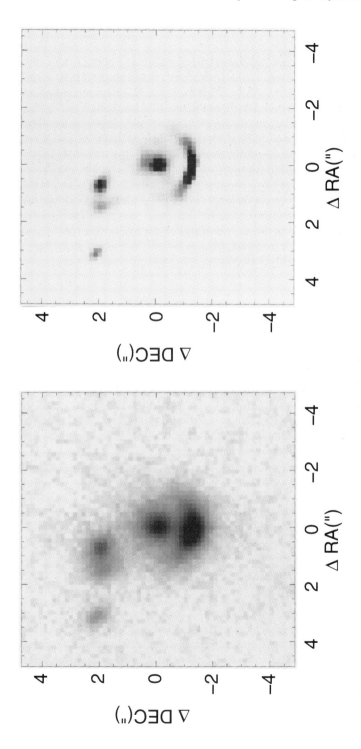

Fig. 11.7. An infrared image obtained with the Keck telescope under outstanding seeing conditions (above) which shows a peculiar arc near the galaxy FSC 10214+4724I. Image reconstruction using MEM (below) reveals that the arc is part of an "Einstein ring", the result of gravitational lensing of the light from a more distant object by the foreground galaxy. Courtesy James Graham.

method compared with others and Fig. 11.7 is an illustration which uses the Maximum Entropy approach.

The Richardson–Lucy iterative inversion scheme was first devised in 1972 by William Hadley Richardson and independently discovered for astronomical applications by Leon Lucy in 1974. It has been used extensively for Hubble Space Telescope image restoration and is available within the STSDAS/IRAF environment. Typical RL restorations require less computer time than Maximum Entropy methods, but more than simpler filtering. The Richardson–Lucy algorithm is a non-linear iterative algorithm based on a linear data gathering model which estimates the prior probability of the image value. Among the most important characteristics of the algorithm are the following:

- image flux is preserved giving the restored image good photometric linearity, even if the iteration is stopped before convergence
- the RL iteration converges to the maximum likelihood solution for Poisson statistics in the data (Schepp and Vardi, 1982) and is therefore suited to astronomical detectors with photon-counting detectors. Note that readout noise from CCDs is a complication
- non-negativity is preserved in the restored image provided that the point spread function and the original image are both non-negative. There is no tendency to "dig" into a zero background
- small errors in the PSF are not significant

Despite these advantages, the RL method suffers from the same generic problem of all maximum likelihood techniques which attempt to fit the data as closely as possible—noise amplification. After many iterations (>50) on an object with an extended light distribution such as a galaxy or nebula, the restored image develops a "speckled" appearance as a result of fitting the noise in the data too closely. Another problem is "ringing", the appearance of dark and light rings at sharp boundaries such as a bright star. Ringing is due to the blurring by the PSF and the loss of information due to the lack of higher spatial frequencies.

SUMMARY

Astronomers use computers for a wide range of purposes. Modern instrumentation is usually controlled by a combination of fast hardware and some form of digital signal processor at the front end and a powerful image display workstation or fast PC at the back end. Large software packages and environments such as IRAF, STARLINK and MIDAS, have been developed by the astronomical community for image analysis and data processing. Commercially available turnkey CCD camera systems, image-processing software, CD-ROMs and the World Wide Web have all had a tremendous impact on electronic imaging in astronomy. Image restoration techniques have been revitalized following the intensive efforts to help the Hubble Space Telescope prior to the successful refurbishment mission.

EXERCISES

1.   Calculate the data rate for a 1024×1024 pixel CCD camera system which digitized each pixel to 16 bits in a time interval of 100 microseconds per pixel. Compare this to an infrared camera system of the same detector format but a pixel rate of only 5 μs per pixel. Suppose the IR detector provided 32 simultaneous outputs, what would the data rate be then?

2.   Explain the difference between image enhancement and image restoration.

3.   Compare and contrast IRAF and IDL for analysis of CCD data.

REFERENCES

Eichhorn, G. (1994) "An overview of the Astrophysics Data System", *Experimental Astronomy*, **5**, 205.

Gull, S.F. and Skilling, J. (1991) *Quantified Maximum Entropy MemSys5 User's Manual*, Royston: Maximum Entropy Data Consultants.

Janes, K.A. and Heasely, J.N. (1987) "Stellar photometry software", *Pub. Astronomical Society of the Pacific*, **99**, 191–222.

Lucy, L.B. (1974) *Astronomical Journal*, **79**, 745.

Puetter, R.C. (1995) "Pixon-based multiresolution image reconstruction and the quantification of picture information content", *Journal of Imaging Systems & Technology*, **6**, 314–331.

Richardson, B.H. (1972) *J. Opt. Soc. Am.*, **62**, 55.

Schild, R.E. (1994) "The many hues of astronomical color imaging", *CCD Astronomy*, **1**, No. 2.

Schepp, L.A. and Vardi, Y. (1982) *IEEE Trans. Medical Imaging*, **MI-1**, 113.

Starlink Bulletin Issue No. 2 (1988), Rutherford Appleton Laboratory, UK.

Stetson, P.B. (1993) "DAOPHOT: A computer program for crowded-field stellar photometry", *Pub. of the Astronomical Society of the Pacific*, **105**, 527–537.

Warmels, R.H. (1991) "The ESO-MIDAS System", *Astronomical Data Analysis Software and Systems I*, PASP Conf. Series, Vol. 25, p. 115.

Wells, D.C., Greisen, E.W. and Harten, R.H. (1981) "FITS: A Flexible Image Transport System", *Astronomy & Astrophysics Supplement*, **363**.

SUGGESTIONS FOR FURTHER READING

Buil, C. (1991) *CCD Astronomy*, Willmann-Bell, Richmond, Virginia.

Computers and the Cosmos (1988) Time-Life Books.

Earnshaw, R.A. and Wiseman, N. (1993), *An Introductory Guide to Scientific Visualization*, Springer, Berlin.

Puetter, R.C. (1996) *Instrumentation for Large Telescopes*, J. M. Rodriguez-Espinosa (Ed.), Cambridge University Press, Cambridge, England.

12

Electronic imaging at other wavelengths

This chapter provides a brief review of the techniques used to obtain images at other wavelengths. Astronomy in the ultraviolet, X-ray and gamma-ray parts of the spectrum can only be carried out from above the Earth's atmosphere using satellites or sounding rockets. Again we come across the CCD, but for the most part, other imaging technologies are needed. At wavelengths longer than that of visible and infrared light the atmosphere provides "windows" to enable ground-based observations over a wide range of radio wavelengths. Most readers will have seen some remarkable and very detailed radio "pictures". How were these images obtained? As mentioned at the beginning of this book, radio astronomy is a vast and mature field which has always been closely linked with electronic and computer technology. Within the scope of this text, we can only include some of the more basic facts and innovations in this field.

12.1 ULTRAVIOLET

In the **ultraviolet** region of the spectrum the Earth's atmosphere is completely opaque due to absorption by ozone in the stratosphere for wavelengths shorter than 300 nm (3000 Å)—even for high altitude sites like Mauna Kea. True UV observations must be carried out from sounding rockets or spacecraft. Ultraviolet space programmes by several nations, in particular Copernicus, TD-1 and ANS laid the groundwork, but in 1978 the launch of the highly successful IUE (International Ultraviolet Explorer) satellite really opened up this vast field to all astronomers. Originally suggested by Bob Wilson as early as 1964, this remarkable NASA/ESA/UK satellite remains one of the longest-running space operations. At the time of writing (mid-1996) IUE continues to operate 24 hours a day, 7 days a week! The satellite, which is located in geosynchronous orbit 36 000 km (22 700 miles) from Earth, carries a telescope with a diameter of 45 cm (18 in) and is equipped with both high- and low-dispersion ultraviolet spectrographs covering the wavelength interval from 1250–3200 Å. With the launch of the Roentgen Satellite (ROSAT) in 1990 and the Extreme Ultraviolet Explorer (EUVE) in 1992, ultraviolet astronomy pushed its boundaries into the 100–1000 Å region. Several UV experiments have utilized the Space Shuttle, such as WUPPE (the Wisconsin Ultraviolet Photo-Polarimeter Experiment) and ORFEUS (Orbiting and Retrievable Far and Extreme Ultraviolet Spectrometer), and of course, the Hubble Space Telescope itself is a superb ultraviolet instrument.

The ultraviolet extends from approximately 70 to 3000 Å, but for practical reasons associated with the technologies used, it is often subdivided into four regions; the **extreme UV (EUV)** from 70 to 900 Å, the **far UV (FUV)** from 912 to 1216 Å, the **UV** from 1216 to 2000 Å and the **near UV (NUV)** from 2000 to 3000 Å. Because of absorption by hydrogen in the interstellar medium below 912 Å, and the difficulty of making telescopes, the EUV region is particularly difficult. Nevertheless, in its first survey the EUVE satellite detected over 400 sources and at least 20 extragalactic objects. Unlike the visible region however, where the silicon CCD has become the unquestioned detector of choice, there are numerous UV detector systems with none being ideally suited for all applications. CCDs and other photoconductive devices are used, but so too are a wide range of photoemissive detectors with various photocathodes tailored to suit certain wavelength ranges. Photoemissive detectors are inherently good UV, FUV and EUV detectors, but some form of intensification is almost always needed. Joseph (1995) has reviewed UV image sensors and associated technologies; see also Carruthers (1993, 1996) and Timothy (1991).

There are several difficulties that UV detectors must overcome. One of the major concerns is that the detector should be "solar blind", that is, it should not be sensitive to visible light photons. The reason for this restriction is simply that many astronomical sources emit 10^4 to 10^8 visible photons for every UV photon in the 1000 to 2000 Å wavelength range, and if the detector actually has its maximum sensitivity at visible light wavelengths, then there will be an enormous background unless the visible light is heavily filtered out. Unfortunately, even blocking to reduce visible light by a factor of 10^5 still isn't sufficient in many cases and such filters, for example the alkali metal or Wood's filter, also absorb UV photons. A CCD with a 20% QE in the ultraviolet would end up with a Detective Quantum Efficiency (DQE) of only 1–4% when it was made solar blind; the DQE takes into account all losses. Another issue arises from the fact that many UV scenes have most of the light concentrated in a few pixels and therefore the local dynamic range or contrast can be more important than the global dynamic range over the entire detector, i.e. the response to a flat field.

Detectors which are operated cold, such as CCDs, will form traces of condensible material on their surfaces, even in the vacuum environment of space, which will destroy UV performance. Using a warm window is an option, but only for wavelengths longer than 1200 Å, and even then the transmission is only about 60–70%.

12.1.1 CCDs as UV detectors
Silicon charge-coupled devices are capable of responding to photons over a huge wavelength range, from 1 to 10 000 Å. Frontside-illuminated CCDs are almost completely insensitive to wavelengths less than 4000 Å and even backside-illuminated CCDs can have severe UV sensitivity problems unless considerable effort and expense is devoted to backside surface control. The toughest wavelength for CCDs is around 2500 Å because the absorption depth is only about 30 Å—much less than the wavelength of the UV light—and so about 70% of the incident radiation is simply reflected! One solution to this problem is to apply a UV phosphor coating to the CCD. An example of such a coating is the organic phosphor, Lumogen, which is also used in yellow "highlighting" pens. This material absorbs all ultraviolet photons with wavelengths shorter than 4200 Å and re-emits the energy at yellow (longer) wavelengths around 5200 Å. This process is called "down converting".

A layer of lumogen about 6000 Å thick can be thermally vacuum deposited onto the CCD and will yield about 15% quantum efficiency from 500 to 4000 Å in wavelength; this is a very big improvement, but it is still only a 3% DQE after the device is rendered "solar blind". Lumogen is very inexpensive compared to backside treatments and can be applied to either the front or the backside of the CCD. Since it is transparent at longer wavelengths it causes no harm and actually improves the quantum efficiency a bit by acting as an anti-reflection coating. The TI 800 × 800 chips for the Hubble Space Telescope WFPC1 instrument were meant to be intrinsically UV-sensitive, but it turned out that they too needed a coating (coronene). Other coatings are also possible and some companies have proprietary coatings such as Metachrome II by Photometrics. Outgassing of the phosphor in sealed systems used in space without re-pumping is a potential problem, both because of contamination and because of changes in the coating's response.

Alternative UV CCD technologies include the use of very thin gate structures (known as "thin-poly" gates) to minimize absorption or the use of virtual phase technology which provides an "open" phase with no overlying gate metal at all. This is a good approach which requires no post-fabrication treatment whatsoever. For thinned CCDs, one can apply a backside charge to drive the photogenerated electrons towards the wells by either creating an electrostatic charge in the native oxide by soaking the CCD in a strong oxidizer such as nitric oxide gas or by use of a thin conducting bias gate on top of a thin insulating layer, or by very shallow ion implants to "build in" an electric field.

Due to the considerable progress on thinning, coating and QE-pinning of CCDs, and with readout noise around 1 e$^-$ rms, CCD detectors have become very important for UV and NUV astronomy, and future UV Space Observatories will almost certainly include UV-sensitive CCDs. Below 2000 Å however, the problems for CCDs are formidable.

12.1.2 Microchannel plates

There are a variety of observing modes, such as high dispersion spectroscopy in the violet and ultraviolet, for which the finite readout noise of currently available CCDs, the presently limited UV response, and the slower data access time of CCDs makes them less satisfactory than photon-counting detectors (or PCDs) employing good UV photoemissive photocathodes and some form of light intensification. One of the most widely used UV detectors is the **microchannel plate** (MCP) which is shown schematically in Fig. 12.1(a). MCPs are used for X-rays, EUV and for FUV/UV. An MCP is essentially a small, thin disk of lead-oxide glass with numerous microscopic channels running parallel to each other from one face of the disk to the other. When an electric potential is applied between the two faces, the MCP becomes an image intensifier. Each tiny channel or pore acts like a photomultiplier tube since electrons hitting the walls eject additional electrons resulting in a cascade of electrons. For one incident photon a charge cloud of 500 000 electrons can emerge from the channel and can therefore be detected or counted. The length of the microchannel is typically 50–100 times the diameter of the channel, which implies a large surface-to-volume ratio and the tendency to trap residual gas unless exceptional measures on cleanliness and plate conditioning are employed. Since MCPs are operated at potentials of a few kiloelectronvolts, residual gases can lead to destructive discharges. The channels have diameters ranging from 8 to 25 μm on 10–40 μm centres, and plates with active areas as large as 100 × 100 mm^2 are available. Note that the response of the MCP is a strong

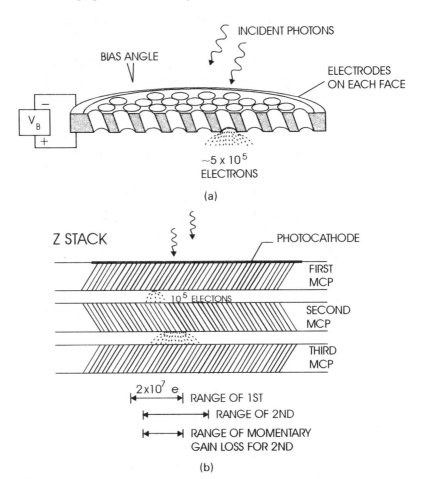

Fig. 12.1. (a) A schematic drawing showing a cross-section through of a microchannel plate (MCP). In practice the length of the channels is about 50–100 times longer than their diameter. Plate dimensions of 40–100 mm diameter are common with of order six million channels. (b) a Z-stack of microchannel plates. Courtesy Chuck Joseph.

function of the angle of incidence of the photons, which is not the case for a CCD detector. Photocathodes can be placed on the top face or on a window in proximity focus immediately above the MCP. Various anode structures or solid-state devices can be placed below the MCP to act as readouts. One possibility is to use a CCD to form a microchannel plate intensified CCD (ICCD). DQEs of 10–20% are routinely obtained with MCPs.

Each conversion between light and electrons causes a degradation in spatial resolution which can be overcome by employing photon counting and centroiding techniques. Photon-counting MCP detectors, which are inherently solar blind, currently have a DQE advantage over direct CCDs at wavelengths shorter than 2000 Å, but lack the dynamic range of a large-pixel CCD. The disadvantage of these photon-counting approaches is the framing rate of the CCD and the phosphor decay effects which can limit the local dynamic range to less than 5 counts/pixel/second. In the extreme UV this is not always a problem.

(a)

(b)

Fig. 12.2. The microchannel plate intensified CCD (MIC). (a) fully assembled (b) schematic. Courtesy John Fordham.

On-going development work on intensified CCD detectors is occurring at University College London and Imperial College (Fig. 12.2). John Fordham and colleagues are developing a large format Microchannel plate Intensified CCD or BIGMIC detector for the X-ray Multi-Mirror Mission (XMM), an ESA "Horizon 2000" project for launch in 1999. Another line of development, being pursued at the Goddard Space Flight Center, is to replace the CCD with a charge-injection device (CID) because the addressable readout capability of the CID can be used to enhance the local dynamic range in selected regions according to the brightness distribution in a given image. CIDs are MOS unit cells, but instead of being charge-coupled, each cell is individually addressable and each has an additional storage well to permit non-destructive reads. This chip architecture results in high capacitance and high noise compared to a CCD, but this disadvantage is eliminated when there is sufficient gain in the intensifier.

Microchannel plate detectors also use a variety of other anode structures. One of the simplest is a single resistive anode in which the location of the event is determined by the amount of charge or current "divided" between amplifiers attached to the corners. Other anode structures include the *wedge-and-strip* anode, the *spiral* anode and the *delay line*, each of which are described as "continuous" anodes. It is also possible to utilize "discrete" anode structures at the expense of many more amplifiers and encode the event location through direct detection. One such system is called the Multi-Anode Microchannel Array (MAMA) and another is the capacitive readout system called Coded Anode Converter (CODACON). A crossed grid system was used on the Einstein and ROSAT missions.

The wedge-and-strip anode (Fig. 12.3) is generally considered the workhorse because it is relatively inexpensive to fabricate—it is used on the Extreme Ultraviolet Explorer (EUVE) satellite for example—despite some fixed pattern noise. Delay-line anodes suffer

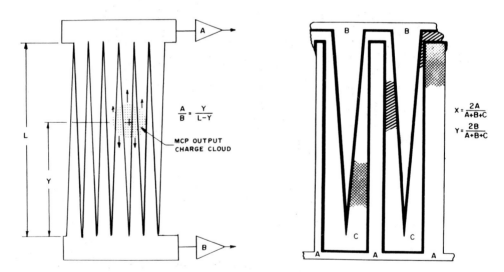

Fig. 12.3. Two readouts for a microchannel plate. Left is a partitioned anode readout and right is a wedge-and-strip readout. The location of the photoevent is determined by the ratio of the amplifier outputs. Only three amplifiers are needed for the wedge-and-strip readout. Courtesy George Carruthers.

much less from the fixed pattern noise caused by stretched and compressed mapping regions of the detector. In a delay line the charge cloud from the MCP strikes an anode structure and two pulses begin to propagate in opposite directions and the event locations is deduced from the difference in arrival times at the ends, in a manner similar to gas proportional counters. Several delay-line technologies have been developed by the Space Astrophysics Group at Berkeley. Two-dimensional planar delay lines on a flat substrate are being developed for the Far Ultraviolet Spectroscopic Explorer (FUSE) mission, and delay lines have been used on the ORFEUS (Orbiting and Retrievable Far and Extreme Ultraviolet Spectrometer) and SOHO (Solar and Heliospheric Observatory) satellites. The joint ESA/NASA SOHO satellite has 12 instruments on board covering a wide range of EUV and UV imagers, spectrographs and coronagraphs, and helioseismology and solar wind experiments. Large format delay lines are readily possible and are well-suited to instruments with large curved focal planes such as the Rowland circle spectrograph on FUSE.

All readout structures which require very large gain usually employ not one, but several MCP in a stack called a "Z-stack" (see Fig. 12.1(b)). A stack of three MCPs can provide

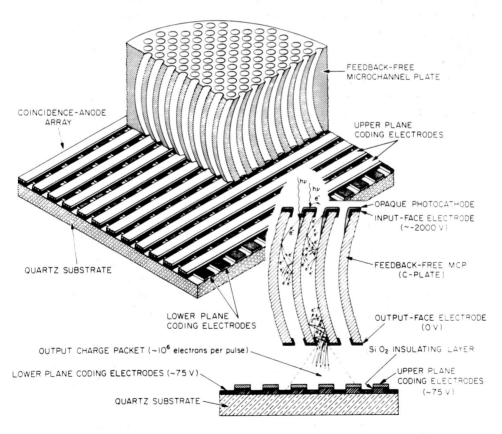

Fig. 12.4. The construction of a microchannel plate detector with a multi-anode microchannel array (MAMA) readout. Courtesy George Carruthers.

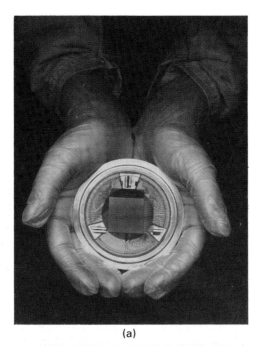

(a)

(b)

Fig. 12.5. (a) One of the MAMA detector constructed for STIS. This device has the equivalent of
1000 × 1000 pixels. (b) The MAMA readout after installation in the STIS Band 2 detector system.
Courtesy Ball Aerospace Corporation.

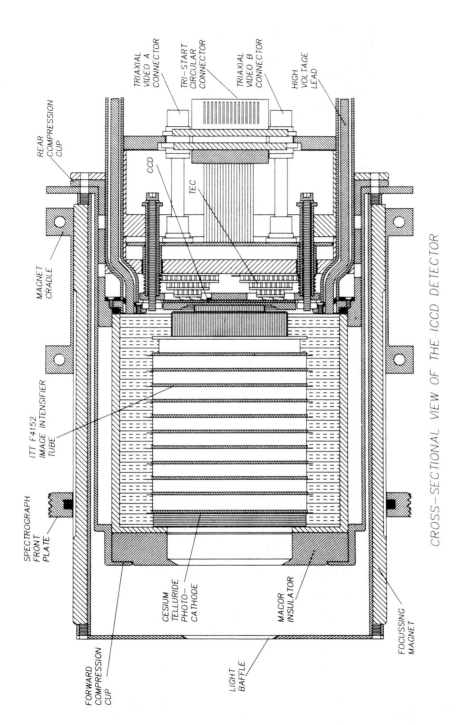

TRIAXIAL VIDEO A CONNECTOR

TRI-START CIRCULAR CONNECTOR

TRIAXIAL VIDEO B CONNECTOR

HIGH VOLTAGE LEAD

REAR COMPRESSION CUP

CCD

TEC

MAGNET CRADLE

ITT F4152 IMAGE INTENSIFIER TUBE

SPECTROGRAPH FRONT PLATE

CESIUM TELLURIDE PHOTO-CATHODE

MACOR INSULATOR

FOCUSSING MAGNET

FORWARD COMPRESSION CUP

LIGHT BAFFLE

CROSS-SECTIONAL VIEW OF THE ICCD DETECTOR

Fig. 12.6. An image-intensified CCD. Courtesy George Carruthers.

a gain of 2×10^7 electrons, which is needed for a Delay Line to give a spatial resolution of about 25 μm, suffers from "gain sag" because numerous MCP channels are activated by charge spreading to achieve this gain. Consequently, a second event arriving within the charge cloud radius and while some of the lower channels are still depleted of charge, will experience less gain. Whether this is a disadvantage or not depends on the application. For example, the Space Telescope Imaging Spectrograph (STIS) project selected the MAMA detector largely for this reason. STIS will fly two MAMA detectors, one with a CsI photo-cathode covering 1150–1750 Å and a second with Cs_2Te covering the 1650–3100 Å band. MAMA detectors are shown schematically in Fig. 12.4 and one of the devices developed by Ball Aerospace for STIS is shown in Fig. 12.5. See Benvenuti et al. (1996) for further details.

12.1.3 Electron-bombarded CCDs

An alternative approach to a UV imaging detector is to accelerate the electrons released from the UV-sensitive photocathode to high energies and allow them to impact a solid state detector. The Digicon detectors used in the Hubble Space Telescope's Faint Object Spectrograph (FOS) and Goddard High-Resolution Spectrograph (GHRS) are in this class. Their disadvantage is the bulky and heavy electromagnetic focussing systems. New technology magnet designs, the availability of CCDs instead of linear arrays of diodes and the fact that certain opaque photocathodes can actually have detective quantum efficien-cies of 70% in the far ultraviolet, has kept this approach competitive. The structure of an EBCCD which has flown on sounding rockets and on the ORFEUS orbital mission is shown in Fig. 12.6. Photoelectrons from the photocathode are accelerated to several thou-sand volts before bombarding the CCD. Each impact on the CCD produces one secondary electron for every 3.6 eV of incident energy.

12.1.4 Telescopes for EUV and X-ray imaging

Telescopes designed for wavelengths below 500 Å (EUV and X-rays) cannot work by reflecting photons from a curved mirror like an optical telescope; EUV and X-ray photons would be absorbed, not reflected! The solution is to use a "grazing incidence" telescope in which the photons are travelling almost parallel to the mirror because at sufficiently low

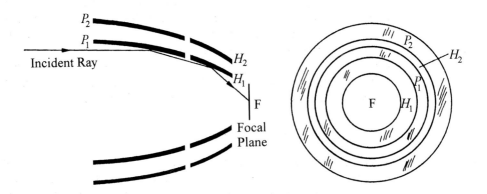

Fig. 12.7. The grazing incidence principle of UV and X-ray telescopes.

angles of incidence most photons will be reflected (Fig. 12.7). The EUVE satellite uses a series of metal grazing-incidence surfaces, whereas glass grazing-incidence optics are often used in X-ray telescopes. Culminating almost 30 years of effort to develop this field, the EUVE science payload was designed and built at the UC Berkeley Space Sciences Laboratory by a team led by Roger Malina. Stuart Bowyer is the EUVE Science Principal Investigator. The imaging detectors devised by Bowyer and Malina use MCP technology in the EUVE focal plane, the detectors produce a "picture" of 600×600 independent pixels. EUVE also has an entirely new kind of spectrometer which is based on gratings with variable spacing—conventional diffraction gratings have a constant spacing. A special grating ruling "engine" by developed by Hitachi was used to make the EUVE gratings. This approach gives high performance and reduced aberrations. The EUVE satellite supplies images in four wavelength regions across the whole EUV band and it also carries three EUV spectrometers.

Circling the Earth once every 96 minutes at an altitude of 530 km, the EUVE can locate sources to an accuracy of 10 arc seconds. For each source, 500–20 000 seconds of observations are accumulated during the time when the satellite is in the Earth's shadow and background EUV radiation from the Sun is minimized. EUVE actually has four telescopes; three point in identical directions, allowing a complete survey of the sky to be made, while the fourth—which is perpendicular to the other three—takes 6 months to scan the sky along the plane of the Earth's orbit. Each quarter of each telescope's focal plane contains different filters, so any source is observed simultaneously in different wavelength ranges. In the fourth telescope, the mirror is subdivided into six segments; three feed radiation into the spectrometers and three produce a single image. This telescope acts as a "deep survey" instrument because it is aligned parallel to the spin axis of the satellite and can therefore observe a given source for 30 minutes along its orbit.

Statistical algorithms are used to discriminate between real EUV sources and spurious background fluctuations. Many hundreds of sources have been identified, some unexpected. EUV emission was detected from hot B stars; helium was detected on Mars for the first time by spectroscopic observations of the 584 Å line; many extragalactic EUV sources were detected, mostly active galaxies of the BL Lac type. The majority of EUVE sources lie within a few hundred light years of the Sun and include such hot, young luminous stars as Eta Canis Majoris, white dwarf stars and cataclysmic variable stars like SS Cygni. Interstellar hydrogen atoms absorb EUV radiation so efficiently that if the density around the solar system was about 100 atoms per cubic centimetre then there would be enough absorption to limit our view to within 10 light years of the Sun. The fact that EUV stars are being discovered at all implies that the density is much lower. The solar neighbourhood may be in a low-density bubble, but EUVE's electronic imaging systems have revealed that there are "tunnels" through the neutral gas in some directions.

12.2 X-RAYS—FANO-NOISE-LIMITED CCDS

One of the major successes of the space era for astronomy has been the ability to study X-ray emissions from cosmic sources. Beginning with brief rocket flights in the late 1940s, X-ray astronomy has flourished through several small satellites in the early 1970s which produced the first all-sky surveys and revealed hundreds of previously unknown sources, including black-hole candidates. This pioneering work led to three large satellites called the High Energy Astrophysical Observatories (HEAOs) launched between 1977 and 1979 which contained several X-ray and gamma-ray experiments. Especially successful among the HEAO missions was the second satellite which was renamed the Einstein Observatory (1978–1981). A big advance came with the launch of ROSAT (in 1990) because this satellite carries an X-ray telescope and a wide-field X-ray camera.

Around a wavelength of 7–10 nm (or 70–100 Å), the extreme ultraviolet merges into the X-ray region which extends to wavelengths as short as 0.01 nm (0.1 Å) beyond which the spectrum is called gamma-rays. At such short wavelengths it becomes more convenient to describe the photon by its energy $E = h\nu = hc/\lambda$ and convert this to electronvolts. The range of energies is from 0.1 keV to 125 keV. Lower energies (< 10 keV) are called "soft" X-rays and the high energy end is called the "hard" X-ray region.

The European Photon Imaging Camera (EPIC) which will fly on board the European Space Agency X-ray Multi-Mirror (XMM) satellite is an X-ray imaging and spectroscopy instrument based on CCD technology. EPIC consists of three CCD cameras at the focus of three grazing-incidence telescopes allowing medium spatial resolution of about 30", but a point source location ability of 5–10", and a broad energy range from 0.1 keV to 10 keV with spectroscopic resolving power $E/\Delta E$ from 5 to 60. Two of the cameras contain a mosaic of seven 600×600 pixel devices and the other is a variation which uses 12 back-illuminated 64×200 pixel CCDs with pn junctions instead of MOS unit cells.

NASA's next major X-ray mission is the Advanced X-ray Astrophysics Facility (AXAF). Initially intended to be one of the "great" observatories with both advanced imaging and spectroscopic capability, AXAF has been scaled-back by budget cuts and will be optimized for X-ray imaging with about 1" resolution; the probable launch date is 1998. A proposed instrument for AXAF is the MIT/Penn State Advanced CCD Imaging Spectrometer (ACIS). This X-ray camera could generate high-resolution images of active galaxies, supernovas and quasars and simultaneously measure the energy of the detected X-rays.

The ACIS instrument will utilize CCDs to obtain images of astronomical sources in the X-ray region of the spectrum. The idea at these wavelengths is to provide images which are spectrally resolved, or in other words, X-ray photons of different energies are separately identified. The best-known form of X-ray imager is, of course, an emulsion film, but this is neither practical nor efficient for astronomy applications. Most approaches to X-ray imaging are via phosphor screens which emit visible light. Evaluation of existing CCDs has led to considerable insights in device design of future UV and X-ray CCDs. Fig. 12.8 shows the approximate efficiency with which photons are converted to electronic charges over the immense spectral range from 1 to 10 000 Å for a backside-illuminated and UV-flooded CCD. The measured data points agree closely with theoretical expectations.

The photoelectric absorption of a "soft" X-ray photon in silicon ejects an electron with energy $E-b$ where $E = h\nu$ is the energy of the X-ray and b is the energy with which it is bound to the atom (typically 1780 electronvolts or 1.78 keV). When free, this energetic electron produces a trail of electron–hole (e–h) pairs by collisions with orbital electrons. On average it takes 3.65 eV of energy to produce a single e–h pair in silicon. If all the energy of an X-ray photon was used to produce e–h pairs directly, there would be no statistical variation in the amount of charge generated by the X-ray event. However, a

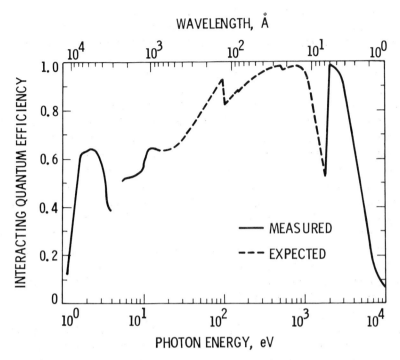

Fig. 12.8. The quantum efficiency of a thinned backside-illuminated CCD over the entire X-ray, ultraviolet, and optical region of the spectrum. Courtesy of James Janesick.

small but significant amount of energy is transferred into the silicon crystal lattice structure, giving a small statistical difference in the number of e–h pairs actually formed. This uncertainty is characterized by the "Fano factor". The Fano[1] factor (F) reflects the ultimate limit of detector energy resolution and relates the measured width or spread in energy (δE) and the average energy (E) needed to produce e–h pairs in silicon by

$$\delta E = 2.355\left(\frac{FE}{3.65}\right)^{\frac{1}{2}} \tag{12.1}$$

where δE is the full width at half maximum energy of the observed spread (in electrons), E is the photon energy in electronvolts and F is the Fano factor. The Fano factor is approximately 0.1 for silicon, and Fano-noise-limited CCDs are now a reality over the entire soft X-ray regime from 100 to 10 000 eV. To achieve this performance the CCD readout noise must be less than two electrons (rms) and any "fat zero" or "preflash" level needed to eliminate any form of charge-trapping must be less than one electron! Several CCD manufacturers now produce devices with such properties. Besides X-ray astronomy, there are several potential applications in high temperature plasma diagnostic instrumentation and

---

[1] The Fano factor was originally formulated by U. Fano in 1947 to describe the uncertainty of the number of ion pairs produced in a volume of gas following the absorption of ionizing radiation.

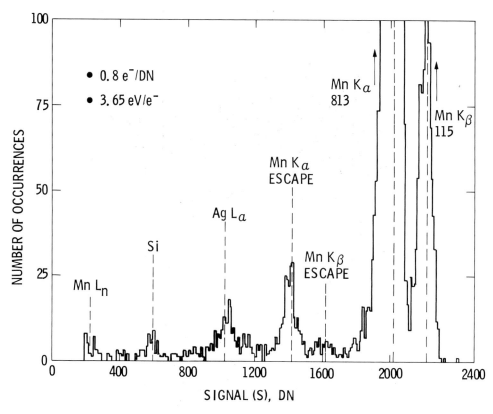

Fig. 12.9. A histogram of X-ray events recorded from a CCD camera revealing its ability to distinguish X-rays of different energies from different atomic species. Courtesy of Jim Janesick.

X-ray experiments on synchrotron sources. An example of the CCD acting as an X-ray detector is shown in Fig. 12.9 for a laboratory energy source of Fe^{55}—an isotope of iron. Notice that Ag (silver) and Mn (magnesium) are also detected in the source. Jim Janesick promoted the use of the X-ray sensitivity of CCDs both as an analyser of X-ray emission and as a diagnostic of CCD performance. In particular, X-rays can be used to determine the charge transfer efficiency of CCDs because a known charge is deposited in a pixel by the X-ray event. For example, the Fe^{55} source yields 5.9 keV X-ray photons (the K_α line) which deposit 1620 electrons each in the CCD. Several references to the technical literature are given at the end of this chapter, and two examples of the usefulness of X-rays for testing CCDs are shown in Fig 12.10.

12.3 GAMMA-RAY ASTRONOMY

The cosmic processes thought to be sufficiently energetic to produce gamma rays include the decay of radioactive nuclei, accelerated particles travelling in strong magnetic fields, electrons colliding with lower energy photons and boosting their energies to gamma ray levels, very energetic nuclei colliding with other nuclei, energetic electrons "braking" as

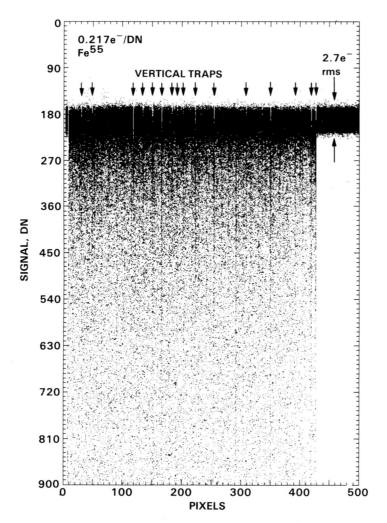

Fig. 12.10. (a) An Fe-55 X-ray line-trace response of a Lincoln CCD which exhibits Fano-noise-limited performance at the 2.7e- read-noise level; vertical single-pixel traps are easily distinguished in this type of trace.

they traverse matter since the deceleration of charged particles results in emission of radiation, and finally the annihilation of matter in the presence of antimatter. Localized sources of gamma rays include supernovas, quasars, the central regions of active galaxies, pulsars and neutron stars, so-called gamma-ray bursters and black holes.

GRO, the Gamma-Ray (or Compton) Observatory, launched in 1990 has a complement of four instruments: the Burst and Transient Source Experiment (BATSE), the Oriented Scintillation Spectrometer Experiment (OSSE), the Imaging Compton Telescope (COMPTEL), and the Energetic Gamma Ray Experiment Telescope (EGRET). These instruments are not like anything we have discussed so far, but they are nevertheless dependent on modern technology and developments in electronics and computers.

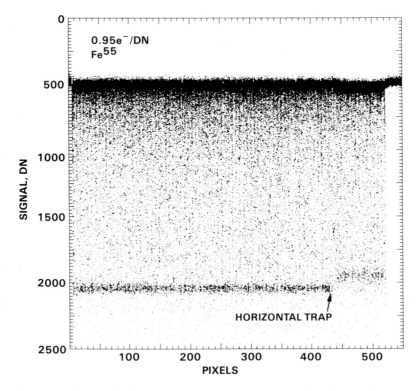

Fig. 12.10. (b) An Fe-55 X-ray line-trace response for a Ford CCD containing a small single-pixel-trap in its horizontal register. The trap displaces about 100e- into deferred charge. The dark line at about 2100e- is due to the 1620e-single-pixel events from the Fe-55 source. Partial events and events split between pixels result in the grey-scale gradation of signal values.

Size is crucial for gamma-ray astronomy because the number of gamma-ray events recorded is directly related to the mass of the detector material used to intercept them. Since the flux of gamma-rays is small, large instruments are needed to detect a significant number in a reasonable time. GRO detects photons with energies from the high end of the X-ray spectrum at 20 000 electronvolts (20 keV) up to a colossal 30 000 million electron-volts (30 GeV). Note that we have completely dispensed with using wavelengths since at photon energies above 10 000 electron volts we have gone below a wavelength of 1 angstrom (Å) unit and are now in a regime of sub-atomic dimensions!

As with X-rays, gamma-rays cannot be focused onto a detector in the usual way for optical astronomy, nor are there any gamma-ray "cameras" which we can use. Neverthe-less, by scanning the telescope/satellite combination across the sky in a systematic way it is possible to "map" the gamma-ray emissions from the whole sky. Each of the GRO instruments involve devices called **scintillators**. In the same way that fluorescent paint converts ultraviolet radiation to visible light (e.g. lumogen-coated CCDs), scintillators change gamma-rays to visible light. When gamma-rays interact with certain types of crys-tals and liquids they produce flashes of light (scintillations) which can be recorded by **photomultiplier tubes** (PMTs) very similar to the tubes which optical astronomers use for direct detection of starlight. In fact, spurious signals from time to time in a photoelectric photometer using poorly shielded PMTs might be due to high-energy particles or gamma-ray photons. Higher-energy gamma-rays produce brighter scintillations.

Since cosmic rays, which are electrically charged particles (not photons) travelling with very high energies, also cause scintillators to produce light, and these particles are much more numerous than gamma-rays, then we have a "background problem" very similar to trying to see normal stars during the daytime or infrared stars against the thermal (heat) emission from the Earth's atmosphere and the telescope. Almost all astronomical measurements must be made against some kind of unwanted background. To reduce the background of cosmic rays, each GRO instrument distinguishes gamma-rays by using a plastic outer detector called an "anti-coincidence counter" which detects cosmic rays and other charged particles by giving off a signal different from gamma-rays. Computer programs compare the signals from the main instrument and the anti-coincidence counter to identify true gamma-rays. The GRO instruments are summarized briefly below.

OSSE: Energy range 0.1–10 MeV. Four identical detectors with a main scintillation crystal made of sodium iodide (NaI) surrounded by other scintillators that absorb gamma radiation arriving from the sides and back. A very massive collimator in front defines a 3.8 × 11.4 degree field of view. Each detector can be rotated to point on and off a source.

BATSE: Energy range 20–600 keV. Observes the entire sky not blocked by the Earth in the field of view of its eight scintillation detectors and looks for changes in gamma-ray intensity in time intervals as short as a fraction of a millisecond.

COMPTEL: Energy range 1–30 MeV. Gamma-rays in this energy range are scattered by electrons in the liquid scintillators in COMPTEL's upper detectors. This process, called Compton scattering, occurs only between electrons and high-energy photons, not between electrons and other particles. The scattered gamma rays are then detected by COMPTEL's lower set of crystal scintillation detectors. COMPTEL has a 60-degree field of view.

EGRET: Energy range >10 MeV. When a gamma ray with energy greater than 10 MeV encounters the metal layers of a detector called a **spark chamber**, it produces an electron–positron pair. The track of the pair and the angle between them is related to the direction and energy of the original gamma-ray. Large electric fields between the plates cause breakdown of the inert gas in the chamber, but only along the track of the charged particles producing a trail of sparks which can be imaged. A massive scintillation detector below the spark chambers helps to determine the energy. EGRET has a 30-degree field of view.

Each instrument measures the energy of individual gamma-ray photons, but BATSE and OSSE can produce spectra of the number of photons at each energy. COMPTEL and EGRET collect spectral data too, but they also produce a map or image of the gamma-ray sky. Each instrument sends data from its own processor to the central onboard computer and finally to the ground station computers. Extensive data processing by computers on the ground is essential to remove all the systematic effects and background problems before the true gamma-ray signal can be known.

12.4 RADIO ASTRONOMY

In 1932, after about four years of work studying background "static" or noise in ship-to-shore communications at a radio wavelength of 15 metres, a young radio engineer named Karl Jansky working at the Bell Telephone Laboratories in Holmdel, New Jersey—the same laboratories from which would come the invention of the transistor, the CCD and the discovery of the cosmic microwave background—realized that a certain kind of radio noise developed a peak approximately once every 23 hours 56 minutes and seemed strongest when the constellation of Sagittarius is high in the sky. Since the centre of the Milky Way galaxy lies in the direction of Sagittarius, Jansky correctly concluded that he was detecting radio waves from outer space. Jansky's work went unnoticed by professional astronomers, but not by an engineer in Illinois, named Grote Reber. During the period 1936–1944 Reber completed a map of the radio emission from the Milky Way using a "backyard" concave dish 9.1 m (about 30 ft) in diameter and "tuned" to a wavelength of 1.87 m. The antennas used by Jansky and Reber are on display at the National Radio Observatory in Green Bank, West Virginia. Of course, the development of radar during World War II (1939–1945) stimulated the technology and very soon after "radio observatories" began to appear all over the world.

Reber's radio dish may seem quite large by comparison with "backyard" optical telescopes—Jansky used a large rotating assembly of aerials rather than a dish—but despite their physical bulk, the angular resolution of these telescopes was worse than the human eye. Since radio wavelengths are approximately the same length as the diameter of the telescope, angular resolution is controlled by the wave phenomenon of diffraction. The Rayleigh criterion gives

$$\theta = 1.22 \frac{\lambda}{D} \approx 70° \frac{\lambda}{D} \tag{12.2}$$

so that for a wavelength of 1 m and a telescope diameter of $D = 10$ m, the angular resolution is only 7° on the sky. At a wavelength of 1 mm the resolution of a 10-m radio telescope improves to 25″ provided the surface of the dish is smooth enough.

Initially, radio astronomy began with equipment for detecting electromagnetic waves with wavelengths of about 1 metre. Gradually, stimulated by the huge military and civilian demand for communications, receiving equipment improved and observations were extended to the centimetre band and most recently to millimetre and sub-millimetre wavelengths. Now a rapidly growing area of astronomy, the sub-millimetre has until relatively recently lagged behind mainstream radio astronomy in general, largely because of lack of technology development outside of astronomy. For example, the communications industry abandoned millimetre waveguides in favour of fibre optics in the seventies. The immense importance of this waveband lies in the fact that numerous molecules have strong emission lines in this region of the spectrum, making it ideal for mapping the molecular gas clouds from which stars form. Actually, the "window" for ground-based radio observations is quite large. For wavelengths less than 2 cm atmospheric water vapour begins to attenuate radio signals and high altitude sites are obligatory. On the other hand, wavelengths longer than 10–20 m suffer absorption and scattering in the Earth's ionosphere. Since the radio waveband is used extensively for a wide range of communication purposes, it has become

necessary to regulate their use and certain wavebands are allocated purely for radio astronomy; a list can be obtained from the Federal Communications Commissions or from the
National Radio Astronomy Observatories. For example, the band from 1400 to 1427 MHz
includes the 21 cm line of hydrogen. This does not prevent people from using these bands
as we discovered while trying to use a 6-ft dish on the roof of our astronomy building to
detect 21 cm emission from hydrogen clouds near the Galactic Centre!

It is customary in radio astronomy to use frequency (ν) rather than wavelength (λ), but
the two are of course easily interchanged using the relationship

$$\nu\lambda = c \equiv 2.9979 \times 10^5 \text{ km/s} \tag{12.3}$$

where c is the speed of light. The metre waveband corresponds to frequencies below 300
megahertz (MHz), the band from 1–10 cm corresponds to frequencies from 3 to 30 giga-
hertz (GHz), and the millimetre and sub-millimetre waveband corresponds to frequencies
above 300 GHz.

12.4.1 Antennas, receivers and noise

Radio wavelengths are about one million times larger than optical/near-infrared wavelengths where, as we have seen, electronic imaging of astronomical sources can be obtained directly using CCD and infrared array detectors comprising of a two-dimensional
grid of pixels, each one of which absorbs photons and converts the energy to electrical
charge. Since this detection process completely ignores the wave nature of light, it is said
to be "incoherent". At sub-millimetre wavelengths it is possible to use discrete bolometers
arranged in a tightly stacked array for imaging. This principle is employed in a monolithic
silicon bolometer array developed for the Caltech Submillimeter Observatory (CSO) on
Mauna Kea. The array has 24 pixels operating in the 350 and 450 µm atmospheric windows and is cooled to 300 mK (0.3 degrees above absolute zero) by a single shot ^{3}He
refrigerator. An even more ambitious system has been developed for the 15-m James
Clerk Maxwell Telescope called SCUBA, the Sub-millimetre Common-User Bolometer
Array. SCUBA is a sub-millimetre camera and photometer. It has two arrays of detectors,
one of 37 pixels optimized for 850 µm and one of 91 pixels optimized for 450 µm. Both
arrays view the same area of sky simultaneously and provide a field of view of approximately 2.3 arcminutes. Each pixel is diffraction limited, which gives 6" at 450 µm and 12"
at 850 µm. There is a filter mechanism to allow the 850 µm array to be used at 750 or 600
µm and the 450 µm array can be used at 350 µm. In addition to the arrays, there are three
separate pixels individually optimized for 1100, 1400 and 2000 µm. Fig. 12.11 shows one
of the bolometer arrays. The focal plane is slightly undersampled because of the use of
detector feed horns. Dithering the telescope between exposures allows the observer to
make a fully sampled image. The bolometer arrays are cooled to 0.1 K. For longer wavelengths and for radio spectral analysis work, other methods are required.

The detection and measurement of radio signals is fundamentally different from that at
shorter wavelengths because it relies primarily on **coherent** processes in which devices
respond directly to the electric field strength in the wave and preserve phase information.
Coherent detectors work by **interference** of two electric fields, that of the incoming wave
and the other from a man-made source called a **local oscillator**. Radio receivers which
function by mixing signals of different frequency are called **heterodyne** receivers. As is

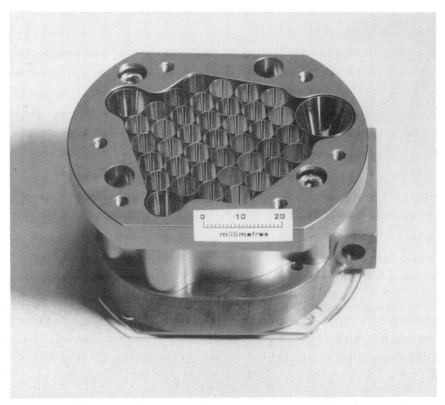

Fig. 12.11. The distribution of silicon bolometers and their Winston cones in one channel of the JCMT Submillimetre Continuum Bolometer Array (SCUBA) instrument. Courtesy Royal Observatory Edinburgh.

well-known from elementary physics, if two signals of different but similar frequencies are added they produce a signal at the "beat" frequency, which is the difference between the two original frequencies and is therefore a much lower frequency. While the resulting mixed signal contains frequencies only from the original two signals, its amplitude is modulated at the difference, or beat frequency. Heterodyne receivers measure this amplitude. The mixed field must be passed through a non-linear circuit element or **mixer** that converts power from the original frequencies to the beat frequency. This element is usually a diode. A typical receiver configuration is shown in Fig. 12.12.

The power (P) in an electromagnetic wave is proportional to the square of the electric field strength (E^2) and the amplitude of the signal voltage (V) is proportional to E. Therefore the power is proportional to V^2 and the ideal characteristic of a mixer would give an output current which is proportional to input power, i.e. I proportional to V^2. Such a device is called a "square law" mixer. In a practical diode mixer, the I–V curve can be represented by a dc term plus a linear term (which gives no net response) and higher-order terms which are quadratic and cubic in the voltage difference ΔV. The output of the mixer is a signal with a frequency of $v_{if} = \pm (v - v_{lo})$, where v_{lo} is the frequency of the local oscillator, which is passed to an **intermediate frequency** (IF) amplifier and then to a rectifying and

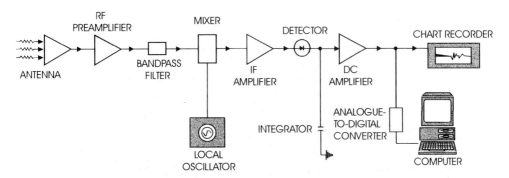

Fig. 12.12. Basic layout of a heterodyne radio detection system showing antenna, mixer, local oscillator, IF amplifier and detector.

smoothing circuit called a "detector". This terminology seems quite strange compared to our previous usage where the detector is the device that receives the photons and converts them to an electrical signal. Rather than a single detector, a "backend" spectrometer often receives the output of the IF amplifier. The spectrometer could consist of a number of electrical "filters" tuned to different frequencies with detectors on their outputs, a digital correlator, or an acousto-optic spectrometer (AOS) such as a Bragg cell, where the frequencies are converted to ultrasonic waves that disperse a monochromatic light beam (e.g. a laser) onto an array of visible light detectors. The principle of the acousto-optic spectrometer is illustrated in Fig. 12.13. The IF signal is applied to a piezo-optic device which sets up an ultrasonic wave in a lithium niobate crystal illuminated by a helium–neon (He–Ne) laser beam. If the IF signal is monochromatic then all the energy is diffracted into one pixel of the CCD or diode array. For a more complex distribution of frequencies in the IF signal the intensity pattern spread over the diodes give the IF spectrum.

Measuring the energy in a radio wave is achieved using a directional **antenna** or **aerial**. The power received at a unit surface element per unit frequency (or wavelength) interval is called the **flux density** S_ν (or S_λ) and is usually measured in W m^{-2} Hz^{-1} (or W m^{-3}); an alternative terminology is spectral irradiance. Radio astronomers also use a flux unit called the **jansky**; 1 Jy = 1 flux unit = 10^{-26} W m^{-2} Hz^{-1}. In the 1940s, a sensitivity of a few jansky was considered good. Nowadays, signal strengths are measured in millionths of a jansky (μJy)! The brightness of the source L_ν is the power arriving per unit frequency interval per unit area of the telescope per unit solid angle and the total power (P) collected by the antenna is given by

$$P = L_\nu A\Omega\Delta\nu \tag{12.4}$$

where A is the collecting area, Ω is the solid angle and $\Delta\nu$ is the bandwidth. For a radio telescope, the A-Omega product or étendue is given by $A\Omega = \lambda^2$ because of diffraction. By using the Rayleigh–Jeans law, any source can be assigned a **brightness temperature** (T_B) given by

$$T_B = \frac{L_\nu \lambda^2}{2k} = \frac{P}{2k\Delta\nu} \tag{12.5}$$

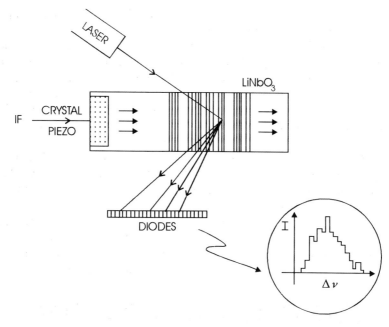

Fig. 12.13. The principle of the acousto-optic spectrometer. An ultrasonic wave in a lithium niobate crystal illuminated by a helium-neon (He-Ne) laser beam diffracts energy into the pixels of a CCD or diode array. The distribution of frequencies in the IF signal is given by the intensity pattern on the diodes.

where k is Boltzmann's constant and λ is the observed wavelength. Electrical currents set up in the antenna due to the passage of the radio wave are fed into a **receiver** which is "tuned" to the frequency (or wavelength) of interest and which amplifies the signal. The receiver system is periodically switched back and forth between the antenna and a stabilized calibration source for comparison and the output signal can be recorded on magnetic tape in analogue form or digitized for storage and processing by computers.

The simplest form of antenna is a half-wave **dipole**, which consists of two aligned metal rods (usually copper) with a small separation between them. Each rod has a length equal to one quarter of the wavelength to be detected and the signal currents are taken by a cable "feed" from each of the closest ends of the pair to the receiver. Although it may seem rather pointless to use a dipole antenna, since it is sensitive to only a rather narrow range of frequencies, some specific frequencies are of great interest such as the 1427 MHz (21 cm) line of hydrogen which has enabled astronomers to map the structure of the Milky Way. The directional sensitivity of a dipole antenna can be illustrated in a **polar diagram** plot which reveals that the dipole is most sensitive when the direction of propagation of the wave is at right angles to the rod and least sensitive to radio waves arriving along the long axis of the antenna (since the electric field in this case cannot drive currents down the length of the rod). In general, the dipole has rather poor directional sensitivity. By adding extra rods which are coplanar with the dipole (usually several in front and one behind) but not actually electrically connected to the dipole or the receiver, and hence are therefore said to be "parasitic", a more directional antenna called a **parasitic antenna** or Yagi is obtained. Apart from the disadvantage caused by the "side lobes" seen in the polar diagram—which imply that strong radio sources not along the main beam will be detected and lead to erroneous fluxes for the object being observed—these antenna are not sensitive enough for radio astronomy. The answer lies in combining many of them into either co-linear (end-to-end) arrays or broadside (parallel formation with half-wavelength separation) arrays. By mounting a combination of both of these arrays on an adjustable platform, and placing a suitable wire mesh reflector to eliminate the backward lobe of the polar diagram, a highly directional antenna with excellent gain can be achieved.

An obvious disadvantage of the directional antennas just described is the fact that the wavelength is at once fixed by the choice of the length of dipole. A much more versatile approach is to use a large parabolic collecting dish to bring the radio waves to a focus where a (selectable) dipole or better, a collecting horn and waveguide, system can be placed. As mentioned previously, to obtain an angular resolution on the sky of only 7° requires that the ratio of the diameter of the dish to the observing wavelength must be at least 10:1. Therefore the world's largest steerable parabolic dishes are very large, for example, the Mark I radio telescope at Jodrell Bank in England has a diameter of 76.2 m (250 ft) and the Effelsberg radio telescope (Fig. 12.14) 40 km south of Bonn in Germany,

Fig. 12.14. The Effelsberg radio telescope. This is the largest steerable dish radio telescope. Courtesy Effelsberg Radio Observatory.

which has a high-quality polished surface of metal plates, is 100 m in diameter and typi-cally works at wavelengths around 6 cm (4996 MHz). At shorter wavelengths the surface accuracy degrades the collecting efficiency. The IRAM Millimetre Radiotelescope at Pico Veleta in Spain is a carbon fibre structure 30 m in diameter with panels machined to an average surface precision of 100 μm, which permits diffraction-limited performance at a wavelength of about 2 mm. Significantly better surface accuracy (30 μm) is obtained by the 15-m James Clerk Maxwell Telescope on Mauna Kea, Hawaii, and better still (15 μm) with the new 10-m dish and panels of the Heinrich Hertz Sub-Millimetre Telescope Ob-servatory on Mount Graham, Arizona—a joint project between the University of Arizona and the Max Planck Institute for Radio Astronomy in Bonn. Made from carbon-fibre rein-forced plastic that is 20 times less sensitive to thermal change than polished metal panels, this dish permits the telescope to work at sub-millimetre wavelengths as short as 0.350 mm. The largest non-steerable radio telescope is the 300-m dish at Arecibo, Puerto Rico.

In the "down conversion" of frequency by the local oscillator and mixer already de-scribed, an ambiguity exists in the sign of the difference signal. It is not possible to tell whether the true frequency was larger or smaller than the local oscillator frequency. Since the IF signal can arise from a combination of two possible inputs it is called a "double sideband" or DSB signal. This is a serious problem for observations of spectral lines at radio frequencies and therefore it is usually desirable to operate in a "single sideband" or SSB configuration if possible by using a narrow bandpass rejection filter in front of the receiver or by tuning the mixer.

In practice, even if there are no sources in the beam, a non-zero current will be mea-sured. This signal arises from various causes such as residual thermal radio emission from the atmosphere, telescope or waveguide; thermal radio emission from the ground detected in the sidelobe pattern of the antenna; thermal noise in the detector itself e.g. fluctuations in the tunnelling current in a gallium arsenide (GaAs) diode. This noise is equivalent to a small amount of power and can be given an effective "noise temperature" which repre-sents the background power against which the signal is to be detected. In the best case the limit is imposed by the detector itself and cannot be lower than the physical temperature of the device. Noise temperatures in the range 10–200 K are typical, with 50 K being typical at wavelengths of 6 to 21 cm.

Any antenna can be considered to have an effective electrical resistance which gener-ates fluctuations or noise due to thermal agitation of the conduction electrons. The power (P) generated by these fluctuations is given by

$$P = kT\Delta v \tag{12.6}$$

where k is Boltzmann's constant, T is the absolute temperature of the resistor and Δv is the **bandwidth** or range of frequencies over which the equipment can respond; typically this would be the bandwidth of the IF amplifier Δv_{IF}. By measuring the power associated with the antenna noise in the relevant bandwidth, radio receiver systems are often characterized by an effective or **noise temperature** (T_N), which of course has nothing to do with the physical temperature of the device. If the receiver operates in the double sideband (DSB) mode then the bandwidth for the noise measurement is generally $\Delta v = 2\Delta v_{IF}$. In most receivers the non-linear element is a significant noise source and amplification of the sig-nal or the IF signal is needed. Various types of amplifiers are possible. Similarly, the power received from a source can be equated with that from a resistor at the appropriate temperature thus providing an equivalent "antenna" temperature (T_A); T_A is related to T_B by integration over the antenna gain pattern.

Parametric amplifiers are essentially RLC circuits with a resistor (R), inductor (L) and capacitor (C). Typical noise temperatures for a cooled parametric amplifier are in the range 10–20 K. Maser amplifiers are based on the use of an external power source to fill the excited electron energy levels of a solid or a gas above the normal population levels expected when the material is in thermal equilibrium. The stimulated emission produced by the incident wave is in phase with it and in the same direction and therefore we have amplification. One common form of maser is the ruby maser which is a single crystal of Al_2O_3 with 0.05% of the Al^{3+} ions replaced by chromium ions Cr^{3+}. Typical noise temperatures of these amplifiers are 15–50 K.

The function of the local oscillator is to provide a coherent, noise-free signal at a frequency very close to the signal frequency, and it must provide sufficient power. Local oscillators employ either klystrons, which is a wave generator for frequencies below 180 GHz which was developed mainly for telecommunications and radar; or the carcinotron, a wave generator which works up to 1000 GHz but is relatively heavy and power-consuming; harmonic generators are frequency multipliers which use a non-linear element like a Schottky diode which radiates harmonics at twice, three times, etc., the frequency when pumped by a klystron. The power obtained reaches several milliwatt up to about 500 GHz and this is sufficient for modern SIS mixers.

12.4.2 Detectors

No quantum detector of the kind we have discussed repeatedly in this book which uses the liberation of a (photo)electron by an incident photon can work at wavelengths longer than about 0.2 mm. Across the entire radio spectrum, the electromagnetic field or the current which it induces in an antenna is applied to a non-linear element (diode) or mixer. The mixer either measures the total power or changes the signal frequency to one which is more easily measured. One of the most frequently used devices is the Schottky diode, which is the same principle as used in the PtSi infrared arrays mentioned in Chapter 8, and another very exciting development, especially at millimetre wavelengths, is the superconducting junction. We will consider each in turn.

When a metal and a semiconductor are brought into contact, the majority charge carriers of the semiconductor leave the contact zone until the Fermi levels of the metal and semiconductor are equalized. As a result, a "barrier" or "depletion region" empty of majority carriers appears in the semiconductor; typical barrier widths are ~0.3 µm. Even without any voltage across the junction a current can flow through it. In practice the semiconductor used is heavily doped gallium arsenide (GaAs) with a very thin lightly doped layer on the surface to reduce quantum mechanical tunneling and a contact layer of metal on top. Cooling the detector to a temperature of about 20 K yields a low noise detector. Above 300 GHz (wavelengths shorter than 1 mm) the capacitance of the diode creates an RC filter which reduces its response.

There has been considerable excitement in the scientific literature in recent years about high-temperature **superconductors**, but other superconductors have made a big impact on detectors for radio astronomy. When a thin insulating barrier is created between a normal

metal and a superconducting metal (note: *not* a semiconductor) or between two superconductors, the structures are called SIN and **SIS junctions**. Non-linear current can flow by
quantum-mechanical tunnelling and hence the device can be used as a mixer. Fig. 12.15
shows the basic structure of the energy levels on each side of the insulator. When the
voltage V is large enough that occupied states on one side are opposite vacant states on the
other side then a tunnelling current can flow. Conversely, no current flows if $eV < 2\Delta$.
Absorption of a photon can excite a charge carrier to the energy where the tunnel effect
occurs. Part (b) of the figure shows the current–voltage behaviour without illumination
(solid line) and with photons present (dashed line). The high-frequency limit is about the
same as the Schottky diode. The Nobeyama Radio Observatory is developing a focal plane
array of 5×5 SIS receivers.

Finally, in very pure (low-doped) InSb at liquid helium temperatures, a fraction of the
conduction electrons stay free and interact only weakly with the crystalline lattice. The
absorption of photons by the electrons raises their temperature and rapidly changes the
resistance of the material. Since this change in conductivity is proportional to incident
energy which in turn is proportional to the square of the electric field in the wave, we again
have a mixer. This device is known as the **hot electron bolometer**. Its bandpass is limited
to about 1 MHz, but it can work at frequencies up to at least 500 GHz.

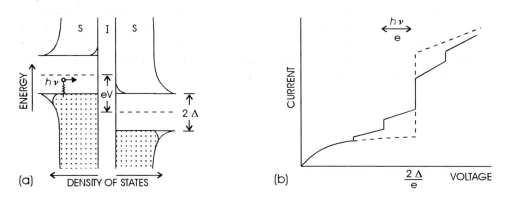

Fig. 12.15. (a) shows the basic structure of the energy levels in an SIS detector. No current flows
if $eV < 2\Delta$. Absorption of a photon can excite a charge carrier to the energy where the tunnel effect
occurs; (b) shows the current–voltage behaviour without illumination (solid line) and with
photons present (dashed line).

12.4.3 Interferometers and aperture synthesis

Radio astronomy faces two serious hurdles. First, the antenna plus detector is a
"single-pixel" device which makes imaging difficult and secondly, because of the much
larger wavelengths involved, high angular resolution can only be achieved by making very
large antenna complexes or very large single dish-type collectors so that the ratio of
wavelength-to-telescope diameter again becomes small. Nevertheless, even the largest radio dish in existence cannot come close to the resolution of the best optical telescopes.

To solve this problem, radio astronomers introduced the technique of **interferometry**
in the late 1940s. In this approach, radio signals are carried over wires between two

well-separated radio telescopes and made to "interfere" or mix together in such a way as to provide the angular resolution (though not the collecting area) of a huge telescope with a diameter equal to the separation of the two smaller telescopes. The concept is the same as the famous 1920s experiment carried out by Michelson with an optical interferometer having variable spacings between the pair of telescopes to measure the diameter of a star. Fig. 12.16 illustrates the principle. A plane wavefront (from a very distant source) arrives at an angle θ to the vertical and intercepts telescope A. At that moment it has an extra distance d to travel before reaching telescope B and therefore there is a "phase lag" between the two detections. When the two signals are combined in the receiver they will interfere according to the phase lag. If the distance d is an exact multiple of the wavelength, $d = n\lambda$ then the waves will be in phase and interfere constructively to reinforce each other; if d is an integer number of half wavelengths then the waves will be completely out of phase and cancel each other. If D is the separation of the radio telescopes, then

$$\sin \theta = n \frac{\lambda}{D}$$ (12.7)

and a large signal is recorded each time a source moving across the sky above a fixed interferometer satisfies the above constraint. Small diameter sources such as the Sun, Cygnus A and Cassiopeia A respond to the interference pattern of an interferometer pair, whereas the diffuse galactic background does not. The resolution is now $\sim\lambda/D$ and is controlled by the separation of the telescopes. Radio telescopes separated by entire continents have now been linked together successfully in a technique known as very-long-baseline interferometry or **VLBI**. In principle, two radio telescopes on opposite sides of the Earth could resolve celestial sources as small as 0.00001 arcseconds apart!

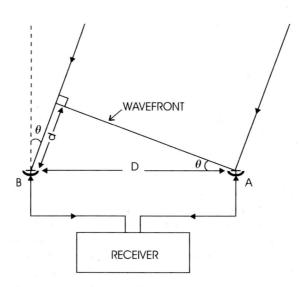

Fig. 12.16. The principle of radio interferometry.

Radio-linked rather than cable-linked interferometers were developed at both Jodrell Bank, England, and in Australia. Among the pioneers were Hanbury Brown, Mills and Palmer. Their success led eventually to the isolation of a class of very compact, star-like (quasi-stellar) radio sources, the quasars. Synthesizing the brightness distribution from interference fringes alone cannot be done without ambiguity. Problems of source identification arise because of the strong secondary side-lobes in the simple two-telescope arrangement mentioned above. By using multi-element interferometers *and* by combining two interferometers at right angles, with special electronic switching and phase shifting techniques, it is possible to create a system with a single narrow beam. This is the principle of the original Mills Cross built near Sydney, Australia. This design is restricted in wavelength coverage and area of sky.

Another approach to high angular resolution with radio telescopes is the method known as **aperture synthesis** which requires a system of two radio telescopes, one at a fixed location and the other "movable" on a railroad track over a large range in positions. As conceived by Martin Ryle and his colleagues, the idea was to observe all the interferometer positional pairs which existed within the equivalent large aperture radio telescope, to effectively recreate or synthesize a uniformly filled aperture. The original "Tee synthesis" telescope at the Mullard Observatory, Cambridge, England, had a multi-element interferometer along an E–W line as the fixed antenna and a small single-dish telescope which moved along a N–S line. The intention was to sample the whole range of spacings; radio astronomers call this "filling the u-v plane". A clever step was to allow the Earth's rotation to sweep out a range of spacings without moving either pair of detectors! Other telescopes began to follow, such as the Westerbork synthesis radio telescope. By 1980 one of the finest systems of radio telescopes in the world began operating in the desert near Socorro, New Mexico. Called the Very Large Array (VLA), it consists of 27 concave dishes, each 26 m (85 ft) in diameter. The 27 telescopes are arranged along the arms of a gigantic Y-shape covering an area 27 km (17 miles) in diameter.

It soon became clear that it was not always going to be possible to devote the time to filling-in the u-v plane. A significant breakthrough was to employ a computational technique called the CLEAN process which was introduced by Jan Hogbom. The principle goes like this. An ideal point source observed by an unfilled aperture (not enough interferometer pairs) will appear as a point of radio emission surrounded by large and widespread sidelobes. If it is the only point source, the pattern will match the theoretical pattern of the unfilled aperture, therefore subtract the ideal pattern and look at what is left. If another weaker radio source is present, then repeat the process. The result will be a list of point sources and a residual map of random noise. The key to the CLEAN process is that for the majority of radio sources, most of the u-v map and consequently the field of view on the sky is *empty* of sources.

When the CLEAN algorithm started running on the radio telescope computers it worked better than anyone expected, even if the source was somewhat extended! The VLA immediately began to produce maps (that is, radio "images") as though it had sampled the u-v plane fully. Finally, the method of "adaptive calibration" was introduced by Readhead and Wilkinson in 1978 in which the amplitude and phase of a diverse set of telescopes was corrected by examining the response to a point source and iterating until it matched the theoretically expected value. Fig. 12.17 is a remarkable radio image of Cygnus A obtained with the VLA. See Verschuur (1987) for more details.

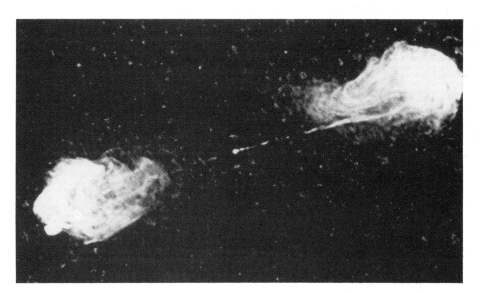

Fig. 12.17. A radio "image" of Cygnus A obtained with the VLA. The double-lobes of radio emission have no optical counterpart. The emission is caused by energetic jets emitted from the peculiar or active galaxy in the centre, which is visible. Courtesy NRAO.

As F. Graham Smith, himself a pioneer in these fields, explains, these various threads of development have now been drawn together in the many high-resolution mapping telescopes, which form a sequence roughly as follows: UK 5km, Westerbork, VLA, MERLIN (UK), VLBI in Europe and the USA, world-wide VLBI. There is a break between MERLIN and VLBI which corresponds to the break between connected elements and the use of tape recorders with accurate time signals imprinted on the recorded radio data. MERLIN has five telescopes outside of Jodrell Bank which provide a maximum baseline of 134 km, and improvements will increase this to 200 km and for some studies will gain in angular resolution over the VLA by a factor of seven. Correlating all of the interferometer signals in a multi-baseline system like MERLIN is very difficult. The system uses 1200 specially designed "correlator chips" and 37 Motorola 68000 microprocessors.

Aperture synthesis arrays are now extending to the sub-millimetre band also, with the first being the Smithsonian Submillimeter Array consisting of six 6-m telescopes on Mauna Kea. Earlier, the JCMT and the CSO on Mauna Kea were linked to form the first sub-millimetre interferometer. These telescopes are separated by 164 m which yields a fringe-spacing of 1.1″ at 345 GHz.

With the introduction of aperture synthesis and very long base-line interferometry, together with antenna and receiver developments and the advent of very fast computers, radio astronomers have been able to completely overcome the initial drawbacks to imaging mentioned earlier and indeed, they have been able to "map" areas of the sky with even higher angular resolution than any optical ground-based telescope.

12.4.4 The microwave background
We close out this look at electronic imaging at other wavelengths with a brief review of the study of the **cosmic microwave background** (CMB) and the highly successful experi-

ments on board the Cosmic Background Explorer (COBE or co-bee) satellite. Microwaves are radio waves with wavelengths of typically a few centimetres, but the total range is often taken to be from about 1 mm to 30 cm, corresponding to frequencies between 300 GHz and 1 GHz.

To understand the importance of the COBE satellite consider the following brief discussion. The Sun consists of about 74% hydrogen, 25% helium and 1% of all other (heavier) elements combined. It is easy to explain that the trace of heavier elements must have been formed by nuclear reactions inside stars which have long ago exploded and enriched the interstellar medium, but it is not so easy to understand the amount of helium in this way. Ralph Alpher and George Gamow were the first to realize that the helium must be primordial, and that the Universe must at one time have been at least as hot as the centre of the Sun, so that helium could form. During this period the Universe would have been filled with many high-energy, short-wavelength photons. The Universe has expanded so much since that era however, that all those short-wavelength photons have been stretched to become low-energy, long-wavelength photons. As a result, the temperature of this cosmic radiation field is now only a few degrees above absolute zero and its spectrum is that of a black body with a corresponding peak emission at a wavelength of a few millimetres. In the early 1960s Robert Dicke and Jim Peebles at Princeton began designing an antenna to look for this radiation. In 1964, just a few miles away, however, at the Bell Telephone Laboratories in Holmdel, New Jersey, Arno Penzias and Robert Wilson were working on a new microwave (7.35 cm) horn antenna designed to relay telephone calls to Earth-orbiting communications satellites. To their surprise, no matter where they pointed their antenna in the sky, they detected a faint background noise. Eventually, Penzias and Wilson were able to show that the microwave radiation was coming from space and that it was isotropic or uniform in all directions. Moreover, the signal strength was matched by that of a black body at a temperature of only 3 K. Through a colleague they learned of the work of Dicke and Peebles and came to realize that they had in fact detected the remnant radiation left over from the Big Bang. The study of the microwave background has continued since 1965 and reached a peak shortly after 1989 with the launch of the COBE satellite. COBE was designed to measure the spectrum and angular distribution of the cosmic microwave background over the wavelength range from 1 μm to 1 cm.

The spacecraft contained three major instruments: the Differential Microwave Radiometers (DMR), the Diffuse Infrared Background Experiment (DIRBE) and the Far Infrared Absolute Spectrometer (FIRAS). In addition to electronics and solar panels, COBE consisted of a large central structure containing a superfluid helium dewar operating at 1.5 K within which was FIRAS and DIRBE. The DMR experiment was composed of three pairs of radiometers around the outside of the dewar.

DIRBE was designed to search for isotropy in the cosmic microwave background over the wavelength range 1–300 μm by making absolute brightness maps in 10 bands: JKLM in the near infrared, the four IRAS bands at 12, 25, 60 and 100 μm, 120–200 μm and 200–300 μm. A small off-axis Gregorian telescope with a 19-cm primary mirror was used. Mike Hauser was the principal investigator for DIRBE. Comparing the spectrum of the CMB to that of a precise on-board black body was the task for the FIRAS experiment led by John Mather. Two spectral ranges were covered, a low frequency range from about 1 cm to 0.5 mm (wavenumbers from 1–20 cm^{-1}) and a high-frequency range from 0.5 mm to 0.1 mm (wavenumbers 20–100 cm^{-1}). A non-imaging parabolic concentrator with a flared aperture to reduce diffraction was used to collect radiation from a 7° diameter beam and spectral resolving power was obtained with a Michelson interferometer. Six differential microwave radiometers formed the DMR experiment led by George Smoot. Two independent radiometers were used at each of three frequencies 31.5, 53 and 90 GHz (9.5, 5.7 and 3.3 mm). Each pair measured the difference between the CMB emission from two parts of the sky separated by 60°, but the combination of spacecraft spin (75 s), orbit (103 min) and orbital precession (1° per day) allowed each sky position to be compared with all others redundantly. Heterodyne receivers were used with the inputs being switched at 100 Hz between two identical corrugated horn antennas.

FIRAS measured the spectrum of the cosmic background radiation over the huge wavelength range indicated above and obtained a near-perfect fit to a black-body curve with a temperature of 2.735 K. The DMR experiment was used to study the uniformity of the microwave background which at first sight appears to be the same in all directions (isotropic). Closer examination shows that it is slightly greater (hotter) in the direction of the constellation Leo and slightly cooler in the opposite direction (Aquarius). This small "dipole" effect is due to the motion of the Earth and Solar System through the cosmos; it is simply a Doppler effect. Our speed toward Leo is about 390 km/s and taking into account the velocity of the Sun around the centre of the Milky Way galaxy, it is found that our entire galaxy is moving in the general direction of the Centaurus Cluster (of galaxies) at 600 km/s, due perhaps to the gravitational pull of a large mass dubbed the Great Attractor. Once the dipole asymmetry in the microwave background due to our local motion is

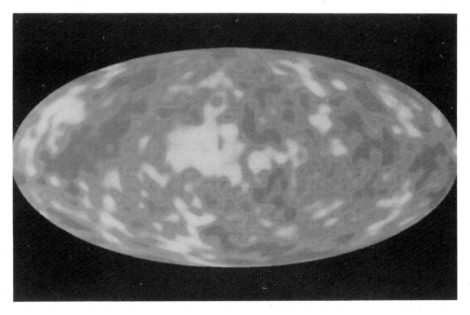

Fig. 12.18. An all-sky image of the fluctuations of the microwave background measured by the DMR experiment on the COBE satellite. Courtesy Ned Wright and COBE team.

eliminated it is possible to begin to study how smooth the true cosmic background is. Just when it was thought that no variations in the incredibly smooth background would be seen, COBE data finally revealed tiny fluctuations of about 1 part in 100 000. The fluctuations were seen after some extremely complex and careful computer processing and modelling of the data by Edward Wright at UCLA (Data team leader) and his colleagues on the DMR experiment (see Fig. 12.18).

SUMMARY

Methods have been found to produce electronic images (or maps) of astronomical sources across the entire electromagnetic spectrum from gamma-rays to radio waves. Charge-coupled devices compete with microchannel plates and other photoemissive detectors in the UV and X-ray domain, whereas bolometer arrays and heterodyne receivers are used with large radio dishes and huge radio interferometers to map the sky at radio and millimetre wavelengths. Semiconductor electronics in silicon and other materials, fibre optics, superconducting junctions, high-speed computers and sophisticated algorithms are among the many technologies used.

EXERCISES

1. Why do CCDs exhibit good response in the X-ray region and in the visible, but poor response in the ultraviolet?
2. Describe the principle of the Microchannel Plate (MCP) and explain what is meant by a "Z-stack". Under what circumstances does a MCP with a Z-stack exhibit "gain sag"?
3. Why must UV detectors be "solar blind"?
4. What is meant by a Fano-noise-limited CCD?
5. How are EUV and X-ray telescopes constructed to provide focal plane images?
6. What are the primary detector technologies used for the detection of gamma-rays?
7. Why must radio telescopes be much larger than optical telescopes to achieve moderate angular resolution on the sky? (a) Calculate the approximate angular resolution of a radio telescope of diameter 26 m (a single dish in the VLA) working at a wavelength of 6 cm. (b) Calculate the diffraction limit of a 4-inch (10 cm) telescope at $\lambda = 500$ nm.
8. Explain the fundamental principle of radio detection and introduce the concepts of a Local Oscillator, mixer and "detector". Why is it convenient to describe the performance of radio receivers in terms of a noise temperature?
9. What is the advantage of combining the signals from two well-separated radio telescopes observing the same source at the same time?
10. Explain the term "aperture synthesis" and the principle behind the CLEAN algorithm.
11. Describe the three main experiments on the COBE satellite and what was learned from them about the cosmic microwave background.

REFERENCES

Bowyer, S. and Malina R. (1994), "Extreme ultraviolet images of the Cosmos", *Physics World*, 7, No. 1, (January), Institute of Physics Publishing, UK.

Carruthers, G.R., (1993), "Ultraviolet and X-ray detectors", in *Electro-Optics Handbook*, Ch. 15, R.W. Waynant & M.N. Ediger (Eds), McGraw-Hill, New York.

Carruthers, G.R. (1996), Private communication.

Elvis, M. (Ed.) (1990), *Imaging X-ray Astronomy—a Decade of Einstein Observatory Achievements*, Cambridge University Press, Cambridge, England.

Fleck, B., Domingo, V. and Poland, A. (Eds) (1995), *The SOHO Mission*, Kluwer Academic Publishers, Dordrecht, Netherlands.

Heap, S.R. (+ 30 co-authors) (1995), "The Goddard High-Resolution Spectrograph: In-orbit performance", *Pub. Astron. Soc. of the Pacific*, 107, 871–887.

Janesick, J., Elliott, T., Collins, S., Daud, T., Campbell, D. and Garmire, G. (1987), "Charge-coupled device advances for X-ray scientific applications in 1986", *Optical Engineering*, 26, No. 2.

Janesick, J., Elliott, T., Bredthauer, R., Chandler, C. and Burke, B., (1988), "Fano-noise-limited CCDs", Optical and Optoelectronic Applied Science and Engineering Symposium; *X-ray Instrumentation in Astronomy*, SPIE, Bellingham, USA.

Joseph, C.L., (1995), "UV image sensors and associated technologies", *Experimental Astronomy*, 6, 97–127, Kluwer Academic Publishers, Dordrecht, Netherlands.

Joseph, C.L. *et al.* (1995), SPIE, Vol. 2551, p.249.

Konda, Y. (Ed.) (1987), *Exploring the Universe with the IUE Satellite*, Kluwer Academic Publishers, Dordrecht, Netherlands.

Kraus, J.D. (1986), *Radio Astronomy*, Cygnus-Quasar Books.

Musso, C., Chiappetti, L, and Bignami, G.F. (1995), "Optimizing mission science: the read-out modes of the EPIC for X-ray Astronomy", *Experimental Astronomy*, 6, 235–248.

Sargent, A.I. and Beckwith, S.V.W. (1993), "The search for forming planetary systems", *Physics Today*, 46, No. 4 (April), 22–29, American Institute of Physics, New York.

Stern, R. A., Liewer, K. and Janesick, J. R. (1983), "Evaluation of a virtual phase charge-coupled device as an imaging x-ray spectrometer", *Review of Scientific Instruments*, 54, 198–205.

Timothy, J.G. (1991), in *Photoelectronic Image Devices 1991*, B.L. Morgan (Ed.), Institute of Physics Conference Series No. 121, UK.

Verschuur, G.L. (1987), *The Invisible Universe Revealed—the Story of Radio Astronomy*, Springer-Verlag, New York.

Wolstencroft, R.D. and Burton, W.B. (Eds) (1987), *Millimetre and Submillimetre Astronomy*, Kluwer Academic Publishers, Dordrecht, Netherlands.

SUGGESTED FURTHER READING

Benvenuti, P., Macchetto, F. and Schreier, E. (Eds) (1996), *Science with the Hubble Space Telescope II*, Space Telescope Science Institute.

Chown, M. (1996), *Afterglow of Creation*, University Science Books, California.

Drew, J.E. (Ed.) (1995), *New Developments in X-ray and Ultraviolet Astronomy*, Pergamon, Oxford.

Fraser, G.W. (1989), *Detectors for X-ray Astronomy*, Cambridge University Press, New York.

Rohlfs, K. (1990), *Tools of Radio Astronomy*, Springer-Verlag, Berlin.

Malina, R. and Bowyer, S. (1991), *Extreme Ultraviolet Astronomy*, Pergamon Press, New York.

Signore, M. and Dupraz, C. (1992), *The Infrared and Submillimeter Sky after COBE*, Kluwer Academic Publishers, Dordrecht, Netherlands.

13

A new generation of telescopes

In addition to the impact of optical and infrared electronic imaging devices, the last decade of the twentieth century has seen the development of a whole new generation of very large ground-based telescopes for optical and IR astronomy. Apertures range from 6.5 m to 10 m and several new technologies have been developed to ensure that the very best image quality accompanies the immense collecting area. Refurbished, the Hubble Space Telescope (HST), the most ambitious of remote-controlled observatories, is fulfilling its initial promise. In this chapter we will review all of these developments.

13.1 THE HUBBLE SPACE TELESCOPE

The tragic loss of the space shuttle "Challenger" in January 1986 delayed the launch of the **Hubble Space Telescope** (Fig. 13.1) by several years and then in April 1990, when this large and complex satellite was finally in orbit, a serious problem was discovered with its image quality! Fortunately however, all was not lost. With an immense effort on image restoration algorithms and computer programs, the telescope was still capable of excellent science while a "rescue" mission was being planned. In January 1994 astronauts installed several new systems in the telescope and subsequent images showed a dramatic improvement (Fig. 13.2 and colour plates). Since then, the images being returned from the Hubble Space Telescope have captured the imagination of astronomers and the general public alike. If you haven't already done so, spend some time "browsing" through the HST library of images on the World Wide Web.

13.1.1 HST instruments
Obviously, the Hubble Space Telescope is a truly remote-controlled 2.4-m (94-inch) telescope and its scientific instrumentation (Fig. 13.3) relies heavily on electronic detectors and computers. Five main instruments were included in the initial complement at launch: the Wide Field Camera / Planetary Camera (WFPC—Principal Investigator (PI), Jim Westphal); Faint Object Camera (FOC—PI, Ducchio Macchetto); Faint Object Spectrograph (FOS—PI, Richard Harms); Goddard High Resolution Spectrograph (GHRS—PI, Jack Brandt); High Speed Photometer/Polarimeter (HSP—PI, Bob Bless). The WFPC is located in a "radial bay" and receives light from a small on-axis "pick-off" mirror,

Fig. 13.1. The Hubble Space Telescope in orbit—a fully remote-controlled observatory with a 2.4m diameter reflecting telescope optimized for wavelengths from 120 nm in the UV to 2400 nm in the infrared.

whereas the other four instruments occupy the "axial bays". The other three radial bays are used for the Fine Guidance Sensors. The HSP was removed in January 1994 to make room for COSTAR, the package of corrective optics, and the Wide Field Planetary Camera was upgraded to WFPC 2 with its own built-in corrective optics. Two new instruments are planned for 1997; the Near-Infrared Camera and Multi-object Spectrometer (NICMOS) and the Space Telescope Imaging Spectrograph (STIS). A third-generation instrument called the Advanced Camera will be installed at the time of orbit re-boost in 1999. Below is a brief summary of these instruments, details can be obtained from the World Wide Web page for HST or from the Space Telescope Science Institute, Baltimore.

Both the old and new Wide Field/ Planetary Camera instruments (the abbreviation WFPC is commonly pronounced "wiff-pick") can be used as a wide-field camera for survey observations or as a high-resolution camera capable of exceptionally fine-detail imaging of, among other things, the planets of the solar system. WFPC 1 used four TI 800 × 800 CCDs arranged in a mosaic. Each detector was separated physically using a shallow "pyramid" reflector in the same manner as the 4-shooter instrument at Palomar Observatory which was indeed the prototype. The four "camera" sections contained small Cassegrain telescopes with the CCD at the focus, but an image of the Hubble's primary mirror at the small secondary mirror of the camera. A square area of 2.67 arcminutes was obtained in

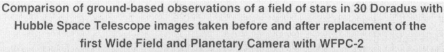

Comparison of ground-based observations of a field of stars in 30 Doradus with Hubble Space Telescope images taken before and after replacement of the first Wide Field and Planetary Camera with WFPC-2

Ground image at 0.6 arcsec resolution | WFPC-1 image (before servicing) | WFPC-2 image (after servicing)

Fig. 13.2. (a) Observations of stars before and after correction for the spherical aberration.

Hubble Space Telescope
Wide Field Planetary Camera 2

Wide Field Planetary Camera 1 | Wide Field Planetary Camera 2

Fig. 13.2. (b) An image of M100 by the HST before and after correction. Courtesy NASA/STScI.

the wide-field (WF) mode and each pixel subtended an angle of 0.1 arcseconds, whereas in the planetary camera (PC) mode the field of view was 68.7 arcseconds and an individual pixel corresponded to a mere 0.043 arcseconds—some ten times smaller than the scales used with ground-based telescopes. Good ultraviolet response from the TI CCDs was obtained by coating the CCD surface with the phosphor coronene. After construction of the

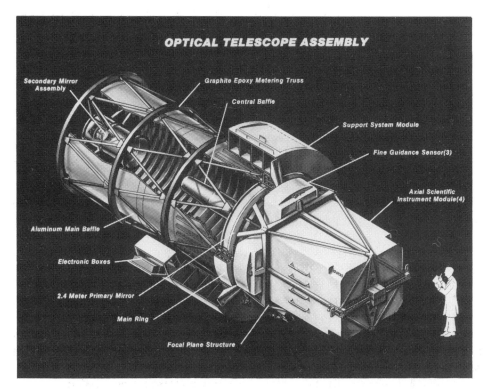

Fig. 13.3. Schematic of the HST showing the location of important components such as the optical telescope assembly, the solar panels and the axial instrument bays. Photo courtesy of Terence Facey, Perkin-Elmer.

spacecraft and instrument it was found that it was necessary to illuminate the CCD with a strong ultraviolet light—a UV flood—to prepare the coronene and ensure good ultraviolet quantum efficiency (see section 7.8). To solve this problem, a hole was drilled in the side of the telescope and a "light-pipe" installed to feed UV light from the Sun to the CCD chamber! Analysis of CCD data from the WFPC 1 was crucial in identifying and understanding the initial image quality problems with HST and deducing that the effect was due to pure spherical aberration.

Unlike the original, WFPC 2 has only one imaging mode. Three CCDs (called WF2, WF3 and WF4) provide the wide field mode with a scale of 0.1 arcsec per pixel, and only one CCD, called PC1, operates in the planetary camera mode with a scale of 0.046 arcsec per pixel. The location of PC1, where WF1 would have been, results in the "steps" seen in many of the public HST images. The CCDs in WFPC 2 are thick front-illuminated chips from Loral Aerospace with 15 μm pixels and the same 800 × 800 format as the TI chips. Therefore, the field of view covered by the three WF chips is about the same as before, whereas the field of PC1 is only 35 × 35 arcseconds. Readout noise is about 7 electrons rms and UV sensitivity is obtained by a coating of lumogen. WFPC 2 is sensitive from 1100 Å to 11 000 Å with a peak total throughput of 13% at about 6500 Å. Blooming, stability and quantum efficiency hysteresis are significantly better compared to WFPC 1.

Cosmic ray hits in orbit are about the same (1.7 hits/s/CCD). Unlike the axial bay instruments, WFPC 2 does not use COSTAR for aberration correction but has its own built-in correction which yields near diffraction-limited performance over the entire field. A set of 48 filters mounted in 12 wheels is available for WFPC 2, including a UBVRI set, a Strömgren series, a Woods filter for solar-blind imaging in the far UV, a series of narrow band filters, methane and polarizing filters.

The Faint Object Camera (FOC) is complementary to the WFPC. It is designed to have higher spatial resolution than the planetary camera and better ultraviolet sensitivity, especially for extremely faint objects. In the FOC, two similar but independent systems are provided to form an image. After installation of COSTAR, there was a magnification factor of about 1.6 applied to each imaging train, but is was decided to retained the original names of the cameras, i.e. $f/96$ and $f/48$. The corresponding new plate scales are 0.014" and 0.028" per pixel and the maximum fields of view are 14" × 14" ($f/96$) and 28" × 28" ($f/48$). Note that the Airy disk has a radius of 0.029" at 250 nm so that FOC remains critically sampled well into the UV. Filters are available for the wavelength range 1150–6500 Å, and there is also some spectroscopic capabilities too. FOC is an efficient imaging system in the ultraviolet and is more sensitive than WFPC 2 below 4500 Å, especially if a Woods filter is required with the CCDs to render them solar blind. In each of the FOC's two camera systems the detector consists of an image photon counting system (IPCS) of the Boksenberg type, i.e., an image intensifier tube with a UV-sensitive photocathode followed by a TV camera for readout and "event-centroiding" electronics to locate individual photon impacts on the detector. As mentioned earlier in the book, this kind of detector saturates at very low flux levels, but it has no readout noise and so it really is intended for faint objects. The FOC also has a means of measuring polarization.

The Faint Object Spectrograph is equipped with two detector systems, one blue-sensitive and one red-sensitive. Both detectors are of the Digicon-type, that is, a photoelectric photocathode as in a photomultiplier tube, with a surrounding magnetic focussing coil which collects and directs the emitted photoelectrons onto a linear array of 512 silicon photodiodes by which means they are detected. Associated with each diode is a discrete amplifier. This is a complex type of detector but it has many of the advantages of the photon-counting detectors and some of the advantages of silicon devices such as large dynamic range. With its two channels, the faint object spectrograph is sensitive to radiation ranging in wavelength from about 1150 Å to 8300 Å at resolving powers of $R = 1300$ and $R = 250$, and in addition it has a polarization measuring capacity (which became somewhat compromised by the pair of additional mirrors of the COSTAR package) and is responsive to fast time-varying astrophysical phenomena. After launch it was found that the magnetic shielding on the detectors was not quite good enough to prevent geomagnetically induced image motion (up to one diode in amplitude) but it was possible to employ a corrective algorithm which works on the electromagnetic optics during actual observations to correct for the image wander.

The Goddard High Resolution Spectrograph also employs Digicon detectors and differs from the FOS largely in terms of the finer spectral resolution it can achieve for relatively brighter sources.

A major component of the HST is the set of "fine guidance" sensors (FGS). The telescope requires accurate target positions relative to an acquired guide star so that light from the target can be fed accurately and continuously into the entrance aperture of the appropriate HST instrument. To do this, a very extensive Guide Star Selection System was developed at the Space Telescope Science Institute in Baltimore. Basically, an image and a catalogue archive has been developed on optical disks using digitized photographic plate surveys from several different Schmidt telescopes (see section 3.2.1). Typically, each survey plate corresponds to an array of 14 000 × 14 000 pixels, with one pixel being equivalent to 1.7 arcseconds on the sky. Four plates (1600 Mbytes) are accommodated on each optical disk and up to 80 disks may be placed in a "juke box", which gives about 20% of the sky "on-line" for image display at any time. In addition, a catalogue with information on approximately 20 million individual objects is available. This catalogue is now widely available on CD-ROM.

Analysis of HST data is generally carried out by "pipeline" reductions developed by the instrument teams and the staff at the primary HST support facilities. HST data can also been reduced using the Space Telescope Science Data Analysis System (STSDAS). This immense software package includes both calibration and analysis software and is fully portable to any host computer that runs the IRAF software system (see Chapter 11) from the National Optical Astronomy Observatories; another form is available for non-IRAF systems.

In reading through the above account of the HST instruments one might be surprised that CCDs feature so little. As already mentioned, however, this state of affairs is not so surprising when one considers the incredibly long "lead-time" needed for space ventures of this scale. The idea of a large Space Telescope was first mooted in 1962, the first group of scientists selected by NASA in 1973 and construction finally authorized in 1977. The first astronomical images with a CCD—see the picture of Uranus obtained by Brad Smith, Jim Janesick and colleagues with the TI 400 × 400 chip—was obtained in 1976, and Jim Westphal and Jim Gunn proposed the Wide Field/Planetary Camera shortly after. Additionally, the Space Telescope was intended to provide excellent coverage in the ultraviolet and photoemissive vacuum-encapsulated TV-type tubes were well-known to be suitable at these wavelengths, and possibly at their peak of development.

13.1.2 COSTAR—fixing the HST

The HST is a classic Ritchey–Chrétien design of two hyperbolic mirrors: an $f/2.35$ concave primary 2.4 m in diameter and a convex secondary of 0.308 m diameter, with the focal plane 1.5 m behind the primary mirror. During manufacture of the primary mirror, a spacing error of a field lens in a measuring device, known as a reflective "null corrector", went unnoticed by contractors which caused an excessive amount of material to be removed from the outer edge of the primary mirror and resulted in an incorrect conic constant. In simplest terms, the shape of the primary mirror became too shallow. Although the total difference from edge to centre was only about 2 μm, this oversight was a disaster because the effect is many times larger than the wavelength of optical and ultraviolet light. Now, instead of 70% of the incident light being focussed into a spot of only 0.1 arcsecond in radius, a mere 15% of the light was able to converge into this spot and the remainder was spread over a large area of a few seconds of arc! The error in the shape of the primary produces a well-known blurring called "spherical aberration" in which the light rays towards the edge of a mirror (or lens) come to a focus which is progressively farther dis-

Fig. 13.4. COSTAR, the corrective optics for the Hubble Space Telescope. Each corrector consists
of two mirrors. Courtesy Ball Aerospace.

placed from that of the central rays; no matter where the detector is placed along the optical axis, the image is blurred.

Over the next several years a huge amount of effort went into computer programs and
algorithms which could "correct" the distorted images and eliminate the fuzzy halo around
the otherwise compact core of light. These efforts were very successful and the HST produced very good science. Of course, this approach cannot put back all that diffuse light
into the original pixels for which it was intended. In other words, the signal-to-noise ratio
of each observation with HST had been seriously reduced because only 15% of the light
was concentrated where it was meant to be. To compensate for this effect, all exposure
times had to be substantially increased, which reduced the number of different observations which could be made, and actually prohibited the longest, most challenging observations. An optical fix was required. That fix arrived on board the shuttle Endeavour in
January 1994 in the form of a curious-looking instrument called COSTAR (Fig. 13.4).

Built by Ball Aerospace and Technology Corporation, COSTAR (Corrective Optics Space Telescope Axial Replacement) is a 7 ft long, box-shaped deployable optical bench made of carbon graphite reinforced epoxy and weighs approximately 640 lbs. At one end, tucked close to a corner, there are four small motor-driven arms made from beryllium which will lower the appropriate corrective optics into the various light paths to the instruments. Before this instrument could be designed, it was essential to derive the erroneous conic constant of the primary to five decimal places. Images from the FOC and WFPC were obtained at many focus settings and analysed by comparing them to computer-generated ray tracings. Many other methods were also used to verify the results. The development of the COSTAR instrument took 28 months and cost almost $50 million. Since the WFPC was a non-axial instrument it could not benefit from COSTAR and therefore it was completely replaced by a new Wide Field/Planetary Camera (WFPC 2) in which the spherical aberration correction was built in.

Each COSTAR corrector consists of two mirrors. The first is a spherical curved field mirror (M1) and a re-imaging mirror (M2) with a complex surface shape and different magnifications in two orthogonal directions; this anamorphic, fourth-order aspheric mirror compensates for the spherical aberration of the primary and for the inherent astigmatism due to the light being directed to a point which is off-axis. COSTAR provides five separately corrected instrument channels distributed among the three instruments: one for the GHRS, two each for the FOC and the FOS. Each instrument has its own dedicated pair of mirrors. One arm carries two M1-type mirrors for the FOC and a second arm carries the matching pair of M2 mirrors. The third arm holds two M2-type mirrors for the FOS and the fourth arm carries the single GHRS M2 mirror. The M1 mirrors for the FOS and GHRS are mounted directly on the deployable optical bench. All the M1 mirrors have biaxial tilt adjustments for on-orbit alignment with the M2 mirrors, and the M1 mirrors for the FOS and the GHRS can move in and out along the telescope's optical axis for focus adjustment. For the FOC, focussing is accomplished by moving the entire optical bench. As illustrated here and in the colour plates, the HST repair mission was an outstanding success. One of the most scientifically valuable images from HST is called the Hubble Deep Field (HDF). The HDF consists of WFPC 2 images in four passbands taken over 10 consecutive days (155 orbits) centred on an unremarkable, relatively "blank" field of low extinction (high galactic latitude) in Ursa Major (RA = 12h 36m, Dec = +62° 13′) which lies in the northern continuous viewing zone of HST. The image reaches limiting magnitudes near 30, and reveals a host of faint galaxies with unexpected colours and shapes.

13.1.3 NICMOS, STIS AND ACS

All of these instruments are axial bay replacements, and are being manufactured by Ball Aerospace. NICMOS (Fig. 13.5) provides access to the near infrared from 0.8 to 2.5 μm and STIS provides a new UV/optical spectroscopic capability. NICMOS contains three near infrared cameras based on 256×256 pixel arrays of HgCdTe from Rockwell International (see Chapter 8) with different magnifications to yield different spatial scales of 0.043″, 0.075″, and 0.2″, and corresponding fields of view of 11×11, 19×19 and 51×51 arcseconds. The image scales of Cameras 1 and 2 provide substantial over-sampling of the diffraction disk. Each camera views a separate part of the field and all can be used simulta-

Fig. 13.5. System integration of NICMOS, the near-infrared camera and multi-object spectrometer built by Ball Aerospace for NASA and the University of Arizona with focal plane infrared arrays by Rockwell. Ball Aerospace photo.

neously and asynchronously in principle. A selection of 20 filters, including grisms and polarizers, is available for each camera. Multi-object spectroscopy is performed using grisms of calcium fluoride in the filter wheel associated with Camera 3; each grism has an "order sorter" interference filter coating on its face. No entrance slit is used and the resolving power is $R = 200$ (per pixel) and three grisms are needed to cover the entire 0.8–2.5 µm range; the bands are 0.8–1.2 µm, 1.1–1.9 µm and 1.4–2.5 µm. At the shorter wavelengths NICMOS benefits immensely from the absence of (variable) OH emission lines in Earth's upper atmosphere and the lack of an entrance slit is not a problem. On the other hand, the fact that the primary mirror of HST is maintained at 20 °C results in thermal emission which restricts performance at the longest wavelength. The detectors are cooled to 58 K and the filters reach 155 K using a solid nitrogen dewar system containing an aluminium "foam" for good conductivity and mechanical stability. The cryogen is expected to last for at least five years.

The Space Telescope Imaging Spectrograph (STIS) is essentially a general-purpose ultraviolet spectrograph to replace the GHRS and the FOS. Both long-slit and cross-dispersed echelle spectroscopy will be available with a wide range of spectral resolutions ($R = 26\,000$–$100\,000$). While the quantum efficiency in these modes is comparable to that of GHRS, the two-dimensional format will dramatically improve the overall efficiency. The far ultraviolet (FUV) and near ultraviolet (NUV) detectors will be similar to those discussed in Chapter 12, namely, Multi Anode Microchannel Array (MAMA) detectors. CCDs will be used for the visible wavelength spectroscopy channel, and both spectrographs include moderate field of view imaging in their respective wave bands.

In 1999 or early 2000, a third service mission to HST will reboost the telescope to a nominal orbit of 320 nautical miles and upgrade several systems. At this time it is likely that the Advanced Camera for Surveys (ACS) will be installed in place of the FOC. This UV/optical imaging instrument will feature three separate camera channels: a "solar blind" channel using a CsI MAMA detector (1024 × 1024 pixels each 0.031″; a "wide field" channel employing a 4096 × 4096 CCD optimized for the I-band and providing 0.05″ pixels; a "high resolution" channel using a SITe 1024 × 1024 CCD designed to provide broad coverage from 2000 Å to 10 000 Å with 0.025″ pixels. With the installation of ACS, the COSTAR package will become dormant since NICMOS, STIS and ACS have "built-in" correctors and so it may be replaced in 2002 with a new instrument.

Finally, it is important to consider the advantages and disadvantages of the Hubble Space Telescope compared with very large ground-based telescopes. Clearly, as a UV telescope there is no competition from ground-based facilities. Also, the improved resolution at visible wavelengths (especially in the blue) resulting from the absence of atmospheric blurring leads to sharper images and much more light per pixel, which in turn means that the HST can image to very faint limits. On the other hand, the HST is very expensive compared to ground-based telescopes, even very large ones. One estimate of the cost of HST is $1.5 billion! In the next few sections we will examine the ground-based competition to achieve HST image resolution.

13.2 VERY LARGE TELESCOPES (VLTS)

In March of 1993, the largest telescope in the world went into operation. It was the first of a pair of 10-metre (10-m) telescopes funded by the W. M. Keck Foundation. The second telescope was inaugurated in May 1996. At that time, at least eight other optical/infrared telescopes with collecting apertures larger than 6.5 metres in diameter were also under construction and several more were being contemplated. Why do we need such large telescopes?

Astronomers have already improved the efficiency of spectroscopy by the use of multislit devices and optical fibres to observe many objects simultaneously. As CCDs get larger and larger more and more area on the sky can be covered and imaged to deeper levels. The advent of infrared arrays has catapulted ground-based infrared astronomy to a position alongside visible light technology. Carefully designed optical cameras and spectrometers now have transmission factors and efficiencies which are virtually impossible to improve due to fundamental effects associated with the refractive and reflective properties of materials and the quantum efficiency of the detector itself. Some improvements are no doubt possible, and should certainly be made even if the improvement is by no more than a factor of 40% or so. Already, however, astronomers are beginning to realize that to probe to the most distant objects in space and the earliest moments in time, to search for other planetary systems and to understand the origin of planets, stars and galaxies will require measurements at least ten times more efficient than ever before.

Once the quantum limits of sensitivity in detectors and instruments have been reached, the only way to gain large factors in efficiency is to construct even larger ground-based telescopes and to develop methods for counteracting the image-blurring effects of turbulence in the Earth's atmosphere.

Table 13.1. The largest telescopes in the pre-Keck era

Telescope name/site	Primary (m)	f/ratio	Mounting
Bol'shoi Teleskop Azimutal'nyi Mt. Pastukhov (Russia)	6.0	f/4.0	alt-az
George Ellery Hale Telescope Mt. Palomar (California, USA)	5.08	f/3.3	equat
Multiple Mirror Telescope (MMT) Mt. Hopkins (Arizona, USA)	4.5	f/2.7	alt-az
William Herschel Telescope (WHT) Roque des los Muchachos (Canary Is.)	4.2	f/2	alt-az
Cerro Tololo Interamerican Observatory (CTIO) Cerro Tololo (Chile)	4.0	f/2.8	equat
Anglo-Australian Telescope (AAT) Siding Spring (NSW, Australia)	3.9	f/3.3	equat
Nicholas Mayall Telescope Kitt Peak (Arizona, USA)	3.8	f/2.8	equat
United Kingdom Infrared Telescope (UKIRT) Mauna Kea (Hawaii)	3.8 (thin)	f/2.5	equat
Canada–France–Hawaii Telescope (CFHT Mauna Kea (Hawaii)	3.6	f/3.8	equat
European Southern Observatory (ESO) Cerro La Silla (Chile)	3.6	f/3.0	equat

Table 13.1 lists the world's largest telescopes prior to the opening of the Keck 10-m telescope in 1993. Interestingly, after construction of the 5-m (200-inch) Hale Telescope on Palomar Mountain in 1949, the trend was for smaller telescopes in what has become known by the imprecise term "4-metre class"; in this definition all telescopes with primary mirror diameters between 3.6 m and 4.5 m are lumped together. The exceptions were the Soviet BTA and the Smithsonian MMT, both of which were unique in their design, and both of which may have been somewhat ahead of their time. What happened of course, was that computers, electronic imaging devices and instrumentation improved greatly. Now, things have come full-circle, and astronomers are once again constructing large telescopes. There are essentially three fundamental issues:

(1) how to achieve a very large collecting aperture (mirror) of the required optical performance

(2) how to support and control in the optimum way such a potentially very heavy mechanical structure

(3)   how to enclose a very large telescope in a cost-effective way with negligible degradation on image quality due to vibration, air disturbance or inadequate environmental protection (wind, dust)

In addition, new telescopes must be designed to capitalize on the very best "seeing" conditions at the world's best sites, and must be designed with "remote control" in mind. As Roger Angel remarks, "the problems of building the new generation of telescopes are compounded because they must not only be bigger, but also must give sharper images than their predecessors". It is not surprising that several different approaches have emerged. Basically, it all comes down to how the mirrors are made and supported. There are three categories:

- **meniscus mirrors**—large monolithic disks of solid glass which are so thin that it must be accepted that they will be flexible and therefore they must be actively controlled to maintain the required shape during operation. Bending by unpredictable forces such as wind gusts requires a rapid servo system of extremely accurate force actuators.
- **segmented mirrors**—smaller monolithic disks of thin polished glass are used to form the surface of an efficient rigid "backing" structure of steel or carbon fibre. Position actuators are still required to make the attachment and to correct for thermal and gravitational effects in the backing frame, but the time scale for such corrections is slow. Each segment is individually supported and global changes are sensed at the gaps between segments.
- **honeycomb mirrors**—a thick mirror is constructed but large pockets of mass are removed from the back to make the mirror lightweight yet very stiff. The method involves a mould and a spinning furnace to form a concave parabolic front face while the ribs and backplate are formed by glass flowing down between the gaps in the mould.

Each of these methods has been applied to build a new generation of very large telescopes for ground-based astronomy at optical and infrared wavelengths (see Table 13.2).

Table 13.2. Telescopes built or planned with primary mirror diameters larger than 6.5 metres

Telescope name	Primary (m)	Mirror technology	Location
Keck I, II (CARA)	10	36 hexagonal segments Zerodur	Mauna Kea, Hawaii
VLT Unit 1, 2, 3, 4 (ESO)	4 × 8.2	thin meniscus	Cerro Paranal, Chile
Subaru (Japan)	8.3	thin meniscus	Mauna Kea, Hawaii
Gemini N & S (GTP)	2 × 8	thin meniscus	Mauna Kea, Hawaii Cerro Pachon, Chile
LBT (former Columbus)	2 × 8	borosilicate honeycomb	Mt. Graham, Arizona
Magellan	2 × 6.5	borosilicate honeycomb	Las Campanas, Chile
MMT upgrade	6.5	borosilicate honeycomb	Mt. Hopkins, Arizona

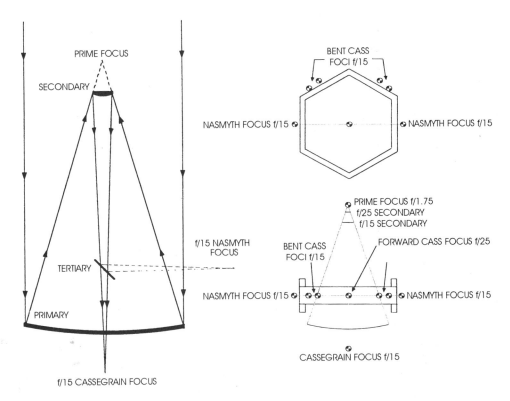

Fig. 13.6. The optical layout of the 10-m W.M. Keck Telescopes showing the *f*/15 foci and the arrangement of the 36 hexagonal segmented mirrors in the primary. A different secondary mirror provides an *f*/25 focus at the intersection of the elevation and optical axes. This "forward Cass" is used for low-background infrared work. Courtesy Jerry Nelson.

13.2.1 The Keck telescopes

In the vanguard of the wave of new large telescopes is the 10 m W. M. Keck Observatory, which is managed by CARA, the California Association for Research in Astronomy, and comprises the California Institute of Technology (CalTech) and the University of California (UC) in collaboration with the University of Hawaii and NASA. The twin Keck Telescopes are already in operation on the summit of Mauna Kea, Hawaii. Each has an effective aperture of 10 m but both possess a unique primary mirror concept championed by Project Scientist Jerry Nelson (now at University of California, Santa Cruz). The primary mirror is not a single piece of glass, instead it is composed of 36 hexagonal "segments" of Zerodur, a special ceramic glass made by the Schott glassworks in Germany; this construction is called **segmented mirror**. Each side of the hexagonal segment is 0.9 m (or 1.8 m across) with each segment being only 7.5 cm thick (Figs 13.6 and 13.7). There are 3×36, or 108, actuators that control the position of those segments to form a surface which is a perfect hyperboloid. The optical design is a Ritchey–Chrétien with a hyberbolic primary and a hyperbolic secondary to give a coma-free field at the Cassegrain focus. Each mirror segment can be moved by the actuators at a rate of twice per second to maintain the global

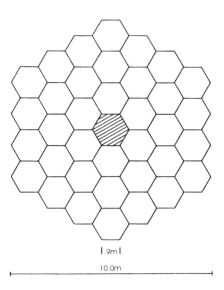

| .9m |

10.0m

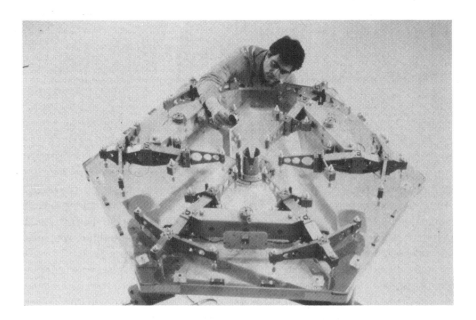

Fig. 13.7. A close-up view of the Zerodur segments and the whiffle-tree warping harness and mounting structure on the rear of the mirror. Photo courtesy CARA.

shape of the hyperboloid to within 50 nanometres; this is known as the Active Control System or ACS.

The history of the Keck Telescope actually dates back to 1977 when a committee of University of California (UC) astronomers met to consider plans for new telescopes and new sites, because the Mt. Hamilton site of the UC/Lick Observatory was becoming too bright. As Jerry Nelson, who was at the Lawrence Berkeley Labs at that time, explained to me, "I volunteered to think about a much bigger telescope"! Jerry began by studying in detail how classical telescope structures were built and by understanding where and why things became hard. He was struck by the fact that the mechanical flexure on the mirror due to gravity was proportional to the fourth power of its diameter (D^4). It seemed to him that this argued for smaller, not larger mirrors! Therefore he began to look into ways of making a continuous mirror surface out of pieces or panels, just like radio telescopes. Jerry soon realized that he could not copy the approach of the huge radio telescopes at optical wavelengths—the surface accuracy is just too small. Deformation of the surface figure would have to be corrected with an "active" control system. Jerry recalled to me that he actually started out with the concept of a relatively large central monolith surrounded by two rings of mirrors, but within a year had dropped that idea in favour of hexagonal segments. Right from the start his vision was of a 10-m telescope. Together with Terry Mast, Jerry worked on all the technical issues, especially the active control system. What should be measured? What do you do with those measurements and how does this relate to the surface accuracy of the mirror? Their breakthrough was the realization that they could perform a "global" best fit to the mirror shape without worrying in detail where one segment was with respect to another. The UC committee approved funds for a full-size prototype segment in 1979 which was to be fully demonstrated. During the period 1979–1984 all the details of the full-scale 36-segmented structure were worked out and led to the famous "blue book" for a Ten-Metre Telescope. The next hurdle was funding.

At first a grant to the University of California of $50 million was received in 1984 from the Hoffman foundation, but this was not enough. The California Institute of Technology (Caltech) was approached as a minor partner who might be able to bring the balance of funding needed. Caltech found the W. M. Keck Foundation who wanted to fund the entire project, not just part of it. The apparent excess of funds actually stimulated the concept of twin 10-metre telescopes working as an optical interferometer. In the end, however, difficulties in reaching legal agreements over the Hoffman gift led to the entire project being funded by the W.M. Keck Foundation, including a second telescope if the first was shown to be successful. In all, the Keck Foundation provided about $145 million. An agreement with the University of Hawaii was successfully negotiated to construct the telescope on Mauna Kea, and all site work was done with two telescopes in mind. By the end of 1991, with 18 segments installed and working beautifully to produce astronomical images, the Keck Foundation was convinced to begin Keck II. The University of California provided funding for the first set of instruments and 25 years of operations costs. In 1996, NASA became an official partner in the Keck Observatory, and the first Scientific Director, Fred Chaffee—formerly Director of the Multiple Mirror Telescope—was appointed.

Fig. 13.8 shows the twin Keck Telescope domes and an interior shot of one of the telescopes. Image quality with Keck I is routinely about 0.5–0.6 arcseconds, and some-

Fig. 13.8. (a) The twin Keck Telescope domes on the summit of Mauna Kea. This picture was taken by the author in June 1996.

(b) An internal view of Keck I. Photo CARA.

times 0.25″. When Keck II was inaugurated in May 1996 the first images were already sub-arcsecond. To the left of the Keck Telescopes lies the Japanese "Subaru" 8-m telescope. Keck I and Keck II are 85 m apart with very large basements connected by a long tunnel to allow for the opportunity to eventually combine the beams to form an optical interferometer.

Both optical and IR instruments are available for the telescopes. In the infrared there is NIRC, the Near Infrared Camera, developed at Caltech (Keith Matthews) and employing a 256×256 InSb array and LWS, the Long Wavelength Spectrometer, developed by UC San Diego (Barbara Jones), which is a 10 and 20 μm imaging and spectrometer system with a 96×96 array. For optical wavelengths the complement is LRIS, the Low Resolution Imaging Spectrograph, also developed at Caltech (Judy Cohen and Bev Oke) which provides multi-slit low resolution CCD spectroscopy and imaging over a 5 arcminute field, and HIRES, the High Resolution Echelle Spectrograph, developed at UC Santa Cruz (Steve Vogt) which is a CCD-based, cross-dispersed echelle spectrograph providing resolving powers of $R > 50\,000$.

Instruments under development at the time of writing include NIRSPEC, the Near Infrared Spectrograph, being developed at UCLA (Ian McLean). This instrument is a cryogenic cross-dispersed echelle spectrograph ($R = 25\,000$) which operates from 1–5 μm and uses a 1024×1024 infrared array. In addition there is ESI, the CCD Echelle Spectrograph and Imager, being developed by UC Santa Cruz (Joe Miller), which is a fixed format cross-dispersed echelle CCD spectrograph, and DEIMOS, the Deep Extragalactic Imaging and Multi-Object Spectrograph, is a very large, wide-field multi-object spectrograph employing a large mosaic of CCDs (Sandra Faber). A "blue" or second channel to the LRIS instrument is being added to make it a double-beam spectrograph; Caltech is developing the blue channel (Jim McCarthy). NIRC 2 is a new diffraction-limited infrared camera (Keith Matthews, Caltech) with multiple image scales to work with the Keck Adaptive Optics system being developed jointly by CARA and the Lawrence Livermore National Laboratory. Additional details are given in Nelson (1995).

The segmented mirror approach works well at longer (radio) wavelengths, as demonstrated by the UK's James Clerk Maxwell Millimetre Telescope completed in 1987, but optical wavelengths are much smaller. There are six different kinds of segments. Each is aspherical and non-symmetrical in shape. The method of manufacture is called stressed-mirror polishing. Forces are applied to the perimeter of a circular blank to distort it in a prescribed way so that it can be polished by conventional means into an axially symmetric shape. When the stresses are released, the segment relaxes elastically into the desired asymmetric shape. Trimming the mirror into its hexagonal form introduces small new strains which add distortions. These effects are removed by a technique called ion-figuring in which a beam of argon ions is used to essentially polish down individual regions. Finally, when each segment is located into the telescope mirror frame, a final fine-tuning of its shape is applied by means of a "warping harness" that exerts a permanent force on the hexagonal segment.

It is one thing to tip and tilt each segment to make all 36 images coincide at a single focus, but this is not the same as making the mirror behave as if it is a single piece of glass!

The light from all the segments must be "phased". What is needed is a set of adjusters attached to each mirror segment which, under computer control, can be manipulated remotely to bring all of the mirror segments into perfect alignment as if they were indeed part of a single giant surface. This is achieved using the Phasing Camera developed by Gary Channan and colleagues at UC Irvine. While the mechanical and optical aspects of this design are challenging enough, it would be impossible to manually set and control this segmented mirror surface; computer control is absolutely essential. Two position sensors are located along every inter-segment edge for a total of 168 sensors. A computer program uses the sensor readings to update 108 position actuators twice every second to keep the mirror in shape as the telescope tracks.

Since about 50% of the cost of any observatory resides with the enclosure or dome, considerable effort is expended to keep the overall length of the telescope to a minimum. In practice this means keeping the focal length of the primary mirror as short as possible despite its diameter, or in other words, making the focal-ratio of the mirror as "fast" as possible; Table 13.1 shows that this ratio is traditionally about 3. The primary mirror of the Keck telescope is an $f/1.75$ hyperbola with a resulting focal length of 17.5 m. Two secondary mirrors are available to provide either an $f/15$ focus or an $f/25$ focus. The latter is quite unusual because it results in a focus location which is before the Cassegrain hole rather than after it; this focus is called the "forward Cass" and is well-suited for infrared work. Incidentally, the dark, infrared-emissive gaps between the segments constitute such a small part of the total area that they have a negligible effect on infrared backgrounds.

The mirror segments were made by Itek and Tinsley, and then ion-figured by Kodak. Will there be more telescopes like the Keck Telescopes? In 1994 contracts were let for the construction of the Spectroscopic Survey Telescope (SST) which is a joint project involving Penn State, University of Texas at Austin, Stanford, Munich and Gottingen. This telescope will have a spherical primary made of 96 small segments, all identical, with an effective diameter of about 9 m and with a Gregorian corrector resulting in a small central obscuration. Moreover, the telescope will remain stationary aimed at a fixed elevation of 35°, rather like the huge 1000 ft wide Arecibo radio dish. Major instruments will be coupled to the telescope by fibre optics. In 1996, Spanish astronomers announced plans for a 10-m telescope almost identical to the Keck telescopes.

13.2.2 The ESO VLT

Equally daring was the proposal announced in 1988 by the European Southern Observatory (ESO) to build an "array" of four independent 8-m telescopes (Fig. 13.9) with the equivalent collecting area of a 16 m telescope on Cerro Paranal in Chile. Cerro Paranal is at an elevation of 2635 m (8645 ft) above sea level and is located in the Atacama Desert about 130 km south of the town of Antofagasta. This is a brand new site, quite distinct from the other locations occupied by ESO (La Silla) and by the Cerro Tololo Inter-American Observatory. Each of the "Unit" telescopes would be alt-az, and each would contain a large monolithic glass meniscus primary mirror ($f/1.8$) which would be so thin for its size that the mirror would require "active" rather than "passive" mechanical support. Hundreds of supporting adjustable "pistons" placed against the back of the mirror exert small forces to compensate for distortions of the figure of the mirror due to gravitational stress, thermal effects, wind-buffeting and slowly varying atmospheric blurring.

Fig. 13.9. The Unit 1 telescope of the ESO VLT project. Four 8-m telescopes with an equivalent collecting area of a 16-m telescope will be located on Cerro Paranal in Chile. Photo courtesy Claus Madsen, ESO.

Forces supplied by the mirror support are updated a few times per second and the entire arrangement is under computer control, responding to an image analyser observing a reference star. The secondary mirror, made of beryllium, is also computer-controlled for focus, centering and tip/tilt.

ESO pioneered many of the required techniques with their 3.5 m New Technology Telescope (NTT) and demonstrated the VLT concept with laboratory trials of a smaller prototype active control system. It would be hard to contemplate such an active control system without the advances in speed and miniaturization of electronic computers. The VLT concept allows for individual operation of each of the Unit Telescopes, incoherent combination to achieve a 16-m light-bucket mode and coherent combination to form an interferometer (the VLTI) with an angular resolution equivalent to a 100–200 m aperture.

Note that each primary is over-sized at 8.2 m and the effective diameter of 8 m is set by the secondary mirrors. The focal ratios are fixed at $f/13.6$ at Cassegrain, $f/15$ at Nasmyth and $f/50$ at coudé. Initially it was believed that removable dome enclosures could be used to improve seeing, but each telescope now has a steel enclosure to minimize buffeting by the wind.

The first 8.6 m diameter mirror blank of 17.5 cm thick Zerodur was successfully delivered by Schott in 1993 to the REOSC Optique factory near Paris where it was ground, figured and polished to its final 8.2 m diameter size before shipping to the VLT site in Chile (Fig. 13.10). VLT Unit Telescope No.1 is expected to go into operation in 1998. A very extensive suite of instruments is planned for each telescope (a total of 12 foci) including medium- and high- resolution CCD spectrographs, near- and mid-infrared cameras and spectrographs and adaptive optics systems. One of the first-light instruments will be a near-infrared imaging system called ISAAC followed by a faint object spectrograph (FORS1). The other instruments are CONICA, NAOS, UVES, VISIR, NIRMOS/WFIS, CRIRES, FORS2, FUEGOS, and SINFONI. Many of the VLT instruments are designed as "multi-mode". Visible light instruments include FORS for imaging and low-resolution spectroscopy, UVES for high-resolution echelle spectroscopy and FUEGOS for multi-

Fig. 13.10. The thin "meniscus" mirror for the Unit I telescope of the VLT. European Southern Observatory photo.

object spectroscopy. At infrared wavelengths ISAAC and VISIR will provide imaging and medium/high-resolution spectroscopy, while CONICA is a high-resolution imaging system with a low-resolution spectrometer mode. Details are given on the ESO World Wide Web pages.

The summit of Cerro Paranal was levelled off and a flat area near the four 8-m telescopes has been retained for later use for an optical interferometer. Smaller telescopes would run along tracks and combine with the four large telescopes.

13.2.3 The Japanese National Large Telescope—Subaru

Subaru—named after the old Japanese word for the young stellar association, the Pleiades (also known to many as the "seven sisters")—is the Japanese National Large Telescope. It too is an 8.2-m alt-azimuth telescope constructed with a meniscus primary mirror fabricated by Corning, USA. Only 20 cm thick, this mirror will be extremely flexible and could never maintain its surface figure against deformations due to gravity as the telescope points to different parts of the sky. By using 264 computer-controlled supporting structures equipped with highly sensitive force sensors, any deformations of the figure of the mirror are actively and automatically compensated for by the control system. Subaru will be located on the summit of Mauna Kea, Hawaii (4200 m elevation) not far from the Keck Telescopes.

Choosing the material for the mirror blank is an important issue. The glass must have good uniformity, high rigidity, and the smallest possible thermal expansion coefficient, otherwise the surface figure will be deformed simply by changes in the ambient temperature. Ultra-low expansion (ULE) glass developed by Corning Glass Works has a linear expansion coefficient of only 10^{-8} per degree Celsius. It requires about 3 years to make approximately 40 boules of ULE glass, each 1.5 m in diameter, and then fuse the boules into a blank of 8.3 m in outer diameter, assembled so as to secure the same expansion properties throughout the blank. That stage of development was completed in mid-1994. Grinding and polishing (by Contraves, USA) requires another 3 years, accompanied of course by numerous measurements of surface shape. The focal length of the primary is 15 m giving a focal ratio of $f/2$ (with a corrector) and a prime focus field of 30 arcminutes. At the Cassegrain focus the focal ratio is $f/12.2$ (or $f/35$ for infrared). Active mirror support employs 264 actuators beneath the primary mirror. Each actuator comprises two parts which provide supporting forces in directions at right-angles to each other. One part is axial support, with a stepper motor and spring, the other part is lateral support which consists of a lever and counterweight. High-precision load sensors are interrogated by the computer-control system every few seconds. This system compensates mainly for gravity, but also for wind loading and residual thermal distortion. The chief technical breakthroughs of this active optics system are the development of the high-sensitivity load sensors, the ability to use computer simulations to study the support and adjustment schemes and engineering experiments on a 60-cm prototype mirror equipped with full-size actuators.

In common with other telescopes, an optical arrangement known as a Hartmann–Shack system analyses stellar images at a rate of about 10 times per second, and a computer calculates the mirror surface deformation and updates the actuators at a maximum rate of once per second. The driving force around both the azimuth and altitude axes will be transferred between huge drums by friction instead of transmission gears to enable, it is hoped, a smoother and accurate drive. Again in common with other telescopes, accurate tracking will employ an automatic guiding system which handles real-time positional measurements of a reference star and feeds back error signals to the drive control system. Since the overall length of the telescope tube structure is about 15 m, several millimetres of deformation are expected due to its own weight. By using the well-known Serrurier truss method, displacements of the primary mirror and the top-end (secondary) ring are parallel to each other.

An important way in which the Subaru differs from other large telescopes is its enclosure or "dome". Most commonly, the classical, semi-spherical dome is used to protect the telescope from the elements, and all domes must be carefully temperature-controlled to keep the mirror and structure in thermal equilibrium with the ambient air. This is quite difficult at most sites where there are large day/night excursions in temperature; it is another attribute of sites like Mauna Kea that these temperature variations are minimal. It is believed, however, that improvements over the dome-shaped enclosure are possible, and detailed studies by the Japanese of a "flushing-type" enclosure have been made using wind-tunnel and water-tunnel modelling. In the end, the design adopted is a cylindrical-shaped enclosure 44 m high that co-rotates with the telescope and which is equipped with a number of active louvres, air filters and air-flushing floor to maintain a fresh, gentle air stream under various outside wind conditions. Using computer models as a basis, the control computer arranges the louvres and air flow to get the best conditions. The Subaru is well-equipped with instruments ranging from the ultraviolet cutoff around 0.3 µm all the way to 30 µm in the infrared. Just one example of developments under way is a mosaic of CCDs comprising 64 devices of 1000×1000 pixels each!

13.2.4 The Gemini Telescopes Project—twin 8-metres

The Gemini Telescopes Project is an international collaboration between the United States, Great Britain, Canada and the South American nations of Chile, Argentina and Brazil. Two identical telescopes will be built, one in the northern hemisphere and one in the southern hemisphere, at Mauna Kea and Cerro Pachon respectively. In the wide-field configuration, image performance goals call for 0.25″ (FWHM) over a 45 arcminute field lying within 20° from the zenith. At 2.2 µm in the infrared the goal is near diffraction-limited imaging using adaptive optics. One of the major drivers for these two telescopes is optimized performance in the infrared. In fact, the goal for the emissivity of the Gemini North telescope on Mauna Kea is a mere 2%. This will be achieved by using silver-coated mirrors which are cleaned regularly with jets of carbon dioxide (CO_2) snow and by using very narrow vanes to support the secondary mirror. Stringent image quality requirements led the Gemini Telescopes designers to opt for a Cassegrain-only arrangement. Since it was no longer required to bring a focus out along the elevation axis, the primary mirror and its cell could be raised so that the centre of mass of the mirror, mirror cell and Cassegrain instrument cluster could be balanced about the elevation axis. This approach results in a minimum mass above the primary mirror which is expected to reduce thermal seeing problems, reduce wind loading, make the primary more accessible for "flushing" to further improve seeing, and provide for a large cluster of four instruments at the Cassegrain focus and an adaptive optics facility.

After some controversy, a review panel in late 1993 selected thin meniscus mirrors for the twin 8-metre telescopes over the competing technology of Roger Angel's honeycombs of borosilicate (see below). Corning was selected to provide the mirror blanks of ULE and REOSC Optique was chosen to do the polishing. The primary mirror support requires 120 actuators and the alignment and figure will be updated via on-board wavefront sensors every few minutes to correct for gravity and thermal deformation. A focal plane sensor will control fast guiding at > 10 Hz via the tip-tilt secondary mirror. The secondary mirror itself is relatively large at a diameter of 1 m, but it will be made from silicon carbide (SiC) to reduce weight (50 kg) and it will have two-axis articulation for fast tip-tilt motion up to 40 Hz. Large panels in the side of the telescope enclosure will open to provide optimum flushing and ventilation, and the primary mirror will be about 20 m above ground level. At the time of writing, the legislated funding for the pair of telescopes was about $176 million and the projected dates for full operation of the telescopes was 1999 for Mauna Kea and 2001 for Cerro Pachon.

13.2.5 The LBT, the Magellan and the MMT Upgrade

Each of these telescopes will benefit from the other major innovation in telescope mirror design, that of honeycomb mirrors, pioneered by Roger Angel of Steward Observatory. To understand his idea we need to recall that for a mirror to focus the light of a distant object it must possess a curved surface—either a spherical or a parabolic surface. The deeper the curve, the shorter the focal length of the primary mirror and therefore the smaller the overall length and the total weight of the telescope. Most classical telescopes have a ratio of focal length-to-diameter of about 3 to 4, which leads to large structures and large buildings (domes). It would be much more economical and provide a stiffer mechanical design if shorter focal lengths could be manufactured. The problem is the depth of the curve and therefore the huge amount of glass which must be "ground and polished away" from the original flat-surface blank. Roger's idea was the following: melt the glass in a rotating furnace and allow it to cool while spinning. During the molten stage the liquid glass will form a deep curve in the shape of a parabola due to the centrifugal force produced by the rotating furnace, and will keep this shape as it cools while spinning. The term for this technology is called "spin-casting". Roger and his team have cast mirror blanks from 3.5 to 7.5 m in size. The mirror can be made very stiff and yet light-weight by using a "honeycomb" construction on the back surface. To do this a mould is made with hexagonal block of ceramic fibre attached to the base of the furnace as shown in Fig. 13.11. As glass is melted into the mould it runs down the gaps between the blocks to form the ribs and backplate. After cooling, the mould is completely removed from the glass using a high-pressure spray of water to break up the ceramic fibre blocks. This casting method produces an internal honeycomb core with 11% of the solid density, with the ribs being 11 mm thick and the faces only 25–28 mm thick. It takes about six weeks to anneal and cool the honeycomb blanks. The honeycomb structure also provides improved thermal performance compared to conventional solid blanks. Although Roger started this amazing endeavour in a backyard furnace in Tucson, there is little argument that it is now a large, sophisticated operation at the Steward Observatory Mirror Lab (SOML), and one which relies heavily on computer-modelling, computer-simulation and computer-control of the furnace environment.

Fig. 13.11. A "honeycomb" mould for a spin cast mirror at Roger Angel's Steward Observatory Mirror Lab is made with hexagonal blocks of ceramic fibre attached to the base of the furnace. As glass is melted into the mould it runs down the gaps between the blocks. Photo courtesy Lori Stiles.

The borosilicate glass suitable for melting and casting may distort because of its larger coefficient of thermal expansion ($\sim10^{-6}$ per °C) compared to materials such as Zerodur and ULE which are highly specialized and partially crystalline. To overcome this effect, the honeycomb cells will be continuously ventilated with air at ambient temperature. Laboratory tests on an earlier 1.8 m mirror showed negligible thermal distortion ($\lambda/50$), and a very rapid thermal time constant of about 35 minutes. Several 3.5 m honeycomb blanks have been made into real mirrors and installed in successful telescopes, e.g. the ARC telescope at Apache Point in New Mexico and the WIYN consortium telescope on Kitt Peak. Two 6.5 m mirrors have also been made, one of these will be used to convert the Multiple-Mirror Telescope (MMT) to a single-mirror telescope and the other is for the Carnegie Institution's Magellan Telescope to be located at Las Campanas, Chile.

The Large Binocular Telescope project (formerly called the Columbus Project) is a unique telescope proposal involving two $f/1.4$, 8.4 m borosilicate honeycomb mirrors on a common mount providing an interferometric baseline of 22.8 m which corresponds to $\lambda/D \sim 5$ milli-arcseconds. Progress on the LBT, which will be located on Mt. Graham in Arizona, will follow the Magellan mirror.

Producing the mirror blank is only the first step. Larger and faster primary mirrors require new polishing methods to achieve their final figure. It is not the deep curvature itself that is the problem, but the asphericity that results in different curvature from place to place and between tangential and radial directions. A rigid pitch lap cannot accommodate the changes of curvature needed as it is stroked over a strongly aspheric surface, unless it is moved only in the tangential direction, but to do this will result in circular grooves or zones. Conventional polishing of 4-m mirrors is typically limited to focal ratios of $f/2$ or slower. To polish the primary mirror of the 4.3 m William Herschel Telescope, David Brown of Grubb Parsons, England (Brown, 1986) used a lap which changed shape as it moved. His method was based on the fact that when a full-sized lap is used to make polishing strokes across a paraboloid, the distortion required to maintain contact is coma. For the same reason, the off-axis aberration of a paraboloid is also coma. There has been considerable development in this field in recent years. The general principle of making the lap change shape as it moves is called "stressed lap polishing".

13.3 REMOTE-CONTROLLED TELESCOPES

Radio astronomers routinely preprogramme their telescopes to perform observations without any further human interaction. Similarly, observatories in space like the Hubble Space Telescope and many others are obviously remotely controlled and once again are programmed to perform a predetermined sequence of observations. Why can't this be done for ground-based telescopes too? The answer is that it can, and it is happening all over the world! Many small telescopes are now in operation which are completely robotic (see the book by Bode in the Wiley–Praxis Astronomy and Astrophysics Series for a review up to 1995 (Bode, 1995)), but scaling this up to the huge telescopes and complex instruments mentioned earlier is quite expensive. Despite this move towards automation and efficiency, there is a degree of reluctance on the part of many astronomers to give up the interactive part of observing which makes it so enjoyable and which often leads to exciting, last-minute changes of plan in response to a recent image or spectrum. So how are these issues being tackled?

13.3.1 Remote observing

With the availability of high-speed, high-density intercontinental links by satellite telecommunications, the possibility for "remote observing" at ground-based observatories, in a manner analogous to remote observing using space facilities, is receiving considerable attention. One of the first demonstrations of remote observing was carried out by the Royal Observatory Edinburgh, Scotland, in the early nineteen-eighties. Using satellite telephone links, the UK Infrared Telescope in Hawaii and an infrared photometer was controlled and data returned to Edinburgh. Other large observatories, notably ESO, and many smaller consortiums of universities, have continued to develop remote observing facilities and methods. The difficulties with full **remote observing** are several:

- All functions must be under computer control, for if even one item requires human presence then the operation is not fully remote. This places very considerable constraints on the observatory and its equipment. Safety of the telescope and instrumentation become a major design issue. There is likely to be a lot more expense arising from the additional equipment needed, and the only way to partially recover such costs is to eliminate staff.
- Reliability and cost of intercontinental links, and their ability to handle large quantities of data such as CCD images. This situation is by no means perfect at the moment, but it is an area of great commercial importance and improvements are likely. Various methods of data compression can be used.

- Proper assessment of climatic conditions at the remote site are needed not only for data quality but for reasons of safety.
- Development of a "mission control centre" or some method of organizing and monitoring access.
- Organization of maintenance, repair and installation at the remote site.

While these problems can all be overcome, given sufficient funding, most large national ground-based facilities have so far opted for a reduced form of remote observing which falls into one of two categories

- remote eavesdropping
- service observing

These techniques are largely self-explanatory. With remote eavesdropping, the distant participant can take part in the observing session by viewing "mimic" displays showing telescope and instrument status, receive data after a short delay and by means of a computer console or telephone can give advice. The actual observing is performed by a qualified Telescope Operator and/or a Staff Scientist. If the guest observers form a large team then one of them can go to the overseas site to represent the team while the others contribute by remote eavesdropping. The Service Observing mode is similar except that the proposers of the scientific experiment are not expected to send a representative, nor are they expected to eavesdrop. The observations are carried out by a qualified staff scientist, the data are transferred to magnetic tape in a standard form and mailed to the proposer. Electronic computer mail systems are used to solicit scientific proposals for Service Observing, the proposals are reviewed by peer review in the normal way, there is a limit of 1–2 hours per proposal, and staff scientists arrange an optimum observing schedule and maintain the selected proposals on-file until completed. Usually the deadline for submission of Service Proposals is only a month in advance of the allotted nights.

These two forms of remote observing are well-established, and more and more effort will be directed toward fully remote-controlled instruments so that, for example, Service Observing can be "pre-programmed" well in advance of the observing run and on the night the operator will simply execute a "command file" or a series of these files. The next step is fully automated control over the telescope and the introduction of "queue scheduling" in a manner similar to radio telescopes. Indeed, this is one of the major goals of the International Gemini Telescopes Project for the 8-m telescopes on Mauna Kea and Cerro Pachon. The idea is that a suite of instruments are attached to the telescope and ready to go at all times, the one in use being selected by the position of a "feed mirror", and the choice of instrument depends on the atmospheric seeing conditions and water vapour content. In other words, several projects are scheduled simultaneously and the one selected depends on the conditions. For example, poorer seeing might be used by a spectrograph for relatively brighter objects, whereas excellent seeing and dark moon conditions would call for a switch to a faint object spectrograph or imager, and a marked improvement in the dryness of the night might argue for a change to a long wavelength infrared instrument. The issues are complex and result in instrumentation and observatory systems which are more highly automated than ever before. Radio observatories and many smaller robotic optical telescopes are helping to lead the way.

13.3.2 Automated imaging telescopes

Complete automation of telescopes and the creation of robotic observatories requiring no human presence except for occasional maintenance is now feasible. Amateur astronomers have played a major role demonstrating the attractions of fully automated small telescopes and observatories. There are now a number of sites where computers and weather monitoring equipment make the decision to open the observatory and begin a pre-programmed set of observations. In fact, there is a network of automated telescopes known as GNAT (Global Network of Automated Telescopes). Initially, the only instrumentation used was a simple photoelectric photometer. Now however, CCDs have led to Automated Imaging Telescopes (AITs).

Another innovative concept, ideal for small teaching observatories and ambitious amateurs, is the "MicroObservatory" idea pioneered by a group at the Harvard-Smithsonian Center for Astrophysics in Cambridge, Massachusetts. The MicroObservatory consists of a cooled slow scan CCD camera attached to a self-contained, weather-protected reflecting telescope on a four-axis mount with stepper motors, encoders and a microcomputer. It is extremely compact and portable. The concept has received support from the National Science Foundation, as well as from Apple Computer Corporation and Kodak.

Unlike remote observing, in which the astronomer controls the telescope and instruments in real time from a distant location, robotic observing, eliminates the astronomer completely! The only task the astronomer must do is submit a list of observations to a computer which controls the telescope instruments and the observatory itself. Provided electronic weather stations signal the computer that all is well, the dome will be opened at the required time, the telescope will be moved to the preset coordinates and the instruments will begin to record data which are transmitted digitally back to the host institute for examination in the next day, or on Monday if the robotic telescope decided to work on the weekend! Of course, this concept is not particularly new. As mentioned earlier, radio astronomers frequently use robotic observing techniques and all of space astronomy is done this way. Optical astronomers with small telescopes have tried this approach also, including many amateur astronomers, but usually with very simple instruments such as photoelectric photometers. In recent years, however, more ambitious "imaging" robotic observing programs have been developed.

The Berkeley Automatic Imaging Telescope, pioneered by Richmond, Treffers and Filippenko in 1993, has now been upgraded, renamed and moved to the Lick Observatory on Mt. Hamilton. This observatory comes equipped with CCD cameras for imaging and guiding, and equally important is the weather station hardware which monitors the temperature, relative humidity (RH) and wind speed. A raindrop sensor ensures that there is no precipitation and an infrared sensor with a 12–13-μm filter pointing at the sky serves as a "cloud" monitor. If the RH or wind speed exceed critical values, or if a single drop of rain hits the raindrop sensor, direct circuits from the weather station to the dome shutter motor over-ride all computer control and close the shutter. In addition, there is a solar cell which prevents opening when the Sun is above 10° altitude and there is a "dead man" timer which closes the observatory if no telescope activity occurs for 20 minutes. All weather data are logged by the computer for later inspection. At the observatory there is a master or host computer which is a Sun SPARCstation 2 running UNIX, and several "slave" PCs all on a local Ethernet-based network running TCP/IP.

Many other robotic observatories are likely to spring up in the next few years, and David Crawford and Eric Craine in Tucson, Arizona, for example, are working to establish a Global Network of Automated Telescopes (GNAT). Their goal is to have six globally distributed sites, each housing several 1-m telescopes. Finding a company to build the desired telescopes within a reasonable budget is the main problem. Past attempts by "small" firms have not been very successful and led to considerable frustration and delays. Informal "networks" of telescopes already exist for certain studies. For example, solar astronomers have established a Global Oscillation Network Group (GONG) to provide much more continuous monitoring of solar oscillations. Together with the SOHO satellite which is positioned at the L1 Lagrange point, this kind of continuous, uninterrupted monitoring will lead to new knowledge about the solar interior and will even force revisions in our understanding of the visible solar atmosphere. Stellar astronomers have also been busy with multi-site campaigns to monitor variable stars of many kinds. Astronomers now talk about the results of "helioseismology" and "asteroseismology".

13.4 SURVEY TELESCOPES

A growing number of specialized telescopes, each highly optimized for a given task and always highly automated, are being designed and built. Most of the these telescopes have moderate apertures (up to about 3.5 m) and they are intended to perform surveys of the sky.

13.4.1 The Sloan Digital Sky Survey and other surveys
This is perhaps the most ambitious of all the "survey" telescopes. This remarkable project, spear-headed by Jim Gunn, uses a dedicated 2.5-m telescope located on Apache Peak, New Mexico, to digitally map half of the northern sky to about 23 magnitude in several bands from U to I and then to select about one million galaxies (about 1%) and 100 000 quasars (about 10%) for spectroscopy using the same wide-field telescope. A five-colour catalogue of all the detected objects will also be produced and the imaging data will be assembled into a high-resolution atlas and a lower resolution map for publication in digital form. As Jim Gunn emphasizes, the scientific value of such a data set is vast, ranging from critical investigations of large-scale structure in three dimensions to the relationships of galaxies with their environments. Funded in part by the Sloan Foundation, the Digital Sky Survey is a joint venture involving Princeton University, the Institute for Advanced Study, Johns Hopkins University, University of Chicago, Fermi National Accelerator Lab, and the Japan Promotion Group.

The detectors (Fig. 13.12) for the photometric survey are an array of thirty (30!) 2048×2048 CCDs manufactured by Morley Blouke's group at SITe in Beaverton, Oregon. The photometric camera for the SDSS consists of two TDI scanning CCD arrays, one uses 30 2048×2048 SITe chips in a 5×6 array for five-colour photometry, and the other using 24 2048×400 chips for astrometry and focus monitoring.

The six wide by five high array is arranged so that two transit swaths cover a stripe 2.53 degrees wide in five independent filters, with a small overlap of about 8.3%. The 22 smaller chips provide astrometric calibration and the remaining two are focus monitors. Scanning is along the five closely spaced columns. The u′, g′, r′, I′, and z′ filters are all custom designed and non-standard but it is optimized for the survey and is a good set for

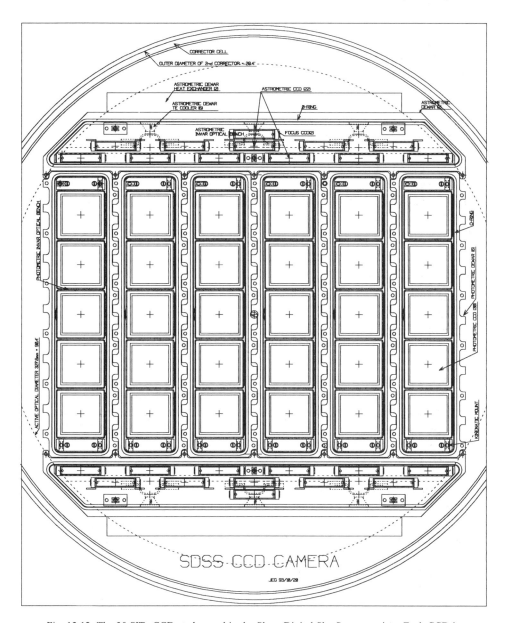

Fig. 13.12. The 30 SITe CCDs to be used in the Sloan Digital Sky Survey project. Each CCD is 2048 × 2048 pixels. Photo courtesy of Jim Gunn.

CCDs. Since the digital sky survey will produce 100 times as much photometry as currently exists, then this system will probably become a standard!

Several other surveys include:

• MACHO and EROS Telescopes: These projects employ large area CCD cameras on small telescopes to systematically survey particular regions of the sky in search of gravitational lensing effects. The EROS project uses a 40-cm $f/10$ reflector with a wide field CCD camera with 16 (2 × 8) three-side buttable $576 × 405$ pixels Thomson THX 31157 chips. The total field is about $1 × 0.5$ degrees oriented so as to follow the bar of the Large Magellanic Cloud. Both the MACHO and the EROS project have already reported brightening of stars with light curves characteristic of micro-lensing by low mass objects in the galactic halo.

- Hobby-Eberly Spectroscopic Survey Telescope: This telescope is a segmented-mirror system similar to Keck which is used at a fixed elevation for faint object imaging and spectroscopy. It is best suited for objects that are uniformly distributed across the sky and/or have low surface densities. It recently achieved "first light".

- 2MASS: This telescope will provide a modern-day equivalent of the 2-μm all-sky infrared survey carried out in the early nineteen-sixties, this time with the benefit of large format infrared array cameras.

SUMMARY

Technological advances involving the way in which telescopes are made and controlled has led to a new generation of very large telescopes, and simultaneously to a new generation of small robotic telescopes. The Hubble Space Telescope and a dozen or so telescopes with apertures from 6.5 to 10 m represent a new era in observational astronomy.

EXERCISES

1. Calculate the diffraction limit for the Hubble Space Telescope ($D = 2.4$ m) at $\lambda = 1$ μm in arcseconds and compare this to the pixel size of the WFPC 2 planetary camera CCD ($0.046''$). Is this satisfactory? What would happen at a wavelength of 500 nm? What strategy would you recommend?

2. Which HST instrument would you use for the following tasks: (i) to obtain images at a wavelength of 2500 Å assuming that you need to at least critically sample the point spread function; (b) to obtain images at a wavelength of 1.6 μm?

3. What was the nature of the aberration in the telescope optics which caused fuzzy images with the Hubble Space Telescope? What were the consequences for imaging with HST? How was this aberration corrected?

4. What are the arguments for constructing a new generation of very large telescopes? Apart from funding, why has it taken so long to consider building telescopes larger than the 5-m (200-inch) Hale Telescope on Mt. Palomar?

5. What are the three main methods for manufacturing very large mirrors for astronomical telescopes? Give an example of a telescope which uses each method.

6. Summarize the advantages and disadvantages of completely remote observing for ground-based astronomy. Why is it harder to come to a clear-cut decision for optical/IR telescopes than for radio telescopes and space telescopes?

7. Calculate the field of view covered by a mosaic of 6 × 5 CCDs with no overlaps if each CCD has 2048 × 2048 pixels mapped to $0.2''$ on the sky. Assuming that each exposure is 20 minutes, how long would it take to cover an area of sky of $10° × 10°$? How long would it take to cover most of one hemisphere of sky?

8.   Essay topic: Discuss the arguments for and against a fully automated, completely robotic observatory with a very large telescope (8–10 metres diameter) for ground-based optical/IR astronomy. Compare and contrast the operational requirements and conditions with the Hubble Space Telescope and the Very Large Array (VLA) radio telescope.

9.   Suppose you plan to construct a robotic observatory that will take observations while you are fast asleep at home or curled up in front of your television set watching re-runs of old science fiction movies. Make a list of the equipment you would need and the steps that the computer program would have to execute from start to finish to ensure both good quality of data and safe operating conditions.

REFERENCES AND SUGGESTED FURTHER READING

Angel, J.R.P. (1989) "The Revolution in Ground-Based Telescopes", *Q. Journal R, astr. Soc.*, **31**, 141–152.

Bode, M.F. (1995) (Ed.) *Robotic Observatories*, Praxis/John Wiley & Sons, Chichester, England.

Brown, D.S. (1986) Technical Digest of OSA Workshop on Optical Fabrication and Test-ing, Seattle, Optical Society of America, Washington, DC.

Crawford, D. and Craine, E.R. (1994) (Eds.) Instrumentation in Astronomy VIII, *Proc. SPIE*, Vol. 2198, Bellingham, USA.

Filippenko, A.V. (1992) (Ed.) *Robotic Telescopes in the 1990's*, Conference Series Vol. 34, Astron. Soc. of the Pacific, San Francisco, CA.

Iye, M, and Nishimura, T. (1995) (Eds.) Scientific Engineering Frontiers for 8–10 m Tele-scopes; Instrumentation for Large Telescopes in the 21st Century, *Frontiers Science Series No. 14*, Universal Academy Press, Tokyo.

Nelson, J. (1995) "The Keck Telescopes and scientific instruments", in *Scientific and Engineering Frontiers for 8–10 m Telescopes*, (Eds.) M. Iye and T. Nishimura, Univer-sal Academy Press, Tokyo, pp.43–50.

Richmond, M.W., Treffers, R.R. and Filippenko, A.V. (1993) "The Berkeley Automated Imaging Telescope", *Pub. Astron. Soc. of the Pacific*, **105**, 1164–1174.

Stepp, L.M. (1994) (Ed.) "Advanced Technology Optical Telescopes V", *SPIE,* Vol. 2199, Bellingham, USA.

14

High angular resolution imaging

The very large collecting apertures of modern ground-based telescopes enable fainter objects to be detected, but astronomers would also like to have the corresponding angular resolution of these giant mirrors. To achieve that goal means overcoming the blurring effects of the Earth's atmosphere. This takes us into the field of adaptive optics and the ultimate prospect of interferometry at very short wavelengths. Computers can now control the actual surface figure of gigantic mirrors, or take a wavefront distorted by its passage through the atmosphere and correct it by means of tiny, rapid changes of a small deformable mirror. It is even possible to create "artificial guide stars" in the sky using laser beacons. In this final chapter we review some of the basics of this rapidly developing field which will ultimately lead to a new level of astronomical imaging.

14.1 ADAPTIVE OPTICS

Active control of telescope optics—generally called **active optics**—to counteract gravity and other deformations is very important (Fig. 14.1), but even at the very best sites, ground-based optical and IR telescopes cannot escape the effects of turbulence in the Earth's atmosphere. In the complete absence of turbulence, the image of a "point source" should be determined only by the quality of the telescope's optics and by the diffraction of light. The characteristic angular size (in radians) of a diffraction-limited image is given by λ/D where λ is the wavelength of the light and D the diameter of the telescope. For example, at a wavelength of 1 μm on a 10-m telescope the value of λ/D is 10^{-7} radians or about 0.02 seconds of arc. This angular resolution corresponds to the orbital radius of the Earth (1 astronomical unit or AU) at a distance of 50 parsecs (pc) or 163 lightyears. Alternatively, this angular resolution corresponds to only 10 times the separation between the Earth and the Moon viewed from the distance of Alpha Centauri (4.2 lightyears). These numbers are very intriguing if only they could be realized in practice. Unfortunately, time-dependent turbulence blurs this image by rapid, random shifts of position resulting in a fuzzy "seeing" disk which can be 10 to 100 times larger in diameter depending on the site and conditions.

Astronomers compare seeing-limited and diffraction-limited images using the **Strehl ratio**—the intensity at the peak of the actual seeing disk divided by the intensity at the

Fig. 14.1. The active mirror support on the ESO New Technology Telescope enables the mirror to maintain perfect shape. This is called active optics. It does not allow high-speed variations in the wavefront due to atmospheric turbulence to be corrected. That requires adaptive optics.

peak of the true Airy diffraction pattern. Typically, the Strehl ratio is ~0.01. If this ratio could be increased to nearer unity, then most of the light would be in the Airy spike and the contrast against the sky background would be increased enormously. Smaller image sizes also mean that narrower slits can be used in spectrographs which in turn implies that the whole spectrometer can be made more compact (see Chapter 5). This is the ultimate goal of **adaptive optics**, a ground-based method of achieving space-based image quality, which has certainly become the major technology push for the future. Several references to this already large and rapidly growing field are given at the end of this chapter and the summary below draws on the work of Babcock, Roddier, Beckers, Fugate, Max and Thompson.

14.1.1 The Fried parameter

What is the effect of the atmosphere? Suppose we have a plane wave arriving at the Earth from a "point" source like a distant star. The wavefronts are distorted randomly by moving cells of air with different indices of refraction. Variations in the refractive index are caused by variations in density which in turn arise from temperature variations in a fully developed turbulent atmosphere. Although there is debate on how well it fits the real atmosphere, the properties of such a turbulent field are usually described statistically in terms of a "structure" function $D(r)$, and the most commonly adopted form is the Kolmogorov turbulence developed in the 1940s which gives the variation in refractive index (n) between two points on the wavefront separated by a distance $(r_1 - r_2)$.

$$D_n(r) = \left\langle \left| n(r_1) - n(r_2) \right|^2 \right\rangle = C_n^2 r^{2/3} \tag{14.1}$$

The stellar wavefront incident on the telescope has spatial variations in both amplitude and phase. Amplitude variations are also called scintillation and contribute much less to image quality and seeing than does phase variations. The phase structure function $D_\phi(r)$ across the entrance of the telescope for Kolmogorov turbulence is given by

$$D_\phi(r) = 6.88 \left(\frac{r}{r_0} \right)^{5/3} \text{rad}^2 \tag{14.2}$$

where

$$r_0(\lambda, \zeta) = 0.185 \lambda^{6/5} \cos^{3/5} \zeta \left(\int C_n^2 \, dh \right)^{-3/5} \tag{14.3}$$

or

$$r(\lambda) = \left(\frac{\lambda}{\lambda_0} \right)^{6/5} \cdot r_0 \tag{14.4}$$

and λ is the wavelength, ζ is the zenith distance and C_n is the structure constant for refractive index variations and is integrated through the atmosphere.

We can therefore characterize the size of the turbulence cells by a length known as the **Fried parameter** (r_0, pronounced r-naught) which is the length over which the wavefront is not significantly perturbed; the larger the better. Typically, $r_0 = 10$ cm at $\lambda_0 = 0.5$ μm. Notice that according to Kolmogorov theory, r_0 is larger at longer wavelengths, i.e. the seeing should be noticeably better at infrared wavelengths than at visible wavelengths. While this is generally observed to be true, it is important to remember that the expressions above are based on theory. Since the turbulent elements responsible for seeing generally last longer than the transit time across their diameters, it is the wind velocities at different heights in the atmosphere that determine the temporal variations in the wavefront. The speed of the turbulence varies with height, but is often described by an average velocity (v) which is about 10 m/s. Typically, the temporal variations across the wavefront are given by

$$\tau_0 \approx 0.314 \frac{r_0}{v} \tag{14.5}$$

which is only a few milliseconds. Substituting the telescope diameter D for r_0 gives an estimate of the slower timescale for seeing image motion. Note that motion in the image plane is independent of wavelength.

Using a "knife-edge" in the focal plane to obscure part of the seeing disk of a star allows the observer to view the illumination of the primary mirror and to detect the direction, speed and size of the changing pattern of turbulent cells. Horace W. Babcock performed pioneering studies of this phenomenon from about 1936 onwards, making many hundreds of visual knife-edge observations at Mt. Hamilton, Mt. Locke, Mt. Wilson, Mt. Palomar and Cerro Las Campanas. The value of r_0 is observed to be site-dependent and also wavelength-dependent, being slightly better in the infrared. At most times at most sites, r_0 is a few centimetres. Values as large as 100 cm at a wavelength of 2.2 μm have been reported for Mauna Kea. If $r_0 \geq D$, the telescope aperture, then seeing effects are negligible. For the very large 8–10 m telescopes described above, even an r_0 of ~1 m is much too small.

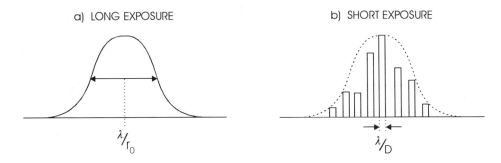

a) LONG EXPOSURE   b) SHORT EXPOSURE

λ/r_0 λ/D

Fig. 14.2. Shows what we could expect when taking a CCD exposure of a star image and then plotting the intensity profiles through the centre of the image.

Fig. 14.2 shows what we could expect when taking a CCD exposure of a star image and then plotting the intensity profiles through the centre of the image. If the exposure time is shorter than $\tau_0 \sim r_0/v$, the time taken for one cell of turbulence to move a distance r_0 across the primary mirror diameter (D), then several small sharp images are seen. The number of small images or "speckles" is approximately D/r_0 and the width of each image is essentially the diffraction width ($\sim \lambda_0/D$). Another short exposure will result in another set of small, sharp speckles but each will be displaced randomly from the first set. When numerous short exposures are taken and added together, or an equivalent long exposure is taken, all the displaced speckles blend together to form a seeing disk of width $\sim \lambda_0/r_0$, instead of an image with diffraction-limited width λ_0/D; unless explicitly stated otherwise, r_0 usually corresponds to $\lambda_0 = 500$ nm. In other words, angular resolution is limited by the size of the turbulence cell (r_0). The larger r_0 can be, the fewer the speckles, the better the long-exposure seeing, the longer the coherence time (τ_0) and the larger the Strehl ratio. Numerous short "speckle" images can be analysed after-the-fact to reconstruct the diffraction-limited image using Fourier analysis techniques. This method is known as **speckle interferometry** and is a very powerful technique. Apart from data handling, the primary drawback of speckle interferometry is that short exposures are required which in turn places a heavy demand on detector performance. Short exposures tend not to be limited by the fundamental photon noise of the background, but rather by the readout noise of the detector (unless this is exceptionally low). This restricts the application of the method to brighter stars. Fig. 14.3 illustrates, however, just how very impressive this technique can be. Two infrared observations by Andrea Ghez of UCLA at a wavelength of 2.2 μm of close binary stars are shown. Normal direct, seeing-limited imaging shows a single

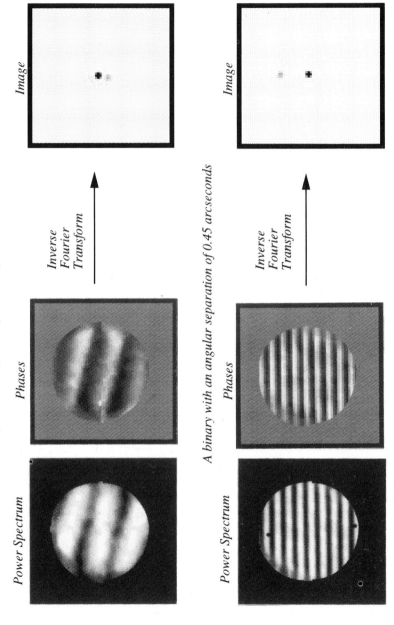

A binary with an angular separation of 0.15 arcseconds

Power Spectrum

Phases

Inverse Fourier Transform

Image

A binary with an angular separation of 0.45 arcseconds

Power Spectrum

Phases

Inverse Fourier Transform

Image

Fig. 14.3. Illustrates the speckle technique from infrared observations with the Keck 10-m telescope. The fringe patterns on the left are the power spectra, or square of the Fourier amplitudes. Note that the spacing of the fringes decreases as the separation of the binary star increases. Next to the power spectra are the Fourier phases. An inverse Fourier transform gives the normal image. A binary of 0.15 arcseconds is cleanly separated from its primary. Courtesy Andrea Ghez, UCLA.

object, but speckle imaging and reconstruction reveals the extremely close compan-
ions.

14.1.2 Correcting the wavefront

So how do we get the value of r_0 to approach D? The first step is to compensate for the
random wandering of the centre (or centroid) of the seeing disk by means of a simple
tip-and-tilt, two-dimensional motion of a mirror, which could even be the secondary mirror of the telescope if it is not too large, and in this way redirect the overall wavefront to
the same (fixed) place in the image. The second step is more complex. An optical arrangement is used to project a real, demagnified image of the primary mirror of the telescope
onto a much smaller mirror whose "shape" can actually be changed or deformed by the
forces applied by numerous small actuators behind it. To do this, a high-speed CCD camera detects the changes in the slope of the coherent areas of turbulence, also referred to as
sub-apertures. By analysing the wavefront distortions rapidly, it is possible to control the
figure of the **deformable mirror** in such a way as to compensate for the wavefront distortions (Fig. 14.4).

In practice, what must be done is to describe the phase variations in the wavefront in
terms of some algebraic quantities called Zernicke polynomials $Z_i(n,m)$ in which n is the

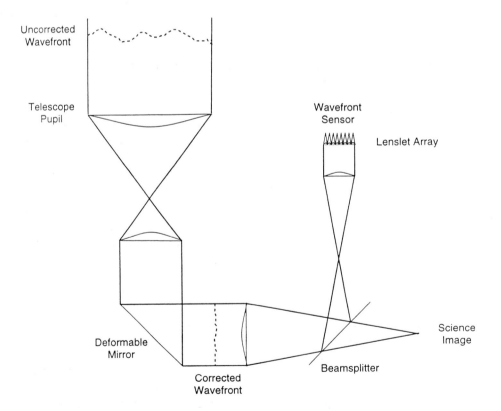

Fig. 14.4. A simple scheme for implementing adaptive optics using a deformable mirror.

degree of a radial polynomial and m is the azimuthal frequency of a sinusoidal term. Noll (1976) gives normalized versions of the Zernicke polynomials such that the rms value of each polynomial over the circle is unity. Table 14.1 lists the low-order terms and explains their meaning. The final column gives the mean square residual amplitude Δ_j in the phase variations at the telescope entrance caused by Kolmogorov turbulence *after* removal of the first j terms; the normalization factor is $S = (D/r_0)^{5/3}$. For large j the approximate value of Δ_j is given by

$$\Delta_j \approx 0.2944\, j^{-0.866} \left(\frac{D}{r_0}\right)^{5/3} \mathrm{rad}^2 \tag{14.6}$$

By substituting the appropriate terms and taking the square root, Table 14.1 yields the rms phase variation for each mode of correction. With no correction, the rms phase variation is $0.162\,(D/r_0)^{5/6}$ waves while after tip-tilt correction the rms phase variation is reduced to $0.053\,(D/r_0)^{5/6}$ waves. If SR is the Strehl ratio, then the Maréchal approximations give

$$1 - SR \approx \Delta \approx 1 - e^{-\Delta} \tag{14.7}$$

for small wavefront disturbances. According to the above theory, it requires a huge number of terms to achieve a Strehl ratio as high as 80% under average seeing conditions on a very large telescope, which in turn implies a large number of actuators on the deformable mirror.

Table 14.1. Modified Zernicke polynomials and the mean square residual amplitude for Kolmogorov turbulence after removal of the first j terms

Z_j	n	m	Expression	Description	Δ_j/S
Z_1	0	0	1	constant	1.030
Z_2	1	1	$2r \sin\phi$	tilt	0.582
Z_3	1	1	$2r \cos\phi$	tilt	0.134
Z_4	2	1	$\sqrt{3}(2r^2-1)$	defocus	0.111
Z_5	2	2	$\sqrt{6}r^2 \sin 2\phi$	astigmatism	0.0880
Z_6	2	2	$\sqrt{6}r^2 \cos 2\phi$	astigmatism	0.0648
Z_7	3	1	$\sqrt{8}(3r^3-2r) \sin\phi$	coma	0.0587
Z_8	3	1	$\sqrt{8}(3r^3-2r) \cos\phi$	coma	0.0525
Z_9	3	3	$\sqrt{8}r^3 \sin 3\phi$	trifoil	0.0463
Z_{10}	3	3	$\sqrt{8}r^3 \cos 3\phi$	trifoil	0.0401
Z_{11}	4	0	$\sqrt{5}(6r^4-6r^2+1)$	spherical	0.0377

14.1.3 The isoplanatic patch

Unfortunately, the correction process described above has several limitations. First, the star being used as a reference for the correction process must be sufficiently bright to provide a good signal-to-noise ratio in each of the sub-apertures of the turbulence pattern across the primary mirror. Generally, the target object being observed will be much too faint to be used as its own reference. Secondly, the effectiveness of the compensation decreases the further the reference star is away, in angular extent, from the object to be observed, simply because the wavefront distortions are not the same. Perfect compensation is limited to an angular patch of radius a few arcseconds around the reference star. This region is known as the **isoplanatic region**.

The isoplanatic angle is usually defined as the radius of a circle over which the wavefront disturbance is essentially identical and is given approximately by

$$\theta_0 = 0.314 \frac{r_0}{H} \tag{14.8}$$

Where H is the average distance of the seeing layer. Sometimes the term is also used to refer to the angular distance over which image motions are practically the same as compared to seeing widths. In this case it is better to use the term "isokinetic patch" which is of order 0.3 D/H. As an example, consider an 8-m telescope at a site for which $r_0 = 13.3$ cm (corresponding to seeing of 0.75″ at $\lambda = 500$ nm) and assume that the distance to the seeing layer is $H = 5000$ m. In this case, the radius of the isoplanatic patch is only 1.7″, whereas the isokinetic patch is much larger at about 100″.

Military applications of this kind of technology have accelerated industrial developments. Deformable mirrors currently use a thin "face" sheet of low-expansion glass sup-

Fig. 14.5. (a) An adaptive optics system developed at the Lawrence Livermore National Labs for the Lick Observatory, and as a prototype for the Keck telescope is shown.

ported on an array of many small, discrete actuators, which consist of several stacked and
bonded layers of lead-magnesium-niobate (PMN) known as electrostatic actuators.
Among the several types of wavefront sensors available, there has been a general trend
towards the Hartmann–Shack type with many sub-apertures. The basic construction is that
of an array of tiny lenses or "lenslets" which is placed near a pupil image to produce a
pattern of many star images on the detector each corresponding to a different part of the
primary aperture. By rapidly finding the centroids of each image it is possible to derive the
slope of the wavefront at that instant.

Many astronomy groups are working on adaptive optics systems. One of the earliest
successes was the COME-ON+ system (now called ADONIS) developed by Meudon and
ESO for the ESO 3.6 m telescope. That system used 52 actuators and a correction rate of
400 Hz. Formerly classified experiments in the USA demonstrated 241 actuators and correction rates of 2 kHz. A system developed at the Lawrence Livermore National Labs for
the Lick Observatory and as a prototype for the Keck telescope is shown in Fig. 14.5 (a,b).
The Hartmann–Shack wavefront sensor is configured with a triangular array of 37 sub-
apertures each of which maps to a diameter of 46 cm at the primary mirror of the 300-cm
telescope. Images from the lenslets are recorded using a special high QE, low-noise, very
fast CCD camera built by Adaptive Optics Associates using a 64 × 64 pixel chip developed by Lincoln Laboratory. The wavefront re-constructor is based on four Intel i860
chips on a Mercury VME board and is controlled by a Unix workstation. The deformable
mirror (or DFM) has a triangular array of 127 PMN actuators with a nominal stroke of
8 μm at 80 volts and the inter-actuator spacing is 11 mm at the DFM which corresponds

Fig. 14.5. (b) A close up of the deformable mirror and its actuators. Courtesy Claire Max, LLNL.

to the 46 cm apertures mapped onto the telescope. A separate tip-tilt system is also in-cluded to conserve the range of the DFM and additionally, it is needed to provide tip-tilt information from a natural guide star during operation of the system with laser guide stars. Note that both sides of the optical bench are used.

All major observatories are now pursuing programs on adaptive optics. For example, at the 4.2-m William Herschel Telescope (WHT), MARTINI is a series of adaptive optics systems located on the Ground-based High Resolution Imaging Lab (GHRIL) optical bench at one of the Nasmyth platforms. Just what can be achieved with these systems is illustrated by the images obtained by Francois Roddier, a pioneer in this field, and his associates using the CFHT (see colour plates). These remarkable images provide direct answers to questions and end years of speculations.

14.1.4 Laser guide stars
When no suitable reference star is found within the isoplanatic patch, then adaptive optics systems cannot function. A remarkable alternative approach is now becoming available

Fig. 14.6. A laser beam emerging from the 120-inch telescope at Lick observatory *en route* to the sodium layer at 92 km. The return beam provides a laser-generated "guide star". Photo courtesy Claire Max, LLNL.

however. The idea is, quite simply, to create an "artificial star" using laser beams (Fig. 14.6). First suggested in non-classified literature by Foy and Labeyrie in 1985, the concept is as follows. A pulsed laser beam tuned to the wavelength of the orange-coloured sodium D resonance line is projected through a telescope and focussed on the so-called sodium (Na) layer in the upper atmosphere at an altitude of 92 km. Resonance fluorescence in this layer produces a glowing, artificial star. There is another kind of laser-generated reference star which employs "sodium" stars excited by "back-scattering" of the laser beam by air molecules in the stratosphere. Fig. 14.17 shows the benefit of the Na laser method. If $r_0 \approx 20$ cm, then the predicted pulsed-laser power required at the zenith is ≈ 6 watts. Outgoing laser pulses return from the Na layer at 92 km in only 700 µs, retracing their path exactly, enabling the adaptive optics system to update the shape of the deformable mirror.

One possible problem with **laser guide stars** is a rapid degradation with angle away from the zenith, especially so for angles $\geq 30°$. Unfortunately, as pointed out by Horace Babcock, restricting one's use of artificial guide stars to the zenith is in direct conflict with the trend toward alt-azimuth mountings for very large telescopes. The required rapid rotation of the azimuth bearing in these telescopes exceeds practical limits for stars whose rising-and-setting arc passes through or near the zenith; the result is an inaccessible zone of a few degrees around the zenith. It is unlikely that this issue will outweigh all the other combined advantages of alt-az telescopes and adaptive optics! When adaptive optics systems become routine, corrections for imperfect seeing—even if only partially

SOR Laser Beacon Adaptive Optics Results: β–Del (0.199 arcsec)

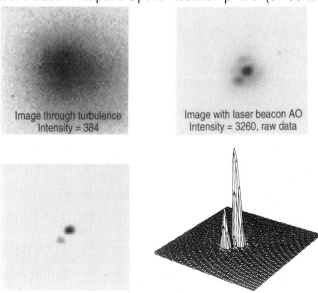

Image through turbulence
Intensity = 384

Image with laser beacon AO
Intensity = 3260, raw data

Processed by iterative deconvolution
60 second exposures, 1.66 arcsec fields, 0.85 µm imaging wavelength, 1.5 m aperture

Fig. 14.17. Image improvements using adaptive optics at the Starfire range. Courtesy Bob Fugate, USAF Phillips Labs.

achieved—will improve the Strehl ratio and significantly revitalize telescopes on poorer sites all around the world.

It should be clear that the term active optics is used to describe ways of controlling the wavefront distortions introduced not by the atmosphere, but by optical, mechanical or thermal deficiencies in the telescope itself. Such distortions vary only slowly with time, compared to the rapid variations caused by the atmosphere. Corrections in the adaptive optics systems must occur at rates of 10–1000 Hz (depending on wavelength), whereas "active" systems like the ESO-NTT work at 1 Hz or slower.

The sodium (and potassium) layer at 92 km is about 11.5 km thick and has a column density of neutral sodium of about 2×10^9 atoms/cm^2. It is believed to be a result of meteoric dust. When illuminated by the laser, Na atoms are radiatively excited to a higher energy level from which they return to the ground state either by spontaneous emission ($\sim 10^{-8}$ s) in all directions or by stimulated emission in which the photons are emitted in the same direction as the incoming beam. Some of the downward-emitted photons are collected by the telescope to form an artificial star. As the laser power increases, the brightness of the laser guide star increases too, up to a saturation point corresponding to the spontaneous emission becoming dominant (so that most of the light goes outward). For a luminous patch of 50 cm, corresponding to an angular size of

$$\frac{0.5 \text{ m}}{92\,000 \text{ m}} \times 206\,265 \approx 1''$$

(14.9)

the pulsed laser power is about 5 kW for saturation.

When the laser transmitter is displaced from the telescope, the configuration is called "bistatic", whereas for the case in which the telescope itself transmits the laser beam the term applied is "monostatic". The positional offset in the bistatic configuration causes the laser guide star to be elongated, which decreases the sensitivity of wavefront corrections. For example, if $d = 10$ m then the elongation is about 3 arcseconds. While most new telescopes install monostatic laser systems, the bistatic configurations are advantageous if it is required to decouple the laser transmitter from the telescope, which is a help in the case of a retro-fit. Bistatic configurations also ensure that back-scattering effects in the lower atmosphere will fall outside the isoplanatic patch. Jacques Beckers has proposed an ingenious scheme for removing the elongation effect. Essentially, a pulsed laser guide star is viewed with microsecond time resolution snapshots. In each snapshot the laser guide star appears as a round spot, but the spot moves across the sodium layer in about 60 μs. Rapid tracking of this motion in the Hartmann–Shack wavefront sensor, by moving the charges along the columns of a CCD at the same rate, should result in a sharp image and good sensitivity.

Robert Fugate and others have used lower atmosphere, Rayleigh scattering laser guide stars successfully for wavefront sensing at the USAF Phillips Laboratory on the 1.5-m telescope at the Starfire Optical Range in Albuquerque, New Mexico. (Examples in Figs 14.7 and 14.8.)

Typically, the laser guide stars have visual magnitudes in the range 12–14, which is bright enough for full adaptive optics at near infrared wavelengths of 1 μm or longer. Claire Max and her team at the Lawrence Livermore National Lab in the USA are one of the groups working on producing a very bright (magnitude 9 or brighter) guide star which

will be suitable for visible wavelength corrections. The LLNL system has been demonstrated on the University of California Observatories (UCO) 3-m Shane telescope at Lick
Observatory. It is based on a 127-actuator continuous-surface deformable mirror, a Hartmann–Shack wavefront sensor equipped with a fast-framing low-noise CCD camera, and
a pulsed solid-state-pumped dye laser tuned to the atomic sodium resonance line at 589
nm. The laser typically generates about 22 W of power and can produce a guide star with

SOR 1.5 m Telescope Images

Trapezium region in H-α light (0.6563 μm)
4 minute exposures, 40 arcsec field

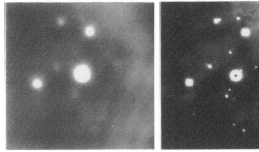

Uncorrected image   Real time image correction
                           with laser beacon adaptive optics

SOR 1.5 m Telescope Image — Laser Beacon Adaptive Optics

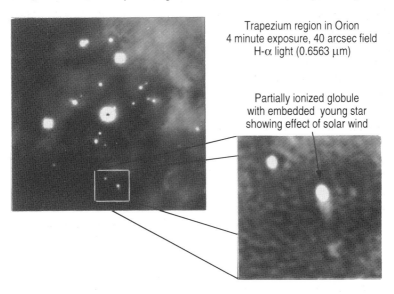

Trapezium region in Orion
4 minute exposure, 40 arcsec field
H-α light (0.6563 μm)

Partially ionized globule
with embedded young star
showing effect of solar wind

Fig. 14.8. A region of the Orion Nebula imaged with the USAF Starfire system. Angular
resolution is comparable to the Hubble Space Telescope. Photo courtesy Bob Fugate.

a V magnitude of about 8 and an image of about 1.0″ in width with seeing conditions of 0.5″; the diffraction-limited size of the guide star in the sodium layer is 0.6″ and of course the beam traverses the atmosphere twice. Strehl ratios as high as 0.7 at a wavelength of 2.2 μm in good seeing are possible. This same system has already demonstrated up to a factor of 12 increase in image peak intensity and a factor of 6.5 reduction in image full width at half maximum (FWHM) using natural guide stars.

14.2 INTERFEROMETRY

We have already discussed **interferometry** very briefly under the topic of imaging at radio wavelengths, but what are the prospects for similar techniques at very short wavelengths? The answer is surprisingly encouraging, given steady improvements in technology.

It was the English physician and physicist Thomas Young who demonstrated the wave nature of light in 1802 and subsequently showed that the spatial structure of an object could be deduced without forming a direct image but by analysis of wave interference effects. In this way, unresolved distant objects such as stars might be studied. In fact, Fizeau proposed an experiment to measure stellar diameters as early as 1868 and Stephan attempted the experiment on Sirius (α CMa) using two apertures separated by 50 cm on the 80-cm refractor at Marseilles, France, in 1873. He saw interference fringes and concluded that much larger separations were needed. Michelson in 1890 suggested using two separate mirrors far apart and then combining the beams for visual inspection. The diameters of several red supergiant stars such as Betelgeuse (α Ori) were measured by Michelson and Pease from 1920 to 1930. Further advances were beyond the technology of the time. Lasers, solid-state detectors, high-speed electronics and computers have changed all that. Long-baseline, large-aperture interferometry at infrared and optical wavelengths has the promise of yielding exceptional spatial resolution.

The push is clearly on towards high-resolution imaging. Speckle interferometry or adaptive optics using deformable mirrors are one way. Both of these approaches make use of "filled-aperture" telescopes. Technology is now available to exploit the aperture synthesis techniques developed by radio astronomers in instruments working at optical and infrared wavelengths to achieve exceptional angular resolution. These unfilled aperture telescopes could yield images with 1 milliarcsecond resolution.

There are essentially two approaches. Aperture synthesis in which the distribution of brightness across a source is the Fourier transform of the complex visibility of the fringes from many baseline pairs observed sequentially in a Michelson stellar interferometer. Alternatively, there is the closure phase method in which simultaneous phase measurements on a triangle of three baselines formed by three apertures yield a summed or closure phase which is independent of unknown atmospheric effects. Jack Baldwin and Craig Mackay were the first to demonstrate optical closure phases and they have developed a prototype instrument known as the Cambridge Optical Aperture Synthesis Telescope (COAST). For optical detection avalanche photodiodes are used, but these investigators plan to use HgCdTe infrared array devices too. The next step is to move COAST to an excellent mountain site.

At the time of writing, many diverse experiments are under way to develop spatial interferometers. For example, the Keck Telescopes and the ESO VLT have both anticipated the need for interferometry. See Breckinridge (1994) and subsequent proceedings in that series for a review.

14.3 THE FUTURE

Among the many possible goals of ground-based and space-based interferometers is the detection and characterization of planets around stars other than our Sun. Whether such ambitious goals are possible or not, only time will tell. All of our knowledge and understanding of the universe revolves around the quality of our measuring apparatus. Certainly, further advances in the kind of electronic imaging technology described in this book will occur. Moreover, fundamentally new approaches will emerge. For example, it is possible to construct a panoramic detector consisting of an array of **superconducting tunnel junctions** (STJ) which are essentially photon-counting detectors with high quantum efficiencies from the visible to the near infrared *and* with the ability to discriminate the energy of individual photons. An array of STJ devices would constitute a combined camera and spectrograph—a three-dimensional camera—involving no gratings or prisms to disperse the light.

A superconducting tunnel junction (Perryman *et al.*, 1994) consists of two films of superconductor sandwiching a thin insulating layer. When operated at temperatures below the critical temperature of 1 K (i.e. below about one degree above absolute zero), the equilibrium state is easily disturbed by an incident photon. By applying a magnetic field and a small bias voltage across the junction, an electrical charge proportional to the energy of the perturbing photon can be extracted. Although STJ devices have been considered for X-ray applications, progress has led to the possibility of using the method for UV and visible photons too.

Another new field which looks very promising is imaging of hard X-rays and gamma-rays with solid state detectors. Neil Gehrels (1995) gives an excellent short review of prospects. Of course, solid state detectors have been flown on gamma-ray astronomy instruments since the 1960s, but the only true "imager" was a small demonstration instrument developed by Skinner and colleagues at the University of Birmingham, England. That device had a 3×3 array of close-packed germanium (Ge) detectors each $1 \times 1 \times 6$ cm, and a passive coded-aperture mask which produced images of the Crab and Cyg X-1. Recent developments in germanium segmented-electrode technology and room temperature solid-state detectors of CdZnTe and purer HgI_2 offer the promise for hard X-ray and gamma-ray imaging. Medical imaging interest in this technology has increased the market and made larger arrays feasible. Arrays of thousands of 1-cm detectors operating in the few keV to few 100 keV range with coded masks giving sub-arcminute angular resolutions may soon be practicable.

The years since the invention of the CCD in 1970 have been a remarkable period of development for astronomy. New technologies have led to new discoveries and the promise of even greater discoveries has stimulated the drive for even better instrumentation. As expected, the access to space has meant that gamma-ray, X-ray, and ultraviolet astronomy have all blossomed. Infrared astronomy has undergone a tremendous surge with the advent of array detectors. New telescopes and new detectors for millimetre wavelengths have finally opened up that part of the radio spectrum too. Conventional ground-based optical astronomy, far from becoming entrenched, has continued to expand with new telescopes and better CCD detectors. Even the very method of building optical reflecting telescopes has undergone radical changes. Throughout this book I have tried to show that the underlying reason for these advances has been a willingness of astronomers to grasp the very latest technologies and push them to their limits. I hope that that will always be true.

SUMMARY

High angular resolution from ground-based telescopes is a top priority and all new facilities will employ both active and adaptive optics in an effort to achieve diffraction-limited performance. Advances in technology have now enabled long-baseline optical/IR interferometers to be developed which could yield spatial information thousands of times better than the very best direct images.

EXERCISES

1.   What is the diffraction-limited angular resolution of two 10-m telescopes separated by 85 m? What is the light-gathering power of the combination? Why is it more important to combine the two telescopes by interferometry rather than simply to "add" their light?

2.   Estimate the angular resolution for a diffraction limited telescope with a collecting aperture of 10-m diameter operating at a wavelength of 2 μm in the near infrared. What is the corresponding linear dimension in miles at the distance of Alpha Centauri (4.2 lightyears)? Suppose your telescope formed an interferometer with a baseline of 100 m, what would the resolution be now? Compare these numbers to distances within the Solar System. Are you impressed? [Assume 1 ly ~ 6×10^{12} miles.]

3.   Explain the terms Fried parameter and isoplanatic patch? What is the order of magnitude of the Fried parameter for an 8-m telescope with seeing of 0.75″ at (a) visible wavelengths (5000 Å) and (b) near-infrared wavelengths (2.2 μm)?

REFERENCES

Babcock, H.W. (1953) *Publ. Astr. Soc. Pacific*, **65**, 229.

Babcock, H.W. (1990) "Adaptive Optics Revisited", *Science*, **249**, 253–257.

Beckers, J.M. (1993) *Ann. Rev. Astron. & Astrophys.*, **31**, 13.

Beckers, J.M. (1996) "Techniques for High Angular resolution Astronomical Imaging", *Instrumentation for Large Telescopes*, J.M. Rodriguez-Espinoza (Ed.), Cambridge University Press.

Breckinridge, J.B. (1994) (Ed.) *Amplitude and Intensity Spatial Interferometry II*, SPIE Vol. 2200.

Fried, D.L. (1965) "Statistics of a geometric representation of wavefront distortion", *J. Opt. Soc. Am.*, **55**, 1427–1435.

Gehrels, N. (1995) "Hard X-ray and gamma-ray imaging with solid state detectors", *Experimental Astronomy*, **6**, 129–135, Kluwer Academic Publishers, Dordrecht, Netherlands.

Noll, R.J. (1976) "Zernicke Polynomials and Atmospheric Turbulence", *J. Opt. Soc. America*, **66**, 207–211.

Olivier, S.S., An, J., Avicola, K., Bissinger, H.D., Brase, J.M., Friedman, H.W., Gavel, D.T., Max, C.E., Salmon, T., and Waltjen, K.E. (1995)?, "Performance of laser guide star adaptive optics at Lick Observatory", SPIE.

Perryman, M.A.C., Peacock, A., Rando, N., van Dordrecht, A., Videler, P. & Foden, C.L. (1994) *Frontiers of space and ground-based astronomy*, W. Wamsteker *et al.* (Eds), Kluwer Academic Publishers, Dordrecht, Netherlands.

Roddier, F. (1981) "The Effects of Atmospheric Turbulence in Optical Astronomy", *Progress in Optics*, **19**, 281–377.

Roddier, F. (1994) *Adaptive Optics for Astronomy*, SPIE Vol. 2201.

Thompson, L.A. (1994) "Adaptive Optics in Astronomy", *Physics Today* **47** (12), 24–31.

Glossary

aberrations Effects associated with the performance of optical components which give rise to imperfect optical images.

absolute zero A temperature of −273 Celsius or 0 kelvins (K). The temperature at which an atom would have no energy of motion.

active optics Controlling the shape of a telescope mirror at a relatively slow rate.

adaptive optics Compensating for atmospheric distortions in a wavefront by high-speed changes in the shape of a small, thin mirror.

ADC (analogue-to-digital converter) An electronic circuit which takes an input voltage in a given range (typically 0–10 volts) and provides a corresponding digital output by setting output lines (bits) high or low. A 16-bit ADC has 16 output lines.

ADU analogue-to-digital units. See DN.

AIPS Astronomical image processing system. Developed by the National Radio Astronomy Observatory.

Airy diffraction disk The central spot in the diffraction pattern of the image of a star at the focus of a telescope. Named for Sir George Airy (1801–1892), seventh Astronomer Royal.

altitude-azimuth (alt-az) mounting A form of mounting similar to that of a radar which allows the telescope tube to be moved horizontally (by rotation in azimuth or compass direction) and vertically (by rotation in altitude or elevation). To follow a star the telescope must be adjusted simultaneously in both axes.

anamorphic magnification The difference in magnification along the spectrum and perpendicular to the spectrum in a spectrograph.

angstrom A unit of length equal to 10^{-10} metres; symbol Å. Named for Swedish physicist Anders Jonas Ångström (1814–1874).

angular dispersion The rate of change of angle (due to refraction or diffraction) with wavelength of the emergent beam in a spectrograph.

antenna (or **aerial**) The part of a radio telescope responsible for detecting an electromagnetic wave.

anti-reflection coating (also **AR coating**) A layer of material of lower refractive index of just the right thickness ($\frac{1}{4}$ wave) is deposited on the optical surface to be coated. More complex coatings are possible which cover a large wavelength range.

aperture photometry Usually refers to magnitude measurements made from digital images by deriving the flux that would have been recorded within a circular aperture large enough to enclose the star's seeing disk.

aperture synthesis The method of combining the signals received by several smaller telescopes distributed over a very large area or baseline to provide the angular resolution of a much large telescope. Used extensively in radio astronomy, e.g. the VLA.

archiving Making a permanent record which can be accessed later at any time.

aspheric An optical surface with departures in shape from a perfect sphere in order to cancel optical imperfections or aberrations.

astigmatism An optical aberration in which off-axis points tend to become elongated ellipses.

astrometry The accurate determination of the positions and motions of celestial objects.

atmospheric dispersion corrector An optical device usually comprising two thin prisms which can rotate to compensate for the elongation of a star image caused by the wavelength dependence of the refractive index of air.

back focal length The distance between the last surface of a compound optical system and the focal plane of the system. This distance may be quite different from the actual focal length.

band gap See forbidden energy gap.

bandwidth Refers to either a wavelength interval (or band) or a frequency interval.

bias (1) The offset from zero in the input of an array detector. (2) A voltage applied across a detector.

BIB Blocked impurity band. See impurity band conduction.

binning On-chip binning. Combining data from a rectangular group of pixels on the CCD itself by appropriate clocking. An improvement in signal-to-noise ratio is achieved at the loss of some picture detail (or spatial resolution).

birefringent (also **birefringence**) Having two values of index of refraction. See refractive index.

blackbody This is the name given to an object which absorbs all radiation incident on it. Such a body is also the most efficient emitter of radiation. The simplest model of a blackbody is a small hole in a matt-black, heated hollow chamber. Since radiation entering the hole has minimal chance of re-emerging, radiation leaving the hole will be characterized by the temperature of the chamber. The amount of radiant energy emitted by a blackbody at any wavelength depends only on the absolute temperature of the body, and is given mathematically by the Planck function. Certain astronomical sources are almost blackbody emitters, but most objects are not quite black enough and therefore have an emissivity of less than 100% of ideal.

blaze angle The tilt of the facets or grooves of a diffraction grating.

blocking layer An undoped layer in an extrinsic infrared detector which converts the action of the photoconductor to a behaviour more like a photovoltaic detector.

BN (Becklin–Neugebauer object) One of the first major discoveries of infrared astronomy. Eric Becklin and Gerry Neugebauer at the California Institute of Technology (Caltech) found a very luminous infrared source with no visible counterpart buried deep in the Orion Molecular Cloud.

bolometer A device for measuring the total amount of radiant energy received from a celestial object.

brightness temperature A statement of the brightness of a radio object in which power is converted to a temperature using the Rayleigh–Jeans law, irrespective of whether or not the source is thermal.

Brown Dwarf A self-gravitating, self-luminous gaseous object which is not sufficiently massive to result in thermonuclear hydrogen fusion reactions in its core and cannot therefore be considered a star. Such objects are expected to have a mass less than 7% of the Sun's mass and represent a "missing link" between low-mass stars and gas giant planets like Jupiter (at 0.1% of the Sun's mass).

buried channel A construction in a silicon CCD which results in a collection zone for photo-generated electrons which is buried well below the surface of the silicon. The zone is produced by a combination of a pn junction within the silicon and a metal electrode on the surface of the silicon.

bus The general term for hardware for dealing with the input–output pathway and back-plane of a computer. There are many types, e.g. IEEE-488 bus, Q-bus, VME-bus.

byte A group of eight "bits" or binary digits (ones or zeros). Two bytes (16 bits) make a word. A megabyte is one million bytes and a gigabyte is one thousand megabytes.

Cassegrain (focus, telescope) Refers to a design of reflecting telescopes in which the light collected and focussed by the large concave primary mirror is refocussed by a smaller convex secondary mirror on the same axis as the primary. The refocussed beam passes through a central hole cut into the primary mirror and emerges behind the primary.

cathode-ray tube (CRT) Basis of the TV tube and the oscilloscope. Electrons emitted by a heated filament are channelled into a very narrow beam and steered by electric fields between charged plates to impact a phosphorescent screen which emits a flash of light for each collision.

charge bleeding The overflow of charge up and down a column in a CCD when the pixel or storage well becomes saturated with photoelectrons.

CCD (charge-coupled device) A small photoelectronic imaging device (typically 1.5 cm square) made from a crystal of semiconductor silicon in which numerous (at least 250 000) individual light-sensitive picture elements (pixels) have been constructed. Each tiny pixel (less than 0.03 mm in size) is capable of storing electronic charges created by the absorption of light. The name derives from the method of extracting the locally stored charges from each pixel which is done by transferring or "coupling" charges from one pixel to the next by the controlled collapse and growth of adjacent storage sites or "potential wells". Each "well" is formed inside the silicon crystal by the electric field generated by voltages applied to tiny, semi-transparent metallic electrodes on the CCD surface.

CD-ROM (compact disk - read only memory) A computer data storage technology. The disk resembles an audio compact disk 120 mm (4.75 in) in diameter, with each platter containing digital information accessible by a laser beam reading system.

CDS (correlated double sampling) A technique used with CCDs to remove an unwanted electrical signal, associated with resetting of the tiny "on-chip" CCD output amplifier, which would otherwise compromise the performance of the detector. It involves making a double measurement of the output voltage before and after a charge transfer and forming a difference to eliminate electrical signals which were the same, i.e., correlated.

Cepheid variables A class of stars named after Delta Cephei which vary in brightness over a regular period of time (typically a few days). The period of change is directly related to the true, average brightness or luminosity of the star. Once the period is known the true brightness can be calculated and the distance estimated by observing the "apparent" brightness of the object as seen from Earth.

channel stops Narrow, heavily doped strips in a silicon CCD which act like walls to prevent sideways movement of charge in a pixel.

charge-coupled device See CCD

charge transfer efficiency See CTE

charge traps See deferred charge.

chopping The method of removing very large background signals at infrared wavelengths by alternating quickly from the object to nearby sky and back using a rocking motion of the telescope's secondary mirror.

chromatic aberration The change in the image size in an optical system due to the wavelength dependence of refractive index of the material.

circularly polarized See polarization.

clocking The process of raising and lowering the voltages between two levels—high and low—on the electrodes or gates of a CCD in order to move charges from one pixel to the next. The voltage levels themselves are often called the "clocks" or "clock levels".

clock frequency The rate at which a CCD is clocked or read out. It is the reciprocal of (or one divided by) the pixel time; e.g. 40 microseconds (μs) per pixel corresponds to 25 kilohertz (kHz).

CMOS (complementary metal oxide semiconductor) Refers to microelectronic logic circuitry which employs both p-type and n-type MOS transistors in a single circuit (see doping). Low power consumption is a feature of CMOS circuits.

coaxial cable A type of electrical wiring. It consists of a copper wire surrounded by an insulator which in turn is surrounded by a braided copper shield which is encased in another plastic insulator.

COBE Cosmic Background Explorer satellite used to study the microwave background.

coherent Involving waves at different locations which have a definite (not random) phase relationship to each other and can therefore exhibit interference effects.

collimate To make parallel, neither diverging nor converging. All rays from a given field point travel in the same direction.

coma An off-axis aberration which produces images with flared tails like comets.

common-user A shared facility or common resource, such as a CCD spectrograph, a computer data reduction program or even a telescope, which has been carefully designed to meet the needs of many research programmes. Common-user facilities are expected to be very reliable and well-supported.

conduction band The combined unoccupied higher energy levels or orbits of all the atoms in a semiconductor crystal.

conic constant   A number used in optics to specify the shape of a surface which is a conic section, i.e. parabolic, hyperbolic, elliptical. Conic sections are obtained by slicing through a cone at the appropriate angle.

convolution   A mathematical combination of two functions which involves multiplying the value of one function at a given point with the value of another function, the weighting function, for a displacement from that point and then integrating over all such displacements. The process is repeated for every point in the image.

coronagraph   A method of obscuring or occulting the direct light from an object such as the Sun or a star to reveal a much fainter surrounding region. Derives from the ease with which the solar corona can be seen during a total eclipse of the Sun.

corrector plates   Thin lens-like optical pieces which remove certain optical aberrations.

coronene   The first ultraviolet phosphor to be tried on the surface of a CCD.

cosmic microwave background (CMB)   The highly uniform radio signal at microwave (cm) frequencies which appears to be the same in all directions and is interpreted as the redshifted remnant of the Big Bang.

cosmic rays   High-speed atomic nuclei or elementary particles travelling through space.

coudé (focus)   A stationary focal point in an equatorial mounted telescope obtained by an arrangement of small auxiliary mirrors in the converging beam which eventually directs the light down the hollow polar axle of the telescope.

CPU (central processing unit)   The part of a digital computer responsible for interpreting and executing instructions.

critical path   A term used in project planning to indicated a segment of the proposed work which if not completed on time will result in one or more other segments being delayed with serious "knock-on" effects for the project.

critical sampling   See Nyquist sampling.

cryostat (also **dewar**)   A vacuum chamber containing a large reservoir of some liquid cryogenic material such as, liquid nitrogen or liquid helium. Any components attached to the cold face of the reservoir will be cooled down.

CTE (charge transfer efficiency)   A term used to characterize the amount of charge successfully moved from pixel to pixel in a CCD. It is usually expressed as a fraction such as $0.99999X$ per pixel transfer, meaning that $99.999X\%$ of the charge is moved on each time; X stands for a number less than nine.

Cu   The chemical symbol for copper.

CVD   Chemical vapour deposition.

dark current   Signals generated in a detector such as a CCD merely by the heat energy of the atoms which, through random jostling, results in the production of free electrons. Dark current is greatly reduced by physically cooling the detector.

Data Number   See DN.

data rate   The number of pieces of digital information being returned each second from an instrument. Measured in bits per second (bps) for serial information or bytes per second for parallel data.

dc   Direct current. Fixed voltages. For example, dc bias in CCD circuits.

declination   Astronomical coordinate. Equivalent to latitude. The angle in degrees above or below the Celestial Equator, i.e. the projection onto the sky of the Earth's equator.

Range of declination is from from zero to ± 90°.

deferred charge   There are a number of phenomena in CCDs which result in charges becoming "trapped" during the charge-coupling process. Some of these traps are due to design faults, others are due to imperfections in the raw materials or difficulties introduced during fabrication. The global effect of charge trapping is poor charge transfer efficiency. Traps often release captured charge at a later time leading to the presence of charge in trailing pixels which should have been empty, that is, the readout of that charge has been delayed or deferred.

deformable mirror   A very thin mirror whose shape can be changed by the force applied by many small pistons behind the mirror.

depletion (region, layer)   The region of a semiconductor in which an electric field has swept out any free charge carriers such as electrons. The field can be applied by a voltage to a metal electrode on the surface of the silicon.

detected quantum efficiency   See DQE.

deuterium   An isotope of the hydrogen atom with one proton and one neutron in the nucleus.

diffraction   A wave-like property of light which allows it to curl around obstacles whose size is about that of the wavelength of the light. The disturbed waves then interfere to produce ripple-patterns.

diffraction grating   An optical device containing thousands of very fine parallel grooves which produce interference patterns in a way which separates out all the components of the light into a spectrum.

digitized   Converting a signal value such as a charge or voltage into a numerical value within a given range. For example, converting the continuous range from 0 to 10 volts into 65 536 steps of 152.5 microvolts each thus digitizes this range into $2^{16} = 65\,536$ bits; usually stated as 16-bit digitization.

dispersion   The separation of a beam of light into the individual wavelengths of which it is composed by means of refraction or diffraction.

DMA (direct memory access)   An efficient electronic method of transferring digital (numerical) information or data to a computer.

DN (data number)   Also called ADUs. One count or one (least significant) bit; the smallest unit of an analogue to digital conversion system. For example, a 10-volt 16-bit system converts the range 0–10 Volts into 65 536 levels or data numbers.

doping   The process of selectively adding known amounts of foreign atoms into the silicon (or other semiconductor) crystal to subtly change its electrical properties. If the doping results in excess electrons then the material is called n-type, if it results in an excess number of positively charged sites (holes) it is called p-type.

Doppler effect   The apparent change of frequency or wavelength of radiation from an object due to its motion toward or away from us. If the object is receding the frequency is decreased and the wavelength is increased, i.e. becomes red-shifted.

down-converter   See fluorescence.

down-loading   The process of passing data or a program from one computer to another, often in a hierarchical structure where the source computer is the more powerful one.

DQE (detective quantum efficiency)   The ratio of the actual number of detected photo-electrons in a detector to the number of incident photons when proper account is taken of noise and other efficiency factors.

drift-scan (or drift-scanning)   See TDI.

dry ice   Solid carbon dioxide (CO_2). An inexpensive cryogen for cooling detectors.

DSP (digital signal processor)   A special kind of computer chip.

dynamic range   The ratio of the maximum possible signal, the saturation level, to the system noise floor.

dynode   Intermediate surface within a photomultiplier tube which emits multiple electrons when bombarded by a single electron.

EBCCD (electron-bombarded CCD)   An imaging device containing a thin target material which emits electrons by the photoelectric effect when illuminated and then magnetically focusses these electrons to impact onto a silicon CCD where they generate a large charge.

echelle   A type of diffraction grating with groove angles of 90°. With the grating at an angle of 45° the grooves resemble a staircase.

electrodes (also called **gates** or **phases**)   Small electrically conducting plates connected to a voltage source (battery or power supply) and arranged in strip patterns to define the picture elements or pixels of a CCD. The plates create an electric field within the semiconductor which therefore forms a storage site for photo-generated charges. Also used as a generic term for any conductor with an applied voltage.

electron–hole pairs   When a photon is absorbed in silicon its energy causes an electron in the valence band to be ejected into the conduction band leaving a (positively charged) vacancy or hole in the valence band.

electronvolt(s)   A unit of energy equal to 1.6×19^{-19} joules. See Appendix.

elevation   The angle in degrees above the horizon toward the zenith or overhead point. Sometimes loosely called the "altitude" of a star, but not to be confused with height above sea level. Elevation angle is 90° minus the zenith distance (or zenith angle).

emissivity   A measure of the efficiency of a source to radiate like a perfect black body; 100% is perfectly black and 0% is perfectly reflecting.

energy levels   The allowed energy states for electrons orbiting the nucleus of an atom

entrance pupil   The real object or image which defines the limit of valid light paths through an optical system.

entropy   A thermodynamic measure of the degree of order in a system. Also used as a measure of information content.

ephemeris   A list or tabulation of astronomical phenomena that *change with time*.

epitaxial   A thin layer of differently doped semiconductor used in the construction of solid-state devices such as the CCD.

EPROM (erasable programmable read only memory)   A small silicon chip containing thousands of individual locations which can be set to either a low or a high voltage level; a 0 or a 1. The settings can be erased by exposure to ultraviolet light.

equatorial mount   The classic type of telescope mount with one axis parallel to the Earth's polar axis (i.e. pointing at the celestial pole) and the other at right angles. Once the object is located, only the polar axis need be driven by a motor to counteract the Earth's rotation.

etalon   Essentially an optical filter that operates by multiple-beam interference of light reflected and transmitted by a pair of parallel flat reflecting plates.

Ethernet   A system for linking computers with a single serial cable.

extinction   The combined effects of absorption and scattering of light, usually by interstellar dust, which dims our view of a distant object.

extrinsic   A semiconductor, such as silicon, which has been doped with impurity atoms to provide smaller energy band gaps for detection of lower-energy photons.

false colour   The use of colours, instead of shades of grey, on a computer image display screen to represent different brightness levels and highlight very small differences in a dramatic way. For example, in an ordinary black-and-white image, objects which differ only slightly in brightness appear as almost the same shade of grey and are hard to distinguish. If instead the numerical brightness values are assigned carefully chosen colours then two objects of almost equal brightness will be strongly distinguished when the image is displayed.

Fano-noise   Fano-noise-limited CCDs used for X-ray detection are limited by intrinsic noise due to the absorption of energy and vibration of the crystal lattice itself.

fast interface states   See traps.

fat zero   See preflash.

FET (field effect transistor)   A tiny transistor amplifier in which the current flow between two terminals, called the source and the drain, is controlled by the electric field generated inside the silicon by an external voltage on a surface called the gate electrode.

fibre optics   A long, thin strand of glass capable of excellent transmission of light over large distances.

FTP   File transfer protocol

field lens   A lens placed in or near the focal plane of a telescope to create an image of the primary mirror inside the instrument.

field of view   The patch of sky or of any image scene which can be seen by an optical system or by one picture element (pixel) of a detector system. Usually expressed in angular measure.

field rotation   The rotation of a star field about the centre which occurs in an alt-az telescope because the motion is not about the polar axis.

FIFO   First-in-first-out buffer. Used in many microprocessor-controlled systems to smooth the flow of data.

finesse   A figure of merit for the reflectance of a Fabry–Perot etalon. It is given by $\pi\sqrt{R}/(1-R)$ where R is the reflectance ($R < 1$).

FITS (flexible image transport system)   A method for saving image data which has become standard in astronomy.

flash gate   An ultra-thin transparent electrode across the entire back surface of certain CCDs used to control the charge on the back surface and hence the QE for blue and UV light.

flat field (or **flat-fielding**)   An extensive, very uniformly illuminated source of light used to determine the relative sensitivity of an imaging detector composed of many picture elements (pixels) each with a slightly different response to light. The correction process is called flat-fielding.

fluorescence   The emission of light at one wavelength, the green say, following absorption of light with a much shorter wavelength such as the ultraviolet. The UV photon parts with its energy by ejecting the electron into a high-energy level from which it cascades back down, releasing photons of lower energy and therefore longer wavelengths. A material which has this property is called a phosphor.

focal plane scale   The relationship between angles on the sky, in seconds of arc, and millimetres of size at the focus of the telescope; i.e. the number of arcsecs per millimetre.

focal ratio   The ratio of the focal length (F) of a mirror or lens to its diameter (D) expressed as a number; $f/\# = F/D$. Also defines the cone angle of the beam. Small focal ratios e.g. $f/\#=F/D=1$ are said to be "fast" and represent a very large cone angle. Large focal ratios e.g. $f/\#=35$ are said to be "slow" and indicate a very small cone angle.

focal reducer   An optical component or system for changing the image scale of a telescope to achieve a better match between the seeing disk and the pixel size. See optical matching.

forbidden energy gap   The unoccupiable interval of electron energy levels which forms in a crystalline substance (that is one having a periodic atomic formation) between the valence band (bound electrons) and the conduction band (free electrons). The forbidden energy gap is very large in insulators, non-existent in metals, and small but finite in a semiconductor.

forward (or **reverse**) **bias**   A term applied to an electronic device known as a diode—usually formed by a junction of p-type and n-type semiconductor material—in which current flows easily if the externally applied voltage has the correct polarity or direction. If the opposite polarity is applied there is almost no current flow and the device is said to be reverse biassed.

frame transfer   A CCD construction in which one half of the imaging area of the device is purposely covered with a mask opaque to light to provide a temporary charge storage section.

free spectral range   A term used in spectrometers to indicate the wavelength interval between occurrences of the same wavelength produced in the next order of interference or diffraction.

frequency   The number of oscillations or wave cycles per second passing a given point. For electromagnetic radiation, the product of the frequency and the wavelength is the speed of light.

Fried parameter   A measure of the scale of the turbulence in the atmosphere. The length over which a disturbance to a wavefront is well-correlated.

fringing   The appearance of complex light and dark contours in a CCD image due to constructive and destructive interference effects of light reflected inside the detector.

FWHM   full width half maximum. The full width of a profile (e.g. the seeing profile or a filter transmission curve) between the two points where the value is 50% of the peak value.

gain   The amplification factor.

gamma-rays   Electromagnetic radiation with a wavelength less than about 1 Å (10^{-10} m); blends from the "hard" X-ray region. Photons of energy greater than about 10 keV.

getter   A chemical absorption method of removing (pumping) gas from a chamber by tying up molecules on a surface. Activated charcoal. Molecular sieve. See zeolite.

gravitational lens   Deflection of electromagnetic radiation from a distant background source by a strong gravitational field associated with a foreground source resulting in more than one image of the original source. Many double-quasars are produced by this phenomenon.

gravitational waves   Disturbances or ripples in spacetime predicted by the General Theory of Relativity due to changing distributions of mass such as the spindown of a neutron star binary system or the implosion of a star during a supernova. No confirmed direct detections so far.

Gregorian   A class of reflecting telescope which uses a concave secondary mirror placed after the prime focus is reached instead of a convex secondary placed before the prime focus.

grism   This is a right-angled glass prism with a transmission diffraction grating deposited on the hypotenuse surface. The spectrum produced by the grating is deflected by the prism to remain on the optical axis and the apex angle of the prism is chosen to get a certain wavelength in the centre of the detector. Grisms can be placed in a filter wheel.

ground loop   A condition in which two pieces of apparatus are connected together while each has a separate earth connection and these are not at identical potentials. A current will flow and small signals will be affected by fluctuations, called noise.

GUI   Graphical user interface.

H   Chemical symbol for hydrogen. The most abundant chemical in the universe. H_2 is the symbol for the molecular hydrogen molecule which is abundant in giant clouds in our galaxy and can be detected by its infrared spectrum. The symbols HI and HII are used to indicate neutral and ionized hydrogen respectively. HII regions are usually associated with star formation, e.g. the Orion Nebula, and are detected by their emission lines such as H-alpha at 656.3 nm. Radio emission at 21 cm wavelength can be detected from neutral hydrogen.

half wave retarder   See retarder.

heliarc welding   A process of joining two metals using an electric arc in an atmosphere of a noble gas.

heterodyne   A detection method used extensively in radio astronomy in which the wave nature of light is used. The method usually involves combining the measured wave with a local oscillator or reference wave and looking for the signal at the difference frequency

hold-time   The time taken to use up all the liquid cryogens, like LN_2, in a cooled CCD cryostat.

hologram   An interferometric method of recording information about the three-dimensional nature of an object which relies on preserving both the amplitudes and phases of the wavefronts which reach the detector, instead of merely the amplitudes. Hologram means "whole record". The basic principle was outlined by D. Gabor in 1948.

honeycomb mirrors   A construction method for a large mirror in which the back is hollowed-out to leave a ribbed structure that resembles a honeycomb.

host computer   The main or master computer in an instrumentation system. The computer responsible for interacting with the user.

HST (Hubble Space Telescope)   A space-based reflecting telescope with a primary mirror diameter of 2.4 m (94 in) capable of high-resolution imaging from the far ultraviolet to the near infrared. A joint NASA/ESA mission. Launched in 1990 with a planned lifetime of 15 years. Encountered reduced performance when the mirror was found to have spherical aberration. Solved by the installation of corrective optics (COSTAR) in 1994.

Hubble expansion ... of the Universe. There are billions of other galaxies and all except the closest ones (the Local Group) are receding from us as deduced from the Doppler redshift effect of their spectra. Thus the Universe appears to be expanding. Moreover, the greater the distance the faster the speed of recession. This is interpreted as the expansion of spacetime itself since an event called the Big Bang.

hybrid array A device in which the roles of radiation (infrared mostly) detector and signal multiplexer are separated. The device is a sandwich of two slabs. Other names include focal plane array (FPA) and sensor chip assembly (SCA).

IC See integrated circuit.

IF (intermediate frequency) The beat frequency between the signal and the local oscillator in a radio detection system.

image intensifier An electronic device for increasing the brightness of a faint optical image. The image is first formed on a thin metallic surface called a photocathode from which electrons are then ejected. The stream of electrons is accelerated and focussed onto a phosphorescent screen which glows brightly as a result of the impact.

imaging spectrometers Refers to a class of instruments which preserve the image field while also determining the spectrum. Integral Field Unit (IFU). Usually implies some kind of image slicing either with facets or fibre optics.

impurity band conduction (IBC) A form of infrared array detector which replaces the photoconductor and provides higher performance.

infrared The region of the electromagnetic spectrum from a wavelength of about 1 μm (10^{-6} m) to about 200 μm. The region from 1 to 5 μm is the near infrared; 5–30 is the mid infrared and 30-200 μm is the far infrared.

InSb (indium antimonide) A compound semiconductor used as an infrared photoelectric detector.

integral field See imaging spectrometers.

integrated circuit A small electronic component made of semiconductor silicon on which an entire electronic circuit of numerous microscopic transistor amplifiers, diodes and resistors has been constructed.

integrating detector Any imaging device, like a photographic emulsion or CCD, which can build up more signal and contrast by a longer exposure to light or other electromagnetic energy.

integration time The interval of time used to collect photons of light on a detector and build-up a strong signal.

interference filters A method of constructing an optical filter to select a particular wavelength band for transmission and reject wavelengths outside this band. Similar to a Fabry–Perot etalon. The construction relies on constructive and destructive interference effects in a multilayer stack of quarter-wave reflective layers and half-wave spacer layers.

interferometry (interferometer) A measurement technique that relies on interference between coherent waves that results in regions of enhanced signal (constructive interference) when the waves are in phase and regions of no signal (destructive interference) when the waves are exactly out of phase. For light, the effect is usually to produce a series of light and dark bands called *fringes*. A record of the fringe pattern is called an *interferogram*.

interline transfer   A CCD construction consisting of vertical strips which are alternately opaque and light sensitive. The opaque strips conceal charge transfer registers.

Internet   A global spider-web-like network of computers and computer systems with no central hub or single point of control.

interstellar dust   Small grains or particles in the interstellar medium.

inverse square law   Decreasing as one over distance squared ($1/r^2$), where r is the distance from the source. Light and gravity both have this property.

inversion   The term used with CCDs to indicate that the applied voltage has not only driven away the majority carriers but has actually attracted the minority carriers of the opposite sign.

ionized   Having lost one or more electrons from an atom.

IPCS (image photon counting system)   A form of very low light level detector used in astronomy. By means of an image intensifier the IPCS is capable of counting individual photons of light.

IR   See infrared.

IRAF (image reduction and analysis facility)   An extensive suite of programs developed for astronomy applications and supported by the US National Optical Astronomy Observatories.

irradiance   The radiant power (watts) received per square metre on a surface.

isoplanatic patch   The angular region on the sky over which the wavefront correction applied by an adaptive optics system remains valid. It is relatively small and therefore a nearby reference star is also required.

isotropy   The same in all directions.

jansky   A unit of flux measurement named after radio astronomy inventor Karl Jansky. See Appendix.

J H K L M N Q   Designations for parts of the infrared waveband transmitted by the Earth's atmosphere in the wavelength range 1 to 20 μm.

JPL   Jet Propulsion Laboratory, Pasadena, California. Funded by NASA and operated by the California Institute of Technology (Caltech).

Julian Day   The number of days since noon on January 1, 4713 BC. The Julian date at noon on January 1, 2000 will be 2 451 545.

kTC noise   See reset noise.

LAN (local area network)   A means of interlinking computers.

laser   An especially intense beam of light at a specific wavelength. The word is derived from Light Amplification by Stimulated Emission of Radiation. Requires over-population of certain energy levels.

LHe   The symbol for liquid helium. The temperature of liquid helium is normally 4 K, that is, four degrees above absolute zero.

linearly polarized   See polarization.

line spectrum A pattern of lines, each corresponding to an image of the entrance slit of the spectrometer, seen when light is either emitted by or interrupted by a hot rarified gas such as hydrogen. The pattern is characteristic of the gas and the wavelength at which the features are observed to occur is indicative of the velocity of the object.

Littrow The configuration of a diffraction grating spectrograph in which the diffracted ray returns along the same direction as the incident ray.

LN_2 The symbol for liquid nitrogen. The temperature of liquid nitrogen is normally 77 K, that is, 77 degrees above absolute zero.

LSB (least significant bit) See also DN.

luminescence In CCDs and infrared arrays—light emission caused by currents through faulty parts of the structure which behave like a light-emitting diode.

Lumogen A material used as a down-converter.

LUT Look-up table

Lyman alpha The spectral line at 1216 Å in the far ultraviolet that corresponds to the transition of an electron in the hydrogen atom between the two lowest energy levels.

magnification The effect of an optical system on the apparent angular size of an object. An increase in angular size occurs if the magnification factor is greater than 1. If the factor is less than 1 then demagnification occurs.

magnitude A logarithmic brightness scale for astronomical objects. See Appendix.

MAMA (Multi Anode Microchannel Analyzer) A detection system used with microchannel plates to detect events. Used as an imaging system in the ultraviolet. See microchannel plates.

Maximum entropy (or **MEM**) An image reconstruction methodology which defines a measure of information content and seeks to maximize it.

mean free path The distance travelled on average by an atom before it collides with another atom.

median Literally the middle value in a sequence of values arranged in increasing size order. A useful mathematical estimator of the true value from a set of values when one of these values is contaminated, i.e. known to be much larger than the average.

MegaFLOPS Millions of floating-point operations per second. A computer benchmark.

meniscus mirror A very thin mirror with a high curvature. A method of constructing very large mirrors which assumes from the outset that the mirror is too thin to hold its shape against gravity and will require an active control system.

Mer–Cad–Tel Mercury–cadmium–telluride (HgCdTe). A semiconductor alloy useful as an infrared photoelectric detector. Also known as CMT (Cad–Mer–Tel).

microchannel plate A compact electrostatic high-voltage electron multiplier with a very large number of narrow pores or channels. A photoelectron generated at the entrance face (photocathode) stimulates a cascade of secondary electrons down the nearest channel to produce a huge cloud of charge at the output face. The output pulse can be used in many different ways to record the event. If it impacts a phosphor screen then light emission can be detected with a CCD. Direct electrical detection can be obtained using a Multi Anode Microchannel Analyzer.

microprocessor A very large silicon integrated circuit with essentially all the functions of a computer on a single chip.

MIDAS   Munich Image Data Analysis System. A suite of programs and a software environment developed at the European Southern Observatory for astronomy applications.

MIPS   Millions of instructions per second. A computer benchmark. In the mid-1980's the DEC Vax 11/780 had a MIPs rating of 1. Desk-top 133 MHz Pentium-based personal computers (PCs) have a MIPs rating of about 80.

mixer   The critical element of a radio detection system which allows the incoming wave to be combined with the reference frequency from the local oscillator. Usually a diode.

MOS   Metal oxide semiconductor. A construction used to fabricate microelectronic components, including CCDs, which consists of three layers, namely a metal conductor, an insulating layer usually made from an oxide of silicon, and a semiconductor such as silicon.

multiplex   Combining many signals into one or a small number of signals.

multi-pinned-phase (also **multi-phase-pinned**)   The method of driving all CCD phases (gates), including the integrating phase, into inversion and thereby greatly reducing the dark current. The penalty is a slightly poorer well-depth for charge storage. The advantage is a much higher operating temperature.

muons   Elementary particles produced when cosmic rays enter the upper atmosphere.

native oxide   The silicon dioxide layer which grows in air on the exposed backside surface of a thinned CCD.

neutrino   A fundamental elementary particle with no electric charge and very small if any rest mass. Believed to be exceedingly abundant in the universe. The neutrino has a very low cross-section for interaction with matter and is almost impossible to detect, hence the uncertainty over its rest mass. The Sun produces neutrinos from thermonuclear reactions in its core, and a large flux of neutrinos carries away most of the energy of a supernova. Neutrinos are one candidate for Dark Matter. Experiments to detect cosmic neutrinos involve large masses of "stopping" material and indirect detection of the effects of neutrino absorption.

Newtonian   A class of reflecting telescope developed by Sir Issac Newton with a paraboloidal primary mirror and a small, plane secondary mirror at 45° to deflect the focus of the primary to a position outside the tube near the top of the telescope.

noise   A term used to describe unwanted electronic signals, sometimes random and sometimes systematic, which contaminate the weak signal from an astronomical source. Types of noise include, *1/f noise* which gets larger at lower frequencies, *white noise* which is independent of frequency, and *reset noise* which is a consequence of device readout architecture but which can be removed by correlated double sampling (see CDS).

non-destructive readout   A means of reading out an infrared array or a CCD with a skipper output in which the output voltage is sampled without resetting the detector. The output voltage can be sampled a large number of times to improve noise performance.

Nyquist sampling   The minimum number of resolution elements required to properly describe or sample a signal, such as a star image, without causing erroneous effects known as aliasing. For electronic imaging, this number is generally taken as 2 pixels across the seeing disk diameter at the half intensity points (FWHM).

Occam   A computer language used with transputers. Also, *Occam's Razor*. A statement that the most likely explanation of a phenomenon is the one that requires the simplest explanation.

occultation   The covering up of one astronomical object be another. For example, the Moon occults many stars each night.

optical fibres   Glass and transparent plastics can be made into very thin wires or fibres. Typical dimensions are 10–50 μm. If a ray enters one end of a fibre at the appropriate angle, it will undergo total internal reflection and travel down the fibre without much loss through the sides.

optical matching   The use of lenses or other optical devices to match the size of the image of the seeing disk, as it appears in the focal plane of the telescope, to the physical size of the CCD pixels. If the telescope yields 10 arcseconds per millimetre and the seeing is 1 arcsecond then the image is 0.1 mm in size. But a typical CCD pixel is 0.022 mm, five times smaller.

optical path difference (OPD)   The difference in path length between the actual wavefront in an optical system and the equivalent spherical wavefront.

out-gassing   The absorbed gases released from the interior walls and components of a vacuum chamber which has already been "roughed-out".

overscan   Additional clock pulses in both the horizontal and vertical directions in excess of the actual number of real pixels in order to sample the electronic offsets in the system.

PDA   Photodiode array.

phase difference   Two identical waves of the same wavelength are said to be "in phase" when the peaks and troughs coincide perfectly. If this is not the case then a phase difference is said to occur. For a phase difference of half a wavelength, the peak of one wave coincides with the trough of the other.

phosphor   See fluorescence.

photocathode   A thin metallic plate housed inside an evacuated tube capable of releasing electrons through the "photoelectric effect" when illuminated by light. These surfaces are best for optical and ultraviolet light.

photoconductivity   Absorption of light increases the number of charge carriers.

photodiode   A light-sensitive device made from the junction of two differently doped species of a semiconductor such as silicon. Also known as a pn junction. An internal electric field is generated at the junction of p and n type material. Photons absorbed in the junction create electron–hole pairs which are separated by the field and create a current.

photoelectric effect   The forced ejection of electrons from certain metals due to the transfer of energy from incident "photons" of light. A demonstration (by Einstein) of the quantum effect. If the ejected electrons are collected as a charge or a current then the intensity of the incident light can be measured. Almost all modern detectors rely on some variant of the photoelectric effect.

photoelectronic devices   Any detector which uses the photoelectric effect to convert photons to electrons.

photographic emulsions   Materials in which the absorption of light leads to a chemical reaction.

photometry (photometer) The quantitative measurement of the intensity of light.

photomultiplier tube   A vacuum encapsulated photocathode from which electrons are ejected by the photoelectric effect followed by multiple cathodes from which many additional electrons are emitted in a cascade. When finally collected, the original single electron may have generated a pulse of over one million electrons.

photon   The term used most often to describe the basic package or "quantum" of electromagnetic energy (light).

photon counting   The detection of individual photons. Photomultiplier tubes are capable of detecting single photons. Photon counting statistics says that if N is the total number of photoelectrons counted then the error in N is $\pm \sqrt{N}$.

photonics   The technology of generating and harnessing light and other forms of radiant energy whose quantum unit is the photon for a range of applications ranging from detection to laser energy production to communications and information processing.

photovoltaic effect   Absorption of a photon leads to the production of a voltage across a junction.

pixel(s)   Derived from picture element. The smallest individual element of an array detector. Note that the size of the detector pixel does not necessarily equate with the resolution of the system.

pixons   A term used to describe pixel-like elements in an image reconstruction algorithm. Pixons range in size and shape depending on the information content in that part of the image.

plate scale   The number of seconds of arc on the sky corresponding to 1 mm in the focal plane of the telescope.

PMT   See photomultiplier tube.

Point spread function   See PSF.

polar axis   The axis of an equatorially mounted telescope that points towards the Celestial Poles and is therefore parallel to the Earth's axis of rotation.

polarization (polarimeter)   The property of transverse electromagnetic waves that describes the plane of vibration of the wave and its behaviour as the wave progresses. *Linearly polarized* light implies that all the waves vibrate in the same plane. *Circular polarization* occurs when the plane of vibration rotates as the wave progresses. A polarimeter is used to measure these properties.

polarization modulator   An optical device sensitive to the plane of vibration of electromagnetic waves, i.e. to their polarization. It is used to convert the polarization of light into a measurable brightness change.

polysilicon   A non-crystalline form of silicon with a high conductivity like a metal; preferred in CCD manufacture to the use of metals because it keeps the entire process in silicon and is more transparent to visible light.

preamplifier   A low noise amplifier designed to be located very close to the source of weak electronic signals, but capable of delivering amplified signals down tens of metres of cables.

precession of equinoxes   The First Point of Aries (0 hr Right Ascension) moves backwards (westward) along the equator at 50.2 arcseconds per year due to the 26 000-year conical motion of the Earth's rotation axis caused by the gravitational pull of the Sun and Moon on the Earth's equatorial bulge. Correcting for this effect yields the mean equator and mean equinox. *Nutation* is the wobble of the Earth's axis as it precesses. Correction for this effect gives the true equator and true equinox.

preflash   The technique of illuminating the CCD with a low light level flash before beginning a long exposure in order to "fill up" any charge traps.

primary mirror   The first mirror encountered by incident light in a telescope system.

prime focus   The focal point of the large primary reflecting mirror in astronomical telescopes when the light source is extremely distant. This focus actually falls at a point just within the upper structure of the telescope itself and is therefore accessible to CCD cameras and other instruments; it provides a large field of view.

printed circuit   A compact double-sided circuit board with no wires but instead fine tracks, etched on a copped-clad board, perform the same function.

PROMS   See EPROMS.

PSF (point spread function)   The size and shape of the actual image of a point source as a result of the combined effects of atmosphere, optics, guiding.

pulse counting   See photon counting.

QEH (quantum efficiency hysteresis)   An increase in QE after exposure to light.

QE pinned   The quantum efficiency of certain CCDs can be driven to their maximum by UV flooding and pinned there by immediate cooling.

quadrature   A means of combining errors or noise by summing the "square" of the values and then taking the "square root".

quanta   Light can carry energy only in specific amounts, proportional to the frequency, as though it came in packets. The term quanta was given to these discrete packets of electromagnetic energy by Max Planck

quantum efficiency (or QE)   The ratio of the number of photoelectrons released for each incident photon of light absorbed by a detector. This ratio cannot exceed unity.

quarter wave retarder   See retarder.

quasar (or QSO)   Compact-looking objects, often radio sources, with emission lines in their spectrum which are displaced by very large amounts towards the red. These redshifts correspond to velocities which are a large fraction of the speed of light, and hence these objects are believed to lie at great distances.

RAM (random access memory)   A silicon micro-chip capable of temporary storage of information in the form of binary digits, either 0 or 1, and enabling rapid access to any part of its storage area.

ray tracing   Computer simulation of light ray paths through an optical system.

receiver   General term for a radio detection system.

reciprocal linear dispersion   The inverse of the linear dispersion of a spectrometer which is the rate of change of position along the spectrum (in millimetres) with wavelength (in angstroms). The reciprocal gives the number of angstroms per millimetre.

reciprocity failure   The non-linear behaviour of a photographic emulsion in which an increase in exposure time does not correspond to an increase in sensitivity by the same factor.

reflectors   Telescopes based on reflections from curved mirrors.

refractive index   A number which characterizes the properties of a material when light is transmitted through it. A vacuum has a refractive index of 1.0; air is fairly close to this also. As the refractive index rises, light waves travel more slowly in the material, more light is reflected and the bending or refraction of light is greater. Some materials have a different refractive index depending on the plane of vibration of the electromagnetic wave, they are called birefringent.

reset noise   The unwanted and uncertain electrical signal transmitted to the output pin of a CCD during the process of recharging, via the reset transistor, the output storage capacitor to its preset value in readiness for the next pixel charge.

reset transistor   See reset noise.

resolution   The ability of an optical system, including detector, to separate two adjacent objects—this is called "spatial resolution"—or two adjacent wavelengths in a spectrometer—this is called "spectral resolution".

resolving power   The ratio of the wavelength of radiation to the smallest interval of wavelength that the instrument can measure.

retarder   A device for introducing a phase delay, such as half-wave or quarter-wave, between two orthogonally polarized components of an electromagnetic wave.

Richardson–Lucy (method)   An image reconstruction algorithm.

Ritchey–Chrétien   A class of reflecting telescope with a hyperbolic primary and secondary.

Schmidt (telescope)   A telescope with a spherical primary mirror and a thin refractive corrector plate with a complex, non-spherical shape. Very wide-field performance for surveys.

Schottky barrier   A metal to semiconductor interface without any insulation layer produces an energy barrier in the semiconductor which can be used like a diode.

scintillator   A detector for high-energy photons such as gamma-rays. The impact of a gamma-ray causes a burst of light which can be observed with a PMT.

secondary mirror   The second reflecting surface encountered by the light in a telescope. The secondary is usually suspended in the beam and therefore obstructs part of the primary.

seeing   Describes the blurring of a stellar (point-like) image due to turbulence in the Earth's atmosphere, both at high altitudes and within the telescope dome. Seeing estimates are often given in terms of the full-width in arcseconds of the image at the points where the intensity has fallen to half its peak value. The typical value at a good site is a little better than 1 arcsecond.

segmented mirrors   A large mirror construction technique in which many smaller elements are built and then actively controlled to conform to the shape of the required large mirror.

semiconductor   A material like silicon or germanium in which the valence band and the conduction band are separated by a small (forbidden) energy gap. Such materials have some of the properties of a good electrical conductor—in which the energy gap is zero—and some of the properties of an insulator—in which the gap is very large.

serial (horizontal) register   The final row of a CCD in which the controlling electrodes are arranged at right angles to those on the rest of the CCD. This enables charges coupled onto this row to be transferred in single-file through the CCD output amplifier.

sequencer   That part of an electronic system responsible for the accurate phasing of time-critical events such as CCD clocking and readout.

SIS junctions   Superconductor–Insulator–Superconductor junctions. Can be used as the mixer in a radio receiver system.

slew   The relatively rapid motion of a telescope (under computer control) as it moves to point at a new position in the sky. Once at the new position the motion of the telescope returns to that required to cancel the effect of the Earth's rotation relative to the stars—the sidereal rate.

solid-state   Usually implies crystalline semiconductor materials used in the electronics industry.

Spark chamber   A means of detecting high energy particles by the trail of ionizations left as they pass through a chamber containing many charge plates.

speckle (interferometry)   The technique of recovering the diffraction-limited angular resolution of a telescope by analysis of images obtained using a very high speed camera system to "freeze" the blurring due to atmospheric turbulence.

spectroscopy (spectrometers)   The quantitative study of the spectral or energy content in a beam of light. Usually achieved by spreading or dispersing the light into its constituent wavelengths by means of a diffraction grating and recording the image with a CCD or other electronic imaging detector. The patterns, wavelengths, intensities and shapes of the spectral features yield considerable physical information.

Stokes parameters   A way of characterizing the polarization state of light which is closely related to actual measurements.

Strehl ratio   The ratio of the peak intensity in the point spread function of an optical system to that of the equivalent diffraction-limited system.

STARLINK   A software environment and suite of programs for astronomical data analysis developed in the UK and supported by the Rutherford-Appleton Labs.

surface channel   A semiconductor device construction in which the electron charges are held or moved near the surface of the silicon crystal.

TDI (time delay and integration or drift-scan)   Methods for averaging the response of a CCD along columns by reading out at the same rate as a mechanical motion is shifting the optical image along the CCD.

tertiary   (mirror) The third mirror to be encountered by the light in a telescope system. A tertiary mirror is required on alt-az telescopes to direct light to the stationary Nasmyth foci.

thermal background   The radiation emitted by the telescope and the atmosphere at infrared wavelengths due to the heat (temperature) of the source.

three-phase   A CCD construction in which three overlapping metal electrodes are used to define a pixel and effect the transfer of charge, in either direction along a column, by the charge-coupling method. If only two electrodes are used then the device is two-phase.

throughput (also **étendue**)   A measure of the efficiency of an optical system. Also étendue. Formally, the product of the solid angle accepted and the effective collecting aperture.

timing waveform   A diagram showing the time sequence and voltage levels of a stream of pulses required, for instance, to perform charge-coupling in a CCD.

transit (telescope)   A stationary support structure for a telescope. Motion is allowed along the meridian from the zenith to the horizon, but stars cannot be tracked east/west. Measurements are only possible when the objects "transit" the meridian due to the Earth's rotation.

transputer   A compact computer chip with a special design for linking to other transputers to make the program run faster.

traps   Irregularities in the silicon crystal lattice which can absorb free charges created in the semiconductor by, for instance, the absorption of light.

twisted-pair   A form of wiring consisting of two strands of single wire twisted together to form a transmission line. One strand carries the signal and the other is grounded to the single "star" ground point of the system.

UBVRI   Designations for parts of the optical waveband, isolated by means of special glass filters which eliminate the unwanted regions, and used for standard astronomical intensity measurements.

UV   Ultraviolet. Wavelengths shorter than about 350 nm.

valence   Also valence band or valence electrons; the electrons in the outermost orbit.

vidicon   General name for the class of vacuum tube imaging devices which employ a scanning electron beam to read out the image.

virtual phase   A type of CCD in which only one electrode is physically outside the silicon and is such as to obscure only half of the pixel. A specially doped layer under the transparent part acts as another or virtual electrode.

VLA (very large array)   A network of 27 radio telescopes in New Mexico, USA.

VLBI   Very long baseline interferometry.

wavefront error   The departure of the true wavefront in an electromagnetic wave propagating through an optical system from the ideal spherical wave at that point.

white dwarf   Faint very-compact stars at the end of their life. Also used as convenient photometric and flux standards.

workstation   A single-user image processing set-up with a powerful small computer, a colour image display and access to a data storage system (a magnetic disk).

World Wide Web   A more systematic way of using the facilities available on the Internet.

X-rays   A large band of electromagnetic radiation with wavelengths smaller than extreme ultraviolet light. A typical X-ray photon has over one thousand times as much energy as a photon of visible light.

zenith   The point in the sky directly overhead.

zeolite   An absorbent material (in the form of small pellets) used in low-temperature cryostats to trap gases released gradually after active pumping has ceased. See outgassing.

zeropoint   The magnitude corresponding to one data number per second from the array detector for a star of zero colour term (like Vega) corrected for absorption in the Earth's atmosphere.

Answers to selected exercises

CHAPTER 1

7. (a) 300 GHz (b) 21 cm.
9. $E = 3.98 \times 10^{-19}$ J = 2.49 eV.
12. Number of electrons at dynodes 1, 2, 3 ... = (3),(9),(27), ... = $3^1, 3^2, 3^3$ ... and after 10 dynodes it is 3^{10}. Q = 9.45×10^{-15} C.
13. (a) ± 1000 counts (b) 0.1% (c) a factor of $\sqrt{4}$ = 2 (to 0.05%).

CHAPTER 3

2. (a) prime focus is faster by $(17/5)^2 = 11.56$ (b) Cass is better in practice because it reduces the level of sky background
3. Tails toward the edges of the field imply coma in the optics.
8. A Schmidt telescope with an objective prism and large photographic plates giving a 7° field.
9. A CCD on a large (D > 4 m) telescope.
10. A polarimeter will show that the emission is highly polarized consistent with the synchrotron process.
11. Speckle interferometry can provide diffraction-limited images.

CHAPTER 4

2. f/number = (206 265)$d_{pix}/D_{tel}\theta_{pix}$ = (206 265"/rad) 24 μm / (10 m × 10^6 μm/m) 0.2" = 2.48. Optics faster than f/2.5 are certainly challenging. 3.4' × 3.4'.
3. $D_{coll} = RD_{tel}(\phi/206\ 265)/2 \tan \theta_B$; R = 20 000, D_{tel} = 10 m, φ = 0.5
   (a) (2 tan θ_B = 0.63) ⇒ D_{coll} = 77 cm
   (b) (2 tan θ_B = 4) ⇒ D_{coll} = 12 cm. The echelle is more practical.
   Focal length = 15 × D_{coll}; (a) 11.55 m (b) 1.8 m.
4. n = 2.4, A = 30°, λ_c = 2.2 μm and R = 500 for two pixels. Pixel size = 27 μm:
   Assuming m = 1 (first order) T = (n−1)$sinA/\lambda_c$ = 318 lines/mm and EFL = 2$d_{pix}R/(n−1) \tan A$ = 33.4 mm.

6. $Q = N(0)-N(45)/N(0)+N(45) = 2000-1000/3000 = +0.33;$
    $U = N(22.5)-N(67.5)/N(22.5)+N(67.5) = 1800-1200/3000 = +0.20.$
    $p = \sqrt{(0.33^2 + 0.20^2)} = 0.39$ (39%), $\theta = \frac{1}{2}\tan^{-1}(U/Q) = 15.6°$
    $\sigma(N)/N = 1/\sqrt{N} = 1/\sqrt{3000} = 1.8\%.$

7. $\Delta\lambda_{FSP} = \lambda^2/2nd$ but $2nd = R\lambda/\mathcal{F} = 250$ μm, giving $\Delta\lambda_{FSP} = 0.001$ μm.

8.   Scan length $= \frac{1}{4}\lambda R = (10$ μm$)(100\,000)/4 = 25$ cm.

CHAPTER 5

2. $\sqrt{n} = 2$. Thickness $= \lambda/4n = 2.2$ μm $/8 = 0.28$ μm (or 2750 Å).

3.   Match the following three detectors to a 0.2-m telescope and then to an 8-m telescope:
    For the small telescope assume seeing of 2″ and 1 pixel = 1″, for large telescope assume these numbers are four times smaller, i.e. 0.25″/pixel. Use Eq. (4.4).
    Kodak KAF-4200 CCD with 9 μm pixels in a 2048 × 2048 format:
        (a) $f/\# = 9.3$, FOV = 34′ (b) $f/\# = 0.93$, FOV = 8.5′.
    SITe CCD with 22 μm pixels in a 1024 × 1024 format:
        (a) $f/\# = 22.7$, FOV = 17′ (b) $f/\# = 2.27$, FOV = 4.27′.
    Hughes-SBRC InSb array with 27 μm pixels in a 1024 × 1024 format.
        (a) $f/\# = 27.8$, FOV = 17′ (b) $f/\# = 2.78$, FOV = 4.27′.
    The Kodak CCD cannot be matched to a large telescope unless the scale is reduced to less than 0.1″/pixel (FOV = 3.4′).

5.   Linear size $= \beta F$. $F = 50$ mm

$$\beta = \frac{1.5(4(1.5)-1)}{128(1.5+2)(1.5-1)^2} \frac{1}{(2)^3} = 0.008 \text{ (rad)}$$

    Linear size = 419 μm >> CCD pixel. Must use a multi-element lens system.

6. F/number for 1″ diameter blur is given by

$$(f/\#)^3 = \frac{206\,265}{128(\beta)} = 1611 \Rightarrow f/\# = 11.7$$

7. Image blur due to coma 1′ off axis (2.9×10^{-4} rad) for an $f/3$ mirror is

$$\beta = \frac{206\,265}{16(3)^2} 2.91 \times 10^{-4} = 0.42″$$

    For $f/1.5$, blur is $(3/1.5)^2$ times greater = 1.67″.

8.   (a) Diameter of diffraction blur = 2.44 (0.5 μm) (2) = 2.44 μm for $f/2$ lens at 500 nm.
    (b) Depth of focus; $\Delta f = \pm\, 2\lambda(f/\#)^2 = \pm\, 4$ μm.

9.   Assuming $\alpha = 24 \times 10^{-6}$ K^{-1}, Young's modulus $E = 10 \times 10^6$ psi and the yield strength of aluminium strut = 40 000 psi.
    Since $F = EA(\Delta L/L)$ and $\Delta L/L = \alpha\Delta T$, therefore $F/A = -\alpha E\Delta T = 50\,400$ psi. The strut will buckle because the stress exceeds the yield strength.

10. $A = 5$ m^2. $\varepsilon = 5\% = 2F_{hc}$, $T_c = 77$ K

(a) for $T_h = 300$ K, $Q_h = 57$ W
(b) for $T_h = 275$ K, $Q_h = 40$ W
Add floating shields or multi-layer insulation.

11. Spectrometer: slit width = w, slit height = h. $A\Omega = \frac{1}{4}\pi D_{coll}^2 (w/F_{coll})(h/F_{coll})$.
Seeing-limited camera: seeing disk diameter = θ. $A\Omega = \frac{1}{4}\pi D_{coll}^2 (\frac{1}{4}\pi\theta^2 /F_{coll}^2)$.

CHAPTER 6

1. See Figs 6.7 and 6.9. Number of electrons stored = CV/e. Draw timing diagram carefully.

4. Inverted operation attracts minority carriers from the channels stops which fill surface traps and eliminate dark current.

6. For germanium $E_G = 0.67$ eV, and longest wavelength (μm) is $\lambda_c = 1.24/E_G = 1.85$ μm.

CHAPTER 7

3. Last pixel has $n = 2048 + 2048 = 4096$ transfers. CTE = 0.999 99, fraction of original charge (Q_0) left after n transfers is $Q' = (CTE)^n Q_0 = (0.999\ 99)^{4096} Q_0 = 0.9599$ or 96%.
Can also use $Q' \approx \{1 - n(1-CTE)\}Q_0 = \{1 - 4096(1\times10^{-5})\}Q_0 = \{1 - 0.040\ 96\}Q_0 = 0.9590$ or 96%.

5. "kTC-noise" = $(1/e)\sqrt{(kTC)}$ electrons. For $T = 150$ K, $C = 0.5 \times 10^{-12}$ F, noise = 201 e$^-$.

7. Shot noise on "pre-flash" = $\sqrt{400} = 20$ electrons adds in quadrature with readout noise of 15 electrons. Final noise = $\sqrt{\{(20)^2 + (15)^2\}} = 25$ e$^-$.

8. Picking up 50 or 60 cycle harmonics from mains; probably due to a ground-loop.

CHAPTER 8

2. Temperature of black body, $T = 1500$ K. Wien's Law gives $\lambda_{max} = 2898/T = 1.93$ μm.

3. Monochromatic flux from a blackbody at 300 K.
$B_\lambda = 1.19 \times 10^{-16} \lambda^{-5} (\exp\{47.97/\lambda(\mu m)\} - 1)^{-1}$.
At 2.2 μm the flux is 7.85×10^{-23} W m^{-2} μm^{-1} per steradian.
$E = hc/\lambda = 9.05 \times 10^{-20}$ J at 2.2 μm.
$N_\lambda = 8.68 \times 10^{-4}$ photons/s/m^2/μm/sr.

5. Charge stored is $Q = CV$, where V is the voltage across the pn junction being discharged by photocurrent, but the capacitance C of the depletion region is also a function of V.

CHAPTER 9

2. Backgrounds are very large due both to variable OH emission at shorter wavelengths and thermal emission from telescope and sky at longer wavelengths. Thermal emission can also change during the night as temperatures, airmass or emissivity changes.

5. (a) all optics. $T = 80$ K (LN$_2$). (b) all optics. $T \sim 20$ K, detector ~ 10 K (LHe).

CHAPTER 10

1. Explain the difference between QE and DQE. Produce additional entries in Table 10.1 for CCDs with 5 and 15 electrons of readout noise.

2. $V = (0.21)S + 41$; linear. Gain factor $g = 1/0.21 = 4.8$ e$^-$/DN.
   Readout noise $R = g\sqrt{41} = 30.5$ e$^-$.

$$A_g = \frac{V_{fs}C}{2^n ge} = \frac{10V \times 0.1 \times 10^{-12}\, F}{2^{16} \times 25 \times 1.6 \times 10^{-19}\, C/e} = 3.81$$

3. $C = 0.1$ pF, $A_{sfd} = 0.75$, $n = 16$, $V_{fs} = 10$ volt, $g = 25$ electrons/DN.
   Dividing A_g by the source follower gain (0.75) gives the gain of the preamp/postamp $= 5.09$.
   The required gain is 4× larger for a 14-bit ADC.

6. Flat-fielding is a multiplicative or "gain" correction. Fringing is an additive term.

8. Zeropoint is the magnitude corresponding to 1 count/s for a star of zero colour above the atmosphere.

11. $m = 24$, V band, $D_{tel} = 4$ m, $\tau\eta = 0.30$, $\theta_{pix} = 0.3''$, $R = 10$ e$^-$ and $I_d \sim 0$.
    F = $(3.92 \times 10^{-12} \times 10^{-0.4(24)}$ W cm^{-2} μm$^{-1})$ (0.55 μm)(0.09 μm)(0.3)(125 600 cm^2)/(1.99 $\times 10^{-19}$ J μm) = 9.2 e-/s
    Assuming $1''$ seeing then star flux is diluted over $n \sim 9$ pixels. Average flux = 1 e$^-$/s/pixel.
    If sky = 22 mag per square arcsecond then $F_{sky} = 6.3 \times F \times (0.3)^2 = 5.2$ e$^-$/s/pixel.
    Background dominates over signal.
    Readout noise = 10 e$^-$, so equivalent flux = 100 e$^-$.
    Must integrate for $t > R^2/F_{sky} = 100/5.2 = 19.2$ s to become background limited; easily achieved, therefore sky-limited.
    Signal-to-noise ratio $\approx F\sqrt{t}/\sqrt{(nF_{sky})}$ ignores signal and readout noise; assumes perfect flats.
    $10 = 9.2\sqrt{t}/\sqrt{(9 \times 5.2)} = 1.34\sqrt{t}$, and $t = 55.3$ s

CHAPTER 11

1. Data rate = number pixels digitized × number of bytes/pixel / frame time
   $= 1M \times 2 / 1024^2 \times 10^{-4} = 0.02$ Mbytes/s
   IR detector data rate = 100/5 = 20 times faster (~0.4 Mbytes/s).
   For 32 simultaneous outputs, data rate is 20/32 times slower again (63%, 0.24 Mbytes/s).

CHAPTER 12

7. Assuming diffraction limited:
   (a) VLA telescope, angular resolution $= 70°\lambda/D = 70°$ (6 cm)/ (26 × 100 cm) = 9.7'.
   (b) small optical telescope, $\theta = (206\,265)(1.22)(5 \times 10^{-5}$ cm)/(10 cm) = 1.25''.

CHAPTER 13

1. Diffraction limit = (206 265) 1.22 λ/D = (206 265)1.22(1 μm)/(2.4 × 10^6 μm) = 0.1″.
    This is 2.28 times the pixel size of WFPC2 planetary camera CCD; satisfactory.
    At 500 nm the diffraction limit is 0.05″, i.e. almost 1:1 sampling. Recommend using
    a dither pattern to sub-sample or using FOC.

7. Calculate the field of view covered by a mosaic of 6 × 5 CCDs with no overlaps if
    each CCD FOV of 1 CCD = 2048 pixels × 0.2″/pixel = 6.83′ × 6.83′.
    FOV of 6 × 5 mosaic = 41′ × 34′ = 0.39 sq. degrees.
    10° × 10° = 100 sq. degrees; 100/0.39 = 257.5 exposures, at 20 min each, gives 85.8
    hours (at least ten clear nights).
    One hemisphere of sky would require about 206 times longer; over 5 years' worth of
    clear skies.

CHAPTER 14

1. Diffraction-limited angular resolution corresponds to D = 85 m: (206 265)1.22λ/D
    Angular resolution is 3 milliarcseconds per micrometre of wavelength.
    Light gathering power is only equivalent to $\sqrt{2}$ × 10-m = 14.1-m telescope (not
    85-m).
    Interferometry gives huge gain in resolution (85/10 = 8.5).

2. θ = 0.05″ at 2 μm. Linear dimension = (0.05/206 265 rad) × (4.2 ly × 6 × 10^{12} miles/ly)
    Linear size = 6.15 million miles.
    Baseline of 100 m = 10× better resolution. Linear size ≈ 615 000 miles or about 2.5×
    the distance to the Moon.

3. Seeing ~ λ_0/r_0.
    (a) r_0 = (0.5 × 10^{-6} m/ 0.75″)206 265 = 0.14 m.
    (b) 2.2/0.5 = 4.4, r_0 = 0.61 m.

Appendix 1: Powers of ten notation

When writing numbers which are very large or very small it is useful to introduce a short-hand notation called "powers of ten". Thus, instead of writing 1 000 000 for one million we write 1×10^6 and understand this to mean 1 followed by six zeros. Similarly, instead of writing 0.000 001 for one millionth we write 1×10^{-6} where the minus sign tells us that this number is one divided by one million. Names and symbols are given to the most frequently used powers of ten as shown below. These names and symbols can be prefixed to any of the units of measurements given below to infer that the basic unit is to be multiplied by that power of ten.

Name	Symbol	Power	Name	Symbol	Power
zepto	z	10^{-21}	zeta	Z	10^{21}
atto	a	10^{-18}	exa	E	10^{18}
femto	f	10^{-15}	peta	P	10^{15}
pico	p	10^{-12}	tera	T	10^{12}
nano	n	10^{-9}	giga	G	10^{9}
micro	μ	10^{-6}	mega	M	10^{6}
milli	m	10^{-3}	kilo	k	10^{3}
centi	c	10^{-2}	hecto	h	10^{2}
deci	d	10^{-1}	deca	da	10^{1}

Notes:
(i) $10^0 = 1$ and in fact $N^0 = 1$ where N is any number.
(ii) $N^a \times N^b = N^{a+b}$ and $N^a \div N^b = N^{a-b}$, e.g. $10^8 \times 10^2 = 10^{10}$.
(iii) $(N^a)^b = N^{ab}$, e.g. $(10^6)^2 = 10^{12}$.
(iv)   We use the definition of 1 billion as 1 000 000 000 (10^9) and 1 trillion is 10^{12}.

If $x = a^y$, then the index y is called the **logarithm** of x to the base a; the logarithm is the number to which the base must be raised to produce the value x. The logarithm (log) of a product is the sum of the individual logs; $\log(ab) = \log(a) + \log(b)$. These rules are needed to understand and manipulate astronomical magnitudes. There are two commonly used bases, the base ten system ($a = 10$) and the natural or Naperian system ($a = e \approx 2.7$); the latter are usually given the symbol 'ln' instead of 'log'.

Decibels (dB) are a logarithmic scale used to represent signal-to-noise ratios. In electronics one is usually concerned with noise power $\Delta P/P$, and 1 dB = 10 log ($\Delta P/P$) or in terms of voltage noise, 1 dB = 20 log ($\Delta V/V$). For example, suppose we want to digitize a voltage to an accuracy of 0.5 bit in 16 bits ($2^{16} = 65\,536$) with a 10 volts ADC, then $\Delta V = 76.25$ μV (from $10V/2 \times 65\,536$) and $\Delta V/V = 0.5/65\,536 = 7.625 \times 10^{-6} = -102.35$ dB. Also, when $\Delta P/P = 0.5$ then dB = −3.0. The level at which the signal has dropped by 50% is called the "3 dB point".

THE GREEK ALPHABET

alpha α,	beta β,	gamma γ,	delta δ, Δ	epsilon ε,	zeta ζ,
eta η	theta θ, Θ,	iota ι,	kappa κ,	lambda λ,	mu μ,
nu ν,	xi ξ,	omicron o,	pi π,	rho ρ,	sigma σ, Σ,
tau τ,	upsilon υ,	phi ϕ,	chi χ,	psi ψ, Ψ,	omega ω, Ω.

Appendix 2: Some units of measurement and useful conversions

Quantity	SI units	Unit symbol
Length	metre	m
Mass	kilogram	kg
Time	second	s
Electric charge	coulomb	C
Thermodynamic temperature	kelvin	K

SI is the Système International d'Unités.

Quantity	Symbol	Derived units	Equivalent
Force	F	newton, N	$1\ N = 1\ kg{\cdot}m/s^2$
Energy	E	joule, J	$1\ J = 1\ N{\cdot}m$
Power	P	watt, W	$1\ W = 1\ J/s$
Electric current	I	ampere, A	$1\ A = 1\ C/s$
Electric potential	V	volt, V	$1\ V = 1\ J/C$
Electric capacitance	C	farad, F	$1\ F = 1\ C/V$
Electric resistance	R	ohm, Ω	$1\ \Omega = 1\ V/A$
Frequency	ν	hertz, Hz	$1\ Hz = 1\ s^{-1}$
Pressure	P	pascal, Pa	$1\ Pa = 1\ N/m^2$
Volume	V	litre, L	$1\ L = 10^{-3}\ m^3$

Other units and conversions:

1 micron = 1 micrometre (μm) = 10^{-6} m

1 inch (in) = 2.54 cm = 25.4 mm (exact)

12 in = 1 ft   5280 ft = 1 mile (mi)

1 km = 0.6214 mi $\approx$ 5/8 mi

1 ounce (oz) = 28.3 g

1 angstrom (Å) = 10^{-10} m = 10^{-8} cm = 0.1 nm

1 cm = 0.3937 in   1 m = 39.37 in

1 mi = 1.6093 km $\approx$ 8/5 km

1 g = 0.0353 oz

1 kg = 2.2046 pound (lb)   1 lb = 0.4536 kg
A temperature change of 1 kelvin (1 K) = 1 °C (Celsius) = 1.8 °F (Fahrenheit)
Freezing point of water = 0 °C = 273 K = 32 °F
1 sidereal year (yr) = 365.256 d = 3.156×10^7 s
1 erg = 10^{-7} joule (J)   1 J = 10^7 erg
1 calorie (cal) = 4.186 J   1 kWh = 3.600×10^6 J
1 horsepower (hp) = 550 ft·lb/s = 745.7 W
1 atmosphere (atm) = 760 mm Hg = 14.70 lb/in^2 = 1.013×10^5 Pa
1 bar = 10^5 Pa = 0.9870 atm   1 torr = 1 mm Hg = 133.3 Pa = 1.33 mbar

Notes:

(i)   One **electronvolt** (eV) is the energy gained by an electron of charge e after being accelerated by a potential (V) of 1 volt, that is, 1 eV = $e \times 1$ V = 1.602×10^{-19} J. Or, 1 joule = 6.242×10^{18} eV.

(ii) The product RC of a resistance R and a capacitance C, has the units of time, and is known as the "RC time constant".

Appendix 3: Physical and astronomical constants

PHYSICAL CONSTANTS

Constant	Symbol	Value
Speed of light in vacuum	c	2.9979×10^8 m/s
Charge on the electron	e	1.6022×10^{-19} C
Planck constant	h	6.6261×10^{-34} J s
Boltzmann constant	k	1.3807×10^{-23} J/K
Electron mass	m_e	9.1094×10^{-31} kg
Proton mass	m_p	1.6726×10^{-27} kg
Avogadro's number	N_A	6.0221×10^{23} mol^{-1}
Gas constant	R	8.3145 J/K·mol
Gravitational constant	G	6.6726×10^{-11} N·m^2/kg^2
Permeability constant	μ_0	$4\pi \times 10^{-7}$ N/A^2
Permittivity constant	ε_0	8.8542×10^{-12} C^2/N·m^2
Stefan–Boltzmann constant	σ	5.669×10^{-8} W·m^{-2}·K^{-4}
Wiens constant	$\lambda_{max}T$	2898 μm·K
Pi	π	3.14159
Base of natural logs	e	2.71828

The constant hc occurs frequently and has the value 1.99×10^{-28} J·km or, expressing 1 km as 10^9 μm (micrometres) then $hc = 1.99 \times 10^{-19}$ J μm.

ASTRONOMICAL DATA

Mean radius of the Earth	$R_\oplus$	6.37×10^6 m
Mass of the Earth	$M_\oplus$	5.98×10^{24} kg
Mean Earth–Sun distance		
= 1 astronomical unit (AU)		1.4960×10^{11} m
		($\approx$ 93 million miles)
Mean radius of the Sun	$R_\odot$	6.96×10^8 m
Mass of the Sun	$M_\odot$	1.989×10^{30} kg
Luminosity of the Sun	$L_\odot$	3.827×10^{26} J/s
1 light year	ly	9.4605×10^{15} m
		($\approx$ 6 trillion miles)
1 parsec	pc	3.0857×10^{16} m
		= 3.2616 ly
		= 206 265 AU

BLACKBODY RADIATION—THE PLANCK FUNCTION

$$B_\lambda = \frac{2hc^2}{\lambda^5} \frac{1}{\left(e^{hc/k\lambda T} - 1\right)} \quad \text{W m}^{-2}\,\text{m}^{-1}\,\text{ster}^{-1}$$

and

$$B_\nu = \frac{2h\nu^3}{c^2} \frac{1}{\left(e^{h\nu/kT} - 1\right)} \quad \text{W m}^{-2}\,\text{Hz}^{-1}\,\text{ster}^{-1}$$

where $2hc^2 = 1.191 \times 10^{-16}$ W m^2 s^{-1} and $hc/k = 1.439 \times 10^{-2}$ m K.

THE RAYLEIGH–JEANS APPROXIMATION ($h\nu \ll kT$)

$$B_\lambda = \frac{2ckT}{\lambda^4} = 8.2782 \times 10^3 \frac{I}{\lambda_{\mu m}^4} \quad \text{W m}^{-2}\,\mu\text{m}^{-1}\,\text{ster}^{-1}$$

$$B_\nu = \frac{2kT\nu^2}{c^2} = 3.0724 \times 10^{-40}\, T\nu^2 \quad \text{W m}^{-2}\,\text{Hz}^{-1}\,\text{ster}^{-1}$$

Note: Irradiance (W m^{-2}) is called "flux" by astronomers.

Appendix 4: Astronomical magnitude scale and its relation to lux and watts

The range of brightnesses of astronomical objects is so great that it is useful to find a way to "compress" that range. This can be done by using a logarithmic scale. Astronomers call their logarithmic scale for brightness a **magnitude** scale. The magnitude, m, of a star of brightness, b, is given by

$$m = constant - 2.5 \log(b) \tag{A4.1}$$

where the "constant" has a value which determines the scale. Notice that the fainter the object the *larger* the value of m. When we have two stars of different brightnesses b_1 and b_2 then we can obtain their magnitude difference $m_1 - m_2$ from

$$m_1 - m_2 = 2.5 \log\left(\frac{b_2}{b_1}\right) \tag{A4.2}$$

In these equations log represents the logarithm to the base ten; a logarithm of a number to the base ten is the power of ten which exactly yields the number. The magnitude difference corresponding to a brightness ratio of 100 is therefore 2.5log (100) = 2.5 × 2 = 5 magnitudes. To find a brightness ratio from a known magnitude difference we can turn the equation around to give

$$\frac{b_2}{b_1} = 10^{0.4(m_1 - m_2)} \tag{A4.3}$$

so that by raising ten to the power of 0.4 × ($m_1 - m_2$) we get the ratio b_2/b_1. The logarithm or the antilogarithm, i.e. the value corresponding to ten to the power of *log(number)*, can be obtained from mathematical tables and from scientific pocket calculators which have "pushbutton" evaluation of the log and antilog of any entered number.

To determine a magnitude from a brightness we need to establish either a relative scale or an absolute scale or both. Astronomers use a relative scale such that the magnitude of the star Vega (Alpha Lyrae) is nearly zero at all wavelengths. On that scale stars fainter than sixth visual magnitude ($m = 6$) cannot be seen by the unaided eye, the brightness of

1 square arcsecond of dark (moonless) sky at an excellent mountain site in the visible (V) waveband is about $m = 21$; CCDs have enabled stars and faint galaxies to be detected to levels fainter than $m = 27$, which is 63 billion times fainter than Vega.

It is also possible to establish an absolute scale. It has been determined by careful experiments that at a wavelength of 0.54 μm in the middle of the visible spectrum that a $m = 0$ star of the same type as Vega produces 3.92×10^{-12} watts per square centimetre per micrometre interval of wavelength just outside the Earth's atmosphere. This is the irradiance per unit wavelength. Notice that the dimensions here are energy per unit time per unit area per unit wavelength interval, and the units are well-known and fundamental. A few other useful terms are as follows, the total amount of energy emitted by a light source per second is its radiant power, measured in watts, and the power emitted in a cone-shape about a particular direction (that is, per unit solid angle) is called the radiant intensity (watt per steradian).

In the electro-optical industry the term illuminance is used to describe the amount of light received per unit surface area and it is measured in a unit called the **lux**. One lux is the illuminance produced by a standard light source of one candela at a distance of one metre and 60 candelas is the luminous intensity of a 1cm^2 blackbody at the temperature of melting platinum (2042 K). One lux is therefore also equivalent to 1 lumen per square metre. Such units are complicated and are never used in astronomy or scientific research in general; they are often found in reference to CCD-based cameras intended for low light level television applications, however. An illuminance of 1 lux is a photon flux of about 3×10^9 photons/s/mm² at $\lambda = 550$ nm, or approximately equivalent to a star of visual magnitude -14. Note the minus sign which indicates that this source is very much brighter than Vega, in fact it is roughly equivalent to the full moon! The brightness in lux of any source in the "visual" waveband is

$$B(\text{lux}) = 10^{-0.4(m_v + 14)} \tag{A4.4}$$

Again note the minus sign. The illuminance of Vega ($m_v = 0$) is only 0.000 002 lux. Ordinary hand-held commercial TV "camcorders", even those with CCDs, operate at frame-rates of 1/60th second and are typically designed with one-inch lenses to image scenes with light levels greater than a few lux (bright moonlight). Some intensified CCD systems are capable of imaging with illumination levels as weak as 0.02 lux. Of course, while a 2-lux CCD camcorder cannot actually take images in a room where the illumination level is equivalent to that from the star Vega, it could actually image the star itself if its standard lens was removed and it was attached to a modest telescope. The reason is simply that the telescope collects more light and, more importantly, focuses all of the available light onto a single (or small number) of pixels so that the illuminance for that pixel is increased by several hundred thousand times. If the CCD is cooled to allow integration times of a few seconds, instead of 1/60th second, then objects can be recorded which are several hundred times fainter still. Cooled, slow-scanned CCD cameras with very long integration times on the world's largest telescopes are therefore very sensitive indeed!

The term **absolute magnitude** (*M*) is used frequently in astronomy, but in fact this is still a relative magnitude; *M* is the magnitude that the star would have at a standard distance of 10 parsec (32.6 ly) in the absence of any interstellar dimming and reddening. For example, Sirius (α CMa A) has an apparent visual (V) magnitude of −1.46 and is the brightest star in the sky. Its distance is 2.6 pc, so at 10 pc its (absolute) magnitude (in the V band) becomes +1.47. In general,

$$M = m - 5 \log \frac{r}{10} \qquad (A4.5)$$

where m is the apparent magnitude and r is the distance in parsecs. The Sun has an absolute visual magnitude of +4.85 but an apparent visual magnitude of −26.72.

The quantity $(m - M)$ is called the distance modulus; if $m - M = 0$ then $r = 10$ pc, for $m - M = 5$, 10, 15, 20, 25 magnitudes then $r = 100$ pc, 1 kpc, 10 kpc, 100 kpc and 1 Mpc (3.26 million light years).

Appendix 5: Basic observational astronomy facts

Most large telescopes must point to within ± 5 arcsec. This means that you must have good "coordinates" for a specific epoch (a date like 1950.0 or 2000.0) and then these coordinates will have to be "precessed" to the current epoch (exact date) to correct for changes in the coordinate system with time.

Coordinates: Positions on the sky are given with reference to the Celestial Equator and the Celestial Poles as shown in Figure A5.1. The measurements are analogous to longitude and latitude on the Earth's surface and are called Right Ascension (RA or α) and Declination (Dec or δ). RA is measured eastwards from the First Point of Aries and Dec is measured ± 90° from the Celestial Equator.

Luni-solar precession: The First Point of Aries moves backwards along the equator at 50.2 arcseconds per year. Correcting for this effect yields the mean equator and mean equinox.

Nutation: The wobble of the Earth's axis as it precesses. Correction for this effect gives the true equator and true equinox.

For the nearest objects, an additional correction for proper motion (the component of the object's true motion in space at right angles to our line of sight) must be made; tabulated in surveys of proper motion. Often, all these corrections are done at the observatory by the telescope control program, but formulas are included in the Astronomical Almanac which is published annually and jointly by the UK and the USA.

To find out when an object is observable you need to convert from RA and Dec to hour angle and zenith distance (or zenith angle) for your latitude. Several basic relationships are given below; it is useful to be familiar with the the Celestial Sphere and basic spherical trigonometry (Green, 1985).

Hour angle: The hour angle (H) is measured either east or west from 0 on the meridian; east is positive and west is negative. In practice, H is given in hours of time and not in degrees because this gives a more intuitive indication of observability of the object, e.g.

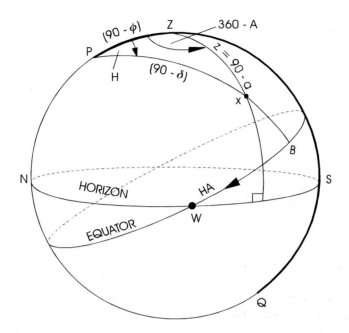

Fig. A5.1. Celestial sphere and spherical triangle relations.

three hours east of the meridian means that it will be three hours until that object crosses the meridian. The usual conversion between degree and hours applies;

$$360° = 24 \text{ hr}, \quad 15° = 1 \text{ hr}, \quad 1° = 4 \text{ min},$$
$$15' = 1 \text{ min}, \quad 1' = 4 \text{ s},$$
$$15'' = 1 \text{ s}, \quad 1'' = 0.067 \text{ s}$$

Zenith distance: ZD or ζ is $90° - a$ (the elevation angle above horizon); a $ZD > 90°$ implies that the object is below the horizon.

Azimuth: Azimuth is the compass bearing in degrees measured from North through East.

Local sidereal time: LST is the hour of right ascension which is on the meridian at that moment. Most observatories have a sidereal clock. There are approximately 23 hours and 56 minutes in one Sidereal Day. The sidereal time at midnight on March 22 is 12 hours and the LST advances about 2 hours per month, so for example, by September 21 the LST is 24 (or 0) hours—this is the RA on the meridian at midnight. In general, *LST = RA + HA*. LST is tabulated at 0 hours Universal Time for every day of the year from the *Astronomical Almanac*.

Universal time is the time on the Greenwich meridian. Most observatories have a clock keeping UT. The exact UT can also be obtained from radio signals issued by the WWV service.

Julian date (no connection to the Julian Calendar) is a numerical day count from an arbitrary zero point. Julian Day number is the number of days that have elapsed since noon Greenwich Mean Time on January 1, 4713 BC plus the decimal fraction of a day since the preceding noon up to the event being recorded. Julian dates are given in the *Astronomical Almanac*.

Coordinate transformations:
Relationship between Hour Angle (H) and Dec (δ) given Azimuth (A) and Elevation (a):

$$\sin \delta = \sin \phi \sin a + \cos \phi \cos a \cos A$$
$$\cos H = (\sin a - \sin \phi \sin \delta) / \cos \delta \cos \phi$$

Relationship between *A* and a given hour angle and zenith distance:

$$\sin a = \cos \phi \cos(90-\delta) + \sin \phi \sin(90-\delta) \cos H$$
$$\cos A = \cos(90-\delta) - \cos \phi \cos(90-a) / \sin \phi \sin (90-a)$$

The altitude (or elevation) of the celestial pole above the horizon in degrees is equal to the latitude of the site.

A plot of airmass (i.e. sec ζ) versus time (or HA) will tell you the best time to observe your source. Airmass cannot be smaller than 1.0 (zenith) and an airmass of 2 corresponds to $\zeta = 60°$ (or the object 30° above the horizon). The times of sunset, sunrise, and the position of the Moon, and many other very useful quantities, are tabulated in the *Astronomical Almanac*. The times of sunset, sunrise and twilight are tabulated at 4-day intervals of time and 10 degree intervals of latitude.

Finding charts can be produced with the help of photographic catalogues such as the Palomar and UK/ESO Schmidt surveys, and the Hubble Space Telescope Guide Star catalogue—which is available on CD-ROM—or images published by previous observers of your sources.

Extensive catalogues of known objects are now available in electronic form on the World Wide Web. Facilities such as ADS, NED, SIMBAD and others can be extremely useful for tracking down information on a wide range of sources over all wavebands.

Older well-known catalogues include:

The *Bright Star Catalogue*, Hoffleit, Yale University Observatory 1982. Magnitudes, B-V colours, MK spectral class, parallax, radial velocity, rotational velocity (if known) of 9096 stars. Limit is about $V = 6.5$ roughly.

The *Smithsonian Astrophysical Observatory Star Catalog* Parts 1–4 gives positions and proper motions of 258 997 stars epoch/equinox 1950.0; limit is about $V = 9.5$ magnitudes.

The *University of Michigan Catalogue of Two-dimensional Spectral Types* for the stars in the Henry Draper (HD) catalogue, Vols 1–3, Houk and Cowley. Gives spectral and luminosity classes of 34 886 stars. Contains plates of comparison spectra.

REFERENCES

Green, R.M. (1985) *Spherical Astronomy*, Cambridge University Press, Cambridge, UK.
Clarke, D. and Roy A.E. (1993) *Astronomy: Principles and Practice*, Adam Hilger, Bristol.

USEFUL STATISTICS

Mean and weighted mean:

$$\bar{x} = \frac{\sum \left(x_i / \sigma_i^2 \right)}{\sum \left(1 / \sigma_i^2 \right)} \rightarrow \frac{1}{N} \sum x_i, \text{ when } \sigma_i = \sigma$$

where σ is the standard deviation of a single observation from the mean and N is the number of observations. The standard error of the mean is $\sigma/\sqrt{N}$.

Variance:

$$\sigma^2 = \frac{\sum \left(x_i - \bar{x} \right)^2}{N - 1}$$

is determined from the sum of the squares of the "residuals" or differences from the mean.

Error in a sum (or difference) of two random variables (A and B):

$$A \pm \sigma_A, \quad B \pm \sigma_B$$

$$C = \left(A + B \right) \pm \sqrt{\sigma_A^2 + \sigma_B^2}$$

The errors add in quadrature.

Error in a ratio of two random variables (A and B):

$$A \pm \sigma_A, \quad B \pm \sigma_B$$

$$C = \frac{A}{B} \pm C \sqrt{\left(\frac{\sigma_A}{A} \right)^2 + \left(\frac{\sigma_B}{B} \right)^2}$$

that is, the fractional errors add in quadrature.

The chi-squared test:

Compares the observed frequency distribution $f(x_i)$ of possible measurements x_i with the predicted distribution $NP(x_i)$ where N is the number of data points and $P(x_i)$ is the theoretical probability distribution.

$$\chi^2 = \sum_{j=1}^{n} \frac{\left[f(x_j) - NP(x_j) \right]^2}{NP(x_i)}$$

The "reduced" chi-squared is χ^2/ν, where ν is called the "degrees of freedom" and is given by $N - p$, where N is the number of data points and p is the number of parameters determined from those data points.

The Gaussian distribution:

For a very large number of observations n of a random variable the probability of obtaining the value x is

$$P(x, \bar{x}, \sigma) = \frac{1}{\sigma\sqrt{2\pi}} \exp\left[-\frac{1}{2} \left(\frac{x - \bar{x}}{\sigma} \right)^2 \right]$$

where the standard deviation σ is related to the full width at half maximum (FWHM) of the distribution by FWHM = 2.354σ, and to the probable error by P.E. = 0.6745σ. The probable error corresponds to the range of the variable which contains 50% of the measurements.

Error bars giving estimates of σ should be applied to all measurements. Assuming that the errors follow a Gaussian distribution, the probability of finding the variable in the interval $\pm 1\sigma$ is 68%. For $\pm 2.5\sigma$ it is 98.7%.

Appendix 6: Typical CCD clock voltages

Typical clock voltages for three types of CCDs as a guide for setting-up a CCD system

Potential (V)	SITe (Tek)	Loral	EEV
V_{RD}	+7.0	+6.5	+4.0
V_{OD}(read)	+16.0	+15.0	+14.0
V_{OD}(integrate)	+7.0	+7.0	+14.0
V_{OG}	−6.0	−6.0	−11.0
V_{SS}	−7.0	−7.0	−7.0
V_{ABG}	−7.0	−6.0	−13.0
V_{BG}	+3.0	+3.0	−13.0
V_{ABD}	+7.0	+7.0	+9.0
V+ (vert. high)	−4.0	−3.0	−3.0
V− (vert. low)	−15.0	−15.0	−13.0
H+ (hor. high)	+3.0	−2.0	−3.0
H− (hor. low)	−9.5	−12.0	−13.5
R+ (reset high)	+3.0	+1.0	−3.0
R− (reset low)	−9.0	−7.0	−13.0

1. A Tek(SITe)1024 thinned chip operated at −90 °C with a constant current load I_{OS} = 1 mA. Nominal readout noise ~4/5 e⁻. Note the non-zero substrate voltage which allows use of an OD that is not too high from the supplies. Transfer gate voltages = parallel clock voltages (V+,V−). Summing well voltages = serial clock voltages (H+,H−). Parallel clocks have 10-μs slopes, but > 300-μs triplet time to allow full overlap and good charge transfer. Serial clocks have ~ 1-μs slopes.
2. Loral 2k × 2k, 15 μm pixels at −103 °C with same current load as above.
3. EEV05-30 and 05-50 (frontside, UV coated) CCDs at −125 °C with 1 mA load. Nominal readout noise ~3/4 e⁻. Note non-zero substrate voltage as above. Parallel clocks have ~2 μs slopes and ~10 μs triplet time. Serial clocks have ~0.5 μs slopes.

The information given above was kindly provided by Paul Jorden of the Royal Greenwich Observatory and Gerry Luppino of the University of Hawaii.

Appendix 7: Suppliers and manufacturers

A7.1 CCD CAMERA SYSTEMS (used in astronomy)

The "*" indicates that complete "turn-key" systems are available including image processing software.

* *AstroCam Limited* (formerly AstroMed), Innovation Centre, Cambridge Science Park, Milton Road, Cambridge CB4 4GS, England.
   Fax: Tel: 01223-316705
* *Axiom Research*, 3340 N. Country Club, #103, Tucson, Arizona 85716, USA.
      Fax: (520) 791 2800   Tel: (520) 791 2864
              http://www.axres.com/axiom/index.html
CompuScope, 3463 State St., Suite 431, Santa Barbara, California 93105, USA.
      Fax. (805) 966 6693   Tel: (805) 966 7179
* *First Magnitude Corporation*, 519 South Fifth Street, Laramie, Wyoming 82070, USA.
      Fax: (307) 745 3743   Tel: (307) 766 6267
      http://www.wyoming.com/~Johnson/fmc.htm
* *Photometrics*, 3440 East Britannia, Tucson, Arizona 85706, USA.
      Fax: (520) 573 1944   Tel: (520) 889 9933
PixelVision Inc., 15250 Greenbrier Parkway, # 1250, Beaverton, Oregon 97006, USA.
   [Partner with SITe. Includes Electron Bombarded CCD systems.]
      Fax: (503) 629 3211   Tel: (503) 629 3210
Princeton Instruments, 3660 Quakerbridge Road, Trenton, New Jersey 08619, USA.
      Fax: (609) 587 1970   Tel: (609) 587 9797
      [Includes Intensified CCD systems.]
PULNiX Inc., 1330 Orleans Drive, Sunnyvale, California 94089, USA.
      Fax: (408) 747 0660   Tel: (408) 747 0300
Santa Barbara Instrument Group (SBIG), 1482 East Valley Road, Suite #33, Santa Barbara, California 93150, USA.
      Fax: (805) 969 4069   Tel: (805) 969 1851
Scanco Inc., P.O Box 987, Kamuela, HI 96743, USA.
                            Tel: (808) 885 6169

* *Software Bisque*, 912 Twelfth St., Suite A, Golden, Colorado 80401, USA.
  Fax: (303) 278 0045 Tel: (303) 278 4478
  [Using SBIG camera systems.]
SpectraSource Instruments, 31324 Via Colinas, # 114, Westlake Village, California 91362, USA.
  Fax: (818) 707 9035 Tel: (818) 707 2655
* *Wright Instruments*, 4 Chalkwell Park Avenue, Enfield, Middlesex EN1 2AJ, England.
  Fax: (44) 181 443 3638 Tel: (44) 181 443 3339

A7.2 INFRARED CAMERA SYSTEMS

Infrared Laboratories Inc., 1808 East 17th Street, Tucson, Arizona 85719, USA.
  Fax: 602 623 0765 Tel: 602 622 7074
System Specialists, 4001 North Runway Dr., Tucson, Arizona 85705, USA.
  Fax: 520 292 9621 Tel: 520 292 9644

A7.3 DETECTOR MANUFACTURERS

Dalsa Inc., 605 McMurray Road, Waterloo, Ontario, Canada N2V 2E9. [Designs CCD systems. Fabricates CCDs via foundries. Large formats and small pixels.]
  Fax: (519) 886 8023 Tel: (519) 886 6000
Eastman Kodak Co., Rochester, New York 14650-2010, USA. [Wide range of devices. Both interline and frame transfer CCDs. Large formats and small pixels. No foundry mode.]
  Fax: (716) 477 4947 Tel: (716) 722 4385
EG&G Reticon, 345 Potrero Avenue, Sunnyvale, California 94086, USA. [Has developed large format CCDs in collaboration with Lick Observatory.]
  Fax: (408) 738 6979 Tel: (408) 738 1009
English Electric Valve, 106 Waterhouse Lane, Chelmsford, Essex CM1 2QU, England. [Wide range of very high-performance scientific CCDs, including devices for X-ray astronomy. Close ties to astronomy. Foundry mode.]
  Fax: (44) 245 492492 Tel: (44) 245 493493
Loral/ Fairchild Imaging Sensors, Ford Road, Newport Beach, California 92658, USA. [Dick Bredthauer. Large format CCDs for astronomy. Foundry service. Comprises former Fairchild (Milipitas, CA) and Ford Aeronutronics (Newport Beach, CA)
  Fax: (714) 720 6741 Tel: (714) 720 6265
MIT/Lincoln Laboratory, P.O. Box 73, Lexington, Massachusetts 02173, USA. [Federally funded R&D centre and part of the Massachusetts Institute of Technology. Innovative devices. High speed. Low-energy X-ray CCD detectors. Not a vendor.]

Orbit Semiconductor, Sunnyvale, California, USA. [CMOS and CCD foundry service. Very large formats produced (e.g. for Dalsa). Provides silicon muxes for IR arrays.]
    Tel: (408) 744 1800

Philips, Eindhoven, Netherlands. [Numerous CCD innovations. Manufactures for internal products.]

David Sarnoff Research Center, 201 Washington Road, Princeton, New Jersey 08543-5300, USA. [Gary Hughes. Former RCA Laboratories. Contract R&D work.]
    Fax: (609) 734 2225   Tel: (609) 734 3056

Scientific Imaging Technologies (SITe), PO Box 569, Beaverton, Oregon 97075-0569, USA. [Morley Blouke. Former Tektronix imaging devices group. Wide range of scientific CCDs. Foundry service.]
    Fax: (503) 644 0798   Tel: (503) 644 0688   http://www.site-inc.com

Texas Instruments, Dallas, Texas, USA. [Does not produce scientific CCDs. Mainly industrial and consumer applications.]

Thomson, Rue de Rocheplaine, BP 128, 38521 Saint Egreve, CEDEX, FRANCE and Thomson TCS Division, 40G Commerce Way, Totowa, New Jersey 07511. [Makes both commercial and scientific image sensors. Large formats developed for ESO.]
    Fax: (201) 812 9050   Tel: (201) 812 9000   USA
    Fax: (33) 76 583406   Tel: (33) 76 583112   France

Infrared array detectors are commercially available from:

Amber, 5756 Thornwood Drive, Goleta, California 93117, USA. [Mainly real time 8–12-μm imagery.]
    Fax: (805) 962 1403   Tel: (805) 692 1200
    http://www.amber-infrared.com

Cincinnati Electronics Corp., 7500 Innovation Way, Mason, Ohio 45040-9699, USA. [Jim Wimmers. Product line. Complete cameras. Mainly real-time thermal imaging.]
    Fax: (513) 573 6290   Tel: (513) 573 6100

Rockwell International Science Center, 1049 Camino Dos Rios, P.O. Box 1085, Thousand Oaks, California 91360, USA. [Kadri Vural. Range of astronomy arrays based on HgCdTe.]

Santa Barbara Research Center, 75 Coromar Drive, Goleta, California 93117, USA. [Alan Hoffman. Product line and foundry service. InSb arrays and extrinsic silicon IBC arrays for astronomy.]
    Fax: (805) 562 4100   Tel: (805) 562 2230

REFERENCES

Janesick, J. (1994) "History and advancements of large area array scientific CCD imagers", SPIE Vol. 2172, *Charge-Coupled Devices and Solid-State Optical Sensors IV*, Bellingham, Washington, USA.

McLean, I.S. (Ed.) (1994) *Infrared Astronomy with Arrays: The Next Generation*, Kluwer Academic Publishers, Dordrecht, Netherlands.

Oates, A.P. and Jorden, P.R. (1993) "New large format CCDs for the ING", ING La Palma Technical Note No. 92, Royal Greenwich Observatory.

Index

WILEY-PRAXIS SERIES IN ASTRONOMY AND ASTROPHYSICS

Forthcoming titles

COSMOLOGY AND PARTICLE ASTROPHYSICS
Lars Bergström and Ariel Goobar, Department of Physics, Stockholm University, Sweden

THE VICTORIAN AMATEUR ASTRONOMER: Independent Astronomical Research in Britain 1820–1920
Allan Chapman, Wadham College, University of Oxford, UK

NEW WORLDS; The Search for Extrasolar Planets
Stuart Clark, Lecturer in Astronomy, University of Hertfordshire, UK

LARGE-SCALE STRUCTURES IN THE UNIVERSE
Anthony P. Fairall, Professor of Astronomy, University of Cape Town, South Africa

URANUS: The Planet, Rings and Satellites, Second edition
Ellis D. Miner, Cassini Project Science Manager, NASA Jet Propulsion Laboratory, Pasadena, California, USA

PROTOSTELLAR DISCS AND PLANETARY SYSTEM FORMATION
John C. B. Papaloizou, Astronomy Unit, Queen Mary and Westfield College, London, UK, and Caroline Terquem, Lick Observatory, University of California, Santa-Cruz, USA

NEW LIGHT ON DARK STARS: Red Dwarfs, Low-Mass Stars, Brown Dwarfs
I. Neill Reid, Senior Research Associate, Caltech, Pasadena, USA
and Suzanne L. Hawley, Assistant Professor, Department of Physics and Astronomy, Michigan State University, USA

COSMIC RAY PHYSICS
Todor S. Stanev, Professor of Physics, Bartol Research Institute, University of Delaware, USA